Statistical Optics

Statistical Optics

JOSEPH W. GOODMAN
Professor of Electrical Engineering
Stanford University

A Wiley-Interscience Publication
John Wiley & Sons
New York / Chichester / Brisbane / Toronto / Singapore

Copyright © 1985 by John Wiley & Sons, Inc.

All rights reserved. Published simultaneously in Canada.

Reproduction or translation of any part of this work beyond that permitted by Section 107 or 108 of the 1976 United States Copyright Act without the permission of the copyright owner is unlawful. Requests for permission or further information should be addressed to the Permissions Department, John Wiley & Sons, Inc.

Library of Congress Cataloging in Publication Data

Goodman, Joseph W.
  Statistical optics.

  (Wiley series in pure and applied optics)
  Includes index.
  1. Optics—Statistical methods. 2. Mathematical statistics. I. Title. II. Series.

QC355.2.G66 1984   535   84-13160
ISBN 0-471-01502-4

Printed in the United States of America

10 9 8 7 6 5 4 3 2 1

*To
Hon Mai,
who has provided
the light.*

Preface

Since the early 1960s it has gradually become accepted that a modern academic training in optics should include a heavy exposure to the concepts of Fourier analysis and linear systems theory. This book is based on the thesis that a similar stage has been reached with respect to the tools of probability and statistics and that some training in the area of statistical optics should be included as a standard part of any advanced optics curriculum. In writing this book I have attempted to fill the need for a suitable textbook in this area.

The subjects covered in this book are very physical but tend to be obscured by mathematics. An author of a book on this subject is thus faced with the dilemma of how best to utilize the powerful mathematical tools available without losing sight of the underlying physics. Some compromises in mathematical rigor must be made, and to the largest extent possible, a repetitive emphasis of the physical meaning of mathematical quantities is needed. Since fringe formation is the most fundamental underlying physical phenomenon involved in most of these subjects, I have tried to stay as close as possible to fringes in dealing with the meaning of the mathematics. I would hope that the treatment used here would be particularly appealing to both optical and electrical engineers, and also useful for physicists. The treatment is suitable for both self-study and for formal presentation in the classroom. Many homework problems are included.

The material contained in this book covers a great deal of ground. An outline is included in Chapter 1 and is not repeated here. The course on which this text is based was taught over the 10 weeks of a single academic quarter, but there is sufficient material for a full 15-week semester, or perhaps even two academic quarters. The problem is then to decide what material to omit in a single-quarter version. If the material is to be covered in one quarter, it is essential that the students have previous exposure to probability theory and stochastic processes as well as a good grasp of Fourier methods. Under these conditions, my suggestion to the instructor is to allow the students to study Chapters 1–3 on their own and to begin the lectures directly with optics in Chapter 4. Later sections that can be omitted or left to optional reading if time is short include Sections 5.6.4, 5.7, 6.1.3, 6.2, 6.3, 7.2.3, 7.5, 8.2.2, 8.6.1, 8.7.2, 8.8.3, 9.4, 9.5, and 9.6. It is perhaps worth mentioning that I have also occasionally used Chapters 2 and 3 as the basis for a full one-quarter course on the fundamentals of probability and stochastic processes.

The book began in the form of rough notes for a course at Stanford University in 1968 and thus has been a long time in the making. In many respects it has been *too* long in the making (as my patient publisher will surely agree), for over a period of more than 15 years any field undergoes important changes. The challenge has thus been to treat the subject matter in a manner that does not become obsolete as time progresses. In an attempt to keep the information as up to date as possible, supplementary lists of recent references have been provided at the ends of various chapters.

The transition from a rough set of notes to a more polished manuscript first began in the academic year 1973–1974, when I was fortunate enough to spend a sabbatical year at the Institute d'Optique, in Orsay, France. The hospitality of my immediate host, Professor Serge Lowenthal, as well as the Institute's Director, Professor André Marechal, was impeccable. Not only did they provide me with all the surroundings needed for productivity, but they were kind enough to relieve me of duties normally accompanying a formal appointment. I am most grateful for their support and advice, without which this book would never have had a solid start.

One benefit from the slowness with which the book progressed was the opportunity over many years to expose the material to a host of graduate students, who have an uncanny ability to spot the weak arguments and the outright errors in such a manuscript. To the students of my statistical optics courses at Stanford, therefore, I owe an enormous debt. The evolving notes were also used at a number of other universities, and I am grateful to both William Rhodes (Georgia Institute of Technology) and Timothy Strand (University of Southern California) for providing me with feedback that improved the presentation.

The relationship between author and publisher is often a distant one and sometimes not even pleasant. Nothing could be further from the truth in this case. Beatrice Shube, the editor at John Wiley & Sons who encouraged me to begin this book 15 years ago, has not only been exceedingly patient and understanding, but has also supplied much encouragement and has become a good personal friend. It has been the greatest of pleasures to work with her.

I owe special debts to K.-C. Chin, of Beijing University, for his enormous investment of time in reading the manuscript and suggesting improvements, and to Judith Clark, who typed the manuscript, including all the difficult mathematics, in an extremely professional way.

Finally, I am unable to express adequate thanks to my wife, Hon Mai, and my daughter Michele, not only for their encouragement, but also for the many hours they accepted being without me while I labored at writing.

JOSEPH W. GOODMAN

Stanford, California
October 1984

Contents

1. **Introduction**   1

   1.1 Deterministic versus Statistical Phenomena and Models   2
   1.2 Statistical Phenomena in Optics   3
   1.3 An Outline of the Book   5

2. **Random Variables**   7

   2.1 Definitions of Probability and Random Variables   7
   2.2 Distribution Functions and Density Functions   9
   2.3 Extension to Two or More Joint Random Variables   12
   2.4 Statistical Averages   15
        2.4.1 Moments of a Random Variable   16
        2.4.2 Joint Moments of Random Variables   17
        2.4.3 Characteristic Functions   19
   2.5 Transformations of Random Variables   21
        2.5.1 General Transformation   21
        2.5.2 Monotonic Functions   23
        2.5.3 Multivariate Probability Transformations   27
   2.6 Sums of Real Random Variables   29
        2.6.1 Two Methods for Finding $p_Z(z)$ 29
        2.6.2 Independent Random Variables   31
        2.6.3 The Central Limit Theorem   31
   2.7 Gaussian Random Variables   33
        2.7.1 Definitions   34
        2.7.2 Special Properties of Gaussian Random Variables   37
   2.8 Complex-Valued Random Variables   40
        2.8.1 General Descriptions   40
        2.8.2 Complex Gaussian Random Variables   41

	2.9	Random Phasor Sums	44
	2.9.1	Initial Assumptions	44
	2.9.2	Calculations of Means, Variances, and the Correlation Coefficient	46
	2.9.3	Statistics of the Length and Phase	48
	2.9.4	A Constant Phasor Plus a Random Phasor Sum	50
	2.9.5	Strong Constant Phasor Plus a Weak Random Phasor Sum	54

3. Random Processes 60

3.1	Definition and Description of a Random Process	60
3.2	Stationarity and Ergodicity	63
3.3	Spectral Analysis of Random Processes	68
3.3.1	Spectral Densities of Known Functions	68
3.3.2	Spectral Density of a Random Process	70
3.3.3	Energy and Power Spectral Densities for Linearly Filtered Random Processes	71
3.4	Autocorrelation Functions and the Wiener–Khinchin Theorem	73
3.5	Cross-Correlation Functions and Cross-Spectral Densities	79
3.6	The Gaussian Random Process	82
3.6.1	Definition	82
3.6.2	Linearly Filtered Gaussian Random Processes	83
3.6.3	Wide-Sense Stationarity and Strict Stationarity	84
3.6.4	Fourth-Order Moments	84
3.7	The Poisson Impulse Process	85
3.7.1	Definitions	85
3.7.2	Derivation of Poisson Statistics from Fundamental Hypotheses	88
3.7.3	Derivation of Poisson Statistics from Random Event Times	90
3.7.4	Energy and Power Spectral Densities of Poisson Processes	91
3.7.5	Doubly Stochastic Poisson Processes	95
3.7.6	Linearly Filtered Poisson Processes	97
3.8	Random Processes Derived from Analytic Signals	99
3.8.1	Representation of a Monochromatic Signal by a Complex Signal	99
3.8.2	Representation of a Nonmonochromatic Signal by a Complex Signal	101
3.8.3	Complex Envelopes or Time-Varying Phasors	103
3.8.4	The Analytic Signal as a Complex-Valued Random Process	104

CONTENTS xiii

    3.9 The Complex Gaussian Random Process   108
    3.10 The Karhunen–Loève Expansion   109

4. Some First-Order Properties of Light Waves   116

    4.1 Propagation of Light Waves   117
        4.1.1 Monochromatic Light   117
        4.1.2 Nonmonochromatic Light   118
        4.1.3 Narrowband Light   120
    4.2 Polarized and Unpolarized Thermal Light   120
        4.2.1 Polarized Thermal Light   121
        4.2.2 Unpolarized Thermal Light   124
    4.3 Partially Polarized Thermal Light   127
        4.3.1 Passage of Narrowband Light Through
                Polarization-Sensitive Instruments   127
        4.3.2 The Coherency Matrix   130
        4.3.3 The Degree of Polarization   134
        4.3.4 First-Order Statistics of the Instantaneous Intensity   136
    4.4 Laser Light   138
        4.4.1 Single-Mode Oscillation   139
        4.4.2 Multimode Laser Light   145
        4.4.3 Pseudothermal Light Produced by Passing Laser
                Light Through a Moving Diffuser   151

5. Coherence of Optical Waves   157

    5.1 Temporal Coherence   158
        5.1.1 The Michelson Interferometer   158
        5.1.2 Mathematical Description of the Experiment   161
        5.1.3 Relationship of the Interferogram to the
                Power Spectral Density of the Light Beam   164
        5.1.4 Fourier Spectroscopy   169
    5.2 Spatial Coherence   170
        5.2.1 Young's Experiment   170
        5.2.2 Mathematical Description of Young's Experiment   173
        5.2.3 Some Geometric Considerations   177
        5.2.4 Interference Under Quasimonochromatic
                Conditions   180
        5.2.5 Effects of Finite Pinhole Size   183
    5.3 Cross-Spectral Purity   187
        5.3.1 Power Spectrum of the Superposition of
                Two Light Beams   187
        5.3.2 Cross-Spectral Purity and Reducibility   189

	5.3.3	Laser Light Scattered by a Moving Diffuser	193
5.4		Propagation of Mutual Coherence	195
	5.4.1	Solution Based on the Huygens–Fresnel Principle	196
	5.4.2	Wave Equations Governing Propagation of Mutual Coherence	199
	5.4.3	Propagation of Cross-Spectral Density	201
5.5		Limiting Forms of the Mutual Coherence Function	202
	5.5.1	A Coherent Field	202
	5.5.2	An Incoherent Field	205
5.6		The Van Cittert–Zernike Theorem	207
	5.6.1	Mathematical Derivation	207
	5.6.2	Discussion	210
	5.6.3	An Example	211
	5.6.4	A Generalized Van Cittert–Zernike Theorem	218
5.7		Diffraction of Partially Coherent Light by an Aperture	222
	5.7.1	Effect of a Thin Transmitting Structure on Mutual Intensity	222
	5.7.2	Calculation of the Observed Intensity Pattern	223
	5.7.3	Discussion	226

6. Some Problems Involving High-Order Coherence — 237

6.1		Statistical Properties of the Integrated Intensity of Thermal or Pseudothermal Light	238
	6.1.1	Mean and Variance of the Integrated Intensity	239
	6.1.2	Approximate Form for the Probability Density Function of Integrated Intensity	244
	6.1.3	Exact Solution for the Probability Density Function of Integrated Intensity	250
6.2		Statistical Properties of Mutual Intensity with Finite Measurement Time	256
	6.2.1	Moments of the Real and Imaginary Parts of $J_{12}(T)$	258
	6.2.2	Statistics of the Modulus and Phase of $J_{12}(T)$ for Long Integration Time and Small μ_{12}	263
	6.2.3	Statistics of the Modulus and Phase of $J_{12}(T)$ Under the Condition of High Signal-to-Noise Ratio	269
6.3		Classical Analysis of the Intensity Interferometer	271
	6.3.1	Amplitude versus Intensity Interferometry	272
	6.3.2	Ideal Output of the Intensity Interferometer	274
	6.3.3	"Classical" or "Self" Noise at the Interferometer Output	277

CONTENTS   xv

7. Effects of Partial Coherence on Imaging Systems   286

- 7.1 Some Preliminary Considerations   287
  - 7.1.1 Effects of a Thin Transmitting Object on Mutual Coherence   287
  - 7.1.2 Time Delays Introduced by a Thin Lens   290
  - 7.1.3 Focal-Plane-to-Focal-Plane Coherence Relationships   292
  - 7.1.4 Object–Image Coherence Relations for a Single Thin Lens   296
  - 7.1.5 Relationship Between Mutual Intensities in the Exit Pupil and the Image   300
- 7.2 Methods for Calculating Image Intensity   303
  - 7.2.1 Integration over the Source   303
  - 7.2.2 Representation of the Source by an Incident Mutual Intensity Function   307
  - 7.2.3 The Four-Dimensional Linear Systems Approach   312
  - 7.2.4 The Incoherent and Coherent Limits   320
- 7.3 Some Examples   324
  - 7.3.1 The Image of Two Closely Spaced Points   324
  - 7.3.2 The Image of a Sinusoidal Amplitude Object   328
- 7.4 Image Formation as an Interferometric Process   331
  - 7.4.1 An Imaging System as an Interferometer   331
  - 7.4.2 Gathering Image Information with Interferometers   335
  - 7.4.3 The Importance of Phase Information   340
  - 7.4.4 Phase Retrieval   343
- 7.5 The Speckle Effect in Coherent Imaging   347
  - 7.5.1 The Origin and First-Order Statistics of Speckle   348
  - 7.5.2 Ensemble Average Coherence   351

8. Imaging in the Presence of Randomly Inhomogeneous Media   361

- 8.1 Effects of Thin Random Screens on Image Quality   362
  - 8.1.1 Assumptions and Simplifications   362
  - 8.1.2 The Average Optical Transfer Function   364
  - 8.1.3 The Average Point-Spread Function   366
- 8.2 Random Absorbing Screens   367
  - 8.2.1 General Forms of the Average OTF and the Average PSF   367
  - 8.2.2 A Specific Example   371
- 8.3 Random-Phase Screens   374
  - 8.3.1 General Formulation   375

		8.3.2 The Gaussian Random-Phase Screen	376
		8.3.3 Limiting Forms for Average OTF and Average PSF for Large Phase Variance	381
	8.4	Effects of an Extended Randomly Inhomogeneous Medium on Wave Propagation	384
		8.4.1 Notation and Definitions	385
		8.4.2 Atmospheric Model	388
		8.4.3 Electromagnetic Wave Propagation Through the Inhomogeneous Atmosphere	393
		8.4.4 The Log-Normal Distribution	399
	8.5	The Long-Exposure OTF	402
		8.5.1 Long-Exposure OTF in Terms of the Wave Structure Function	402
		8.5.2 Near-Field Calculation of the Wave Structure Function	407
	8.6	Generalizations of the Theory	414
		8.6.1 Extension to Longer Propagation Paths—Amplitude and Phase Filter Functions	415
		8.6.2 Effects of Smooth Variations of the Structure Constant C_n^2	427
		8.6.3 The Atmospheric Coherence Diameter r_0	429
		8.6.4 Structure Function for a Spherical Wave	432
	8.7	The Short-Exposure OTF	433
		8.7.1 Long versus Short Exposures	433
		8.7.2 Calculation of the Average Short-Exposure OTF	436
	8.8	Stellar Speckle Interferometry	441
		8.8.1 Principle of the Method	442
		8.8.2 Heuristic Analysis of the Method	446
		8.8.3 A More Complete Analysis of Stellar Speckle Interferometry	450
		8.8.4 Extensions	455
	8.9	Generality of the Theoretical Results	457
9.	**Fundamental Limits in Photoelectric Detection of Light**		**465**
	9.1	The Semiclassical Model for Photoelectric Detection	466
	9.2	Effects of Stochastic Fluctuations of the Classical Intensity	468
		9.2.1 Photocount Statistics for Well-Stabilized, Single-Mode Laser Radiation	470
		9.2.2 Photocount Statistics for Polarized Thermal Radiation with a Counting Time Much Shorter Than the Coherence Time	472
		9.2.3 Photocount Statistics for Polarized Thermal Light and an Arbitrary Counting Interval	475

	9.2.4 Polarization Effects	477
	9.2.5 Effects of Incomplete Spatial Coherence	479
9.3	The Degeneracy Parameter	481
	9.3.1 Fluctuations of Photocounts	481
	9.3.2 The Degeneracy Parameter for Blackbody Radiation	486
9.4	Noise Limitations of the Amplitude Interferometer at Low Light Levels	490
	9.4.1 The Measurement System and the Quantities to Be Measured	491
	9.4.2 Statistical Properties of the Count Vector	493
	9.4.3 The Discrete Fourier Transform as an Estimation Tool	494
	9.4.4 Accuracy of the Visibility and Phase Estimates	496
9.5	Noise Limitations of the Intensity Interferometer at Low Light Levels	501
	9.5.1 The Counting Version of the Intensity Interferometer	502
	9.5.2 The Expected Value of the Count-Fluctuation Product and Its Relationship to Fringe Visibility	503
	9.5.3 The Signal-to-Noise Ratio Associated with the Visibility Estimate	506
9.6	Noise Limitations in Speckle Interferometry	510
	9.6.1 A Continuous Model for the Detection Process	511
	9.6.2 The Spectral Density of the Detected Imagery	512
	9.6.3 Fluctuations of the Estimate of Image Spectral Density	517
	9.6.4 Signal-to-Noise Ratio for Stellar Speckle Interferometry	519
	9.6.5 Discussion of the Results	521

Appendix A. The Fourier Transform — **528**

A.1	Fourier Transform Definitions	528
A.2	Basic Properties of the Fourier Transform	529
A.3	Table of One-Dimensional Fourier Transforms	531
A.4	Table of Two-Dimensional Fourier Transform Pairs	532

Appendix B. Random Phasor Sums — **533**

Appendix C. Fourth-Order Moment of the Spectrum of a Detected Speckle Image — **539**

Index — **543**

Statistical Optics

1
Introduction

Optics, as a field of science, is well into its second millennium of life; yet in spite of its age, it remains remarkably vigorous and youthful. During the middle of the twentieth century, various events and discoveries have given new life, energy, and richness to the field. Especially important in this regard were (1) the introduction of the concepts and tools of Fourier analysis and communication theory into optics, primarily in the late 1940s and throughout the 1950s, (2) the discovery and successful realization of the laser in the late 1950s, and (3) the origin of the field of nonlinear optics in the 1960s. It is the thesis of this book that a less dramatic but equally important change has taken place gradually, but with an accelerating pace, throughout the entire century, namely, the infusion of statistical concepts and methods of analysis into the field of optics. It is to the role of such concepts in optics that this book is devoted.

The field of statistical optics has a considerable history of its own. Many fundamental statistical problems were solved in the late nineteenth century and applied to acoustics and optics by Lord Rayleigh. The need for statistical methods in optics increased dramatically with the discovery of the quantized nature of light, and particularly with the statistical interpretation of quantum mechanics introduced by Max Born. The introduction by E. Wolf in 1954 of an elegant and broad framework for considering the coherence properties of waves laid a foundation within which many of the important statistical problems in optics could be treated in a unified way. Also worth special mention is the semiclassical theory of light detection, pioneered by L. Mandel, which tied together (in a comparatively simple way) knowledge of the statistical fluctuations of classical wave quantities (fields, intensities) and fluctuations associated with the interaction of light and matter. This history is far from complete but is dealt with in more detail in the individual chapters that follow.

1.1 DETERMINISTIC VERSUS STATISTICAL PHENOMENA AND MODELS

In the normal course of events, a student of physics or engineering first encounters optics in an entirely deterministic framework. Physical quantities are represented by mathematical functions that are either completely specified in advance or are assumed to be precisely measurable. These physical quantities are subjected to well-defined transformations that modify their form in perfectly predictable ways. For example, if a monochromatic light wave with a known complex field distribution is incident on a transparent aperture in a perfectly opaque screen, the resulting complex field distribution some distance away from the screen can be calculated precisely by using the well-established diffraction formulas of wave optics.

The students emerging from such an introductory course may feel confident that they have grasped the basic physical concepts and laws and are ready to find a precise answer to almost any problem that comes their way. To be sure, they have probably been warned that there are certain problems, arising particularly in the detection of weak light waves, for which a statistical approach is required. But a statistical approach to problem solving often appears at first glance to be a "second-class" approach, for statistics is generally used when we lack sufficient information to carry out the aesthetically more pleasing "exact" solution. The problem may be inherently too complex to be solved analytically or numerically, or the boundary conditions may be poorly defined. Surely the preferred way to solve a problem must be the deterministic way, with statistics entering only as a sign of our own weakness or limitations. Partially as a consequence of this viewpoint, the subject of statistical optics is usually left for the more advanced students, particularly those with a mathematical flair.

Although the origins of the above viewpoint are quite clear and understandable, the conclusions reached regarding the relative merits of deterministic and statistical analysis are very greatly in error, for several important reasons. First, it is difficult, if not impossible, to conceive of a real engineering problem in optics that does not contain some element of uncertainty requiring statistical analysis. Even the lens designer, who traces rays through application of precise physical laws accepted for centuries, must ultimately worry about quality control! Thus statistics is certainly not a subject to be left primarily to those more interested in mathematics than in physics and engineering.

Furthermore, the view that the use of statistics is an admission of one's limitations and thus should be avoided is based on too narrow a view of the nature of statistical phenomena. Experimental evidence indicates, and indeed the great majority of physicists believe, that the interaction of light and matter is *fundamentally* a statistical phenomenon, which cannot in principle be predicted with perfect precision in advance. Thus statistical phenomena play a role of the greatest importance in the world around us, independent of our particular mental capabilities or limitations.

Finally, in defense of statistical analysis, we must say that, whereas both deterministic and statistical approaches to problem solving require the construction of mathematical models of physical phenomena, the models constructed for statistical analysis are inherently more general and flexible. Indeed, they invariably contain the deterministic model as a special case! For a statistical model to be accurate and useful, it should fully incorporate the current state of our knowledge regarding the physical parameters of concern. Our solutions to statistical problems will be no more accurate than the models we use to describe both the physical laws involved and the state of knowledge or ignorance.

The statistical approach is indeed somewhat more complex than the deterministic approach, for it requires knowledge of the elements of probability theory. In the long run, however, statistical models are far more powerful and useful than deterministic models in solving physical problems of genuine practical interest. Hopefully the reader will agree with this viewpoint by the time this book is completed.

1.2 STATISTICAL PHENOMENA IN OPTICS

Statistical phenomena are so plentiful in optics that there is no difficulty in compiling a long list of examples. Because of the wide variety of these problems, it is difficult to find a general scheme for classifying them. Here we attempt to identify several broad aspects of optics that require statistical treatment. These aspects are conveniently discussed in the context of an optical imaging problem.

Most optical imaging problems are of the following type. Nature assumes some particular state (e.g., a certain collection of atoms and/or molecules in a distant region of space, a certain distribution of reflectance over terrain of unknown characteristics, or a certain distribution of transmittance in a sample of interest). By operating on optical waves that arise as a consequence of this state of Nature, we wish to deduce exactly what that state is.

Statistics is involved in this task in a wide variety of ways, as can be discovered by reference to Fig. 1-1. First, and most fundamentally, the state of Nature is known to us a priori only in a statistical sense. If it were known exactly, there would be no need for any measurement in the first place. Thus the state of Nature is random, and in order to properly assess the performance of the system, we must have a statistical model, ideally representing

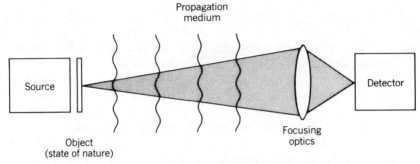

Figure 1-1. An optical imaging system.

the set of possible states, together with their associated probabilities. Usually, a less complete description of the statistical properties of the object will suffice.

Our measurement system operates not on the state of Nature per se, but rather on an optical representation of that state (e.g., radiated light, transmitted light, or reflected light). The representation of the state of Nature by an optical wave has statistical attributes itself, primarily as a result of the statistical or random properties of all real light waves. Because of the fundamentally statistical nature of the interaction of light and matter, all optical sources produce radiation that is statistical in its properties. At one extreme we have the chaotic and unordered emission of light by a thermal source, such as an incandescent lamp; at the other extreme we have the comparatively ordered emission of light by a continuous-wave (CW) gas laser. Such light comes close to containing a single frequency and traveling in a single direction. Nonetheless, any real laser emits light with statistical properties, in particular random fluctuations of both the amplitude and phase of the radiation. Statistical fluctuations of light are of great importance in many optical experiments and indeed play a central role in determining the character of the image produced by the system depicted in Fig. 1-1.

After interacting with the state of Nature, the radiation travels through an intervening medium until it reaches our measurement instrument. The parameters of that medium may or may not be well known. If the medium is a perfect vacuum, it introduces no additional statistical aspects to the problem. On the other hand, if the medium is the Earth's atmosphere and the optical path is a few meters or more in length, the random fluctuations of the atmospheric index or refraction can have dramatic effects on the wave and can seriously degrade the image obtained by the system. Statistical methods are required to quantify this degradation.

The light eventually reaches our measurement apparatus, which performs some desired operations on it before it is detected. For example, the light beam may pass through an interferometer, as in Fourier spectroscopy, or through a system of lenses, as in aerial photography. How well are the exact parameters of our measurement instrument known? Any lack of knowledge of these parameters must be taken into account in our statistical model for the measurement process. For example, there may be unknown errors in the wavefront deformation introduced by passage through the lens system. Such errors can often be modeled statistically and should be taken into account in assessment of the performance of the system.

The radiation finally reaches an optical detector, where again there is an interaction of light and matter. Random fluctuations of the detected energy are readily observed, particularly at low light levels, and can be attributed to a variety of causes, including the discrete nature of the interaction between light and matter and the presence of internal electronic detector noise (thermal noise). The result of the measurement is related in only a statistical way to the image falling on the detector.

At all stages of the optical problem, including illumination, transmission, image formation, and detection, therefore, statistical treatment is needed in order to fully assess the performance of the system. Our goal in this book is to lay the necessary foundation and to illustrate the application of statistics to the many diverse areas of optics where it is needed.

1.3 AN OUTLINE OF THE BOOK

Eight chapters follow this Introduction. Since many scientists and engineers working in the field of optics may feel a need to sharpen their abilities with statistical tools, Chapter 2 presents a review of probability theory, and Chapter 3 contains a review of the theory of random processes, which are used as models for many of the statistical phenomena described in later chapters. The reader already familiar with these subjects may wish to proceed directly to Chapter 4, using the earlier material primarily as a reference resource.

Discussion of optical problems begins in Chapter 4, which deals with the "first-order" statistics (i.e., the statistics at a single point in space and time) of several kinds of light waves, including light generated by thermal sources and light generated by lasers. Also included is an introduction to a formalism that allows characterization of the polarization properties of an optical wave.

Chapter 5 introduces the concepts of time and space coherence (which are "second-order" statistical properties of light waves) and deals at length with the propagation of coherence under various conditions. Chapter 6 extends this theory to coherence of order higher than 2 and illustrates the need for fourth-order coherence functions in a variety of optical problems, including classical analysis of the intensity interferometer.

Chapter 7 is devoted to the theory of image formation with partially coherent light. Several analytical approaches to the problem are introduced. The concept of interferometric imaging, as widely practiced in radio astronomy, is also introduced in this chapter and is used to lend insight into the character of optical imaging systems. The phase retrieval problem is introduced and discussed.

Chapter 8 is concerned with the effects of random media, such as the Earth's atmosphere, on the quality of images formed by optical instruments. The origin of random refractive-index fluctuations in the atmosphere is reviewed, and statistical models for such fluctuations are introduced. The effects of these fluctuations on optical waves are also modeled, and image degradations introduced by the atmosphere are treated from a statistical viewpoint. Stellar speckle interferometry, a method for partially overcoming the effects of atmospheric turbulence, is discussed in some detail.

Finally, Chapter 9 treats the semiclassical theory of light detection and illustrates the theory with analyses of the sensitivity limitations of amplitude interferometry, intensity interferometry, and stellar speckle interferometry.

Appendixes A through C present supplemental background material and analysis.

2

Random Variables

Since this book deals primarily with statistical problems in optics, it is essential that we start with a clear understanding of the mathematical methods used to analyze random or statistical phenomena. We shall assume at the start that the reader has been exposed previously to at least some of the basic elements of probability theory. The purpose of this chapter is to provide a review of the most important material, to establish notation, and to present a few specific results that will be useful in later applications of the theory. The emphasis is *not* on mathematical rigor, but rather on physical plausibility. For more rigorous treatment of the theory of probability, the reader may consult texts on statistics (e.g., Refs. 2-1 and 2-2). In addition, there are many excellent engineering-oriented books that discuss the theory of random variables and random processes (e.g., Refs. 2-3 through 2-8).

2.1 DEFINITIONS OF PROBABILITY AND RANDOM VARIABLES

By a *random experiment* we mean an experiment with an outcome that cannot be predicted in advance. Let the collection of possible outcomes be represented by the set of events $\{A\}$. For example, if the experiment consists of the tossing of two coins side by side, the possible "elementary events" are HH, HT, TH, TT, where H indicates "heads" and T denotes "tails." However, the set $\{A\}$ contains more than four elements, since events such as "at least one head occurs in the two tosses" (HH or HT or TH) are included. If A_1 and A_2 are any two events, the set $\{A\}$ must also contain A_1 and A_2, A_1 or A_2, not A_1 and not A_2. In this way, the complete set A is derived from the underlying elementary events.

If we repeat the experiment N times and observe the specific event A to occur n times, we define the relative frequency of the event A to be the ratio n/N. It is then appealing to attempt to define the probability of the event A as the limit of the relative frequency as the number of trials N increases without bound,

$$P(A) = \lim_{N \to \infty} \frac{n}{N}. \tag{2.1-1}$$

Unfortunately, although this definition of probability has physical appeal, it is not entirely satisfactory. Note that we have assumed that the relative frequency of each event will indeed approach a limit as N increases, an assumption we are by no means prepared to prove. Furthermore, we can never really measure the exact value of $P(A)$, for to do so would require an infinite number of experimental trials. As a consequence of these difficulties and others, it is preferable to adopt an axiomatic approach to probability theory, assuming at the start that probabilities obey certain axioms, all of which are derived from corresponding properties of relative frequencies. The necessary axioms are as follows:

(1) Any probability $P(A)$ obeys $P(A) \geq 0$.
(2) If S is an event certain to occur, then $P(S) = 1$.
(3) If A_1 and A_2 are *mutually exclusive* events, that is, the occurrence of one guarantees that the second does not occur, the probability of the event A_1 *or* A_2 satisfies

$$P(A_1 \text{ or } A_2) = P(A_1) + P(A_2).$$

The theory of probability is based on these axioms.

The problem of assigning specific numerical values to the probabilities of various events is not addressed by the axiomatic approach, but rather is left to our physical intuition. Whatever number we assign for the probability of a given event must agree with our intuitive feeling for the limiting relative frequency of that event. In the end, we are simply building a statistical model that we hope will represent the experiment. The necessity to hypothesize a model should not be disturbing, for every deterministic analysis likewise requires hypotheses about the physical entities concerned and the transformations they undergo. Our statistical model must be judged on the basis of its accuracy in describing the behavior of experimental results over many trials.

We are now prepared to introduce the concept of a *random variable*. To every possible elementary event A of our underlying random experiment we assign a real number $u(A)$. The random variable[†] U consists of all possible

---

[†] Here and in Chapter 3 we consistently represent random variables by capital letters and specific values of random variables by lowercase letters.

DISTRIBUTION FUNCTIONS AND DENSITY FUNCTIONS

$u(A)$, together with an associated measure of their probabilities. Note especially that the random variable consists of *both* the set of values *and* their associated probabilities and hence encompasses the entire statistical model that we hypothesize for the random phenomenon.

2.2 DISTRIBUTION FUNCTIONS AND DENSITY FUNCTIONS

A random variable U is called *discrete* if the experimental outcomes consist of a discrete set of possible numbers. A random variable is called *continuous* if the experimental results can lie anywhere on a continuum of possible values. Occasionally, a *mixed* random variable is encountered, with outcomes that lie on either a discrete set (with certain probabilities) or on a continuum.

In all cases it is convenient to describe the random variable U by a probability distribution function $F_U(u)$, which is defined by[†]

$$F_U(u) = \text{Prob}\{U \leq u\}, \qquad (2.2\text{-}1)$$

or in other words, the probability that the random variable U assumes a value less than or equal to the specific value u. From the basic axioms of probability theory we can show that $F_U(u)$ must have the following properties:

(1) $F_U(u)$ is nondecreasing to the right.
(2) $F_U(-\infty) = 0$.
(3) $F_U(+\infty) = 1$.

Figure 2-1 shows typical forms for $F_U(u)$ in the discrete, continuous, and mixed cases. Note that the probability that U lies between the limits $a < U \leq b$ can be expressed as

$$\text{Prob}\{a < U \leq b\} = F_U(b) - F_U(a). \qquad (2.2\text{-}2)$$

Of more importance to us in practical applications will be the probability density function, represented by $p_U(u)$ and defined by

$$p_U(u) \triangleq \frac{d}{du} F_U(u). \qquad (2.2\text{-}3)$$

[†] The symbol Prob{ } means the probability that the event described within the brackets occurs.

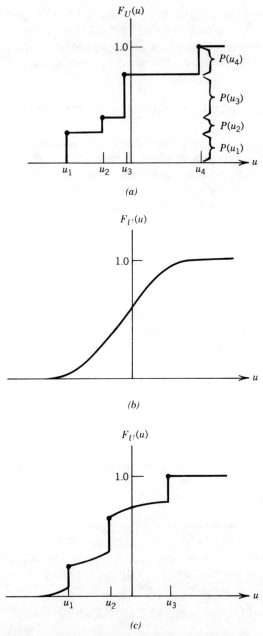

Figure 2-1. Typical probability distribution functions for (a) discrete, (b) continuous, and (c) mixed random variables.

DISTRIBUTION FUNCTIONS AND DENSITY FUNCTIONS

For a continuous random variable U, there is no difficulty in applying this definition, for $F_U(u)$ is everywhere differentiable. Noting from the definition of a derivative that

$$p_U(u) = \lim_{\Delta u \to 0} \frac{F_U(u) - F_U(u - \Delta u)}{\Delta u},$$

we see that for sufficiently small Δu,

$$p_U(u)\Delta u \cong F_U(u) - F_U(u - \Delta u) = \text{Prob}\{u - \Delta u < U \le u\},$$

or in words, $p_U(u)\Delta u$ is the probability that U lies in the range $u - \Delta u < U \le u$. From the fundamental properties of $F_U(u)$, it follows that $p_U(u)$ must have the basic properties

$$p_U(u) \ge 0; \qquad \int_{-\infty}^{\infty} p_U(u)\,du = 1. \qquad (2.2\text{-}4)$$

The probability that U assumes a value between the limits a and b can be expressed in terms of the probability density function by

$$\text{Prob}\{a < U \le b\} = \int_a^b p_U(u)\,du. \qquad (2.2\text{-}5)$$

When U is a discrete random variable, $F_U(u)$ is discontinuous, and hence $p_U(u)$ does not exist in the usual sense. By introducing Dirac δ functions (Ref. 2-9, Chapter 5), however, we can include this case within our framework. The probability density function becomes

$$p_U(u) = \sum_{k=1}^{\infty} P(u_k)\delta(u - u_k), \qquad (2.2\text{-}6)$$

where $\{u_1, u_2, \ldots, u_k, \ldots\}$ represents the discrete set of possible numerical values, and the δ function is defined to have the properties[†]

$$\delta(u - u_k) = 0; \qquad u \ne u_k,$$

$$\int_{-\infty}^{\infty} g(u)\delta(u - u_k)\,du = g(u_k^-). \qquad (2.2\text{-}7)$$

The density function for a mixed random variable contains both a continu-

---

[†] The symbol $g(u_k^-)$ signifies the limit of $g(u)$ as u approaches u_k from the left. For a continuous $g(u)$, $g(u_k^-) = g(u_k)$.

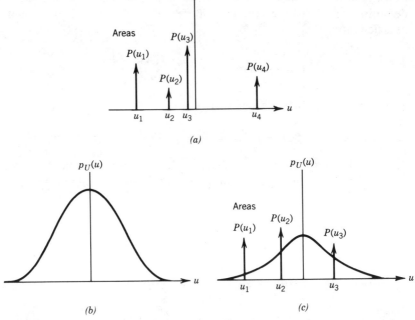

Figure 2-2. Typical probability density functions for (*a*) discrete, (*b*) continuous, and (*c*) mixed random variables.

ous component and δ-function components. Figure 2-2 illustrates the character of probability density functions in these three cases.

Two specific probability density functions will illustrate the continuous and discrete cases; both are important to us for later work:

Gaussian density $\quad p_U(u) = \dfrac{1}{\sqrt{2\pi}\,\sigma} \exp\left\{ -\dfrac{(u - \bar{u})^2}{2\sigma^2} \right\}$

Poisson density $\quad p_U(u) = \displaystyle\sum_{k=0}^{\infty} \dfrac{(\bar{k})^k}{k!} e^{-\bar{k}}\, \delta(u - k),$

where $\bar{u}$, σ, and $\bar{k}$ are parameters.

2.3 EXTENSION TO TWO OR MORE JOINT RANDOM VARIABLES

Consider two random experiments with sets of possible events $\{A\}$ and $\{B\}$. If the events are taken in pairs, one from each set, we define a new set of possible joint outcomes, which we denote $\{A \times B\}$. The relative frequency

EXTENSION TO TWO OR MORE JOINT RANDOM VARIABLES

with which the specific event A occurs jointly with the specific event B is denoted n/N, where N is the number of joint experimental trials, whereas n is the number of times A and B occur as joint results of the two experiments. We assign a joint probability $P(A, B)$ to this pair of outcomes, and the specific value of this probability is determined by our intuitive notions concerning the limiting value of the relative frequency n/N. Since it is a probability, $P(A, B)$ must satisfy the axioms given in Section 2.1.

To each outcome A of the first experiment we assign a numerical value $u(A)$ and to each outcome B of the second experiment a value $v(B)$. The *joint random variable UV* is defined to be the collection of all possible joint numbers (u, v), together with an associated measure of probability.

The *probability distribution function* $F_{UV}(u, v)$ for the joint random variable UV is defined as

$$F_{UV}(u, v) \triangleq \text{Prob}\{U \leq u \text{ and } V \leq v\} \tag{2.3-1}$$

and the *probability density function* $p_{UV}(u, v)$ by

$$p_{UV}(u, v) \triangleq \frac{\partial^2}{\partial u \, \partial v} F_{UV}(u, v). \tag{2.3-2}$$

Here the partial derivatives must be interpreted as existing either in the usual sense or in a δ-function sense, depending on whether F_{UV} is or is not continuous. The density function $p_{UV}(u, v)$ must have unit volume, that is,

$$\iint_{-\infty}^{\infty} p_{UV}(u, v) \, du \, dv = 1. \tag{2.3-3}$$

If we know the joint probabilities of all specific events A and B, we may wish to determine the probability that a specific event A occurs, regardless of the particular event B that accompanies it. Reasoning directly from relative frequency concepts, we can show that

$$P(A) = \sum_{\substack{\text{all} \\ B}} P(A, B)$$

and similarly

$$P(B) = \sum_{\substack{\text{all} \\ A}} P(A, B).$$

Values $P(A)$ and $P(B)$ so defined are referred to as the *marginal probabilities* of A and B, respectively.

In a similar fashion, the marginal probability density functions of the random variables U and V derived from the two random experiments are defined by

$$p_U(u) \triangleq \int_{-\infty}^{\infty} p_{UV}(u,v)\, dv$$

$$p_V(v) \triangleq \int_{-\infty}^{\infty} p_{UV}(u,v)\, du. \qquad (2.3\text{-}4)$$

These functions are the density functions of one random variable when the particular value assumed by the second random variable is of no concern.

The probability of observing the event B in one experiment, given that the event A has already been observed in the other experiment, is called the *conditional probability* of B given A and is written $P(B|A)$. Note that the relative frequency of the joint event (A, B) can be written

$$\frac{n}{N} = \frac{n}{m} \cdot \frac{m}{N}$$

where n is the number of times the joint event (A, B) occurs in N trials, whereas m is the number of times A occurs in N trials, regardless of the particular value of B. But m/N represents the (marginal) relative frequency of A, whereas n/m represents the (conditional) relative frequency of B, given that A has occurred. It follows that the probabilities of concern must satisfy

$$P(A, B) = P(A) P(B|A)$$

or

$$P(B|A) = \frac{P(A, B)}{P(A)}. \qquad (2.3\text{-}5)$$

Similarly,

$$P(A|B) = \frac{P(A, B)}{P(B)} \qquad (2.3\text{-}6)$$

Taken together, (2.3-5) and (2.3-6) imply that

$$P(B|A) = \frac{P(A|B) P(B)}{P(A)}, \qquad (2.3\text{-}7)$$

which is known as *Bayes' rule*.

Following the above reasoning, the conditional probability density functions of U and V are defined by

$$p_{V|U}(v|u) = \frac{p_{UV}(u,v)}{p_U(u)},$$

$$p_{U|V}(u|v) = \frac{p_{UV}(u,v)}{p_V(v)}. \qquad (2.3\text{-}8)$$

Finally, we introduce the concept of statistical independence. Two random variables U and V are called *statistically independent* if knowledge about the value assumed by one does not influence the probabilities associated with possible outcomes of the second. It follows that for statistically independent random variables, we have

$$p_{V|U}(v|u) = p_V(v). \qquad (2.3\text{-}9)$$

This fact, in turn, implies that

$$p_{UV}(u,v) = p_U(u) p_{V|U}(v|u) = p_U(u) p_V(v), \qquad (2.3\text{-}10)$$

or, in words, the joint probability density function of two independent random variables factors into the product of their two marginal density functions.†

2.4 STATISTICAL AVERAGES

Let $g(u)$ be a function that for every real number u assigns a new real number $g(u)$. If u represents the value of a random variable, $g(u)$ is also the value of a random variable.

We define the *statistical average* (mean value, expected value) of $g(u)$ by

$$\bar{g}(u) = E[g(u)] \triangleq \int_{-\infty}^{\infty} g(u) p_U(u) \, du. \qquad (2.4\text{-}1)$$

For a discrete random variable, $p_U(u)$ is of the form

$$p_U(u) = \sum_k P(u_k) \delta(u - u_k) \qquad (2.4\text{-}2)$$

†More generally, two events A and B are statistically independent if $P(A, B) = P(A) P(B)$.

with the result that

$$\bar{g}(u) = \sum_k P(u_k) g(u_k). \qquad (2.4\text{-}3)$$

For a continuous random variable, however, the average must be found by integration.

2.4.1 Moments of a Random Variable

The simplest average properties of a random variable are its *moments*, which (if they exist) are obtained by setting

$$g(u) = u^n$$

in Eq. (2.4-1). Of particular importance is the first moment (mean value, expected value, average value),

$$\bar{u} = \int_{-\infty}^{\infty} u\, p_U(u)\, du, \qquad (2.4\text{-}4)$$

and the *second moment* (mean-square value),

$$\overline{u^2} = \int_{-\infty}^{\infty} u^2 p_U(u)\, du. \qquad (2.4\text{-}5)$$

Often, the fluctuations of a random variable about its mean are of greatest interest, in which case we deal with the *central moments*, obtained with

$$g(u) = (u - \bar{u})^n. \qquad (2.4\text{-}6)$$

Of most importance is the second central moment, or *variance*, defined by

$$\sigma^2 = \int_{-\infty}^{\infty} (u - \bar{u})^2 p_U(u)\, du. \qquad (2.4\text{-}7)$$

As a simple exercise (see Problem 2-1), the reader can prove the following relationship between the moments of any random variable:

$$\overline{u^2} = (\bar{u})^2 + \sigma^2.$$

The square root of the variance, σ, is called the *standard deviation* and is a measure of the dispersion or spread of values assumed by the random variable U.

2.4.2 Joint Moments of Random Variables

Let U and V be random variables jointly distributed with probability density function $p_{UV}(u,v)$. The joint moments of U and V are defined by

$$\overline{u^n v^m} \triangleq \iint_{-\infty}^{\infty} u^n v^m p_{UV}(u,v)\, du\, dv. \tag{2.4-8}$$

Of particular importance are the *correlation* of U and V,

$$\Gamma_{UV} = \overline{uv} = \iint_{-\infty}^{\infty} uv\, p_{UV}(u,v)\, du\, dv, \tag{2.4-9}$$

the *covariance* of U and V,

$$C_{UV} = \overline{(u-\bar{u})(v-\bar{v})} = \Gamma_{UV} - \bar{u}\bar{v} \tag{2.4-10}$$

and the *correlation coefficient*

$$\rho = \frac{C_{UV}}{\sigma_U \sigma_V}. \tag{2.4-11}$$

The correlation coefficient is a direct measure of the similarity of the fluctuations of U and V. As we show in the argument to follow, the modulus of ρ always lies between zero and one. The argument begins with Schwarz's inequality, which states that for any two (real or complex-valued) functions[†] $\mathbf{f}(u,v)$ and $\mathbf{g}(u,v)$,

$$\left| \iint_{-\infty}^{\infty} \mathbf{f}(u,v)\mathbf{g}(u,v)\, du\, dv \right|^2 \leq \iint_{-\infty}^{\infty} |\mathbf{f}(u,v)|^2\, du\, dv \iint_{-\infty}^{\infty} |\mathbf{g}(u,v)|^2\, du\, dv$$

$$\tag{2.4-12}$$

with equality if and only if

$$\mathbf{g}(u,v) = \mathbf{a}\mathbf{f}^*(u,v). \tag{2.4-13}$$

where $\mathbf{a}$ is a complex constant and * indicates a complex conjugate. Making the specific choices

$$\mathbf{f}(u,v) = (u-\bar{u})\sqrt{p_{UV}(u,v)}$$

$$\mathbf{g}(u,v) = (v-\bar{v})\sqrt{p_{UV}(u,v)}, \tag{2.4-14}$$

[†] We shall consistently use boldface characters to indicate quantities that are or could be complex valued.

we obtain

$$\left| \iint_{-\infty}^{\infty} (u - \bar{u})(v - \bar{v}) p_{UV}(u,v)\, du\, dv \right|^2$$

$$\leq \iint_{-\infty}^{\infty} (u - \bar{u})^2 p_{UV}(u,v)\, du\, dv \iint_{-\infty}^{\infty} (v - \bar{v})^2 p_{UV}(u,v)\, du\, dv,$$

(2.4-15)

or equivalently $|C_{UV}| \leq \sigma_U \sigma_V$, thus proving that

$$0 \leq |\rho| \leq 1. \tag{2.4-16}$$

If $\rho = 1$, we say that U and V are *perfectly correlated*, meaning that their fluctuations are essentially identical, up to possible scaling factors. If $\rho = -1$, we say that U and V are *anticorrelated*, meaning that their fluctuations are identical but in an opposite sense (again up to scaling factors), with a large positive excursion of U accompanied by a large negative excursion of V, for example.

When ρ is identically zero, U and V are said to be *uncorrelated*. The reader can easily show (see Problem 2-2) that *two statistically independent random variables are always uncorrelated*. However, the converse is not true; that is, lack of correlation does not necessarily imply statistical independence. A classic illustration is provided by the random variables

$$U = \cos \Theta$$

$$V = \sin \Theta \tag{2.4-17}$$

with Θ a random variable uniformly distributed on $(-\pi/2, \pi/2)$, that is,

$$p_\Theta(\theta) = \begin{cases} \dfrac{1}{\pi} & -\dfrac{\pi}{2} < \theta \leq \dfrac{\pi}{2} \\ 0 & \text{otherwise.} \end{cases}$$

Knowledge of the value of V uniquely identifies the value of U, and hence the two random variables are statistically *dependent*. Nonetheless, the reader can verify (see Problem 2-3) that U and V are uncorrelated random variables.

2.4.3 Characteristic Functions

The *characteristic function* of a random variable U is defined as the expected value of $\exp(j\omega u)$,

$$\mathbf{M}_U(\omega) \triangleq \int_{-\infty}^{\infty} \exp(j\omega u) p_U(u)\, du. \qquad (2.4\text{-}18)$$

Thus the characteristic function is the Fourier transform[†] of the probability density function of U. If this integral exists, at least in the sense of δ functions, the relationship is an invertible one, and the probability density function is expressible as

$$p_U(u) = \frac{1}{2\pi} \int_{-\infty}^{\infty} \mathbf{M}_U(\omega) \exp(-j\omega u)\, d\omega. \qquad (2.4\text{-}19)$$

The characteristic function thus contains all information about the first-order statistical properties of the random variable U.

Under certain circumstances it is possible to obtain the characteristic function (and hence the probability density function by 2.4-19) from knowledge of the nth-order moments for all n. To demonstrate this fact, we expand the exponential in Eq. (2.4-18) in a power series,

$$\exp(j\omega u) = \sum_{n=0}^{\infty} \frac{(j\omega u)^n}{n!}$$

If we assume that the orders of summation and integration can be interchanged, we obtain

$$\mathbf{M}_U(\omega) = \sum_{n=0}^{\infty} \frac{(j\omega)^n}{n!} \int_{-\infty}^{\infty} u^n p_U(u)\, du = \sum_{n=0}^{\infty} \frac{(j\omega)^n}{n!} \overline{u^n} \qquad (2.4\text{-}20)$$

[As a result of conditions required for validity of the interchange of orders of integration and summation given above, this result is valid only if all the moments are finite and the resulting series converges absolutely (Ref. 2-3).]

In addition, if the nth absolute moment $\int_{-\infty}^{\infty} |u|^n p_U(u)\, du$ exists, then the nth moment of U can be found from

$$\overline{u^n} = \left. \frac{1}{j^n} \frac{d^n}{d\omega^n} \mathbf{M}_U(\omega) \right|_{\omega=0}, \qquad (2.4\text{-}21)$$

as is made plausible by Eq. (2.4-20).

[†] For a brief review of Fourier transforms, see Appendix A.

The characteristic functions of the Gaussian and Poisson random variables are readily shown to be

$$\text{Gaussian:} \quad \mathbf{M}_U(\omega) = \exp\left(-\frac{\sigma^2\omega^2}{2}\right)\exp(j\omega\bar{u})$$

$$\text{Poisson:} \quad \mathbf{M}_U(\omega) = \sum_{k=0}^{\infty} \frac{(\bar{k})^k}{k!} e^{-\bar{k}} \exp(j\omega k)$$

$$= \exp\{\bar{k}(e^{j\omega} - 1)\}.$$

On occasion we shall have use for the *joint characteristic function* of two random variables U and V, defined by

$$\mathbf{M}_{UV}(\omega_U, \omega_V) = \iint_{-\infty}^{\infty} \exp[j(\omega_U u + \omega_V v)] p_{UV}(u,v) \, du \, dv.$$

(2.4-22)

The joint density function is recoverable from $\mathbf{M}_{UV}(\omega_U, \omega_V)$ by a two-dimensional Fourier inversion. In addition, joint moments of U and V are expressible in the form (see Problem 2-5)

$$\overline{u^n v^m} = \frac{1}{(j)^{n+m}} \frac{\partial^{n+m}}{\partial \omega_U^n \partial \omega_V^m} \mathbf{M}_{UV}(\omega_U, \omega_V) \bigg|_{\omega_U = \omega_V = 0} \quad (2.4\text{-}23)$$

provided $|\overline{u^n v^m}| < \infty$.

Finally, the nth-order joint characteristic function of the random variables $U_1, U_2, \ldots, U_n$ is defined by

$$\mathbf{M}_U^{(n)}(\omega_1, \omega_2, \ldots, \omega_n) \triangleq E\{\exp[j(\omega_1 u_1 + \omega_2 u_2 + \cdots + \omega_n u_n)]\}.$$

(2.4-24)

Equivalently, in matrix notation we can write

$$\mathbf{M}_U(\underline{\omega}) \triangleq E\{\exp[j\underline{\omega}^t \underline{u}]\}, \quad (2.4\text{-}25)$$

where $\underline{\omega}$ and $\underline{u}$ are column matrices,

$$\underline{\omega} = \begin{bmatrix} \omega_1 \\ \omega_2 \\ \vdots \\ \omega_n \end{bmatrix} \quad \underline{u} = \begin{bmatrix} u_1 \\ u_2 \\ \vdots \\ u_n \end{bmatrix}, \quad (2.4\text{-}26)$$

and the superscript t indicates a matrix transpose operation. The nth-order joint probability density function $p_U(\underline{u})$ can be obtained from $\mathbf{M}_U(\underline{\omega})$ by an nth-order Fourier inversion.

2.5 TRANSFORMATIONS OF RANDOM VARIABLES

It is important in practical applications to be able to determine the probability density function of a random variable after it has been subjected to a linear or nonlinear transformation. Generally we know the probability density function $p_U(u)$ of the random variable U, and U is subjected to a transformation

$$z = f(u). \tag{2.5-1}$$

The problem is then to find the probability density function $p_Z(z)$. Different approaches to this problem are possible, depending on the nature of the function $f(u)$.

2.5.1 General Transformation

We first treat the most general case, in which we assume only that $f(u)$ is single valued; thus each value of u maps into only one value of z. (For each z, however, there may be many values of u.) Figure 2-3 illustrates one possible function $f(u)$.

To find $p_Z(z)$, the most general approach is to first find the distribution function $F_Z(z)$ and then differentiate it with respect to z. Again referring to Fig. 2-3, we choose a specific value of z and let the symbol L_z represent the set of all points on the u axis that map into values less than or equal to that z (L_z is the crosshatched region of the u axis.) The region L_z is, of course, a function of the particular value of z chosen. Now the probability that $Z \leq z$ can be expressed as

$$F_Z(z) = \text{Prob}\{U \text{ lies in } L_z\}. \tag{2.5-2}$$

The density function $p_Z(z)$ is then given by

$$p_Z(z) = \frac{d}{dz} \text{Prob}\{U \text{ lies in } L_z\}. \tag{2.5-3}$$

The application of this formalism is best understood with the aid of an example. Let U be a random variable with known density function $p_U(u)$

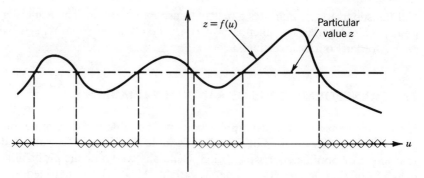

Figure 2-3. The crosshatched line segments represent the values of u for which the random variable Z will be less than or equal to the particular value of z shown.

and let $z = au^2$. The problem is to find $p_Z(z)$. We first plot the function $z = au^2$ in Fig. 2-4. Then we choose a particular value of z and identify the region L_z, as shown in the figure. Clearly,

$$F_Z(z) = \text{Prob}\{Z \le z\} = \text{Prob}\left\{-\sqrt{\frac{z}{a}} < U \le +\sqrt{\frac{z}{a}}\right\}. \quad (2.5\text{-}4)$$

This expression can be restated in the form

$$F_Z(z) = \int_{-\infty}^{+\sqrt{z/a}} p_U(u)\, du - \int_{-\infty}^{-\sqrt{z/a}} p_U(u)\, du. \quad (2.5\text{-}5)$$

To find the density function $p_Z(z)$, it remains to differentiate (2.5-5) with respect to z. As an aid in this task, we make use of the general relation (which will be useful several times in the future)

$$\frac{d}{dz}\int_{-\infty}^{g(z)} p_U(u)\, du = p_U[g(z)]\frac{dg}{dz}. \quad (2.5\text{-}6)$$

In this particular example we have

$$g(z) = \sqrt{\frac{z}{a}}, \qquad \frac{dg}{dz} = \frac{1}{2\sqrt{az}}$$

for one integral and

$$g(z) = -\sqrt{\frac{z}{a}}, \qquad \frac{dg}{dz} = -\frac{1}{2\sqrt{az}}$$

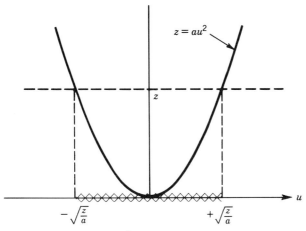

Figure 2-4. The transformation $z = au^2$. The crosshatched region, bounded by $u = \pm \sqrt{z/a}$, is the region L_z.

for the second. It follows that

$$p_Z(z) = \frac{p_U\left(\sqrt{\frac{z}{a}}\right) + p_U\left(-\sqrt{\frac{z}{a}}\right)}{2\sqrt{az}}. \qquad (2.5\text{-}7)$$

The reader may wish to try other examples suggested in Problem 2-7.

2.5.2 Monotonic Functions

If the transformation $z = f(u)$ is a one-to-one mapping and thus invertible (each value of u maps into one value of z and each value of z arises from a unique value of u), a simpler procedure can be used to find $p_Z(z)$. Such a transformation is shown in Fig. 2-5. Consider a small increment Δz about the point z. If we map this incremental region back through the transformation, we obtain an increment Δu about the point $u = f^{-1}(z)$, where f^{-1} is the inverse of f. Now we use the fact that

$$\text{Prob}\{Z \text{ in } \Delta z\} = \text{Prob}\{U \text{ in } \Delta u\}. \qquad (2.5\text{-}8)$$

For small Δu and Δz, this equality can be stated approximately as

$$p_Z(z)\,\Delta z \cong p_U(u)\,\Delta u, \qquad (2.5\text{-}9)$$

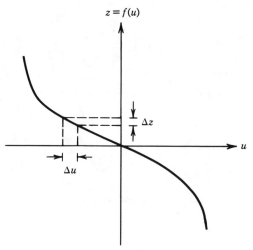

Figure 2-5. Example of a one-to-one probability transformation.

where $u = f^{-1}(z)$. Furthermore, again for small Δu and Δz,

$$\Delta u \cong \left|\frac{du}{dz}\right| \Delta z. \qquad (2.5\text{-}10)$$

Substituting (2.5-10) in (2.5-9) and canceling Δz, we obtain a relationship that becomes exact as Δu and Δz approach zero,

$$p_Z(z) = p_U[f^{-1}(z)]\left|\frac{du}{dz}\right|. \qquad (2.5\text{-}11)$$

Since $du/dz = (dz/du)^{-1}$, we can equivalently write

$$p_Z(z) = \frac{p_U[f^{-1}(z)]}{\left|\frac{dz}{du}\right|}, \qquad (2.5\text{-}12)$$

where $|dz/du|$ must be expressed in terms of the variable z. With either (2.5-11) or (2.5-12), $p_Z(z)$ can easily be calculated in any specific case.

Interpreting Eq. (2.5-12) in a physical way, we note that the slope dz/du of the transformation controls the manner in which probability density in the u domain is spread over the z domain. If $|dz/du|$ is large, a small region of u is mapped into a large region of z; hence the probability density is spread thinly in the z domain. On the other hand, if $|dz/du|$ is small, a large region of the u axis is mapped into a small region of the z axis, and probability density is accordingly mounded high in that region.

As an example of the application of this method, consider the transformation

$$z = \cos u \qquad (2.5\text{-}13)$$

and a probability density function

$$p_U(u) = \begin{cases} \dfrac{1}{\pi} & 0 < u \leq \pi \\ 0 & \text{otherwise.} \end{cases} \qquad (2.5\text{-}14)$$

This transformation is invertible over the region of u for which $p_U(u)$ is nonzero; the inverse function is

$$u = \cos^{-1} z. \qquad (2.5\text{-}15)$$

The required derivative is

$$\left| \frac{du}{dz} \right| = \frac{1}{\sqrt{1-z^2}},$$

and thus

$$p_Z(z) = \begin{cases} \dfrac{1}{\pi} \left| \dfrac{du}{dz} \right| = \dfrac{1}{\pi\sqrt{1-z^2}} & -1 < z \leq 1 \\ 0 & \text{otherwise.} \end{cases} \qquad (2.5\text{-}16)$$

Figure 2-6 shows a plot of $p_U(u)$, the transformation $z = \cos u$, and the resulting $p_Z(z)$.

When the function $z = f(u)$ is not invertible but does consist of invertible segments, a procedure similar to that used above can be employed. If on the nth segment the function can be represented by the invertible function $f_n(u)$, the probability density function of z can be written

$$p_Z(z) = \sum_n p_U\left[u = f_n^{-1}(z)\right] \left| \frac{df_n^{-1}(z)}{dz} \right|. \qquad (2.5\text{-}17)$$

As a specific example we again take the square law characteristic $z = au^2$,

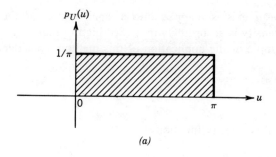

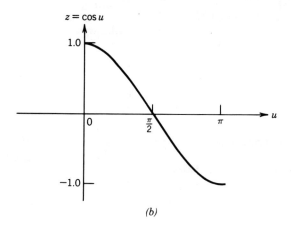

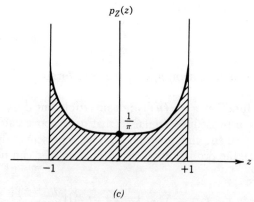

Figure 2-6. Plots of (a) the probability density before transformation, (b) the transformation law, and (c) the probability density after transformation.

TRANSFORMATIONS OF RANDOM VARIABLES

which can be inverted in segments as follows:

$$u = +\sqrt{\frac{z}{a}} \qquad 0 < u \le \infty$$

$$u = -\sqrt{\frac{z}{a}} \qquad -\infty < u \le 0.$$

On both segments we have

$$\left|\frac{du}{dz}\right| = \frac{1}{2\sqrt{az}},$$

and thus

$$p_Z(z) = \frac{p_U\left(\sqrt{\frac{z}{a}}\right) + p_U\left(-\sqrt{\frac{z}{a}}\right)}{2\sqrt{az}} \qquad (2.5\text{-}18)$$

in agreement with Eq. (2.5-7).

2.5.3 Multivariate Probability Transformations

Consider two jointly distributed random variables W and Z that are functionally related to two underlying jointly distributed random variables U and V by

$$w = f(u, v)$$

$$z = g(u, v). \qquad (2.5\text{-}19)$$

We assume that the joint density function $p_{UV}(u, v)$ is given, and we wish to find the joint density function $p_{WZ}(w, z)$.

In the most general case of interest, the mapping [Eq. (2.5-19)] is single valued [i.e., a given pair (u, v) maps into only one pair (w, z)], but not necessarily one to one and invertible. By analogy with Eqs. (2.5-2) and (2.5-3), we must find the joint distribution function $F_{WZ}(w, z)$ and then differentiate it with respect to w and z. Let A_{wz} represent the region of the (u, v) plane for which the inequalities $W \le w$ and $Z \le z$ are *both* satisfied. Then

$$F_{WZ}(w, z) = \text{Prob}\{(u, v) \text{ lies in } A_{wz}\} \qquad (2.5\text{-}20)$$

and

$$p_{WZ}(w, z) = \frac{\partial^2}{\partial w\, \partial z} \text{Prob}\{(u, v) \text{ lies in } A_{wz}\}. \qquad (2.5\text{-}21)$$

Since this most general approach will not be required in our later considerations, we defer an example to the problems (see Problem 2-8).

If the mappings $f(u, v)$ and $g(u, v)$ are one to one and have inverses, a simpler approach is possible. We write u and v in terms of w and z as follows:

$$u = F(w, z)$$

$$v = G(w, z). \qquad (2.5\text{-}22)$$

The probability that the values of u and v lie in an incremental area $\Delta u\, \Delta v$ is equal to the probability that w and z lie in the elementary area $\Delta w \Delta z$, representing the projection of $\Delta u\, \Delta v$ through the inverse transformation. Thus

$$p_{WZ}(w, z)\, \Delta w \Delta z = p_{UV}(u, v)\, \Delta u\, \Delta v. \qquad (2.5\text{-}23)$$

But for small $(\Delta u, \Delta v)$ we have

$$\Delta u\, \Delta v \cong |J|\, \Delta w \Delta z, \qquad (2.5\text{-}24)$$

where $|J|$ is the Jacobian of the inverse transformation

$$|J| = \begin{Vmatrix} \dfrac{\partial F}{\partial w} & \dfrac{\partial F}{\partial z} \\ \dfrac{\partial G}{\partial w} & \dfrac{\partial G}{\partial z} \end{Vmatrix} \qquad (2.5\text{-}25)$$

and the $\|\cdot\|$ signs indicate the modulus of the determinant. If Δu and Δv are allowed to become arbitrarily small, the approximation (2.5-24) becomes arbitrarily good. Substituting (2.5-24) in (2.5-23) and canceling $\Delta w \Delta z$, we obtain

$$p_{WZ}(w, z) = |J|\, p_{UV}[u = F(w, z), v = G(w, z)], \qquad (2.5\text{-}26)$$

which represents our final result.

2.6 SUMS OF REAL RANDOM VARIABLES

Attention is now turned to the important problem of finding the probability density function of a random variable that is itself the sum of two other random variables. Let the random variable Z be defined as

$$Z = U + V, \qquad (2.6\text{-}1)$$

where U and V are random variables with joint probability density function $p_{UV}(u,v)$. Knowing $p_{UV}(u,v)$, we wish to find $p_Z(z)$. For illustration purposes we shall find the solution by two different methods.

2.6.1 Two Methods for Finding $p_Z(z)$

As our first method for finding $p_Z(z)$, we calculate $F_Z(z)$ and differentiate with respect to z. Figure 2-7 illustrates the calculation; choosing a particular value of z we draw the line $z = u + v$ and identify the region within which the random variable Z is less than or equal to z. The distribution function

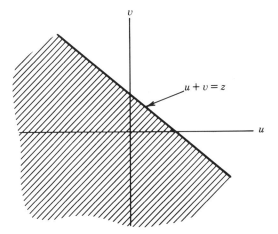

Figure 2-7. The shaded region represents the area within which $Z \leq z$.

$F_Z(z)$ represents the probability that (u, v) falls within this region. Calculating this probability, we write

$$F_Z(z) = \int_{-\infty}^{\infty} dv \int_{-\infty}^{z-v} du\, p_{UV}(u, v). \qquad (2.6\text{-}2)$$

With the help of Eq. (2.5-6), $F_Z(z)$ is differentiated with respect to z, yielding

$$p_Z(z) = \int_{-\infty}^{\infty} p_{UV}(z - v, v)\, dv. \qquad (2.6\text{-}3)$$

Thus from a given joint density function p_{UV} we can now calculate p_Z.

Next an alternative method for calculating the same result is illustrated. Here the result (2.5-26) for multivariate transformations is used. Since we have only one equation relating z, u, and v, a second transformation must be invented to suit our purposes. Exactly what the second transformation should be is not obvious, but it can be found with some trial-and-error experience. We choose the simple transformation $w = v$, yielding the pair of transformations

$$z = u + v$$

$$w = v. \qquad (2.6\text{-}4)$$

This transformation pair is invertible as follows:

$$u = z - w$$

$$v = w. \qquad (2.6\text{-}5)$$

The Jacobian of the inverse transformation is

$$|J| = \begin{Vmatrix} \dfrac{\partial u}{\partial z} & \dfrac{\partial u}{\partial w} \\ \dfrac{\partial v}{\partial z} & \dfrac{\partial v}{\partial w} \end{Vmatrix} = \begin{Vmatrix} 1 & -1 \\ 0 & 1 \end{Vmatrix} = 1. \qquad (2.6\text{-}6)$$

With the use of Eq. (2.5-26), therefore, the joint density function of w and z is

$$p_{WZ}(w, z) = p_{UV}(z - w, w). \qquad (2.6\text{-}7)$$

But we are interested only in the marginal density function $p_Z(z)$, which we obtain by integrating p_{WZ} with respect to w,

$$p_Z(z) = \int_{-\infty}^{\infty} p_{UV}(z - w, w)\, dw, \qquad (2.6\text{-}8)$$

which is identical with the previous result (2.6-3).

2.6.2 Independent Random Variables

If the random variable Z represents the sum of two *independent* random variables U and V, the function $p_Z(z)$ acquires a particularly simple relation to the probability density functions of U and V. For independent U and V, the integrand of Eq. (2.6-8) factors,

$$p_{UV}(z-w,w) = p_U(z-w)p_V(w) \qquad (2.6\text{-}9)$$

yielding

$$p_Z(z) = \int_{-\infty}^{\infty} p_U(z-w)p_V(w)\,dw. \qquad (2.6\text{-}10)$$

Such an integral is recognized to be a convolution and arises so frequently that we use a special notation for it. In shorthand notation, we write the convolution (2.6-10) as

$$p_Z = p_U * p_V. \qquad (2.6\text{-}11)$$

The fact that p_Z is the convolution of p_U and p_V can also be derived in another way by using characteristic functions. Because of the brevity of this proof we present it here. The characteristic function of Z is by definition

$$\mathbf{M}_Z(\omega) = \overline{\exp(j\omega z)} = \overline{\exp[j\omega(u+v)]}. \qquad (2.6\text{-}12)$$

But since U and V are independent, the last average can be split into the product of two averages,

$$\mathbf{M}_Z(\omega) = \overline{\exp(j\omega u)} \cdot \overline{\exp(j\omega v)} = \mathbf{M}_U(\omega)\mathbf{M}_V(\omega). \qquad (2.6\text{-}13)$$

Thus the characteristic function of Z is the product of the characteristic functions of u and v. To find $p_Z(z)$ we must inverse Fourier transform $\mathbf{M}_Z(\omega)$. But the inverse Fourier transform of a product of two functions is equal to the convolution of their individual inverse Fourier transforms. Hence we again obtain the result that $p_Z(z)$ is equal to the convolution of p_U and p_V. This proof is a good indication of the simplifications that can often be obtained by reasoning with characteristic functions rather than directly with probability density functions.

2.6.3 The Central Limit Theorem

A basic theorem of enormous importance to us in later applications of statistics is the *central limit theorem*. In our discussion we first state the theorem in a form useful to us, then mention a set of sufficient conditions that assure that it holds, and finally present an intuitive and nonrigorous "proof."

Let $U_1, U_2, \ldots, U_n$ be *independent* random variables with arbitrary probability distributions (not necessarily the same), means $\bar{u}_1, \bar{u}_2, \ldots, \bar{u}_n$, and variances $\sigma_1^2, \sigma_2^2, \ldots, \sigma_n^2$. Furthermore, let the random variable Z be defined by

$$Z = \frac{1}{\sqrt{n}} \sum_{i=1}^{n} \frac{U_i - \bar{u}_i}{\sigma_i}. \qquad (2.6\text{-}14)$$

(Note that for every n, Z has zero mean and unit standard deviation.) Then under certain conditions that are often met in practice and are discussed below, as the number n of random variables tends to infinity, the probability density function $p_Z(z)$ approaches a *Gaussian* density,

$$\lim_{n \to \infty} p_Z(z) = \frac{1}{\sqrt{2\pi}} e^{-z^2/2}. \qquad (2.6\text{-}15)$$

There exists a large body of statistical literature on the conditions required for this theorem to hold. Here we are satisfied to state a set of sufficient conditions as follows (Ref. 2-6, p. 201):[†] there must exist two positive numbers p and q such that

$$\left.\begin{array}{l} \sigma_i^2 > p > 0 \\ E\left[|u_i - \bar{u}_i|^3\right] < q \end{array}\right\} \text{ for all } i. \qquad (2.6\text{-}16)$$

Finally, a brief and nonrigorous "proof" of the central limit theorem is presented. Let $\mathbf{M}_i(\omega)$ represent the characteristic function of the random variable $U_i - \bar{u}_i$; we assume that all such characteristic functions exist. It follows from (2.6-13) that the characteristic function of Z is

$$\mathbf{M}_Z(\omega) = \prod_{i=1}^{n} \mathbf{M}_i\left(\frac{\omega}{\sqrt{n}\,\sigma_i}\right). \qquad (2.6\text{-}17)$$

According to the first condition of (2.6-16), the σ_i are bounded from below. Hence for any given ω it is always possible to find an n large enough that

---

[†] Less stringent sufficient conditions can be stated (see Ref. 2-1, pp. 431–433). If the U_i have identical distributions, it suffices that the mean and variance of that distribution be finite. If the U_i have different distributions, it suffices that they have finite means and finite $(2 + \delta)$th absolute central moment for some $\delta > 0$ and that they satisfy the so-called Lyapunov condition.

the argument of $\mathbf{M}_i$ is extremely small. The second condition of (2.6-16) guarantees that for small argument, $M_i(\omega/\sqrt{n}\,\sigma_i)$ is convex and parabolic [cf. Eq. (2.4-20)],

$$\mathbf{M}_i\left(\frac{\omega}{\sqrt{n}\,\sigma_i}\right) \cong 1 - \frac{\omega^2}{2n}. \qquad (2.6\text{-}18)$$

Thus for sufficiently large n, the characteristic function of Z behaves as

$$\mathbf{M}_Z(\omega) \cong \prod_{i=1}^{n}\left(1 - \frac{\omega^2}{2n}\right) = \left(1 - \frac{\omega^2}{2n}\right)^n. \qquad (2.6\text{-}19)$$

Letting n grow without bound, we find

$$\lim_{n\to\infty}\mathbf{M}_Z(\omega) = \lim_{n\to\infty}\left(1 - \frac{\omega^2}{2n}\right)^n = \exp\left(-\frac{\omega^2}{2}\right); \qquad (2.6\text{-}20)$$

a Fourier transform of this result yields

$$\lim_{n\to\infty} p_Z(z) = \frac{1}{\sqrt{2\pi}}\exp\left\{-\frac{z^2}{2}\right\}. \qquad (2.6\text{-}21)$$

Thus the density function of Z is asymptotically Gaussian.

A word of caution should be injected here. Whereas $p_Z(z)$ is asymptotically Gaussian, the Gaussian density function may or may not be a good approximation to $p_Z(z)$ for a *finite* n. The quality of the approximation depends on just how large n may be and how far out in the "tails" of $p_Z(z)$ we wish to work. Results of questionable accuracy may be obtained if the Gaussian approximation is used to calculate probabilities of extremely large and improbable excursions of Z. Nonetheless, the central limit theorem is of great utility when applied to problems that contain enormous numbers of independent contributions.

2.7 GAUSSIAN RANDOM VARIABLES

In many problems in physics and engineering we encounter random phenomena that are the result of many additive and independent random events. By virtue of the central limit theorem, Gaussian statistics accordingly play a role of unsurpassed importance in the statistical analysis of physical phenomena. In this section we summarize the most important properties of Gaussian random variables.

2.7.1 Definitions

A random variable U is called *Gaussian* (or normal) if its characteristic function is of the form

$$\mathbf{M}_U(\omega) = \exp\left[j\omega\bar{u} - \frac{\omega^2\sigma^2}{2}\right]. \quad (2.7\text{-}1)$$

By appropriate differentiations of $\mathbf{M}_U(\omega)$, we can show that $\bar{u}$ and σ are indeed the mean and standard deviation of the random variable U. More generally, the nth central moment is found to be

$$\overline{(u - \bar{u})^n} = \begin{cases} 1 \cdot 3 \cdot 5 \cdot \cdots \cdot (n-1)\sigma^n & n \text{ even} \\ 0 & n \text{ odd} \end{cases} \quad (2.7\text{-}2)$$

A Fourier inversion of $\mathbf{M}_U(\omega)$ shows that the probability density function of U is

$$p_U(u) = \frac{1}{\sqrt{2\pi}\,\sigma} \exp\left\{-\frac{(u-\bar{u})^2}{2\sigma^2}\right\}. \quad (2.7\text{-}3)$$

A plot of this density function is shown in Fig. 2-8.

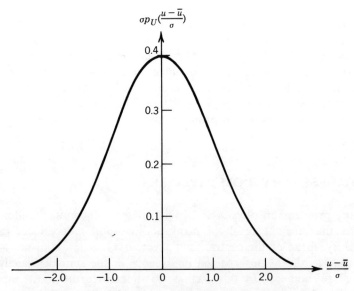

Figure 2-8. The Gaussian probability density function.

GAUSSIAN RANDOM VARIABLES

Furthermore, n random variables $U_1, U_2, \ldots, U_n$ are said to be *jointly Gaussian* if their joint characteristic function is of the form

$$\mathbf{M}_U(\underline{\omega}) = \exp\{ j\underline{\bar{u}}'\underline{\omega} - \tfrac{1}{2}\underline{\omega}'\underline{C}\underline{\omega} \} \tag{2.7-4}$$

where

$$\underline{\bar{u}} = \begin{bmatrix} \bar{u}_1 \\ \bar{u}_2 \\ \vdots \\ \bar{u}_n \end{bmatrix}, \quad \underline{\omega} = \begin{bmatrix} \omega_1 \\ \omega_2 \\ \vdots \\ \omega_n \end{bmatrix} \tag{2.7-5}$$

and $\underline{C}$ is an $n \times n$ covariance matrix, with element σ_{ik}^2 in the ith row and kth column defined by

$$\sigma_{ik}^2 = E[(u_i - \bar{u}_i)(u_k - \bar{u}_k)]. \tag{2.7-6}$$

The corresponding nth-order probability density function can be shown to be

$$p_U(\underline{u}) = \frac{1}{(2\pi)^{n/2}|\underline{C}|^{1/2}} \exp\left\{ -\frac{1}{2}(\underline{u} - \underline{\bar{u}})'\underline{C}^{-1}(\underline{u} - \underline{\bar{u}}) \right\} \tag{2.7-7}$$

where $|\underline{C}|$ and $\underline{C}^{-1}$ are the determinant and matrix inverse of $\underline{C}$, respectively, and $\underline{u}$ is a column matrix of the u values.

Of most importance for our future work is the form of (2.7-7) when we have *two* jointly distributed Gaussian random variables U and V, each having zero mean, and with $\sigma_U^2 = \sigma_V^2 = \sigma^2$. In this case (2.7-7) becomes

$$p_{UV}(u, v) = \frac{\exp\left[-\dfrac{u^2 + v^2 - 2\rho uv}{2(1 - \rho^2)\sigma^2} \right]}{2\pi\sigma^2\sqrt{1 - \rho^2}}. \tag{2.7-8}$$

where

$$\rho \triangleq \frac{\overline{uv}}{\sigma^2}. \tag{2.7-9}$$

Figure 2-9 shows contours of constant probability density in the (u, v) plane for the cases $\rho = 0$, $0 < \rho < 1$, and $\rho = 1$. As the correlation coefficient

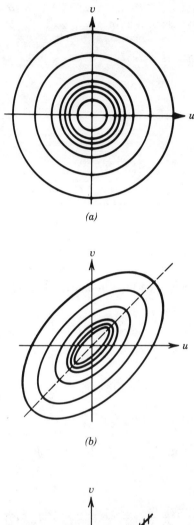
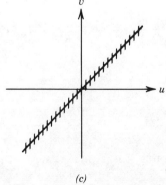

Figure 2-9. Contours of constant probability density for a joint Gaussian density with $\bar{u} = \bar{v} = 0$, $\sigma_U^2 = \sigma_V^2 = \sigma^2$, and (a) $\rho = 0$, (b) $0 < \rho < 1$, (c) $\rho \cong 1$.

GAUSSIAN RANDOM VARIABLES

increases in a positive sense, the density function passes from circular symmetry to elliptical shape, with major axis along the line $u = v$. For negative correlation coefficient, the major axis is the line $u = -v$.

2.7.2 Special Properties of Gaussian Random Variables

In addition to arising with great frequency in practical problems, the Gaussian random variable is notable for its many special properties that make it particularly easy to deal with. Here we summarize these properties, in most cases with at least an intuitive kind of proof.

(a) Two Uncorrelated Jointly Gaussian Random Variables Are Also Statistically Independent. As pointed out in Section 2.4.2, lack of correlation rarely implies statistical independence. In the case of jointly distributed Gaussian random variables, however, the two properties are synonymous. To demonstrate this fact, we let the correlation coefficient ρ in Eq. (2.7-8) be identically zero, in which case the joint density function becomes

$$p_{UV}(u,v) = \frac{\exp\left\{-\dfrac{u^2+v^2}{2\sigma^2}\right\}}{2\pi\sigma^2}$$

$$= \frac{1}{\sqrt{2\pi}\,\sigma}\exp\left(-\frac{u^2}{2\sigma^2}\right) \cdot \frac{1}{\sqrt{2\pi}\,\sigma}\exp\left(-\frac{v^2}{2\sigma^2}\right)$$

$$= p_U(u)p_V(v).$$

Since the joint density function factors into the product of the two marginal density functions, U and V are independent.

(b) The Sum of Two Statistically Independent Jointly Gaussian Random Variables Is Itself Gaussian. Suppose that U and V are Gaussian and independent, with characteristic functions

$$\mathbf{M}_U(\omega) = \exp\left[j\omega\bar{u} - \frac{\omega^2\sigma_U^2}{2}\right]$$

$$\mathbf{M}_V(\omega) = \exp\left[j\omega\bar{v} - \frac{\omega^2\sigma_V^2}{2}\right].$$

Let Z be the sum of U and V. Then by Eq. (2.6-13) we have

$$\mathbf{M}_Z(\omega) = \mathbf{M}_U(\omega)\mathbf{M}_V(\omega)$$

$$= \exp\left[j\omega(\bar{u} + \bar{v}) - \frac{\omega^2}{2}(\sigma_U^2 + \sigma_V^2)\right].$$

Thus Z is a Gaussian random variable with mean $\bar{u} + \bar{v}$ and variance $\sigma_U^2 + \sigma_V^2$.

(c) The Sum of Two Dependent (Correlated) Gaussian Random Variables Is Itself Gaussian. Let U and V be jointly Gaussian random variables with correlation coefficient $\rho \neq 0$. In addition, for simplicity let $\bar{u} = \bar{v} = 0$ and $\sigma_U^2 = \sigma_V^2 = \sigma^2$. Then

$$p_{UV}(u,v) = \frac{\exp\left[-\dfrac{u^2 + v^2 - 2\rho uv}{2(1-\rho^2)\sigma^2}\right]}{2\pi\sigma^2\sqrt{1-\rho^2}}.$$

Let the random variable Z again be the sum of U and V. From Eq. (2.6-3), we obtain

$$p_Z(z) = \int_{-\infty}^{\infty} p_{UV}(z-v, v)\, dv$$

$$= \int_{-\infty}^{\infty} \frac{\exp\left[-\dfrac{(z-v)^2 + v^2 - 2\rho(z-v)v}{2(1-\rho^2)\sigma^2}\right]}{2\pi\sigma^2\sqrt{1-\rho^2}}\, dv.$$

We next complete the square in the exponent of the integrand, giving

$$p_Z(z) = \frac{\exp\left[-\dfrac{z^2}{4(1+\rho)\sigma^2}\right]}{2\pi\sigma^2\sqrt{1-\rho^2}} \int_{-\infty}^{\infty} \exp\left[-\frac{\left(v - \dfrac{z}{2}\right)^2}{(1-\rho)\sigma^2}\right] dv.$$

The integral can be performed to yield

$$p_Z(z) = \frac{\exp\left[-\dfrac{z^2}{4(1+\rho)\sigma^2}\right]}{\sqrt{2\pi}\sqrt{2(1+\rho)\sigma^2}}. \tag{2.7-10}$$

GAUSSIAN RANDOM VARIABLES

Thus Z is Gaussian, with zero mean and variance

$$\sigma_Z^2 = 2(1 + \rho)\sigma^2. \qquad (2.7\text{-}11)$$

When $\rho \to 0$, $\sigma_Z^2 \to 2\sigma^2$, whereas when $\rho \to 1$, $\sigma_Z^2 \to 4\sigma^2$.

(d) Any Linear Combination of Jointly Gaussian Random Variables, Dependent or Independent, Is a Gaussian Random Variable. Let Z be defined by

$$Z = \sum_{i=1}^{n} a_i U_i,$$

where the a_i are known constants and the U_i are jointly Gaussian. By repeated application of the result (2.7-10), Z is readily seen to be Gaussian.

(e) For Jointly Gaussian Random Variables $U_1, U_2, \ldots, U_n$, Joint Moments of Order Higher than 2 Can Always Be Expressed in Terms of the First- and Second-Order Moments. A moment of the form $u_1^p u_2^q \cdots u_n^k$ can be obtained by partial differentiation of the characteristic function as follows [cf., Eq. (2.4-23)]:

$$\overline{u_1^p u_2^q \cdots u_n^k} = \frac{1}{(j)^{p+q+\cdots+k}} \frac{\partial^{p+q+\cdots+k}}{\partial \omega_1^p \partial \omega_2^q \cdots \partial \omega_n^k} [M_U(\underline{\omega})]_{\underline{\omega}=\underline{0}}.$$

Since the only parameters appearing in the characteristic function are means and covariances, the $(p + q + \cdots + k)$th-order moment must be expressible in terms of these first- and second-order moments.

By differentiating the characteristic function an appropriate number of times, it is possible to prove the following basic property of zero-mean Gaussian random variables:

$$\overline{u_1 u_2 \cdots u_{2k+1}} = 0,$$

$$\overline{u_1 u_2 \cdots u_{2k}} = \sum_P \left(\overline{u_j u_m} \, \overline{u_l u_p} \cdots \overline{u_q u_s} \right)_{\substack{j \neq m \\ l \neq p \\ q \neq s}}, \qquad (2.7\text{-}12)$$

where $\sum_P$ indicates the summation over all possible distinct groupings of the $2k$ variables in pairs. It can be shown that there are $(2k)!/2^k k!$ such distinct groupings. For the most important case of $k = 2$, we have

$$\overline{u_1 u_2 u_3 u_4} = \overline{u_1 u_2}\,\overline{u_3 u_4} + \overline{u_1 u_3}\,\overline{u_2 u_4} + \overline{u_1 u_4}\,\overline{u_2 u_3}. \qquad (2.7\text{-}13)$$

This relationship is known as the *moment theorem* for real Gaussian random variables.

2.8 COMPLEX-VALUED RANDOM VARIABLES

In the previous sections we have studied the properties of random variables that take on real values. Frequently in the study of waves it is necessary to consider random variables that take on complex values. Accordingly, it will be helpful to explore briefly the methods that are used to describe complex-valued random variables.

2.8.1 General Descriptions

Underlying the definition of every random variable there is a space of events $\{A\}$ and a set of associated probabilities $P(A)$. If to each event A we assign a complex number $\mathbf{u}(A)$, the set of possible complex numbers, together with their associated probability measures, define a complex-valued random variable $\mathbf{U}$.

To describe mathematically the statistical properties of the random variable $\mathbf{U}$, it is usually most convenient to describe the joint statistical properties of its real and imaginary parts. Thus if $\mathbf{U} = R + jI$ represents a complex random variable that can take on specific complex values $\mathbf{u} = r + ji$, a complete description of $\mathbf{U}$ entails specification of either the joint distribution function of R and I,

$$F_{\mathbf{U}}(\mathbf{u}) \triangleq F_{RI}(r,i) \triangleq \text{Prob}\{R \le r \text{ and } I \le i\}, \qquad (2.8\text{-}1)$$

or the joint density function of R and I,

$$p_{\mathbf{U}}(\mathbf{u}) \triangleq p_{RI}(r,i) = \frac{\partial^2}{\partial r\, \partial i} F_{RI}(r,i), \qquad (2.8\text{-}2)$$

or, alternatively, the joint characteristic function of R and I,

$$\mathbf{M}_{\mathbf{U}}(\omega^r, \omega^i) \triangleq E\left[\exp\left[j(\omega^r r + \omega^i i)\right]\right]. \qquad (2.8\text{-}3)$$

For n joint complex random variables $\mathbf{U}_1, \mathbf{U}_2, \ldots, \mathbf{U}_n$, which take on specific values $\mathbf{u}_1 = r_1 + ji_1$, $\mathbf{u}_2 = r_2 + ji_2$, and so on, the joint distribution function may be written

$$F_{\mathbf{U}}(\underline{\mathbf{u}}) \triangleq \text{Prob}\{R_1 \le r_1, R_2 \le r_2, \ldots, R_n \le r_n, I_1 \le i_1, I_2 \le i_2, \ldots, I_n \le i_n\}$$

$$(2.8\text{-}4)$$

COMPLEX-VALUED RANDOM VARIABLES

where the probability in question is the joint probability that all the events indicated occur, and the argument of F_U is regarded as a matrix with n complex elements,

$$\underline{u} = \begin{bmatrix} u_1 \\ u_2 \\ \vdots \\ u_n \end{bmatrix}. \qquad (2.8\text{-}5)$$

Corresponding to the distribution function $F_U(\underline{u})$ is a joint probability density function of the $2n$ real variables $\{r_1, r_2, \ldots, r_n, i_1, i_2, \ldots, i_n\}$,

$$p_U(\underline{u}) \triangleq \frac{\partial^{2n}}{\partial r_1 \cdots \partial r_n \, \partial i_1 \cdots \partial i_n} F_U(\underline{u}). \qquad (2.8\text{-}6)$$

Finally, it is possible to describe the joint statistics by means of a characteristic function defined by

$$\mathbf{M}_U(\underline{\omega}) \triangleq E\big[\exp(j\underline{\omega}^t \underline{u})\big] \qquad (2.8\text{-}7)$$

where $\underline{\omega}$ and $\underline{u}$ are column matrices with $2n$ real-valued entries,

$$\underline{u} = \begin{bmatrix} r_1 \\ \vdots \\ r_n \\ i_1 \\ \vdots \\ i_n \end{bmatrix} \quad \underline{\omega} = \begin{bmatrix} \omega_1^r \\ \vdots \\ \omega_n^r \\ \omega_1^i \\ \vdots \\ \omega_n^i \end{bmatrix}. \qquad (2.8\text{-}8)$$

2.8.2 Complex Gaussian Random Variables

The n complex random variables $U_1, U_2, \ldots, U_n$ are said to be jointly Gaussian if their characteristic function is of the form

$$\mathbf{M}_U(\underline{\omega}) = \exp\{j\underline{\bar{u}}^t \underline{\omega} - \tfrac{1}{2}\underline{\omega}^t \underline{C}\,\underline{\omega}\} \qquad (2.8\text{-}9)$$

where $\underline{\omega}$ is again given by (2.8-8), $\underline{\bar{u}}$ is a column matrix with $2n$ real-valued elements that are the mean values of the elements of $\underline{u}$, and $\underline{C}$ is a $2n \times 2n$ covariance matrix, with real-valued elements, defined by

$$\underline{C} = E\big[(\underline{u} - \underline{\bar{u}})(\underline{u} - \underline{\bar{u}})^t\big]. \qquad (2.8\text{-}10)$$

By means of a $2n$-dimensional Fourier transformation of $\mathbf{M}_U(\underline{\omega})$, the corresponding probability density function is found to be

$$p_U(\underline{u}) = \frac{1}{(2\pi)^n |\underline{C}|^{1/2}} \exp\left\{-\frac{1}{2}(\underline{u} - \underline{\bar{u}})^t \underline{C}^{-1}(\underline{u} - \underline{\bar{u}})\right\} \quad (2.8\text{-}11)$$

where $|\underline{C}|$ and $\underline{C}^{-1}$ are, respectively, the determinant and inverse of the $2n \times 2n$ covariance matrix $\underline{C}$.

For future reference, it is useful to define a special class of complex Gaussian random variables. But to do so we must first define some new symbols. Let $\underline{r}$ and $\underline{i}$ be n-element column matrices of the real parts and imaginary parts, respectively, of the n complex random variables $\mathbf{U}_k$ ($k = 1, 2, \ldots, n$); thus

$$\underline{r} \triangleq \begin{bmatrix} r_1 \\ r_2 \\ \vdots \\ r_n \end{bmatrix}, \quad \underline{i} \triangleq \begin{bmatrix} i_1 \\ i_2 \\ \vdots \\ i_n \end{bmatrix}. \quad (2.8\text{-}12)$$

Further, let the following covariance matrices be defined:

$$\underline{C}^{(rr)} \triangleq E\left[(\underline{r} - \underline{\bar{r}})(\underline{r} - \underline{\bar{r}})^t\right], \quad \underline{C}^{(ii)} \triangleq E\left[(\underline{i} - \underline{\bar{i}})(\underline{i} - \underline{\bar{i}})^t\right]$$

$$\underline{C}^{(ri)} \triangleq E\left[(\underline{r} - \underline{\bar{r}})(\underline{i} - \underline{\bar{i}})^t\right], \quad \underline{C}^{(ir)} \triangleq E\left[(\underline{i} - \underline{\bar{i}})(\underline{r} - \underline{\bar{r}})^t\right].$$

We call the complex $\mathbf{U}_k$ ($k = 1, 2, \ldots, n$) jointly *circular* complex random variables if the following special relations hold:

(1)
$$\underline{\bar{r}} = \begin{bmatrix} 0 \\ 0 \\ \vdots \\ 0 \end{bmatrix}, \quad \underline{\bar{i}} = \begin{bmatrix} 0 \\ 0 \\ \vdots \\ 0 \end{bmatrix}. \quad (2.8\text{-}14)$$

(2) $\quad\quad\quad\quad \underline{C}^{(rr)} = \underline{C}^{(ii)}, \quad \underline{C}^{(ri)} = -\underline{C}^{(ir)}. \quad (2.8\text{-}15)$

The origin of the term "circular" is perhaps best understood by considering the simple case of a single circular complex Gaussian random variable.

We have

$$\bar{r} = \bar{r}, \qquad \bar{i} = \bar{i}$$

$$\underline{C}^{(rr)} = \sigma_r^2, \qquad \underline{C}^{(ii)} = \sigma_i^2$$

$$\underline{C}^{(ir)} = \sigma_i \sigma_r \rho, \qquad \underline{C}^{(ri)} = \sigma_r \sigma_i \rho, \qquad (2.8\text{-}16)$$

where σ_r^2 and σ_i^2 are the variances of the real and imaginary parts of **U**, whereas ρ is the correlation coefficient of the real and imaginary parts. Imposition of the circularity conditions (2.8-14) and (2.8-15) yields the requirements

$$\bar{r} = \bar{i} = 0$$

$$\sigma_r^2 = \sigma_i^2 = \sigma^2$$

$$\rho = 0. \qquad (2.8\text{-}17)$$

Thus the 2×2 covariance matrix $\underline{C}$ is given by

$$\underline{C} = \begin{bmatrix} \sigma^2 & 0 \\ 0 & \sigma^2 \end{bmatrix} \qquad (2.8\text{-}18)$$

and for the case of Gaussian statistics, the probability density function of **U** becomes

$$p_\mathbf{U}(\mathbf{u}) = \frac{1}{2\pi\sigma^2} \exp\left(-\frac{r^2 + i^2}{2\sigma^2}\right). \qquad (2.8\text{-}19)$$

Contours of constant probability are circles in the (r, i) plane, and hence **U** is called a *circular* complex Gaussian random variable.

Note that the real and imaginary parts of a circular complex Gaussian random variable are uncorrelated and hence independent. If $\mathbf{U}_1$ and $\mathbf{U}_2$ are two such joint random variables, however, the real part of $\mathbf{U}_1$ may have an arbitrary degree of correlation with the real and imaginary parts of $\mathbf{U}_2$, provided only that the conditions

$$\overline{r_1 r_2} = \overline{i_1 i_2}$$

$$\overline{r_1 i_2} = -\overline{i_1 r_2} \qquad (2.8\text{-}20)$$

are satisfied, in accord with (2.8-15).

Circular complex Gaussian random variables are frequently encountered in practice. An important property of such random variables is the *complex Gaussian moment theorem*, which can be derived from the real Gaussian moment theorem (2.7-13), together with the conditions (2.8-14) and (2.8-15) for circularity. Let $U_1, U_2, \ldots, U_{2k}$ be zero-mean jointly circular complex Gaussian random variables. Then

$$\overline{u_1^* \cdots u_k^* u_{k+1} \cdots u_{2k}} = \sum_\pi \overline{u_1^* u_p} \, \overline{u_2^* u_q} \cdots \overline{u_k^* u_r} \qquad (2.8\text{-}21)$$

where $\sum_\pi$ denotes a summation over the $k!$ possible permutations $(p, q, \ldots, r)$ of $(1, 2, \ldots, k)$. For the simplest case of $k = 2$, we have

$$\overline{u_1^* u_2^* u_3 u_4} = \overline{u_1^* u_3} \, \overline{u_2^* u_4} + \overline{u_1^* u_4} \, \overline{u_2^* u_3}. \qquad (2.8\text{-}22)$$

2.9 RANDOM PHASOR SUMS

In many areas of physics, and particularly in optics, we must deal with complex-valued random variables that arise as a sum of many small "elementary" complex-valued contributions. The complex numbers of concern are often *phasors*, representing the amplitude and phase of a monochromatic or nearly monochromatic wave disturbance. A complex addition of many small independent phasors results, for example, when we calculate the total complex amplitude of the wave that arises as a result of scattering by a collection of small, independent scatterers. More generally, such complex sums occur whenever we add a number of complex-valued analytic signals, which are defined and discussed in detail in Section 3.8. Sums of complex-valued random variables are referred to here as *random phasor sums*, and their properties are discussed in this section.

2.9.1 Initial Assumptions

Consider a sum of a very large number N of complex phasors, the kth phasor having random length $\alpha_k/\sqrt{N}$ and random phase ϕ_k. The resultant phasor, with length a and phase θ, is defined by

$$\mathbf{a} = ae^{j\theta} = \frac{1}{\sqrt{N}} \sum_{k=1}^{N} \alpha_k e^{j\phi_k} \qquad (2.9\text{-}1)$$

and is illustrated in Fig. 2-10.

RANDOM PHASOR SUMS

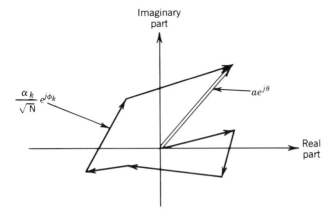

Figure 2-10. Random phasor sum.

For simplicity of analysis, we make a number of assumptions about the statistical properties of the elementary phasors composing the sum, properties that are generally satisfied in practical problems of interest:

(1) The amplitude $\alpha_k/\sqrt{N}$ and phase ϕ_k of the kth elementary phasor are statistically independent of each other and of the amplitudes and phases of all other elementary phasors.
(2) The random variables α_k are identically distributed for all k, with mean $\bar{\alpha}$ and second moment $\overline{\alpha^2}$.
(3) The phases ϕ_k are all uniformly distributed on $(-\pi, \pi)$.

Of the various assumptions, 1 is the most important, whereas 2 and 3 can both be relaxed, with some changes in the results [see, e.g., Ref. 2-10, pp. 119–137, and Appendix B].

Let the real and imaginary parts r and i of the resultant phasor be defined by

$$r \triangleq \text{Re}\{ae^{j\theta}\} = \frac{1}{\sqrt{N}} \sum_{k=1}^{N} \alpha_k \cos\phi_k$$

$$i \triangleq \text{Im}\{ae^{j\theta}\} = \frac{1}{\sqrt{N}} \sum_{k=1}^{N} \alpha_k \sin\phi_k. \qquad (2.9\text{-}2)$$

Noting that both r and i are sums of many independent random contributions, we conclude that by virtue of the central limit theorem, both r and i

will be approximately *Gaussian* random variables for large N.[†] To specify in detail the joint density function for r and i, we must first calculate $\bar{r}$, $\bar{i}$, σ_r^2, σ_i^2, and their correlation coefficient ρ.

2.9.2 Calculations of Means, Variances, and the Correlation Coefficient

The mean values of the real and imaginary parts r and i are calculated as follows:

$$\bar{r} = \frac{1}{\sqrt{N}} \sum_{k=1}^{N} \overline{\alpha_k \cos\phi_k} = \frac{1}{\sqrt{N}} \sum_{k=1}^{N} \overline{\alpha_k} \, \overline{\cos\phi_k} = \sqrt{N}\, \overline{\alpha}\, \overline{\cos\phi}$$

$$\bar{i} = \frac{1}{\sqrt{N}} \sum_{k=1}^{N} \overline{\alpha_k \sin\phi_k} = \frac{1}{\sqrt{N}} \sum_{k=1}^{N} \overline{\alpha_k} \, \overline{\sin\phi_k} = \sqrt{N}\, \overline{\alpha}\, \overline{\sin\phi}.$$

Here we have explicitly used the facts that α_k and ϕ_k are independent and identically distributed for all k. But in addition, by assumption 3, the random variable ϕ is uniformly distributed on $(-\pi, \pi)$, with the result $\overline{\cos\phi} = \overline{\sin\phi} = 0$ and hence

$$\bar{r} = \bar{i} = 0. \tag{2.9-3}$$

Thus both the real and the imaginary parts have zero means.

To evaluate the variances σ_r^2 and σ_i^2, we can equivalently evaluate the second moments $\overline{r^2}$ and $\overline{i^2}$ (since $\bar{r} = \bar{i} = 0$). Using the independence of the amplitudes and phases, we write

$$\overline{r^2} = \frac{1}{N} \sum_{k=1}^{N} \sum_{n=1}^{N} \overline{\alpha_k \alpha_n} \, \overline{\cos\phi_k \cos\phi_n}$$

$$\overline{i^2} = \frac{1}{N} \sum_{k=1}^{N} \sum_{n=1}^{N} \overline{\alpha_k \alpha_n} \, \overline{\sin\phi_k \sin\phi_n}.$$

But in addition:

$$\overline{\cos\phi_k \cos\phi_n} = \overline{\sin\phi_k \sin\phi_n} = \begin{cases} 0 & k \neq n \\ \frac{1}{2} & k = n, \end{cases}$$

[†] A subtlety has been avoided in this argument. Although the marginal statistics of r and i clearly are asymptotically Gaussian, we have not proved that the two random variables are *jointly* Gaussian. Such a proof is provided in Appendix B.

again due to the uniform distribution of the phases. Thus we have

$$\overline{r^2} = \overline{i^2} = \frac{\overline{\alpha^2}}{2} \triangleq \sigma^2. \quad (2.9\text{-}4)$$

Finally we evaluate the correlation between r and i,

$$\overline{ri} = \frac{1}{N} \sum_{k=1}^{N} \sum_{n=1}^{N} \overline{\alpha_k \alpha_n \cos\phi_k \sin\phi_n}.$$

Noting that $\cos\phi \sin\phi = \frac{1}{2}\sin 2\phi$, we have

$$\overline{\cos\phi_k \sin\phi_n} = \begin{cases} \overline{\cos\phi}\,\overline{\sin\phi} = 0 & k \neq n \\ \frac{1}{2}\overline{\sin 2\phi} = 0 & k = n. \end{cases}$$

Thus the real and imaginary parts of the resultant are *uncorrelated*. Note that the zero means, equality of variances, and lack of correlation are true for any N, finite or infinite.

To summarize our results, we now know that in the limit of very large N, the joint density function of the real and imaginary parts of the random phasor sum is asymptotically ($N \to \infty$)

$$p_{RI}(r,i) = \frac{1}{2\pi\sigma^2} \exp\left\{-\frac{r^2 + i^2}{2\sigma^2}\right\}, \quad (2.9\text{-}5)$$

where

$$\sigma^2 = \frac{\overline{\alpha^2}}{2}. \quad (2.9\text{-}6)$$

In the terminology given in Section 2.8, the random variable **a** representing the resultant is a circular complex Gaussian random variable. Figure 2-11 shows contours of constant probability density in the (r, i) plane.

The reader will find in Appendix B that when a distribution $p_\Phi(\phi)$ other than uniform is chosen for the phase of an elementary phasor, the resulting two-dimensional joint density function will in general not have zero means, equal variances, and zero correlation coefficient. Rather, the contours of constant probability density will be ellipses in the complex plane (see, e.g., Problem 2-10).

2.9.3 Statistics of the Length and Phase

In the previous section we found the joint statistics of the real and imaginary parts of a random phasor sum. In many applications it is desired to know instead the statistics of the length a and phase θ of the resultant, where

$$a = \sqrt{r^2 + i^2}$$

$$\theta = \tan^{-1}\frac{i}{r}. \qquad (2.9\text{-}7)$$

The change from rectangular to polar coordinates is a one-to-one mapping, and hence we can use the methods given in Section 2.5.3. to find the joint statistics of a and θ. The inverse functions are

$$r = a\cos\theta$$
$$i = a\sin\theta, \qquad (2.9\text{-}8)$$

and the corresponding Jacobian is

$$\|J\| = \left\| \begin{array}{cc} \dfrac{\partial r}{\partial a} & \dfrac{\partial r}{\partial \theta} \\ \dfrac{\partial i}{\partial a} & \dfrac{\partial i}{\partial \theta} \end{array} \right\| = \left\| \begin{array}{cc} \cos\theta & -a\sin\theta \\ \sin\theta & a\cos\theta \end{array} \right\| = a. \qquad (2.9\text{-}9)$$

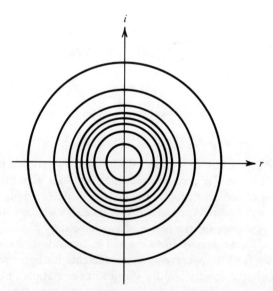

Figure 2-11. Contours of constant probability density in the (r, i) plane.

Thus we have a joint density function

$$p_{A\Theta}(a,\theta) = p_{RI}(r = a\cos\theta, i = a\sin\theta) \cdot a \qquad (2.9\text{-}10)$$

which becomes, with the help of (2.9-5),

$$p_{A\Theta}(a,\theta) = \begin{cases} \dfrac{a}{2\pi\sigma^2}\exp\left\{-\dfrac{a^2}{2\sigma^2}\right\} & \begin{array}{l} -\pi < \theta \leq \pi \\ a > 0 \end{array} \\ 0 & \text{otherwise.} \end{cases} \qquad (2.9\text{-}11)$$

The marginal densities of the length and phase can now be found. Integrating first with respect to angle θ, we have

$$p_A(a) = \int_{-\pi}^{\pi} p_{A\Theta}(a,\theta)\, d\theta = \begin{cases} \dfrac{a}{\sigma^2}\exp\left\{-\dfrac{a^2}{2\sigma^2}\right\} & a > 0 \\ 0 & \text{otherwise.} \end{cases}$$

$$(2.9\text{-}12)$$

This density function is known as a *Rayleigh* density function and is plotted in Fig. 2-12. Its mean and variance are

$$\bar{a} = \sqrt{\dfrac{\pi}{2}}\,\sigma$$

$$\sigma_a^2 = \left[2 - \dfrac{\pi}{2}\right]\sigma^2. \qquad (2.9\text{-}13)$$

To find the probability density function of the phase θ, we integrate Eq. (2.9-11) with respect to a,

$$p_\Theta(\theta) = \begin{cases} \dfrac{1}{2\pi}\displaystyle\int_0^\infty \dfrac{a}{\sigma^2}\exp\left\{-\dfrac{a^2}{2\sigma^2}\right\} da, & -\pi < \theta \leq \pi \\ 0 & \text{otherwise.} \end{cases} \qquad (2.9\text{-}14)$$

But the integral is precisely the integral of a Rayleigh density function and hence must be unity. We conclude that the phase θ of the resultant is

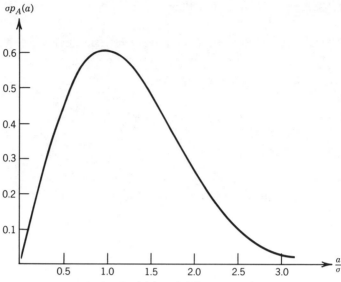

Figure 2-12. Rayleigh probability density function.

uniformly distributed on $(-\pi, \pi)$,

$$p_\Theta(\theta) = \begin{cases} \dfrac{1}{2\pi} & -\pi < \theta \leq \pi \\ 0 & \text{otherwise.} \end{cases} \qquad (2.9\text{-}15)$$

Note that the joint density function $p_{A\Theta}(a, \theta)$ can be expressed as a simple product of the marginal densities $p_A(a)$ and $p_\Theta(\theta)$. Thus A and Θ are *independent* random variables, as were the real and imaginary parts R and I described in Section 2.9.2.

2.9.4 A Constant Phasor Plus a Random Phasor Sum

We consider next the statistical properties of the sum of a constant known phasor plus a random phasor sum. Without loss of generality, the known phasor can be taken to be entirely real and positive with length s (this simply amounts to choosing a phase reference that coincides with the phase of s). Figure 2-13 illustrates the complex sum of interest.

RANDOM PHASOR SUMS

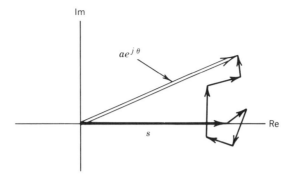

Figure 2-13. Sum of a constant phasor and a random phasor sum.

The real part of the resultant phasor is readily expressible as

$$r = s + \frac{1}{\sqrt{N}} \sum_{k=1}^{N} \alpha_k \cos \phi_k, \qquad (2.9\text{-}16)$$

whereas the imaginary part remains as before,

$$i = \frac{1}{\sqrt{N}} \sum_{k=1}^{N} \alpha_k \sin \phi_k. \qquad (2.9\text{-}17)$$

Thus the only effect of adding the known phasor has been to add a bias to the real part of the resultant phasor. In the limit of large N, the joint statistics of R and I remain approximately Gaussian, but with a modified mean,

$$p_{RI}(r, i) = \frac{1}{2\pi\sigma^2} \exp\left\{-\frac{(r-s)^2 + i^2}{2\sigma^2}\right\}. \qquad (2.9\text{-}18)$$

Again our chief interest is often in the statistics of the length a and phase θ of the resultant phasor. Since the transformation to polar coordinates is identical to that considered earlier, the Jacobian remains a, and

$$p_{A\Theta}(a, \theta) = \begin{cases} \dfrac{a}{2\pi\sigma^2} \exp\left\{-\dfrac{(a\cos\theta - s)^2 + (a\sin\theta)^2}{2\sigma^2}\right\} & a > 0 \\ & -\pi < \theta \leq \pi \\ 0 & \text{otherwise.} \end{cases}$$

$$(2.9\text{-}19)$$

To find the marginal density function for A, we must evaluate

$$p_A(a) = \int_{-\pi}^{\pi} p_{A,\Theta}(a,\theta)\,d\theta$$

$$= \frac{a}{2\pi\sigma^2} \exp\left(-\frac{a^2+s^2}{2\sigma^2}\right) \int_{-\pi}^{\pi} \exp\left(\frac{as}{\sigma^2}\cos\theta\right) d\theta.$$

The integral can be expressed as $2\pi I_0(as/\sigma^2)$, where I_0 is a modified Bessel function of the first kind, zero order. Thus

$$p_A(a) = \begin{cases} \dfrac{a}{\sigma^2} \exp\left(-\dfrac{a^2+s^2}{2\sigma^2}\right) I_0\left(\dfrac{as}{\sigma^2}\right) & a > 0 \\ 0 & \text{otherwise,} \end{cases} \qquad (2.9\text{-}20)$$

which is known as a *Rician* density function.

Figure 2-14 plots $\sigma p_A(a)$ against a/σ for various values of the parameter $k = s/\sigma$. As the strength of the known phasor increases, the shape of the probability density function changes from that of a Rayleigh density to what will be seen in the next section to be approximately a Gaussian density with mean equal to s.

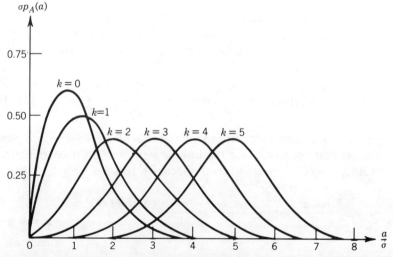

Figure 2-14. Probability density function of the amplitude A of the sum of a constant phasor (length s) and a random phasor sum (variance σ^2). Parameter $k = s/\sigma$. (After J. B. Thomas, Ref. 2-6, p. 163. Reprinted with the permission of the author and John Wiley and Sons, Inc.)

RANDOM PHASOR SUMS

Two moments of the density function (2.9-20) will be of use to us in later chapters. These are the mean value,

$$\bar{a} = \int_0^\infty \frac{a^2}{\sigma^2} \exp\left(-\frac{a^2 + s^2}{2\sigma^2}\right) I_0\left(\frac{as}{\sigma^2}\right) da \qquad (2.9\text{-}21)$$

and the second moment,

$$\overline{a^2} = \int_0^\infty \frac{a^3}{\sigma^2} \exp\left(-\frac{a^2 + s^2}{2\sigma^2}\right) I_0\left(\frac{as}{\sigma^2}\right) da. \qquad (2.9\text{-}22)$$

These integrals can be evaluated and yield (see Ref. 2-6, Section 4.8)

$$\bar{a} = \sqrt{\frac{\pi}{2}}\,\sigma e^{-k^2/4}\left[\left(1 + \frac{k^2}{2}\right) I_0\left(\frac{k^2}{4}\right) + \frac{k^2}{2} I_1\left(\frac{k^2}{4}\right)\right] \qquad (2.9\text{-}23)$$

$$\overline{a^2} = \sigma^2[2 + k^2], \qquad (2.9\text{-}24)$$

where I_0 and I_1 are modified Bessel functions of the first kind, orders zero and one, respectively.

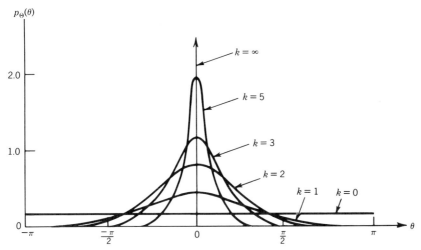

Figure 2-15. The probability density function $p_\Theta(\theta)$ for a constant phasor plus a random phasor sum. Again the parameter k is s/σ. (After J. B. Thomas, Ref. 2-6, p. 167. Reprinted with the permission of the author and John Wiley and Sons, Inc.)

To find the marginal density function $p_\Theta(\theta)$ for the phase, we must evaluate

$$p_\Theta(\theta) = \int_0^\infty p_{A\Theta}(a,\theta)\, da.$$

The integration is a difficult one, so we present only the result here (see Ref. 2-6, Section 4.8 again):

$$p_\Theta(\theta) = \frac{e^{-k^2/2}}{2\pi} + \frac{k\cos\theta}{\sqrt{2\pi}} \exp\left[-\frac{k^2\sin^2\theta}{2}\right] \Phi(k\cos\theta) \quad (2.9\text{-}25)$$

where

$$\Phi(b) = \frac{1}{\sqrt{2\pi}} \int_{-\infty}^b e^{-y^2/2}\, dy. \quad (2.9\text{-}26)$$

A plot of $p_\Theta(\theta)$ is shown in Fig. 2-15 for various values of $k = s/\sigma$. When $k = 0$, the distribution is uniform, whereas with increasing k the density function becomes more narrow, converging toward a δ function at $\theta = 0$, the phase of the constant phasor.

2.9.5 Strong Constant Phasor Plus a Weak Random Phasor Sum

When the known phasor is much stronger than the random phasor sum, the results obtained in the previous section simplify considerably. Thus we wish to consider the approximate form of the expressions for $p_A(a)$ and $p_\Theta(\theta)$ when $s \gg \sigma$, or equivalently $k \gg 1$. One approach is to apply the condition $s \gg \sigma$ to equations (2.9-20) and (2.9-25) and to discover the approximate forms through mathematical approximation. However, we choose here a more physical approach that yields exactly the same results in a more appealing way.

Our approximation is based on the observation that when $s \gg \sigma$, we are dealing with a tiny probability "cloud" centered on the tip of an extremely long, known phasor, as shown in Fig. 2-16. In such a case, with extremely high probability, the resultant of the random phasor sum is much smaller than the length of the known phasor. As a consequence, variations in the length a of the total resultant are caused primarily by the *real part* of the random phasor sum, whereas variations of the phase θ of the resultant are caused primarily by the *imaginary part* of the random phasor, which is

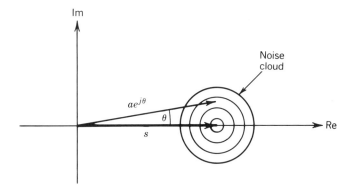

Figure 2-16. Large constant phasor s plus a small "noise cloud."

orthogonal to the known phasor. Since the real part of the random phasor sum is Gaussian with zero mean, we have that

$$p_A(a) \cong \frac{1}{\sqrt{2\pi}\,\sigma} \exp\left\{ -\frac{(a-s)^2}{2\sigma^2} \right\}, \qquad s \gg \sigma. \qquad (2.9\text{-}27)$$

As for the phase θ, with $s \gg \sigma$ its fluctuations will be small about zero, and

$$\theta \cong \tan\theta \cong \frac{i}{s} \qquad (2.9\text{-}28)$$

Therefore

$$p_\Theta(\theta) \cong s p_I(i = s\theta) \qquad (2.9\text{-}29)$$

or

$$p_\Theta(\theta) \cong \frac{k}{\sqrt{2\pi}} \exp\left\{ -\frac{k^2\theta^2}{2} \right\}. \qquad (2.9\text{-}30)$$

We conclude that both A and Θ are approximately Gaussian for $s \gg \sigma$. For the amplitude we have mean $\bar{a} = s$ and variance $\sigma_a^2 = \sigma^2$, whereas for the phase we have $\bar{\theta} = 0$ and $\sigma_\theta^2 = 1/k^2 = \sigma^2/s^2$. These results provide useful approximations when the condition $s \gg \sigma$ is met.

REFERENCES

2-1 E. Parzen, *Modern Probability Theory and Its Applications*, John Wiley & Sons, New York (1960).

2-2 W. Feller, *An Introduction to Probability Theory and Its Applications*, Vol. 1, John Wiley & Sons, New York (1957).

2-3 A. Papoulis, *Probability, Random Variables, and Stochastic Processes*, McGraw-Hill Book Company, New York (1965).

2-4 A. Papoulis, *Systems and Transforms with Applications in Optics*, McGraw-Hill Book Company, New York (1968).

2-5 D. Middleton, *An Introduction to Statistical Communication Theory*, McGraw-Hill Book Company, New York (1960).

2-6 J. B. Thomas, *An Introduction to Statistical Communication Theory*, John Wiley & Sons, New York (1969).

2-7 W. B. Davenport and W. L. Root, *An Introduction to the Theory of Random Signals and Noise*, McGraw-Hill Book Company, New York (1958).

2-8 P. Beckmann, *Probability in Communication Engineering*, Harcourt, Brace and World, Inc., New York (1967).

2-9 R. Bracewell, *The Fourier Transform and Its Applications*, McGraw-Hill Book Company, New York (1965).

2-10 P. Beckmann and A. Spizzichino, *The Scattering of Electromagnetic Waves from Rough Surfaces*, Pergamon Press, Oxford (1963).

PROBLEMS

2-1 Show that for any random variable U,

$$\overline{u^2} = \sigma^2 + (\bar{u})^2.$$

2-2 Show that any two statistically independent random variables have a correlation coefficient that is zero.

2-3 Given the random variables

$$U = \cos\Theta$$
$$V = \sin\Theta$$

with

$$p_\Theta(\theta) = \begin{cases} \dfrac{1}{\pi} & -\dfrac{\pi}{2} < \theta \leq \dfrac{\pi}{2} \\ 0 & \text{otherwise,} \end{cases}$$

show that $\rho = 0$.

PROBLEMS

2-4 Prove the following properties of characteristic functions:
  (a) Every characteristic function has value unity at zero argument.
  (b) The second-order characteristic function $\mathbf{M}_{UV}(\omega_U, \omega_V)$ with $\omega_V = 0$ is equal to the characteristic function $\mathbf{M}_U(\omega)$ of the random variable U alone.
  (c) For two independent random variables U and V,

$$\mathbf{M}_{UV}(\omega_U, \omega_V) = \mathbf{M}_U(\omega_U)\mathbf{M}_V(\omega_V)$$

2-5 Show that the moment $\overline{u^n v^m}$, if it exists, can be found from the joint characteristic function $\mathbf{M}_{UV}(\omega_U, \omega_V)$ by the formula

$$\overline{u^n v^m} = \frac{1}{j^{n+m}} \frac{\partial^{n+m}}{\partial \omega_U^n \, \partial \omega_V^m} \mathbf{M}_{UV}(\omega_U, \omega_V) \bigg|_{\omega_U = \omega_V = 0}.$$

2-6 (a) Show that a sum of two statistically independent Poisson-distributed random variables is Poisson distributed.
  (b) Show that if K is Poisson distributed, then

$$\overline{K(K-1)\cdots(K-k+1)} = (\overline{K})^k.$$

2-7 Find the probability density function of the random variable Z in terms of the known density function $p_U(u)$ when
  (a) $z = au + b$
  (b) $z = \begin{cases} |u| & -1 < u \le 1 \\ 1 & \text{otherwise.} \end{cases}$

2-8 Using the method given in Eq. (2.5-20), find the joint probability density function $p_{WZ}(w, z)$ when

$$w = u^2$$
$$z = u + v$$

and $p_{UV}(u, v) = \text{rect } u \text{ rect } v$, where $\text{rect } x = 1$ for $|x| \le \frac{1}{2}$ and is zero otherwise.

2-9 Consider two independent, identically distributed random variables Θ_1 and Θ_2, each of which obeys a probability density function

$$p_\Theta(\theta) = \begin{cases} \dfrac{1}{2\pi}, & -\pi < \theta \le \pi \\ 0 & \text{otherwise.} \end{cases}$$

(a) Find the probability density function of the random variable Z defined by

$$Z = \Theta_1 + \Theta_2$$

(b) If Z represents a phase angle that can only be measured modulo 2π, show that, despite the result of (a), Z is uniformly distributed on $(-\pi, \pi)$.

2-10 Consider the random phasor sum in Section 2.9.1 with the single change that the phases ϕ_k are uniformly distributed on $(-\pi/2, \pi/2)$. Find the following quantities: $\bar{r}, \bar{i}, \sigma_r^2, \sigma_i^2$, and ρ_{ri}. Make a rough plot of the contours of constant probability in the complex plane.

2-11 Let the random variables U_1 and U_2 be jointly Gaussian, with zero means, equal variances, and correlation coefficient $\rho \neq 0$. Consider new random variables V_1 and V_2 defined by a rotational transformation about the origin of the (u_1, u_2) plane,

$$\begin{bmatrix} v_1 \\ v_2 \end{bmatrix} = \begin{bmatrix} \cos\phi & \sin\phi \\ -\sin\phi & \cos\phi \end{bmatrix} \begin{bmatrix} u_1 \\ u_2 \end{bmatrix}$$

where ϕ is the rotation angle. Show that if ϕ is chosen to be $45°$, V_1 and V_2 are *independent* random variables. What are the means and variances of V_1 and V_2 in this case?

2-12 Consider n independent random variables $U_1, U_2, \ldots, U_n$, each of which obeys a Cauchy density function,

$$p_U(u) = \frac{1}{\pi\beta\left[1 + \left(\dfrac{u}{\beta}\right)^2\right]}$$

(a) Show that this density function violates one of the conditions (2.6-16) associated with the validity of the central limit theorem.
(b) Show that the random variable

$$Y = \frac{1}{n}\sum_{i=1}^{n} U_i$$

obeys a Cauchy distribution for all n.

2-13 A certain computer contains a random number generator that generates numbers with uniform relative frequencies (or probability density) on the interval $(0, 1)$. Suppose, however, that it is desired to simulate trials of a random variable Z with density function $p_Z(z)$ that is not uniform.

(a) If the values generated by the computer are represented by u, with

$$p_U(u) = \begin{cases} 1 & 0 < u \le 1 \\ 0 & \text{otherwise,} \end{cases}$$

show that by means of a monotonic transformation $z = g(u)$ it is possible to obtain the desired $p_Z(z)$, and that if $u = G(z)$ represents the inverse of $g(\cdot)$, then G should be chosen to satisfy

$$G(z) = \pm \int p_Z(z)\, dz$$

where $\int$ is an indefinite integral.

(b) Show that to generate a random variable with probability density

$$p_Z(z) = \begin{cases} e^{-z} & 0 \le z < \infty \\ 0 & \text{otherwise,} \end{cases}$$

either of the following transformations could be used:

$$z = -\ln u$$
$$z = -\ln(1 - u)$$

3

Random Processes

A natural generalization of the concept of a random variable is a *random process*, for which the basic unpredictable or random events are *functions* (usually of time and/or space) rather than numbers. The theory of random processes thus deals with the mathematical description of functions having a structure that cannot be predicted in detail in advance. Such functions play a role of great importance in optics; for example, the wave amplitude emitted by any real source has properties that change with time in an unpredictable way to some degree. In this chapter we review the basic concepts underlying the theory of such random phenomena. Emphasis is placed here on functions of *time*. However, generalizations to functions of *space* are straightforward.

3.1 DEFINITION AND DESCRIPTION OF A RANDOM PROCESS

Underlying the concept of a random process is again a random experiment, with a set of possible events $\{A\}$ and an associated probability measure. To define a random variable, we assigned a real-valued number $u(A)$ to each elementary event A. To define a random process, we assign a real-valued function $u(A; t)$, with independent variable t, to each elementary event A. The collection of possible "sample functions" $u(A; t)$, together with their associated probability measure, constitute a random process.

In general, the explicit dependence of the random process on the underlying set of events $\{A\}$ is not indicated in notation, with the random process represented by the symbol $U(t)$ and the specific sample functions indicated by lowercase letters $u(t)$. It should be remembered, however, that $U(t)$ consists of an entire ensemble of possible $u(t)$, together with a measure of their probabilities.

There are various possible ways to describe a random process mathematically. Most general is a complete denumeration of all sample functions composing the random process, together with a specification of their probabilities. We illustrate this complete description with the following example.

DEFINITION AND DESCRIPTION OF A RANDOM PROCESS

Let the underlying random experiment consist of two tosses of a "fair" coin, that is, a coin that is equally likely to land heads or tails. The "elementary events" in the set $\{A\}$ are $A_1 = HH$, $A_2 = HT$, $A_3 = TH$, and $A_4 = TT$. To each elementary event we assign a sample function as shown below:

$$u(A_1; t) = \exp(t)$$

$$u(A_2; t) = \exp(2t)$$

$$u(A_3; t) = \exp(3t)$$

$$u(A_4; t) = \exp(4t). \tag{3.1-1}$$

In each case the probability associated with the corresponding event must be calculated. Note that if several different events generate the same sample function, all possible ways of generating each sample function must be discovered, and the probability that any of these events occurs becomes the probability associated with that sample function. Thus, with much labor we arrive at a denumeration of all sample functions in the ensemble, together with their probabilities; a complete description of the random process is then in hand.

Such a complete description is seldom possible or even desirable. In most practical applications only a partial description of the random process is needed for calculation of the quantities of physical interest. Various different kinds of partial descriptions are possible. In some applications it may suffice to view the parameter t as fixed and to specify the *first-order* probability density function of the random variable $U(t)$, which we denote by $p_U(u; t)$. From such a description we can specify $\bar{u}, \overline{u^2}$ and other moments of U for any value of t.

More commonly, the *second-order* probability density function of U with parameter values t_1 and t_2 is required. Figure 3-1 illustrates the ensemble of waveforms and a pair of parameter values t_1 and t_2. The second-order density function is the joint density function of the random variables $U(t_1)$ and $U(t_2)$. In general this density function depends on both t_1 and t_2 and hence is denoted $p_U(u_1, u_2; t_1, t_2)$, where $u_1 = u(t_1)$, $u_2 = u(t_2)$. From such a description we can calculate joint moments, such as

$$\overline{u_1 u_2} = \iint_{-\infty}^{\infty} u_1 u_2 \, p_U(u_1, u_2; t_1, t_2) \, du_1 \, du_2. \tag{3.1-2}$$

In some cases, even higher order density functions may be required. To *completely* describe the random process $U(t)$, it must be possible to specify the kth-order density function $p_U(u_1, u_2, \ldots, u_k; t_1, t_2, \ldots, t_k)$ *for all k*. Such a description is equivalent to the complete description discussed previously and generally is just as difficult to state. In practice, a complete description is never needed.

In closing, we note that a random process is a mathematical model that is useful to us only before the exact sample function $u(t)$ is determined by measurement. Before the measurement, the random process represents our a priori state of knowledge. After $u(t)$ has been determined by measurement, only one sample function remains of interest, namely, the one that was observed.

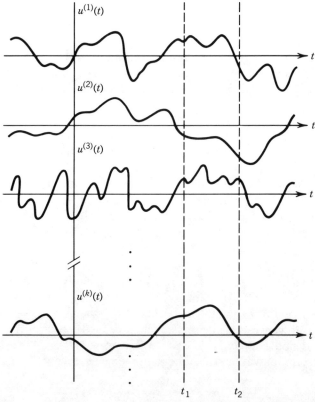

Figure 3-1. An ensemble of sample functions, where t_1 and t_2 are the parameter values for which the joint density function $p_U(u_1, u_2; t_1, t_2)$ is specified.

3.2 STATIONARITY AND ERGODICITY

Of the infinite variety of random process models that could in principle be constructed, only certain restricted types are of great importance in physical applications. Various restricted classes are defined and discussed in this section. This classification is by no means complete or exhaustive but merely identifies certain types of models we deal with in the future.

A random process is called *strictly stationary* if the kth-order joint probability density function $p_U(u_1, u_2, \ldots, u_k; t_1, t_2, \ldots, t_k)$ is independent of the choice of time origin *for all k*. Stating this definition mathematically, we require that

$$p_U(u_1, u_2, \ldots, u_k; t_1, t_2, \ldots, t_k)$$
$$= p_U(u_1, u_2, \ldots, u_k; t_1 - T, t_2 - T, \ldots, t_k - T) \quad (3.2\text{-}1)$$

for all k and all T. For such a process, the first-order density function is independent of time and hence can be written $p_U(u)$. Similarly, the second-order density function depends only on the time difference $\tau = t_2 - t_1$ and can be written $p_U(u_1, u_2; \tau)$.

A random process is called *wide-sense* stationary if the following two conditions are met:

(i) $E[u(t)]$ is independent of t.
(ii) $E[u(t_1)u(t_2)]$ depends only on $\tau = t_2 - t_1$.

Every strictly stationary random process is also wide-sense stationary; however, a wide-sense stationary process need not be strictly stationary.

If the difference $U(t_2) - U(t_1)$ is strictly stationary for all t_2 and t_1, $U(t)$ is said to have *stationary increments*.[†] If $\Phi(t)$ is a strictly stationary random process, the new random process

$$U(t) = U(t_0) + \int_{t_0}^{t} \Phi(\xi)\, d\xi \quad (t > t_0) \quad (3.2\text{-}2)$$

[constructed from integrals of the sample functions of $\Phi(t)$] is nonstationary but does have stationary increments. Such random processes play an important role in certain practical problems.

[†] We should differentiate here between strictly stationary increments and wide-sense stationary increments. For simplicity, however, we try to avoid using too many qualifiers and assume the kind of stationarity actually needed in each case.

Since a full description of a random process is seldom needed or even possible, we are generally satisfied with descriptions of finite-order (especially first- and second-order) statistics. In such cases it is necessary only to know the stationarity properties of the random process to finite order. (For example, are the second-order statistics strictly stationary, wide-sense stationary, or stationary in increments?) When in the future we refer to a random process simply as *stationary*, without specifying the type or order of stationarity, we mean that the particular statistical quantities necessary for use in our calculation are assumed to be independent of the choice of time origin. Depending on just what calculations are to be performed, the term may mean different types of stationarity in different cases. When there is danger of confusion, the exact type of stationarity assumed is stated precisely.

The most restrictive class of random processes and the class used most frequently in practice is the class of *ergodic* random processes. In this case we are interested in a comparison of the properties of an individual sample function as it evolves along the time axis, with the properties of the entire ensemble at one or more specific instants of time. We may state this in the form of a question by asking whether each sample function is in some sense typical of the entire ensemble.

For a more precise definition, a random process is called *ergodic* if every sample function (except possibly a subset with zero probability) takes on values along the time axis (i.e., "horizontally") with the same joint relative frequencies observed across the ensemble at any instant or collection of instants (i.e., "vertically").

For a random process to be ergodic, it is necessary that it be strictly stationary. This requirement is perhaps best understood by considering an example of a random process that is nonergodic by virtue of nonstationarity. Sample functions of such a process are shown in Fig. 3-2. Suppose that all sample functions have exactly the same relative frequency distributions along the time axis. Now clearly the relative frequencies observed across the process at time instants t_1 and t_2 will not be the same since the fluctuations of all sample functions are greater at t_2 than at t_1. Thus there is no unique distribution of relative frequencies across the process. Hence the relative frequencies observed across the process and along the process cannot be equal for all time. The process is thus nonergodic.

Although a process must be strictly stationary to be ergodic, not all strictly stationary processes are necessarily ergodic. We illustrate this fact with a specific example. Let $U(t)$ be the random process

$$U(t) = A\cos(\omega t + \Phi) \qquad (3.2\text{-}3)$$

where ω is a known constant, whereas A and Φ are independent random

STATIONARITY AND ERGODICITY

variables with probability density functions

$$p_A(a) = \tfrac{1}{2}\delta(a-1) + \tfrac{1}{2}\delta(a-2)$$

$$p_\Phi(\phi) = \begin{cases} \dfrac{1}{2\pi} & -\pi < \phi \leq \pi \\ 0 & \text{otherwise.} \end{cases} \quad (3.2\text{-}4)$$

Because of the uniform distribution of Φ on $(-\pi, \pi)$, this random process is strictly stationary. However, as illustrated in Fig. 3-3, a single sample

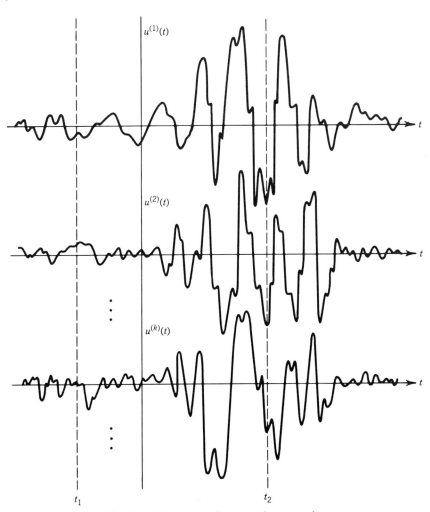

Figure 3-2. Sample functions of a nonstationary random process.

function is not typical of the entire process. Rather, there are two classes of sample functions; one class has amplitude 1 and the other, amplitude 2. Each class occurs with probability $\frac{1}{2}$. Clearly the relative frequencies observed along a sample function with amplitude 1 are different from the relative frequencies when the amplitude is 2. Thus not all sample functions have the same relative frequencies in time as those observed across the process.

If a random process is ergodic, any average calculated along a sample function (i.e., a time average) must equal the same average calculated across the ensemble (i.e., an ensemble average). Thus if $g(u)$ is the quantity to be averaged, we have that the time average,

$$\langle g \rangle = \lim_{T \to \infty} \frac{1}{T} \int_{-T/2}^{T/2} g[u(t)] \, dt \qquad (3.2\text{-}5)$$

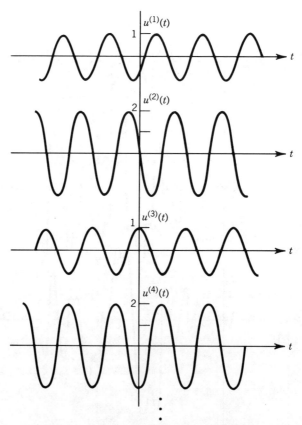

Figure 3-3. A stationary process that is nonergodic.

STATIONARITY AND ERGODICITY

must equal the ensemble average

$$\bar{g} = \int_{-\infty}^{\infty} g(u) p_U(u) \, du. \tag{3.2-6}$$

For an ergodic random process, time and ensemble averages are equal and interchangeable.

An important question remains. How can we methodically determine whether a certain random process model, which we believe accurately represents the random phenomenon under study, is ergodic? To establish ergodicity, it is necessary to consider the entire ensemble of sample functions. This ensemble can be said to be ergodic provided (Ref. 2-5, p. 56):

(a) The ensemble is strictly stationary.
(b) The ensemble contains no strictly stationary subensembles that occur with probability other than zero or one.

It should be noted that some random phenomena require a nonergodic ensemble for accurate modeling.

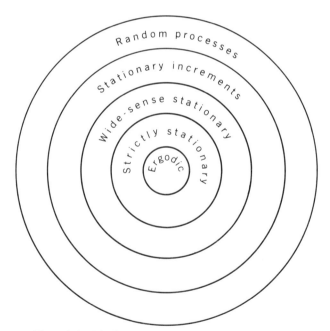

Figure 3-4. The hierarchy of classes of random processes.

The hierarchy of types of random processes is illustrated in Fig. 3-4, which shows the progression from the broad collection of all random processes to the far narrower class of ergodic random processes. The circles within circles represent subsets of the broader collections in each case.

3.3 SPECTRAL ANALYSIS OF RANDOM PROCESSES

Let $u(t)$ be a known function of time. Two different classes of time functions can be distinguished. If $u(t)$ has the property that

$$\int_{-\infty}^{\infty} |u(t)|\, dt < \infty, \qquad (3.3\text{-}1)$$

we say that $u(t)$ is Fourier transformable. On the other hand, it may be that $u(t)$ does not satisfy (3.3-1) but does satisfy

$$\lim_{T \to \infty} \frac{1}{T} \int_{-T/2}^{T/2} u^2(t)\, dt < \infty, \qquad (3.3\text{-}2)$$

in which case we say that $u(t)$ has *finite average power*. In each case it is important in practice to be able to specify the distribution of energy [when Eq. (3.3-1) is satisfied] or average power [when Eq. (3.3-2) is satisfied] over frequency. Such descriptions are called, respectively, the *energy spectral density* (energy spectrum) and the *power spectral density* (power spectrum) of the function $u(t)$.

Similarly, if $U(t)$ is a random process with sample functions satisfying (3.3-1) or (3.3-2), it is important to be able to characterize the manner in which energy or average power is distributed over frequency, not just for one sample function but for the entire random process. Since the particular sample function that will occur experimentally is unknown in advance, the logical quantity to be concerned with is the *expected* distribution of energy or average power over frequency. These expected or mean distributions are called, respectively, the *energy spectral density* and the *power spectral density* of the random process $U(t)$. The distinction between spectral densities of known functions and of random processes is an important one and is developed in further detail in the following section.

3.3.1 Spectral Densities of Known Functions

If $u(t)$ is a Fourier transformable function, then

$$\mathcal{U}(\nu) = \int_{-\infty}^{\infty} u(t) e^{j2\pi\nu t}\, dt \qquad (3.3\text{-}3)$$

SPECTRAL ANALYSIS OF RANDOM PROCESSES

always exists. Further, according to Parseval's theorem (Ref. 2-6, p. 380), the area under $|\mathscr{U}(v)|^2$ is equal to the total energy contained in $u(t)$; that is,

$$\int_{-\infty}^{\infty} u^2(t)\, dt = \int_{-\infty}^{\infty} |\mathscr{U}(v)|^2\, dv. \tag{3.3-4}$$

Thus the quantity

$$\mathscr{E}(v) = |\mathscr{U}(v)|^2 \tag{3.3-5}$$

has the dimensions of energy per unit frequency, and we accordingly refer to it as the energy spectral density of $u(t)$.

On the other hand, suppose that $u(t)$ is not Fourier transformable but does have finite average power. Then, in general, the integral (3.3-3) does not exist. However, the truncated function

$$u_T(t) = \begin{cases} u(t) & -\dfrac{T}{2} \le t < \dfrac{T}{2} \\ 0 & \text{otherwise} \end{cases} \tag{3.3-6}$$

does have a transform, which we denote by $\mathscr{U}_T(v)$. Furthermore, the quantity $|\mathscr{U}_T(v)|^2$ represents the distribution of energy over frequency for the truncated waveform $u_T(t)$. Thus the normalized energy spectrum

$$\mathscr{G}_T(v) = \frac{|\mathscr{U}_T(v)|^2}{T}$$

has the dimension of power per unit frequency, and we are logically led to define the power spectral density of $u(t)$ by

$$\mathscr{G}(v) \triangleq \lim_{T \to \infty} \frac{|\mathscr{U}_T(v)|^2}{T}.$$

Such a definition works adequately for some functions. For example, the reader may wish to prove, by means of the limiting process above, that the function

$$u(t) = 1 \quad (\text{all } t)$$

has a power spectral density

$$\mathscr{G}(v) = \delta(v).$$

Therefore, whereas strictly speaking, the limit above does not exist in this case, it does exist in the sense of δ functions.

Unfortunately, however, *there are also many functions for which the limit does not even exist in the sense of δ functions*. Rather, the value of $\mathcal{G}_T(\nu)$ fluctuates erratically at each ν as T is increased without bound. Such is often the case when $u(t)$ represents a sample function of a stationary random process.

In addition, note that the above definitions of $\mathcal{E}(\nu)$ and $\mathcal{G}(\nu)$ apply only for a single function $u(t)$, but a random process contains an entire ensemble of different functions. Clearly a different definition of power spectral density is needed for a random process.

3.3.2 Spectral Density of a Random Process

There exists a simple and logical modification of the definitions of energy and power spectral densities that proves quite satisfactory in practice. Since we wish to find a spectral distribution that characterizes an entire random process, it is logical to define such quantities in terms of averages over the entire random process. Accordingly, we define the energy and power spectral densities, respectively, by

$$\mathcal{E}_U(\nu) \triangleq E\big[|\mathcal{U}(\nu)|^2\big] \qquad (3.3\text{-}7a)$$

$$\mathcal{G}_U(\nu) \triangleq \lim_{T \to \infty} \frac{E\big[|\mathcal{U}_T(\nu)|^2\big]}{T}. \qquad (3.3\text{-}7b)$$

The latter limit does indeed exist in most cases of practical interest.

Several basic properties of spectral density functions follow directly from the definitions (3.3-7):

(i) $\mathcal{E}_U(\nu) \geq 0$, $\mathcal{G}_U(\nu) \geq 0$; energy and power spectral densities are nonnegative (and real-valued).

(ii) $\mathcal{E}_U(-\nu) = \mathcal{E}_U(\nu)$, $\mathcal{G}_U(-\nu) = \mathcal{G}_U(\nu)$; energy and power spectral densities are even functions of ν, provided $U(t)$ is a real-valued random process.

(iii)
$$\int_{-\infty}^{\infty} \mathcal{E}_U(\nu)\, d\nu = \int_{-\infty}^{\infty} \overline{u^2(t)}\, dt,$$

$$\int_{-\infty}^{\infty} \mathcal{G}_U(\nu)\, d\nu = \begin{cases} \overline{u^2} & \text{for } U(t) \text{ stationary} \\ \langle \overline{u^2(t)} \rangle & \text{for } U(t) \text{ nonstationary}. \end{cases}$$

SPECTRAL ANALYSIS OF RANDOM PROCESSES

Proofs of these properties are straightforward. Property (i) follows directly from the positivity of the right-hand sides of Eq. (3.3-7). Property (ii) follows from the hermitian character of $\mathcal{U}(\nu)$ and $\mathcal{U}_T(\nu)$ [i.e., $\mathcal{U}(-\nu) = \mathcal{U}^*(\nu)$, $\mathcal{U}_T(-\nu) = \mathcal{U}_T^*(\nu)$] for any real-valued $u(t)$. Property (iii) for the energy spectral density follows from Parseval's theorem and an interchange of orders of averaging and integration. Property (iii) for power spectral densities can be proved by noting

$$\int_{-\infty}^{\infty} \mathcal{G}_U(\nu) \, d\nu = \int_{-\infty}^{\infty} \lim_{T \to \infty} \frac{E[|\mathcal{U}_T(\nu)|^2]}{T} \, d\nu$$

$$= \lim_{T \to \infty} \frac{1}{T} E\left[\int_{-\infty}^{\infty} |\mathcal{U}_T(\nu)|^2 \, d\nu \right] = \lim_{T \to \infty} \frac{1}{T} E\left[\int_{-\infty}^{\infty} u_T^2(t) \, dt \right]$$

where Parseval's theorem was used in the last step. Continuing,

$$\lim_{T \to \infty} \frac{1}{T} E\left[\int_{-\infty}^{\infty} u_T^2(t) \, dt \right] = \lim_{T \to \infty} \frac{1}{T} \int_{-T/2}^{T/2} E\left[u_T^2(t) \right] dt$$

$$= \begin{cases} \overline{u^2} & \text{if } U(t) \text{ is stationary} \\ \langle \overline{u^2(t)} \rangle & \text{if } U(t) \text{ is nonstationary.} \end{cases}$$

Thus the basic properties have been proved.

3.3.3 Energy and Power Spectral Densities for Linearly Filtered Random Processes

Let the random process $V(t)$ consist of sample functions that result from passing all sample functions of the random process $U(t)$ through a known linear filter.[†] Then $V(t)$ is called a *linearly filtered* random process. In the case of a random process with Fourier transformable sample functions, we wish to find the relationship between the energy spectral densities $\mathcal{E}_V(\nu)$ of the filter output and $\mathcal{E}_U(\nu)$ of the filter input. If the sample functions of $U(t)$ are not Fourier transformable but do have finite average power, the desired relationship is between the power spectral densities $\mathcal{G}_V(\nu)$ and $\mathcal{G}_U(\nu)$.

The case of Fourier transformable waveforms is considered first. The linear filter is assumed to be time invariant, in which case a single output sample function is related to the corresponding input sample function $u(t)$

[†] For a review of the properties of linear filters, see, for example, Ref. 2-9, Chapter 9.

by a convolution

$$v(t) = \int_{-\infty}^{\infty} h(t - \xi) u(\xi) \, d\xi, \tag{3.3-8}$$

where $h(t)$ represents the known response of the filter at time t to a unit impulse applied at time $t = 0$ (i.e., $h(t)$ is the "impulse response" of the filter). In the frequency domain, this relationship becomes a simple multiplicative one,

$$\mathscr{V}(\nu) = \mathscr{H}(\nu) \mathscr{U}(\nu) \tag{3.3-9}$$

where $\mathscr{V}(\nu)$ and $\mathscr{U}(\nu)$ are the Fourier transforms of $v(t)$ and $u(t)$ and $\mathscr{H}(\nu)$ is the Fourier transform of $h(t)$ (called the *transfer function*). The definition given in Eq. (3.3-7a) is now used for $\mathscr{E}_V(\nu)$,

$$\mathscr{E}_V(\nu) = E\left[|\mathscr{H}(\nu) \mathscr{U}(\nu)|^2\right] = |\mathscr{H}(\nu)|^2 \mathscr{E}_U(\nu). \tag{3.3-10}$$

Thus the average spectral distribution of energy in the random process is modified by the simple multiplicative factor $|\mathscr{H}(\nu)|^2$.

For the case of finite-average-power processes, the relationship between the power spectral densities $\mathscr{G}_V(\nu)$ and $\mathscr{G}_U(\nu)$ must be found by a more subtle argument. In this case the Fourier transforms $\mathscr{V}(\nu)$ and $\mathscr{U}(\nu)$ generally do not exist. However, the truncated waveforms $v_T(t)$ and $u_T(t)$ do have transforms $\mathscr{V}_T(\nu)$ and $\mathscr{U}_T(\nu)$. Furthermore, although the relationship is not exact due to "end effects," we can nonetheless write

$$v_T(t) \cong \int_{-\infty}^{\infty} h(t - \xi) u_T(\xi) \, d\xi \tag{3.3-11}$$

with an approximation that becomes arbitrarily good as T increases.[†] Subject to the same approximation, we have the frequency domain relationship

$$\mathscr{V}_T(\nu) \cong \mathscr{H}(\nu) \mathscr{U}_T(\nu).$$

[†] The approximation arises because the response of the filter to a truncated excitation is in general not itself truncated. As T grows, however, these end effects eventually have negligible consequence.

The power spectral density of $v(t)$ can now be written

$$\mathcal{G}_V(\nu) = \lim_{T\to\infty} \frac{E[|\mathcal{V}_T(\nu)|^2]}{T} = \lim_{T\to\infty} \frac{E[|\mathcal{H}(\nu)|^2|\mathcal{U}_T(\nu)|^2]}{T}$$

$$= |\mathcal{H}(\nu)|^2 \lim_{T\to\infty} \frac{E[|\mathcal{U}_T(\nu)|^2]}{T},$$

or equivalently

$$\mathcal{G}_V(\nu) = |\mathcal{H}(\nu)|^2 \mathcal{G}_U(\nu). \tag{3.3-12}$$

Thus the power spectral density of the output random process is simply the squared modulus of the transfer function of the filter times the power spectral density of the input random process.

3.4 AUTOCORRELATION FUNCTIONS AND THE WIENER–KHINCHIN THEOREM

In the theory of coherence (Chapter 5) a role of great importance is played by correlation functions. In preparation for these discussions, we accordingly introduce the concept of an autocorrelation function.

Given a single known time function $u(t)$, which may be one sample function of a random process, the *time autocorrelation function* of $u(t)$ is defined by

$$\tilde{\Gamma}_U(\tau) \triangleq \langle u(t+\tau)u(t) \rangle$$

$$= \lim_{T\to\infty} \frac{1}{T} \int_{-T/2}^{T/2} u(t+\tau)u(t)\, dt. \tag{3.4-1}$$

Closely related, but a property of an entire random process $U(t)$, is the *statistical autocorrelation function*, defined by

$$\Gamma_U(t_2, t_1) \triangleq \overline{u(t_2)u(t_1)}$$

$$= \iint_{-\infty}^{\infty} u_2 u_1 p_U(u_1, u_2; t_2, t_1)\, du_1\, du_2. \tag{3.4-2}$$

From a physical point of view, the time autocorrelation function measures the structural similarity of $u(t)$ and $u(t + \tau)$, averaged over all time, whereas the statistical autocorrelation function measures the statistical similarity of $u(t_1)$ and $u(t_2)$ over the ensemble.

For a random process with at least wide-sense stationarity, Γ_U is a function only of the time difference $\tau = t_2 - t_1$. For the more restrictive class of ergodic random processes, the time autocorrelation functions of all sample functions are equal to each other and are also equal to the statistical autocorrelation function. For ergodic processes, therefore,

$$\tilde{\Gamma}_U(\tau) = \Gamma_U(\tau) \quad \begin{pmatrix} \text{all sample} \\ \text{functions} \end{pmatrix}. \qquad (3.4\text{-}3)$$

It is thus pointless to distinguish between the two types of autocorrelation function for such processes.

Two important properties of autocorrelation functions of processes that are at least wide-sense stationary follow directly from the definition:

(i) $\Gamma_U(0) = \overline{u^2}$
(ii) $\Gamma_U(-\tau) = \Gamma_U(\tau)$. $\qquad (3.4\text{-}4)$

A third property, $\qquad (3.4\text{-}5)$

(iii) $|\Gamma_U(\tau)| \leq \Gamma_U(0)$,

can be proved using Schwarz's inequality [cf. argument leading to Eq. (2.4-16)].

However, a major practical importance of autocorrelation functions lies in the very special relationship they enjoy with respect to power spectral density. In the derivation to follow we shall show that, for a process that is at least wide-sense stationary, the autocorrelation function and power spectral density form a Fourier transform pair,

$$\mathcal{G}_U(\nu) = \int_{-\infty}^{\infty} \Gamma_U(\tau) e^{j2\pi\nu\tau} \, d\tau$$

$$\Gamma_U(\tau) = \int_{-\infty}^{\infty} \mathcal{G}_U(\nu) e^{-j2\pi\nu\tau} \, d\nu. \qquad (3.4\text{-}6)$$

This very special relationship is known as the *Wiener–Khinchin theorem*.

To prove the above relationship, we begin with the definition of power spectral density,

$$\mathcal{G}_U(\nu) = \lim_{T \to \infty} \frac{E[\mathcal{U}_T(\nu)\mathcal{U}_T^*(\nu)]}{T}. \qquad (3.4\text{-}7)$$

AUTOCORRELATION FUNCTIONS AND THE WIENER-KHINCHIN THEOREM

Since $u(t)$ is real valued, we have $\mathcal{U}_T^*(\nu) = \mathcal{U}_T(-\nu)$, and we further note that[†]

$$\mathcal{U}_T(\nu) = \int_{-\infty}^{\infty} \text{rect}\frac{\xi}{T} u(\xi) \exp(j2\pi\nu\xi) \, d\xi$$

$$\mathcal{U}_T(-\nu) = \int_{-\infty}^{\infty} \text{rect}\frac{\eta}{T} u(\eta) \exp(-j2\pi\nu\eta) \, d\eta. \quad (3.4\text{-}8)$$

Substituting (3.4-8) in (3.4-7), we find

$$\frac{E[|\mathcal{U}_T(\nu)|^2]}{T} = \frac{1}{T} \iint_{-\infty}^{\infty} \text{rect}\frac{\xi}{T} \text{rect}\frac{\eta}{T} E[u(\xi)u(\eta)]$$

$$\cdot \exp[j2\pi\nu(\xi-\eta)] \, d\xi \, d\eta.$$

The expectation is recognized as the statistical autocorrelation function of $U(t)$. For the sake of generality, we allow $\Gamma_U(\xi, \eta)$ to depend on both ξ and η, deferring our assumption of stationarity until a later point. Thus we obtain

$$\frac{E[|\mathcal{U}_T(\nu)|^2]}{T} = \frac{1}{T} \iint_{-\infty}^{\infty} \text{rect}\frac{\xi}{T} \text{rect}\frac{\eta}{T} \Gamma_U(\xi, \eta) \exp[j2\pi\nu(\xi-\eta)] \, d\xi \, d\eta.$$

Now with a simple change of variables, with ξ replaced by $t + \tau$ and η by t, the integral becomes

$$\frac{E[|\mathcal{U}_T(\nu)|^2]}{T} = \frac{1}{T} \iint_{-\infty}^{\infty} \text{rect}\frac{t+\tau}{T} \text{rect}\frac{t}{T} \Gamma(t+\tau, t) \exp(j2\pi\nu\tau) \, dt \, d\tau.$$

The power spectral density $\mathcal{G}_U(\nu)$ is the limit of this quantity as $T \to \infty$. Interchanging orders of integration with respect to τ and the limit, and noting that for any fixed τ

$$\lim_{T \to \infty} \frac{1}{T} \int_{-\infty}^{\infty} \text{rect}\frac{t+\tau}{T} \text{rect}\frac{t}{T} \Gamma(t+\tau, t) \, dt = \langle \Gamma(t+\tau, t) \rangle,$$

we obtain

$$\mathcal{G}_U(\nu) = \int_{-\infty}^{\infty} \langle \Gamma_U(t+\tau, t) \rangle e^{j2\pi\nu\tau} \, d\tau, \quad (3.4\text{-}9)$$

where the angle brackets, as usual, signify a time averaging operation.

[†] Here and throughout, the function rect x is defined to be unity for $|x| \leq \frac{1}{2}$ and zero otherwise.

The result [Eq. (3.4-9)] shows that the power spectral density of any random process, stationary or nonstationary, can be found from a Fourier transform of a suitably averaged autocorrelation function. When the random process is at least wide-sense stationary, we have $\Gamma_U(t + \tau, t) = \Gamma_U(\tau)$ and

$$\mathcal{G}_U(\nu) = \int_{-\infty}^{\infty} \Gamma_U(\tau) \exp(j2\pi\nu\tau) \, d\tau, \qquad (3.4\text{-}10)$$

which is the relationship that was to be proved. Provided this transform exists, at least in the sense of δ functions, the inverse relationship

$$\Gamma_U(\tau) = \int_{-\infty}^{\infty} \mathcal{G}_U(\nu) \exp(-j2\pi\nu\tau) \, d\nu \qquad (3.4\text{-}11)$$

follows from the basic properties of Fourier transforms.

The importance of autocorrelation functions stems from two sources. First (and of particular relevance in Fourier spectroscopy), the autocorrelation of a signal can often be directly measured, thereby providing an experimental means for ultimately determining the power spectral density of the signal. The experimentally measured autocorrelation function is Fourier transformed by either digital or analog means to provide a distribution of power over frequency.

Second, the autocorrelation function often provides an analytic means for calculating the power spectral density of a random process model described only in statistical terms. Often it is much easier to calculate the autocorrelation function of Eq. (3.4-2) than to directly calculate the power spectral density using (3.3-7). However, once the autocorrelation function is found, the power spectral density is easily obtained by means of a Fourier transformation.

To illustrate with a simple example, consider a random process $U(t)$ with a typical sample function as shown in Fig. 3-5. The value of $u(t)$ jumps between $+1$ and -1. Assume that our statistical model, based on physical intuition about the phenomenon underlying the process, is that the number n of jumps occurring in a $|\tau|$ second interval obeys Poisson statistics,

$$P(n; |\tau|) = \frac{(k|\tau|)^n}{n!} e^{-k|\tau|}, \qquad (3.4\text{-}12)$$

where k is the mean rate (jumps per second). The autocorrelation function $\Gamma_U(t_2, t_1)$ is given by

$$\Gamma_U(t_2, t_1) = \overline{u(t_2) u(t_1)} = 1 \cdot \text{Prob}\{u(t_1) = u(t_2)\}$$
$$- 1 \cdot \text{Prob}\{u(t_1) \neq u(t_2)\}.$$

AUTOCORRELATION FUNCTIONS AND THE WIENER–KHINCHIN THEOREM

But

$$\text{Prob}\{u(t_1) = u(t_2)\} = \text{Prob}\begin{pmatrix}\text{even number of}\\ \text{jumps in }|\tau|\end{pmatrix}$$

$$\text{Prob}\{u(t_1) \neq u(t_2)\} = \text{Prob}\begin{pmatrix}\text{odd number of}\\ \text{jumps in }|\tau|\end{pmatrix}.$$

Thus

$$\Gamma_U(t_2, t_1) = \sum_{m\text{ even}} \frac{(k|\tau|)^m}{m!} e^{-k|\tau|} - \sum_{m\text{ odd}} \frac{(k|\tau|)^m}{m!} e^{-k|\tau|}$$

$$= e^{-k|\tau|} \sum_{m=0}^{\infty} \frac{(-k|\tau|)^m}{m!}.$$

The series is simply equal to $e^{-k|\tau|}$, so

$$\Gamma_U(t_2, t_1) = \Gamma_U(\tau) = \exp(-2k|\tau|). \qquad (3.4\text{-}13)$$

We see that the process is wide-sense stationary, and on Fourier transformation of $\Gamma_U(\tau)$ we find the power spectral density

$$\mathcal{G}_U(\nu) = \frac{1/k}{\left[1 + \left(\dfrac{\pi\nu}{k}\right)^2\right]}. \qquad (3.4\text{-}14)$$

Both the autocorrelation function and the power spectral density are illustrated in Fig. 3-6. To find the power spectral density directly from the definition would require appreciably more work than that involved in the preceding calculation.

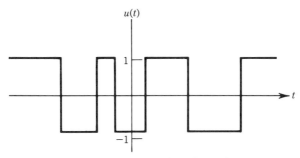

Figure 3-5. Sample function of a random process.

For our later use, it will be convenient to define some additional quantities closely related to the autocorrelation function. First, we define the *autocovariance* function,

$$C_U(t_2, t_1) \triangleq \overline{[u(t_2) - \bar{u}(t_2)][u(t_1) - \bar{u}(t_1)]}. \qquad (3.4\text{-}15)$$

Thus

$$C_U(t_2, t_1) = \Gamma_U(t_2, t_1) - \bar{u}(t_2)\bar{u}(t_1) \qquad (3.4\text{-}16)$$

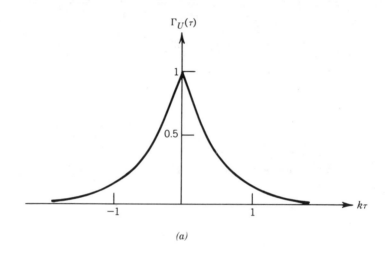

(a)

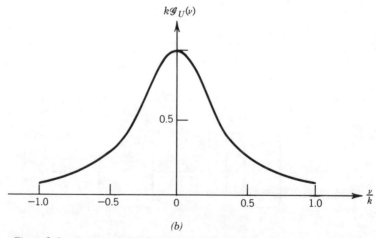

(b)

Figure 3-6. Autocorrelation function and corresponding power spectral density.

specifies the close relationship between autocovariance functions and autocorrelation functions.

A second quantity of considerable utility is the *structure function* $D_U(t_2, t_1)$ of the random process $U(t)$, as defined by

$$D_U(t_2, t_1) \triangleq \overline{[u(t_2) - u(t_1)]^2}. \tag{3.4-17}$$

The structure function and the autocorrelation function are related by

$$D_U(t_2, t_1) = \overline{u^2(t_2)} + \overline{u^2(t_1)} - 2\Gamma_U(t_2, t_1). \tag{3.4-18}$$

The structure function has the advantage that it depends only on the delay $\tau = t_2 - t_1$ even for some random processes that are not wide-sense stationary. For example, it is easy to show that a random process that is nonstationary but is stationary in increments has a structure function that depends only on τ. Of course, $D_U(t_2, t_1)$ depends only on τ for stronger types of stationarity, too. If $U(t)$ is wide-sense stationary, $D_U(\tau)$ and $\Gamma_U(\tau)$ are related by

$$D_U(\tau) = 2\Gamma_U(0) - 2\Gamma_U(\tau). \tag{3.4-19}$$

In addition, $D_U(\tau)$ can be expressed in terms of the power spectral density $\mathscr{G}_U(\nu)$,

$$D_U(\tau) = 2\int_{-\infty}^{\infty} \mathscr{G}_U(\nu)[1 - \cos 2\pi\nu\tau]\, d\nu. \tag{3.4-20}$$

3.5 CROSS-CORRELATION FUNCTIONS AND CROSS-SPECTRAL DENSITIES

A natural generalization of the concept of an autocorrelation function is the cross-correlation function of two random processes $U(t)$ and $V(t)$, as defined by

$$\Gamma_{UV}(t_2, t_1) \triangleq E[u(t_2)v(t_1)]. \tag{3.5-1}$$

In addition to the ensemble-average definition above, we can define the time-average cross-correlation function,

$$\tilde{\Gamma}_{UV}(\tau) \triangleq \langle u(t+\tau)v(t) \rangle. \tag{3.5-2}$$

The random processes $U(t)$ and $V(t)$ are said to be jointly wide-sense stationary if $\Gamma_{UV}(t_2, t_1)$ depends only on the time difference $\tau = t_2 - t_1$, in which case

$$\Gamma_{UV}(t_2, t_1) = \Gamma_{UV}(\tau). \qquad (3.5\text{-}3)$$

For such processes, the cross-correlation function exhibits the following properties

(i) $\qquad \Gamma_{UV}(0) = \overline{uv}$

(ii) $\qquad \Gamma_{UV}(-\tau) = \Gamma_{VU}(\tau)$

(iii) $\qquad |\Gamma_{UV}(\tau)| \leq [\Gamma_U(0)\Gamma_V(0)]^{1/2}. \qquad (3.5\text{-}4)$

The first two properties follow directly from the definition. Proof of the third property requires the help of Schwarz's inequality.

Closely related to cross-correlation functions are *cross-spectral density functions*, defined by

$$\mathscr{G}_{UV}(\nu) \triangleq \lim_{T \to \infty} \frac{E[\mathscr{U}_T(\nu)\mathscr{V}_T^*(\nu)]}{T}$$

$$\mathscr{G}_{VU}(\nu) \triangleq \lim_{T \to \infty} \frac{E[\mathscr{V}_T(\nu)\mathscr{U}_T^*(\nu)]}{T}. \qquad (3.5\text{-}5)$$

The functions $\mathscr{G}_{UV}(\nu)$ and $\mathscr{G}_{VU}(\nu)$ may be regarded as measures of the statistical similarity of the random processes $U(t)$ and $V(t)$ at each frequency ν. The cross-spectral density is in general a complex-valued function. In addition, it has the following basic properties

(i) $\qquad \mathscr{G}_{VU}(\nu) = \mathscr{G}_{UV}^*(\nu) \qquad$ for any real-valued random processes

(ii) $\qquad \mathscr{G}_{UV}(-\nu) = \mathscr{G}_{UV}^*(\nu) \qquad U(t)$ and $V(t)$. $\qquad (3.5\text{-}6)$

By an argument strictly similar to that leading to Eq. (3.4-10), we can prove the important fact that for jointly wide-sense stationary random processes $U(t)$ and $V(t)$, $\mathscr{G}_{UV}(\nu)$ and $\Gamma_{UV}(\nu)$ are a Fourier transform pair; that is,

$$\mathscr{G}_{UV}(\nu) = \int_{-\infty}^{\infty} \Gamma_{UV}(\tau) e^{j2\pi\nu\tau} d\tau$$

$$\Gamma_{UV}(\tau) = \int_{-\infty}^{\infty} \mathscr{G}_{UV}(\nu) e^{-2\pi\nu\tau} d\nu. \qquad (3.5\text{-}7)$$

CROSS-CORRELATION FUNCTIONS AND CROSS-SPECTRAL DENSITIES

In addition, using a derivation analogous to that used in Section 3.3.4, we can discover the effect of linear filtering on the cross-spectral density. Referring to Fig. 3-7, let the random process $U(t)$ be passed through a linear, time-invariant filter with transfer function $\mathcal{H}_1(\nu)$ to produce a random process $W(t)$, and let $V(t)$ be passed through a linear, time-invariant filter with transfer function $\mathcal{H}_2(\nu)$ to produce a random process $Z(t)$. By straightforward extension of the arguments in Section 3.3.3, we can show that

$$\mathcal{G}_{WZ}(\nu) = \mathcal{H}_1(\nu)\mathcal{H}_2^*(\nu)\mathcal{G}_{UV}(\nu). \tag{3.5-8}$$

The reader may well be wondering what the utility of the concepts introduced in this section might be. Cross-correlation functions and cross-spectral density functions are found to play extremely important roles in the theory of optical coherence, for they are directly related to the fringe-forming capabilities of light beams. For the present it suffices to point out that these concepts arise quite naturally when we consider a random process $Z(t)$ having sample functions $z(t)$ that are sums of the sample functions $u(t)$ and $v(t)$ of two jointly wide-sense stationary random processes $U(t)$ and $V(t)$; thus

$$z(t) = u(t) + v(t).$$

For such a process, the power spectral density is easily seen to be

$$\mathcal{G}_Z(\nu) = \lim_{T \to \infty} \frac{E\left[\mathcal{Z}_T(\nu)\mathcal{Z}_T^*(\nu)\right]}{T}$$

$$= \lim_{T \to \infty} \frac{E\left[(\mathcal{U}_T(\nu) + \mathcal{V}_T(\nu))(\mathcal{U}_T^*(\nu) + \mathcal{V}_T^*(\nu))\right]}{T}.$$

(3.5-9)

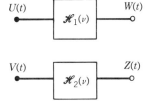

Figure 3-7. Transformation of cross-spectral density under linear filtering.

Expanding the argument of the expectation and averaging the four resulting terms individually, we obtain

$$\mathscr{G}_Z(\nu) = \mathscr{G}_U(\nu) + \mathscr{G}_V(\nu) + \mathscr{G}_{UV}(\nu) + \mathscr{G}_{VU}(\nu). \quad (3.5\text{-}10)$$

The corresponding relationship for the autocorrelation function of $Z(t)$ is

$$\Gamma_Z(\tau) = \Gamma_U(\tau) + \Gamma_V(\tau) + \Gamma_{UV}(\tau) + \Gamma_{VU}(\tau). \quad (3.5\text{-}11)$$

Clearly the autocorrelation function and power spectral density of $Z(t)$ depend not only on the corresponding properties of $U(t)$ and $V(t)$ individually, but also on the statistical relationship *between* these two latter processes, through the cross-correlation functions and the corresponding cross-spectral densities.

3.6 THE GAUSSIAN RANDOM PROCESS

Just as Gaussian random variables represent the most important kind of random variable in physical applications, so too the Gaussian random process plays a role of major importance. The underlying reason for this importance is again the fact that many physical phenomena are composed of a multitude of independent additive contributions, which, as a result of the central limit theorem leads to Gaussian statistics. Here we briefly review the most important properties of the Gaussian random process.

3.6.1 Definition

A random process $U(t)$ is said to be a *Gaussian random process* if the random variables $U(t_1), U(t_2), \ldots, U(t_k), \ldots$ are jointly Gaussian random variables for all finite sets of time instants. For n time instants $t_1, t_2, \ldots, t_n$, the joint probability density function is thus of the form

$$p_U(\underline{u}) = \frac{1}{(2\pi)^{n/2}|\underline{C}|^{1/2}} \exp\left\{-\frac{1}{2}(\underline{u} - \underline{\bar{u}})^t \underline{C}^{-1}(\underline{u} - \underline{\bar{u}})\right\}, \quad (3.6\text{-}1)$$

where

$$\underline{u} = \begin{bmatrix} u(t_1) \\ u(t_2) \\ \vdots \\ u(t_n) \end{bmatrix}, \quad \underline{\bar{u}} = \begin{bmatrix} \bar{u}(t_1) \\ \bar{u}(t_2) \\ \vdots \\ \bar{u}(t_n) \end{bmatrix} \quad (3.6\text{-}2)$$

and $\underline{C}$ is a covariance matrix with element in the ith row and jth column defined by

$$\sigma_{ij}^2 = E\big[[u(t_i) - \bar{u}(t_i)][u(t_j) - \bar{u}(t_j)]\big]. \qquad (3.6\text{-}3)$$

Corresponding to the density function of Eq. (3.6-1) is the joint characteristic function of the n jointly Gaussian random variables

$$\mathbf{M}_U(\underline{\omega}) = \exp\{j\underline{\omega}'\bar{\underline{u}} - \tfrac{1}{2}\underline{\omega}'\underline{C}\underline{\omega}\}, \qquad (3.6\text{-}4)$$

where

$$\underline{\omega} = \begin{bmatrix} \omega_1 \\ \omega_2 \\ \vdots \\ \omega_n \end{bmatrix}.$$

3.6.2 Linearly Filtered Gaussian Random Processes

The Gaussian random process possesses many unique properties that make it particularly simple to deal with. One such property is the following: *a linearly filtered Gaussian random process is also a Gaussian random process.*

A rigorous proof of this fact is beyond the scope of our treatment (see, e.g., Ref. 2-7, pp. 155–157). However, the following loose argument makes the result plausible. If $V(t)$ is a linearly filtered random process, each sample function $v(t)$ can be related to an input sample function $u(t)$ by means of a superposition integral,

$$v(t) = \int_{-\infty}^{\infty} h(t, \xi) u(\xi) \, d\xi, \qquad (3.6\text{-}5)$$

where $h(t, \xi)$ is response of the filter at time t to a unit impulse applied at time ξ. The integral can be written as a limit of approximating sums,

$$v(t) = \lim_{\Delta\xi \to 0} \sum_{k=-\infty}^{\infty} h(t, \xi_k) u(\xi_k) \Delta\xi,$$

where ξ_k is a point in the kth subinterval of width $\Delta\xi$. Over the ensemble of input sample functions, the value $u(\xi_k)$ is Gaussianly distributed, by assumption. Since $h(t, \xi_k)$ is simply a known real number, each term of the sum obeys Gaussian statistics over the ensemble. Finally, the sum of any number of Gaussian random variables, dependent or independent, is itself Gaussian. Hence the first-order statistics of $v(t)$ are Gaussian.

Thus the Gaussian random process has a certain unique kind of permanence. Although passage through a linear filter may change the parameters of the distribution (i.e., means, variances, covariances), the Gaussian character of the random process is retained.

3.6.3 Wide-Sense Stationarity and Strict Stationarity

A final unusual and important property of the Gaussian process is the following: *a Gaussian random process that is stationary in the wide sense is also strictly stationary*. The proof of this fact is straightforward. The nth-order probability density function of Eq. (3.6-1) depends only on the means and covariances of the n sampled values. If the random process $U(t)$ is wide-sense stationary, the mean is independent of time and the covariances depend only on the time differences between the instants involved. It follows directly that the nth-order density function is independent of the time origin for all n, and hence $U(t)$ is strictly stationary. When dealing with Gaussian random processes, therefore, we rarely specify the type of stationarity possessed by the process, since the two most important kinds of stationarity are equivalent.

3.6.4 Fourth-Order Moments

In some applications it is desirable to know the fourth-order moment of the form $\overline{u^2(t)u^2(t+\tau)}$ of a stationary, zero-mean, real-valued Gaussian random process. Such a moment is needed, for example, in calculating the autocorrelation function at the output of a square-law device, for which the output $v(t)$ and input $u(t)$ are related by

$$v(t) = u^2(t). \qquad (3.6\text{-}6)$$

This moment can readily be found with the help of Eq. (2.7-13), valid for zero-mean, real-valued, Gaussian random variables. Applying this equation, we find that

$$\Gamma_V(\tau) = \overline{v(t)v(t+\tau)} = \overline{u^2(t)u^2(t+\tau)}$$

$$= \Gamma_U^2(0) + 2\Gamma_U^2(\tau). \qquad (3.6\text{-}7)$$

THE POISSON IMPULSE PROCESS

More generally, for a moment of the form $\overline{u(t_1)u(t_2)u(t_3)u(t_4)}$, we have

$$\overline{u(t_1)u(t_2)u(t_3)u(t_4)} = \Gamma_U(t_2,t_1)\Gamma_U(t_4,t_3)$$
$$+ \Gamma_U(t_3,t_1)\Gamma_U(t_4,t_2) + \Gamma_U(t_4,t_1)\Gamma_U(t_3,t_2).$$
(3.6-8)

In optical applications, such moments are often of interest, but generally for complex-valued random processes. The relationships are somewhat different in this case, as we shall demonstrate in Section 3.9.

3.7 THE POISSON IMPULSE PROCESS

Of great importance in many optical problems is the Poisson impulse process. In this section we develop some of the basic properties of such processes in preparation for later consideration of various problems associated with the detection of light.

3.7.1 Definitions

Consider a random process $U(t)$ with sample functions $u(t)$ that consist of a multitude of Dirac delta (or impulse) functions, as illustrated in Fig. 3-8a. This random process will be called a *Poisson impulse process* or, for brevity, simply a *Poisson process*, if the following two conditions are satisfied:

(1) The probability $P(K; t_1, t_2)$ that K impulses fall within the time interval $\{t_1 < t \leq t_2\}$ is given by

$$P(K; t_1, t_2) = \frac{\left(\int_{t_1}^{t_2} \lambda(t)\, dt\right)^K}{K!} \exp\left(-\int_{t_1}^{t_2} \lambda(t)\, dt\right), \quad (3.7\text{-}1)$$

where $\lambda(t) \geq 0$ is called the *rate* of the process.
(2) The numbers of impulses falling in any two nonoverlapping time intervals are statistically independent.

A typical rate function $\lambda(t)$ is illustrated in Fig. 3-8b for the sample function shown. From (3.7-1) it can readily be shown that, for a given $\lambda(t)$, the mean and second moment of the number of impulses (or "events") in the time interval ($t_1 < t \le t_2$) are given by

$$\overline{K} = \int_{t_1}^{t_2} \lambda(t)\, dt,$$

$$\overline{K^2} = \overline{K} + (\overline{K})^2. \tag{3.7-2}$$

In addition, the following moment theorem can be shown to hold (see Problem 2-6):

$$\overline{K(K-1)\cdots(K-k+1)} = (\overline{K})^k. \tag{3.7-3}$$

Two important cases can be distinguished. First, the rate function $\lambda(t)$ may be a known (i.e., deterministic) function, in which case all randomness

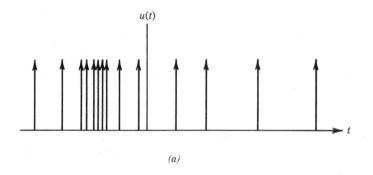

(a)

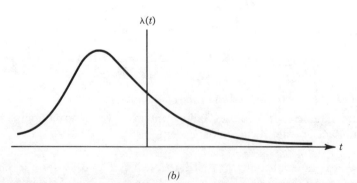

(b)

Figure 3-8. (a) A sample function of a Poisson impulse process, together with (b) the corresponding rate function.

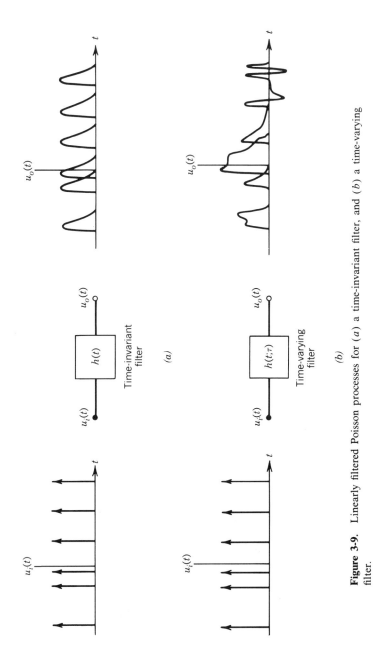

Figure 3-9. Linearly filtered Poisson processes for (*a*) a time-invariant filter, and (*b*) a time-varying filter.

associated with the process $U(t)$ arises from transformation of a given $\lambda(t)$ into a sample function $u(t)$ of the Poisson process. Alternatively, the rate $\lambda(t)$ may itself be a sample function of a random process $\Lambda(t)$, in which case $U(t)$ is often called a *doubly stochastic* Poisson process. In this latter case, some of the randomness of $U(t)$ can be attributed to the transformation of a specific $\lambda(t)$ into a sample function $u(t)$ and some to the statistical uncertainties about $\lambda(t)$ itself.

Finally, we note that in most practical applications of the theory, the random process $U(t)$ consists not of perfect unit-area impulses, but rather of a multitude of finite-width pulses. Thus each impulse $\delta(t - t_k)$ is replaced by a finite pulse $h(t - t_k)$. In some cases the pulses may all have identical shape and area. Such a process may be regarded as being generated by passage of a Poisson impulse process through a linear, time-invariant filter with impulse response $h(t)$, as illustrated in Fig. 3-9a.

Alternatively, some phenomena (e.g., the output of a photomultiplier tube) require modeling by a process characterized by random changes of pulse shape and area from pulse to pulse. Such a process may be regarded as arising from passage of a Poisson impulse process through a randomly time-varying linear filter with an impulse response $h(t; \tau)$ that is a sample function of a random process, as illustrated in Fig. 3-9b. Both Poisson processes described in the preceding paragraphs are referred to as *linearly filtered Poisson processes*.

In order to develop some physical intuition regarding the reasons why the Poisson process is so important in practice, we devote the next two sections to discussion of equivalent conditions that lead to Poisson statistics.

3.7.2 Derivation of Poisson Statistics from Fundamental Hypotheses

It is possible to arrive at the statistical model described in the previous section from a number of different sets of hypotheses (see Ref. 2-3, Chapter 16). The set to be discussed in this section is perhaps the most fundamental and most meaningful physically. Our derivation is a slight generalization of that found in Ref. 2-7 (Section 7.2). Throughout this and the following section, the rate function $\lambda(t)$ is assumed to be known. The case of a stochastic $\lambda(t)$ is deferred to Section 3.7.5.

We begin with the following three basic hypotheses:

(1) For sufficiently small Δt, the probability of a single impulse occurring in the time interval t to $t + \Delta t$ is equal to the product of Δt and a real nonnegative function $\lambda(t)$; thus

$$P(1; t, t + \Delta t) = \lambda(t)\Delta t. \tag{3.7-4}$$

(2) For sufficiently small Δt, the probability that more than one impulse occurs in Δt is negligibly small (i.e., there are no "multiple" events); hence

$$P(0; t, t + \Delta t) = 1 - \lambda(t)\Delta t. \quad (3.7\text{-}5)$$

(3) The numbers of impulses in nonoverlapping time intervals are statistically independent.

With these assumptions we can now ask, what is the probability $P(K; t, t + \tau + \Delta\tau)$ that K impulses occur in the time interval t to $t + \tau + \Delta\tau$? If $\Delta\tau$ is small, there are only two ways that we could obtain K impulses in $(t, t + \tau + \Delta\tau)$. Specifically, we could have K impulses in $(t, t + \tau)$ and no impulses in $(t + \tau, t + \tau + \Delta\tau)$, or we could have $K - 1$ impulses in $(t, t + \tau)$ and one impulse in $(t + \tau, t + \tau + \Delta\tau)$. Employing all three hypotheses above, we write

$$P(K; t, t + \tau + \Delta\tau) = P(K; t, t + \tau)[1 - \lambda(t + \tau)]\Delta\tau$$
$$+ P(K - 1; t, t + \tau)[\lambda(t + \tau)\Delta\tau]. \quad (3.7\text{-}6)$$

Rearranging terms and dividing by $\Delta\tau$, we obtain

$$\frac{P(K; t, t + \tau + \Delta\tau) - P(K; t, t + \tau)}{\Delta\tau}$$
$$= \lambda(t + \tau)[P(K - 1; t, t + \tau) - P(K; t, t + \tau)].$$

Now letting $\Delta\tau$ go to zero, we find that $P(K; t, t + \tau)$ must satisfy the differential equation

$$\frac{dP(K; t, t + \tau)}{d\tau} = \lambda(t + \tau)[P(K - 1; t, t + \tau) - P(K; t, t + \tau)].$$
$$(3.7\text{-}7)$$

By using standard methods for the solution of linear differential equations, coupled with the boundary condition $P(0; t, t) = 1$, we are led uniquely to the solution

$$P(K; t, t + \tau) = \frac{\left[\int_t^{t+\tau} \lambda(\xi)\, d\xi\right]^K}{K!} \exp\left\{-\int_t^{t+\tau} \lambda(\xi)\, d\xi\right\} \quad (3.7\text{-}8)$$

in agreement with Eq. (3.7-1).

When dealing with Poisson processes in the future, we shall feel free to use the three fundamental hypotheses above whenever it proves convenient.

3.7.3 Derivation of Poisson Statistics from Random Event Times

An alternative model that leads to the same type of Poisson process is based on certain assumptions about the statistical distribution of event times t_k.

Suppose that we have a collection of a large number N of "events," which we drop onto the infinite time interval. A random process can then be constructed by inserting a unit impulse function at the location of each event. We assume that the N events are dropped onto the time axis in accord with the following hypotheses: the N event times t_k ($k = 1, 2, \ldots, N$) are (1) statistically independent and (2) identically distributed with probability density function $p(t_k)$.

Using the two properties above, we readily conclude that the number K of events lying in any subinterval (t_1, t_2) obeys a binomial distribution,

$$P(K; t_1, t_2) = \frac{N!}{K!(N-K)!} \left[\int_{t_1}^{t_2} p(\xi)\, d\xi \right]^K \left[1 - \int_{t_1}^{t_2} p(\xi)\, d\xi \right]^{N-K}.$$

Now suppose that we let $N \to \infty$ and $p(t) \to 0$, subject to the restriction that

$$Np(t) = \lambda(t) \qquad (3.7\text{-}9)$$

remains fixed for each t. The probability of obtaining K events or impulses in (t_1, t_2) becomes, for any fixed N,

$$P(K; t_1, t_2) = \frac{N(N-1) \cdots (N-K+1)}{K! N^K}$$

$$\times \left[\int_{t_1}^{t_2} \lambda(\xi)\, d\xi \right]^K \left[1 - \frac{1}{N} \int_{t_1}^{t_2} \lambda(\xi)\, d\xi \right]^{N-K}.$$

Letting N become large, we have

$$\left[1 - \frac{1}{N} \int_{t_1}^{t_2} \lambda(\xi)\, d\xi \right]^{N-K} \cong \left[1 - \frac{1}{N} \int_{t_1}^{t_2} \lambda(\xi)\, d\xi \right]^{N} \to \exp\left\{ -\int_{t_1}^{t_2} \lambda(\xi)\, d\xi \right\},$$

$$\frac{N(N-1) \cdots (N-K+1)}{N^K} \to 1.$$

Thus

$$\lim_{N \to \infty} P(K; t_1, t_2) = \frac{\left[\int_{t_1}^{t_2} \lambda(\xi)\, d\xi\right]^K}{K!} \exp\left\{-\int_{t_1}^{t_2} \lambda(\xi)\, d\xi\right\},$$

which is again the Poisson distribution. In addition, since the event times t_k are statistically independent and there is an inexhaustible supply of events ($N \to \infty$), the number of events occurring in one interval conveys no information about the number occurring in a second disjoint interval. Hence the numbers of events in nonoverlapping intervals are statistically independent.

Thus we have arrived at the same random process model from two different sets of hypotheses. In the future we shall use the set of hypotheses that best suits our purpose.

3.7.4 Energy and Power Spectral Densities of Poisson Processes

In this section we investigate the energy spectral density and power spectral density of Poisson impulse processes. Note that because such processes are composed of ideal δ functions and because an ideal δ function contains infinite energy, it might seem that only power spectral densities are of interest in this case. However, we shall see that the energy spectral density is of utility when the rate function $\lambda(t)$ is Fourier transformable, that is, when $\int_{-\infty}^{\infty} |\lambda(t)|\, dt < \infty$. On the other hand, when the rate function is not Fourier transformable but does have finite average power, that is, when $\int_{-\infty}^{\infty} |\lambda(t)|\, dt = \infty$ but $\lim_{T \to \infty} (1/T) \int_{-T/2}^{T/2} \lambda^2(t)\, dt < \infty$, the power spectral density is the quantity of most interest. Again we assume that $\lambda(t)$ is an entirely deterministic function and defer generalization to Section 3.7.5.

Let $\lambda(t)$ be a Fourier-transformable function. A sample function of the corresponding Poisson impulse process can be expressed as

$$u(t) = \sum_{k=1}^{K} \delta(t - t_k), \qquad (3.7\text{-}10)$$

which depends on the $K + 1$ random variables $t_1, t_2, \ldots, t_K$ and K. This sample function has a Fourier transform

$$\mathcal{U}(\nu) = \sum_{k=1}^{K} \exp(j2\pi \nu t_k). \qquad (3.7\text{-}11)$$

The corresponding energy spectrum for this one sample function is

$$|\mathscr{U}(\nu)|^2 = \sum_{k=1}^{K} \sum_{q=1}^{K} \exp\left[j2\pi\nu(t_k - t_q)\right].$$

The energy spectral density for the random process $U(t)$ is thus

$$\mathscr{E}_U(\nu) = E\left[|\mathscr{U}(\nu)|^2\right] = E\left\{\sum_{k=1}^{K} \sum_{q=1}^{K} \exp\left[j2\pi\nu(t_k - t_q)\right]\right\}.$$

(3.7-12)

Now the expectation with respect to $t_1, t_2, \ldots, t_K$ and K can be performed in two steps. First we average with respect to times t_k under the assumption that K is given and then average with respect to K. This procedure is justified by noting that

$$p(t_1, t_2, \ldots, t_K, K) = p(t_1, t_2, \ldots, t_K | K) P(K).$$

Thus we rewrite (3.7-12) as

$$\mathscr{E}_U(\nu) = E_K\left\{\sum_{k=1}^{K} \sum_{q=1}^{K} E_{t|K}\left\{\exp\left[j2\pi\nu(t_k - t_q)\right]\right\}\right\}, \quad (3.7\text{-}13)$$

where E_K signifies expected value with respect to K, whereas $E_{t|K}$ means the expected value with respect to the times t_k, given the value of K.

Recall that the times t_k are identically distributed, independent random variables. Furthermore, from the proportionality (3.7-9) between $p(t)$ and $\lambda(t)$, we must have

$$p(t_k) = \frac{\lambda(t_k)}{\int_{-\infty}^{\infty} \lambda(t)\, dt}, \quad (3.7\text{-}14)$$

with the normalization chosen to assure unit area. To perform the averaging operation, it is helpful to consider two different sets of terms. There exist K separate terms for which $k = q$, and each such term contributes unity. In addition there are $K^2 - K$ terms having $k \neq q$. Using (3.7-14) and the independence of t_k and t_q, we find

$$E_{t|K}\{\exp[j2\pi\nu(t_k - t_q)]\} = \frac{\int_{-\infty}^{\infty} \lambda(t_k)e^{j2\pi\nu t_k} dt_k}{\int_{-\infty}^{\infty} \lambda(t) dt} \cdot \frac{\int_{-\infty}^{\infty} \lambda(t_q)e^{-j2\pi\nu t_q} dt_q}{\int_{-\infty}^{\infty} \lambda(t) dt}$$

$$= \frac{|\mathscr{L}(\nu)|^2}{(\overline{K})^2} = \frac{\mathscr{E}_\lambda(\nu)}{(\overline{K})^2} \qquad (k \neq q), \qquad (3.7\text{-}15)$$

where $\mathscr{L}(\nu)$ is the Fourier transform of $\lambda(t)$, $\mathscr{E}_\lambda(\nu)$ is the energy spectral density of $\lambda(t)$, and we have used the fact (see Eq. 3.7-2) that

$$\int_{-\infty}^{\infty} \lambda(t) dt = \overline{K}. \qquad (3.7\text{-}16)$$

Performing the final expectation over K, we obtain

$$\mathscr{E}_U(\nu) = \overline{K} + \frac{\overline{K^2} - \overline{K}}{(\overline{K})^2} \mathscr{E}_\lambda(\nu). \qquad (3.7\text{-}17)$$

But for a Poisson distributed K, $\overline{K^2} = (\overline{K})^2 + \overline{K}$, and hence

$$\mathscr{E}_U(\nu) = \overline{K} + \mathscr{E}_\lambda(\nu). \qquad (3.7\text{-}18)$$

Thus the energy spectral density of a Poisson impulse process consists of a constant $\overline{K}$ plus the energy spectral density of the rate function. Note that, because of the constant $\overline{K}$, the total energy associated with $U(t)$ is infinite even though $\lambda(t)$ has finite energy.

When the rate function is not Fourier transformable but does have finite average power, some change in the argument must be made. First, we truncate the random process $U(t)$ so that it is identically zero outside the interval $(-T/2, T/2)$. Then a single sample function $u_T(t)$ can again be written in terms of $K + 1$ random variables,

$$u_T(t) = \sum_{k=1}^{K} \delta(t - t_k) \qquad (3.7\text{-}19)$$

and the corresponding Fourier transform is given by

$$\mathscr{U}_T(\nu) = \sum_{k=1}^{K} \exp(j2\pi\nu t_k). \qquad (3.7\text{-}20)$$

The probability density function of the times t_k must now be taken to be

$$p_T(t_k) = \begin{cases} \dfrac{\lambda(t_k)}{\int_{-T/2}^{T/2} \lambda(t)\, dt} & -\dfrac{T}{2} < t_k \leq \dfrac{T}{2} \\ 0 & \text{otherwise.} \end{cases} \qquad (3.7\text{-}21)$$

The power spectral density is found from the definition

$$\mathcal{G}_U(\nu) = \lim_{T \to \infty} \frac{1}{T} E\left[|\mathcal{U}_T(\nu)|^2\right]$$

$$= \lim_{T \to \infty} \frac{1}{T} E\left[\sum_{k=1}^{K} \sum_{q=1}^{K} \exp[j2\pi\nu(t_k - t_q)]\right].$$

The average is performed as before. Denoting by $\overline{K}_T$ the average number of events in the T-second interval, we find

$$\mathcal{G}_U(\nu) = \lim_{T \to \infty} \left[\frac{\overline{K}_T}{T} + \frac{|\mathcal{L}_T(\nu)|^2}{T}\right].$$

Now

$$\lim_{T \to \infty} \frac{\overline{K}_T}{T} = \lim_{T \to \infty} \frac{1}{T} \int_{-T/2}^{T/2} \lambda(t)\, dt \triangleq \langle \lambda \rangle \qquad (3.7\text{-}22)$$

and

$$\lim_{T \to \infty} \frac{|\mathcal{L}_T(\nu)|^2}{T} = \mathcal{G}_\lambda(\nu), \qquad (3.7\text{-}23)$$

where $\langle \lambda \rangle$ is the time-averaged rate of the process and $\mathcal{G}_\lambda(\nu)$ is the power spectral density of the rate function $\lambda(t)$. Thus

$$\mathcal{G}_U(\nu) = \langle \lambda \rangle + \mathcal{G}_\lambda(\nu) \qquad (3.7\text{-}24)$$

provides the desired relationship between the power spectral densities of $U(t)$ and $\lambda(t)$. Figure 3-10 shows the relationship between $\mathcal{G}_U(\nu)$ and $\mathcal{G}_\lambda(\nu)$ in pictorial form. Note that $U(t)$ contains infinite total average power, even though the average power content of $\lambda(t)$ is finite. Note also that the limits appearing in Eqs. (3.7-22), (3.7-23) have tacitly been assumed to exist, at least in the sense of δ functions.

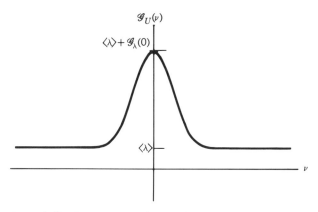

Figure 3-10. Power spectral density of a Poisson impulse process.

3.7.5 Doubly Stochastic Poisson Processes

Suppose that $\lambda(t)$ is not a known function, but rather is a single sample function of a random process $\Lambda(t)$. The various moments of the random process $U(t)$ calculated earlier may now be regarded as conditional moments, conditioned by a particular realization $\lambda(t)$. Moments for the doubly stochastic Poisson process can be calculated simply by averaging the earlier results over the statistics of the random process $\Lambda(t)$.

We illustrate with some simple examples. It was previously stated [Eq. (3.7-2)] that, for a known $\lambda(t)$, the mean number of events in the interval (t_1, t_2) is

$$E_{K|\lambda}(K) = \int_{t_1}^{t_2} \lambda(t)\, dt.$$

If $\lambda(t)$ is a sample function of a stationary random process $\Lambda(t)$, we must additionally average over Λ to obtain

$$\overline{K} = \int_{t_1}^{t_2} \overline{\lambda(t)}\, dt = \overline{\lambda}\tau, \qquad (3.7\text{-}25)$$

where stationarity of $\Lambda(t)$ is used in the last step, and $\tau = t_2 - t_1$.

As for the second moment of K, for a given sample function $\lambda(t)$ we have

$$E_{K|\lambda}[K^2] = \int_{t_1}^{t_2} \lambda(t)\, dt + \iint_{t_1}^{t_2} \lambda(\xi)\lambda(\eta)\, d\xi\, d\eta. \qquad (3.7\text{-}26)$$

Averaging over the ensemble $\Lambda(t)$, we obtain

$$\overline{K^2} = \bar{\lambda}\tau + \iint\limits_{t_1}^{t_2} \Gamma_\Lambda(\xi - \eta)\, d\xi\, d\eta, \qquad (3.7\text{-}27)$$

where Γ_Λ is the autocorrelation function of $\Lambda(t)$, which has been assumed wide-sense stationary. Using arguments similar to those preceding Eq. (3.4-9), we can reduce the double integral to a single integral,

$$\iint\limits_{t_1}^{t_2} \Gamma_\Lambda(\xi - \eta)\, d\xi\, d\eta = 2\tau \int_0^\tau \left(1 - \frac{\zeta}{\tau}\right) \Gamma_\Lambda(\zeta)\, d\zeta. \qquad (3.7\text{-}28)$$

Noting that $\Gamma_\Lambda(\zeta) = (\bar{\lambda})^2 + C_\Lambda(\zeta)$, where $C_\Lambda(\zeta)$ is the autocovariance function of $\Lambda(t)$, we obtain

$$\overline{K^2} = \bar{K} + (\bar{K})^2 + 2\tau \int_0^\tau \left(1 - \frac{\zeta}{\tau}\right) C_\Lambda(\zeta)\, d\zeta. \qquad (3.7\text{-}29)$$

Equivalently, the variance σ_K^2 of K is given by

$$\sigma_K^2 = \bar{K} + 2\tau \int_0^\tau \left(1 - \frac{\zeta}{\tau}\right) C_\Lambda(\zeta)\, d\zeta, \qquad (3.7\text{-}30)$$

which exceeds the variance $\sigma_K^2 = \bar{K}$ associated with a Poisson impulse process having a known rate function $\lambda(t)$. The higher variance is due to the statistical fluctuations associated with the random process $\Lambda(t)$. We defer a further discussion of this fact, as well as a more detailed evaluation of σ_K^2, to Chapter 9.

Finally, we consider modifications of Eqs. (3.7-18) and (3.7-24) for energy spectral density and power spectral density when $\lambda(t)$ is a sample function of a random process. By definition, the energy spectral density and power spectral density are given by

$$\mathcal{E}_U(\nu) = \lim_{T \to \infty} E\left[|\mathcal{U}_T(\nu)|^2\right]$$

$$\mathcal{G}_U(\nu) = \lim_{T \to \infty} \frac{E\left[|\mathcal{U}_T(\nu)|^2\right]}{T}, \qquad (3.7\text{-}31)$$

where $\mathcal{U}_T(\nu)$ is given explicitly by Eq. (3.7-20). Evaluation of the expectations of the sums involved in a manner identical to that already used for the

THE POISSON IMPULSE PROCESS

case of deterministic rate functions, leads us to

$$\mathcal{E}_U(\nu) = \lim_{T \to \infty} \left\{ \overline{K}_T + E\left[|\mathcal{L}_T(\nu)|^2\right] \right\}$$

$$\mathcal{G}_U(\nu) = \lim_{T \to \infty} \left\{ \frac{\overline{K}_T}{T} + \frac{E\left[|\mathcal{L}_T(\nu)|^2\right]}{T} \right\}.$$

Finally, allowing T to become arbitrarily large, we obtain

$$\mathcal{E}_U(\nu) = \overline{K} + \mathcal{E}_\Lambda(\nu)$$

$$\mathcal{G}_U(\nu) = \overline{\lambda} + \mathcal{G}_\Lambda(\nu) \qquad (3.7\text{-}32)$$

where $\overline{\lambda} \triangleq \langle E[\lambda(t)] \rangle$. Thus we see that in the case of a stochastic rate function, both spectral densities consist or a constant plus the corresponding spectral density of the stochastic rate process.

3.7.6 Linearly Filtered Poisson Processes

Finally, we consider the case of a linearly filtered Poisson process, and in particular the energy or power spectral density of such a process. First it is assumed that the process consists of pulses of identical shape and area; thus any truncated sample function is of the form

$$u_T(t) = \text{rect}\frac{t}{T} \cdot \sum_{k=1}^{K} h(t - t_k). \qquad (3.7\text{-}33)$$

As illustrated in Fig. 3-9a, such a process may be regarded as arising from passage of a Poisson impulse process through a linear time-invariant filter. If $\mathcal{H}(\nu)$ represents the transfer function of the required filter, i.e.

$$\mathcal{H}(\nu) = \int_{-\infty}^{\infty} h(t) e^{j2\pi\nu t} dt, \qquad (3.7\text{-}34)$$

then Eqs. (3.3-10) and (3.3-12) allow us to express the spectral density of the linearly filtered Poisson process as the product of $|\mathcal{H}(\nu)|^2$ and the spectral density of the underlying Poisson impulse process. The results are

$$\mathcal{E}_U(\nu) = \overline{K}|\mathcal{H}(\nu)|^2 + |\mathcal{H}(\nu)|^2 \mathcal{E}_\Lambda(\nu) \qquad (3.7\text{-}35)$$

and

$$\mathcal{G}_U(\nu) = \overline{\lambda}|\mathcal{H}(\nu)|^2 + |\mathcal{H}(\nu)|^2 \mathcal{G}_\Lambda(\nu) \qquad (3.7\text{-}36)$$

for the energy spectral density and the power spectral density, respectively.

If the pulses composing $U(t)$ have random shape and area, a modification is necessary. Illustrating for the case of an energy spectral density, we have

$$U(t) = \sum_{k=1}^{K} h(t; t_k) \qquad (3.7\text{-}37)$$

and

$$|\mathcal{U}(\nu)|^2 = \sum_{k=1}^{K} \sum_{q=1}^{K} \mathcal{H}(\nu; t_k) \mathcal{H}^*(\nu; t_q) \exp[j2\pi\nu(t_k - t_q)], \qquad (3.7\text{-}38)$$

where $\mathcal{H}(\nu; t_k)$ is the Fourier transform of the pulse shape associated with the kth pulse,

$$\mathcal{H}(\nu; t_k) = \int_{-\infty}^{\infty} h(t; t_k) e^{j2\pi\nu(t-t_k)} d(t - t_k). \qquad (3.7\text{-}39)$$

The expectation applied to Eq. (3.7-38) must now be taken over the statistics of $t_1, t_2, \ldots, t_K$, K and the statistics of the $\mathcal{H}(\nu; t_k)$.

Again it is helpful to consider separately the K terms for which $k = q$ and the $K^2 - K$ terms for which $k \neq q$. For the former terms, our previous contribution $\overline{K}$ to the energy spectrum must be multiplied by $\overline{|\mathcal{H}(\nu; t_k)|^2}$, which we assume to be the same for all t_k and hence representable as $\overline{|\mathcal{H}(\nu)|^2}$. For the $K^2 - K$ terms with $k \neq q$, we must multiply by $\overline{\mathcal{H}(\nu; t_k) \mathcal{H}^*(\nu; t_q)}$. If the statistics of different pulses are independent, this multiplier reduces to $[\overline{\mathcal{H}(\nu)}]^2$. Thus we obtain for the energy and power spectral densities

$$\mathcal{E}_U(\nu) = \overline{K} \overline{|\mathcal{H}(\nu)|^2} + [\overline{\mathcal{H}(\nu)}]^2 \mathcal{E}_\Lambda(\nu) \qquad (3.7\text{-}40)$$

and

$$\mathcal{G}_U(\nu) = \overline{\lambda} \overline{|\mathcal{H}(\nu)|^2} + [\overline{\mathcal{H}(\nu)}]^2 \mathcal{G}_\Lambda(\nu). \qquad (3.7\text{-}41)$$

In closing we note that implicit in these results is the assumption that the statistics of $\mathcal{H}(\nu; t_k)$ are independent of the statistics of $\Lambda(t)$.

3.8 RANDOM PROCESSES DERIVED FROM ANALYTIC SIGNALS

It is common practice in physics and engineering to represent real-valued signals by related complex-valued signals. The complex representation is chosen such that its real part is the original real-valued signal; thus provided that only linear operations are performed on the complex signal, the original signal can be specified at any stage simply by taking the real part of the complex waveform.

The reason for preferring a complex representation, rather than the real-valued signal itself, can be traced to a fundamental property of linear, time-invariant systems. Specifically, the eigenfunctions of such a system are complex exponentials of the form $\exp(-j2\pi\nu t)$. Thus if we represent the linear, time-invariant system by an operation $\mathscr{L}\{\ \}$, we can show that

$$\mathscr{L}\{\exp(-j2\pi\nu t)\} = \mathscr{H}(\nu)\exp(-j2\pi\nu t),$$

where $\mathscr{H}(\nu)$ is the transfer function of the system, evaluated at frequency ν (for a proof of this fact, see Ref. 2-9, p. 186). Passage of a real-valued signal through the system requires operations on both positive- and negative-frequency complex exponentials and thus entails a greater amount of algebra.

With these comments as motivation, we turn to an examination of complex signal representation in greater mathematical detail.

3.8.1 Representation of a Monochromatic Signal by a Complex Signal

Consider a monochromatic (i.e., single-frequency) real-valued signal $u^{(r)}(t)$ described by

$$u^{(r)}(t) = A\cos(2\pi\nu_0 t - \phi), \tag{3.8-1}$$

where A, ν_0, and ϕ represent constant amplitude, frequency and phase, respectively. The complex representation of this signal is

$$\mathbf{u}(t) = A\exp\{-j(2\pi\nu_0 t - \phi)\}, \tag{3.8-2}$$

which has a real part equal to the original $u^{(r)}(t)$. Related to this complex representation is the *phasor amplitude* of $\mathbf{u}(t)$, defined by

$$\mathbf{A} \triangleq A\exp(+j\phi) \tag{3.8-3}$$

and representing the amplitude and phase of the monochromatic signal.

Note that the imaginary part of the complex representation has not been chosen arbitrarily, but rather is closely related to the original real-valued signal.

Exactly what operations are involved in arriving at the specific complex representation of (3.8-2)? The question is most readily answered by frequency-domain reasoning. Let the real-valued signal be expanded in complex exponential components,

$$u^{(r)}(t) = \frac{A}{2}e^{-j\phi}e^{j2\pi\nu_0 t} + \frac{A}{2}e^{j\phi}e^{-j2\pi\nu_0 t}.$$

Representing the Fourier transform operation by an operator $\mathscr{F}\{\ \}$, we further note that

$$\mathscr{F}\{e^{j2\pi\nu_0 t}\} = \delta(\nu + \nu_0)$$

$$\mathscr{F}\{e^{-j2\pi\nu_0 t}\} = \delta(\nu - \nu_0).$$

Therefore, the Fourier spectrum of $u^{(r)}(t)$ is

$$\mathscr{U}(\nu) = \frac{A}{2}e^{-j\phi}\delta(\nu + \nu_0) + \frac{A}{2}e^{j\phi}\delta(\nu - \nu_0).$$

For the complex representation $\mathbf{u}(t)$, however, we have

$$\mathscr{F}\{\mathbf{u}(t)\} = Ae^{j\phi}\delta(\nu - \nu_0). \qquad (3.8\text{-}4)$$

Thus the relationship between $u^{(r)}(t)$ and $\mathbf{u}(t)$ can be stated as follows: *in*

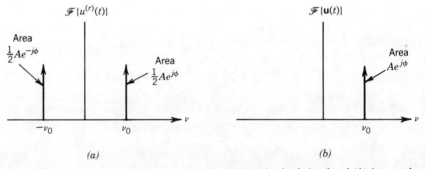

Figure 3-11. Fourier spectra of (*a*) a monochromatic real-valued signal and (*b*) its complex representation.

passing from $u^{(r)}(t)$ to $\mathbf{u}(t)$, we double the strength of the positive frequency component and entirely remove the negative frequency component. This operation is illustrated for the monochromatic case in Fig. 3-11. It is this very specific operation that imposes a fixed relationship between the real and imaginary parts of $\mathbf{u}(t)$.

3.8.2 Representation of a Nonmonochromatic Signal by a Complex Signal

Suppose we are given a real-valued nonmonochromatic signal $u^{(r)}(t)$ with Fourier transform $\mathcal{U}(\nu)$. How can we represent $u^{(r)}(t)$ by a complex signal $\mathbf{u}(t)$? We can follow exactly the same procedure used in the monochromatic case, doubling the positive frequency components and removing the negative frequency components. Thus our definition is

$$\mathbf{u}(t) \triangleq 2\int_0^\infty \mathcal{U}(\nu)e^{-j2\pi\nu t}\,d\nu. \tag{3.8-5}$$

The function $\mathbf{u}(t)$ is called the *analytic signal* representation of $u(t)$. For a comprehensive discussion of the properties of analytic signals, see Refs. 3-1 and 3-2.

Before turning attention to the properties of the analytic signal, one mathematical fine point should be clarified. This fine point concerns exactly what is done to the spectrum at $\nu = 0$ in passing from $u^{(r)}(t)$ to $\mathbf{u}(t)$. The question is immaterial if $u^{(r)}(t)$ contains no δ-function component at $\nu = 0$, for changing the spectrum by a finite amount at a single point will not affect $\mathbf{u}(t)$. If $u^{(r)}(t)$ does contain a δ-function component at $\nu = 0$, the convention will be adopted that the δ-function component remains unchanged. This convention allows us to represent in the frequency domain the operation of passing from $u^{(r)}(t)$ to $\mathbf{u}(t)$ by

$$\mathcal{U}(\nu) \to [1 + \operatorname{sgn}\nu]\mathcal{U}(\nu), \tag{3.8-6}$$

where

$$\operatorname{sgn}\nu \triangleq \begin{cases} +1 & \nu > 0 \\ 0 & \nu = 0 \\ -1 & \nu < 0 \end{cases}. \tag{3.8-7}$$

Thus

$$\mathbf{u}(t) = \int_{-\infty}^\infty [1 + \operatorname{sgn}\nu]\mathcal{U}(\nu)e^{-j2\pi\nu t}\,d\nu. \tag{3.8-8}$$

The Fourier integral representation of $\mathbf{u}(t)$ above allows us to discover some important properties of the analytic signal. Representing the inverse Fourier transform operation by an operator $\mathscr{F}^{-1}\{\ \}$, we see that $\mathbf{u}(t)$ can be expressed as the sum of two terms,

$$\mathbf{u}(t) = \mathscr{F}^{-1}\{\mathscr{U}(\nu)\} + \mathscr{F}^{-1}\{\operatorname{sgn}\nu\,\mathscr{U}(\nu)\}.$$

The first term is simply $u^{(r)}(t)$, the original signal. With use of the convolution theorem, the second term can be expressed by

$$\mathscr{F}^{-1}\{\operatorname{sgn}\nu\,\mathscr{U}(\nu)\} = \mathscr{F}^{-1}\{\operatorname{sgn}\nu\} * \mathscr{F}^{-1}\{\mathscr{U}(\nu)\}.$$

Noting that $\mathscr{F}^{-1}\{\operatorname{sgn}\nu\} = (-j/\pi t)$ (see Appendix A), we find

$$\mathbf{u}(t) = u^{(r)}(t) + \frac{j}{\pi}\int_{-\infty}^{\infty}\frac{u^{(r)}(\xi)}{\xi - t}\,d\xi, \tag{3.8-9}$$

where the symbol $\int_{-\infty}^{\infty}$ indicates that the *Cauchy principal value* of the integral must be taken. That is

$$\frac{1}{\pi}\int_{-\infty}^{\infty}\frac{u^{(r)}(\xi)}{\xi - t}\,d\xi \triangleq \frac{1}{\pi}\lim_{\varepsilon \to 0}\left[\int_{-\infty}^{t-\varepsilon}\frac{u^{(r)}(\xi)}{\xi - t}\,d\xi + \int_{t+\varepsilon}^{\infty}\frac{u^{(r)}(\xi)}{\xi - t}\,d\xi\right].$$

$$\tag{3.8-10}$$

The integral transformation of (3.8-10) is known as the *Hilbert transform* of $u^{(r)}(t)$ (for a more detailed discussion of Hilbert transforms, see also Ref. 2-9, pp. 267–272).

The important properties of the analytic signal can now be stated on the basis of Eqs. (3.8-8) and (3.8-9):

(i) $$u^{(r)}(t) \triangleq \operatorname{Re}\{\mathbf{u}(t)\} \tag{3.8-11}$$

(ii) $$u^{(i)}(t) \triangleq \operatorname{Im}\{\mathbf{u}(t)\} = \frac{1}{\pi}\int_{-\infty}^{\infty}\frac{u^{(r)}(\xi)}{\xi - t}\,d\xi \tag{3.8-12}$$

(iii) $$\mathscr{F}\{u^{(i)}(t)\} = -j\operatorname{sgn}\nu \cdot \mathscr{F}\{u^{(r)}(t)\}$$
$$= -j\operatorname{sgn}\nu \cdot \mathscr{U}(\nu). \tag{3.8-13}$$

Thus the real part of the analytic signal is indeed the original real-valued signal we started with. The imaginary part of the analytic signal is simply

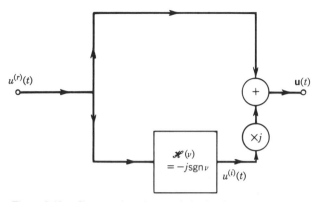

Figure 3-12. Construction of an analytic signal from a real signal.

the Hilbert transform of the original signal. Finally, the spectrum of the imaginary part of the analytic signal can be obtained by multiplying the spectrum of the real part by $-j\,\text{sgn}\,\nu$.

The last property, represented by Eq. (3.8-13), lends itself to a useful interpretation. The imaginary part of the analytic signal can be obtained by passing the real part through a linear, time-invariant filter with transfer function

$$\mathcal{H}(\nu) = -j\,\text{sgn}\,\nu. \qquad (3.8\text{-}14)$$

We refer to such a filter as a "Hilbert transforming" filter. The construction of the analytic signal $\mathbf{u}(t)$ from the real signal $u^{(r)}(t)$ can thus be represented diagrammatically as shown in Fig. 3-12.

3.8.3 Complex Envelopes or Time-Varying Phasors

Consider a real-valued waveform $u^{(r)}(t)$ that is nonmonochromatic but nonetheless possesses a "narrowband" power spectrum. As illustrated in Fig. 3-13, if $\Delta\nu$ represents the nominal width of the spectrum about its center frequency ν_0, we require that $\Delta\nu \ll \nu_0$.

Such a signal may be written in terms of a slowly varying envelope $A(t)$ and a slowly varying phase $\phi(t)$ as follows:

$$u^{(r)}(t) = A(t)\cos[2\pi\nu_0 t - \phi(t)]. \qquad (3.8\text{-}15)$$

To a good approximation, doubling the positive frequency components and

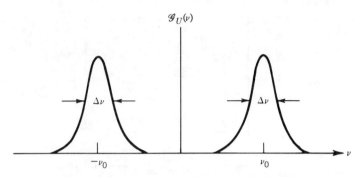

Figure 3-13. Power spectrum of a narrowband signal.

removing the negative frequency components yields an analytic signal with only one exponential component of (3.8-15),

$$\mathbf{u}(t) = A(t)e^{j\phi(t)}e^{-j2\pi\nu_0 t}. \qquad (3.8\text{-}16)$$

By analogy with the monochromatic case, we define the *time-varying phasor amplitude*, or the *complex envelope*, of $\mathbf{u}(t)$ by

$$\mathbf{A}(t) \triangleq A(t)e^{j\phi(t)}. \qquad (3.8\text{-}17)$$

For any signal (wideband or narrowband), we can write the analytic signal representation in the form

$$\mathbf{u}(t) = \mathbf{A}(t)e^{-j2\pi\nu_0 t}. \qquad (3.8\text{-}18)$$

If the signal is narrowband, the complex envelope $\mathbf{A}(t)$ varies much more slowly than the complex carrier $\exp(-j2\pi\nu_0 t)$, and $|\mathbf{A}(t)|$ is approximately the same as $A(t)$ in Eq. (3.8-15).

3.8.4 The Analytic Signal as a Complex-Valued Random Process

If the real signal $u^{(r)}(t)$ is a sample function of a random process $U(t)$, the analytic signal may be regarded as a sample function of a complex-valued random process $\mathbf{U}(t)$. We consider some of the basic properties of such a random process in this section.

The reader may be concerned that we have defined the analytic signal in terms of the Fourier transform of the real-valued signal and that such a spectrum does not exist for a random process. However, we could define the analytic signal alternatively as

$$\mathbf{u}(t) \triangleq \left[\delta(t) - \frac{j}{\pi t}\right] * u^{(r)}(t), \qquad (3.8\text{-}19)$$

in complete accord with the definition (3.8-8), but without introducing Fourier transforms at all. Thus the analytic signal representing a sample function of a random process is indeed well defined.

For a complete description of the random process $\mathbf{U}(t)$, it is necessary to specify the joint statistics of the real and imaginary parts of the process for all possible collections of time instants. However, specification of the statistics of $\mathbf{U}(t)$ at even a single point in time is in general difficult, for the joint statistics of the real and imaginary parts must be found, based on knowledge of the statistics of only the real part and the Hilbert transform relationship

$$u^{(i)}(t) = \frac{1}{\pi} \int_{-\infty}^{\infty} \frac{u^{(r)}(\xi)}{\xi - t} d\xi. \tag{3.8-20}$$

The problem is a difficult one except in the case of a Gaussian $U(t)$ treated in the section to follow.

Nonetheless, of general interest, regardless of the probability density functions involved, are the autocorrelation functions and cross-correlation functions of the real and imaginary parts of $\mathbf{U}(t)$. To find these functions, we use the linear filtering interpretation of the Hilbert transform operation, as implied by Eq. (3.8-14). Let the function $\Gamma_U^{(r,r)}(\tau)$ represent the autocorrelation function of the real process $U(t)$, which is assumed to be at least wide-sense stationary but otherwise arbitrary. The corresponding power spectral density of the real process is

$$\mathscr{G}_U^{(r,r)}(\nu) = \int_{-\infty}^{\infty} \Gamma_U^{(r,r)}(\tau) e^{j2\pi\nu\tau} d\tau. \tag{3.8-21}$$

The power spectral density of the imaginary part of the process is represented by $\mathscr{G}_U^{(i,i)}(\nu)$. Using Eqs. (3.3-12) and (3.8-14), we find that

$$\mathscr{G}_U^{(i,i)}(\nu) = |-j \operatorname{sgn} \nu|^2 \mathscr{G}_U^{(r,r)}(\nu).$$

Furthermore, provided the random process $U^{(r)}(t)$ has zero mean (its power spectral density has no δ-function component at $\nu = 0$), we can conclude that

$$|-j \operatorname{sgn} \nu|^2 = 1. \tag{3.8-22}$$

As a consequence,

$$\mathscr{G}_U^{(i,i)}(\nu) = \mathscr{G}_U^{(r,r)}(\nu) \tag{3.8-23}$$

and thus
$$\Gamma_U^{(i,i)}(\tau) = \Gamma_U^{(r,r)}(\tau). \qquad (3.8\text{-}24)$$

As for the cross-correlation functions
$$\Gamma_U^{(r,i)}(\tau) = \overline{u^{(r)}(t+\tau)u^{(i)}(t)}$$
$$\Gamma_U^{(i,r)}(\tau) = \overline{u^{(i)}(t+\tau)u^{(r)}(t)}, \qquad (3.8\text{-}25)$$

we use Eqs. (3.5-8) and (3.8-14), with one filter having transfer function unity and the second having transfer function $-j\operatorname{sgn} \nu$. The result is
$$\mathscr{G}_U^{(r,i)}(\nu) \triangleq \mathscr{F}\{\Gamma_U^{(r,i)}(\tau)\}$$
$$= 1 \cdot (+j\operatorname{sgn} \nu)\mathscr{G}_U^{(r,r)}(\nu) = j\operatorname{sgn} \nu\, \mathscr{G}_U^{(r,r)}(\nu) \qquad (3.8\text{-}26)$$

and similarly
$$\mathscr{G}_U^{(i,r)}(\nu) = \mathscr{F}\{\Gamma_U^{(i,r)}(\tau)\}$$
$$= (-j\operatorname{sgn} \nu) \cdot 1 \cdot \mathscr{G}_U^{(r,r)}(\nu) = -j\operatorname{sgn} \nu\, \mathscr{G}_U^{(r,r)}(\nu). \qquad (3.8\text{-}27)$$

From these results we conclude first that
$$\Gamma_U^{(i,r)}(\tau) = -\Gamma_U^{(r,i)}(\tau) \qquad (3.8\text{-}28)$$

and in addition, from (3.8-27), that
$$\Gamma_U^{(i,r)}(\tau) = \frac{1}{\pi}\int_{-\infty}^{\infty} \frac{\Gamma_U^{(r,r)}(\xi)}{\xi - \tau}\, d\xi. \qquad (3.8\text{-}29)$$

It is convenient for future applications to define the autocorrelation function of a complex-valued random process by
$$\Gamma_U(t_2, t_1) = \overline{\mathbf{u}(t_2)\mathbf{u}^*(t_1)}. \qquad (3.8\text{-}30)$$

When the real and imaginary parts of $U(t)$ are at least wide-sense stationary, $\Gamma_U(\tau)$ so defined has the following basic properties:

(i) $\Gamma_U(0) = \overline{[u^{(r)}(t)]^2} + \overline{[u^{(i)}(t)]^2}$

(ii) $\Gamma_U(-\tau) = \Gamma_U^*(\tau) \qquad (3.8\text{-}31)$

(iii) $|\Gamma_U(\tau)| \leq |\Gamma_U(0)|.$

By expanding the analytic signals in terms of their real and imaginary parts, we can readily show that (3.8-30) becomes

$$\Gamma_U(\tau) = \left[\Gamma_U^{(r,r)}(\tau) + \Gamma_U^{(i,i)}(\tau)\right] + j\left[\Gamma_U^{(i,r)}(\tau) - \Gamma_U^{(r,i)}(\tau)\right]. \quad (3.8\text{-}32)$$

Using Eqs. (3.8-24) and (3.8-28), we see directly that

$$\Gamma_U(\tau) = 2\Gamma_U^{(r,r)}(\tau) + j2\Gamma_U^{(i,r)}(\tau). \quad (3.8\text{-}33)$$

Thus the real part of the complex autocorrelation function is just twice the autocorrelation function of the original real-valued random process. Furthermore, using (3.8-29), we see that the imaginary part of $\Gamma_U(\tau)$ is just twice the Hilbert transform of the autocorrelation function of the real random process.

We consider next the Fourier transform of $\Gamma_U(\tau)$, which we call the *power spectral density* of the complex-valued random process $U(t)$. Proceeding directly, we have

$$\mathscr{G}_U(\nu) \triangleq \mathscr{F}\{\Gamma_U(\tau)\} = 2\mathscr{F}\{\Gamma_U^{(r,r)}(\tau)\} + 2j\mathscr{F}\{\Gamma_U^{(i,r)}(\tau)\}$$

$$= 2\mathscr{G}_U^{(r,r)}(\nu) + 2\,\mathrm{sgn}\,\nu\,\mathscr{G}_U^{(r,r)}(\nu)$$

$$= \begin{cases} 4\mathscr{G}_U^{(r,r)}(\nu) & \nu > 0 \\ 0 & \nu < 0. \end{cases} \quad (3.8\text{-}34)$$

Thus the autocorrelation function $\Gamma_U(\tau)$ of an analytic signal has a one-sided Fourier spectrum and *is itself an analytic signal*.

Finally, we consider the cross-correlation function of two jointly wide-sense stationary analytic signals, defined by

$$\Gamma_{UV}(\tau) \triangleq E\left[\mathbf{u}(t+\tau)\mathbf{v}^*(t)\right]. \quad (3.8\text{-}35)$$

This particular function plays a role of central importance in the theory of partial coherence. With the notation

$$\mathbf{u}(t) = u^{(r)}(t) + ju^{(i)}(t)$$

$$\mathbf{v}(t) = v^{(r)}(t) + jv^{(i)}(t),$$

direct substitution in Eq. (3.8-35) yields an expanded form of $\Gamma_{UV}(\tau)$,

$$\Gamma_{UV}(\tau) = \left[\Gamma_{UV}^{(r,r)}(\tau) + \Gamma_{UV}^{(i,i)}(\tau)\right]$$

$$+ j\left[\Gamma_{UV}^{(i,r)}(\tau) - \Gamma_{UV}^{(r,i)}(\tau)\right]. \quad (3.8\text{-}36)$$

In a manner identical to that used in arriving at (3.8-33), we can readily reduce this equation to the simpler form

$$\Gamma_{UV}(\tau) = 2\Gamma_{UV}^{(r,r)}(\tau) + j2\Gamma_{UV}^{(i,r)}(\tau). \qquad (3.8\text{-}37)$$

As with the case of the autocorrelation function, the cross-correlation function of two analytic signals has a one-sided Fourier spectrum and thus is itself an analytic signal, as can be demonstrated with the help of Eq. (3.5-8).

3.9 THE COMPLEX GAUSSIAN RANDOM PROCESS

In most general terms, a complex random process $U(t)$ is called a complex *Gaussian* random process if its real and imaginary parts are joint Gaussian processes. Consider a real-valued Gaussian random process $U^{(r)}(t)$ and the corresponding complex random process $U(t)$ consisting of the analytic signal representations of the real sample functions of $U^{(r)}(t)$. Since Gaussian statistics are preserved under linear operations of the form of Eq. (3.6-5), for a Gaussian $u^{(r)}(t)$ the imaginary part $u^{(i)}(t)$ defined by Eq. (3.8-20) also obeys Gaussian statistics. Thus the real and imaginary parts of $U(t)$ are both Gaussian random processes. We conclude that the analytic signal representation of a Gaussian random process is a complex Gaussian random process. However, not every complex Gaussian random process has sample functions that are analytic signals.

In later chapters, we shall occasionally be interested in calculating fourth-order moments $\mathbf{u}^*(t_1)\mathbf{u}^*(t_2)\mathbf{u}(t_3)\mathbf{u}(t_4)$ of a complex Gaussian random process. Such calculations can be performed with the help of the complex Gaussian moment theorem, provided $\mathbf{u}(t_1)$, $\mathbf{u}(t_2)$, $\mathbf{u}(t_3)$ and $\mathbf{u}(t_4)$ obey *circular* joint Gaussian statistics, that is, provided

(i) $\qquad \overline{u^{(r)}(t_m)} = \overline{u^{(i)}(t_m)} = 0 \qquad m = 1,2,3,4$

(ii) $\quad \overline{u^{(r)}(t_m)u^{(r)}(t_n)} = \overline{u^{(i)}(t_m)u^{(i)}(t_n)}; \qquad m,n = 1,2,3,4$

and

(iii) $\quad \overline{u^{(r)}(t_m)u^{(i)}(t_n)} = -\overline{u^{(r)}(t_n)u^{(i)}(t_m)}; \qquad m,n = 1,2,3,4 \quad (3.9\text{-}1)$

A random process satisfying (3.9-1) is said to be a *circular* complex random process. Fortunately, an analytic signal representation of a zero-mean random process does indeed satisfy the circularity conditions, as evidenced by Eqs. (3.8-24) and (3.8-28). With these conditions satisfied, the fourth-order moment is given by

$$E\left[\mathbf{u}^*(t_1)\mathbf{u}^*(t_2)\mathbf{u}(t_3)\mathbf{u}(t_4)\right] = \Gamma_U(t_3, t_1)\Gamma_U(t_4, t_2) + \Gamma_U(t_3, t_2)\Gamma_U(t_4, t_1).$$

(3.9-2)

Of special interest in the later work will be the case $t_3 = t_1$, $t_4 = t_2$, for which

$$E\left[|\mathbf{u}(t_1)|^2|\mathbf{u}(t_2)|^2\right] = \Gamma_U(t_1, t_1)\Gamma_U(t_2, t_2) + |\Gamma_U(t_2, t_1)|^2, \quad (3.9\text{-}3)$$

where we have used the fact that $\Gamma_U(t_2, t_1)$ equals $\Gamma_U^*(t_1, t_2)$. The reader is reminded again that these relationships hold only for *circular* complex Gaussian random processes.

3.10 THE KARHUNEN–LOÈVE EXPANSION

In certain applications to be encountered in later chapters, it will be helpful to be able to expand the sample functions $\mathbf{u}(t)$ of a complex random process $U(t)$ in terms of a set of functions orthonormal on the interval $(-T/2, T/2)$. Even greater benefits will accrue if, over the ensemble, the expansion coefficients are uncorrelated random variables, and we now attempt to find such an expansion.

Let the set of functions $\{\phi_1(t), \phi_2(t), \ldots, \phi_n(t), \ldots\}$ be orthonormal and complete on the interval $(-T/2, T/2)$. Then any reasonably well behaved sample function $\mathbf{u}(t)$ can be expanded in the form

$$\mathbf{u}(t) = \sum_{n=0}^{\infty} \mathbf{b}_n \phi_n(t) \qquad |t| \leq \frac{T}{2}, \qquad (3.10\text{-}1)$$

where

$$\int_{-T/2}^{T/2} \phi_m(t)\phi_n^*(t)\, dt = \begin{cases} 1 & n = m \\ 0 & n \neq m, \end{cases} \qquad (3.10\text{-}2)$$

and the expansion coefficients $\mathbf{b}_n$ are given by

$$\mathbf{b}_n = \int_{-T/2}^{T/2} \mathbf{u}(t)\phi_n^*(t)\, dt \qquad n = 0, 1, 2, \ldots. \qquad (3.10\text{-}3)$$

Now we ask whether, for a random process with a given autocorrelation function $\Gamma_U(t_2, t_1)$, it is possible to choose a particular set of orthonormal functions such that the expansion coefficients $\{b_n\}$ are uncorrelated.

For simplicity we assume that the random process $U(t)$ has zero mean for all time; the process may be nonstationary in other respects, however. The mean value of every expansion coefficient can now be seen to be zero, since

$$E[\mathbf{b}_n] = \int_{-T/2}^{T/2} E[\mathbf{u}(t)] \phi_n^*(t) \, dt = 0. \qquad (3.10\text{-}4)$$

Thus for the expansion coefficients to be uncorrelated, we require that

$$E[\mathbf{b}_n \mathbf{b}_m^*] = \begin{cases} \lambda_m & m = n \\ 0 & m \neq n. \end{cases} \qquad (3.10\text{-}5)$$

For satisfaction of the uncorrelated condition (3.10-5), the orthonormal set $\{\phi_m(t)\}$ must be properly chosen. To discover the conditions imposed on $\{\phi_m(t)\}$, we substitute Eq. (3.10-3) directly into (3.10-5), yielding

$$E[\mathbf{b}_n \mathbf{b}_m^*] = \int_{-T/2}^{T/2} \int_{-T/2}^{T/2} E[\mathbf{u}^*(t_1)\mathbf{u}(t_2)] \phi_n^*(t_2) \phi_m(t_1) \, dt_1 \, dt_2$$

$$= \int_{-T/2}^{T/2} \left[\int_{-T/2}^{T/2} \Gamma_U(t_2, t_1) \phi_m(t_1) \, dt_1 \right] \phi_n^*(t_2) \, dt_2.$$

(3.10-6)

Suppose now that the set $\{\phi_m(t)\}$ is chosen to satisfy the integral equation

$$\int_{-T/2}^{T/2} \Gamma_U(t_2, t_1) \phi_m(t_1) \, dt_1 = \lambda_m \phi_m(t_2). \qquad (3.10\text{-}7)$$

The correlation of the expansion coefficients then becomes

$$E[\mathbf{b}_n \mathbf{b}_m^*] = \int_{-T/2}^{T/2} \lambda_m \phi_m(t_2) \phi_n^*(t_2) \, dt_2 = \begin{cases} \lambda_m & m = n \\ 0 & m \neq n \end{cases} \qquad (3.10\text{-}8)$$

as required.

The requirement placed on the set of functions $\{\phi_m(t)\}$ by the integral equation (3.10-7) can be stated in mathematical language familiar to some: the required set of functions $\{\phi_m(t)\}$ is the set of eigenfunctions of an integral equation having $\Gamma_U(t_2, t_1)$ as its kernel, and the set of coefficients $\{\lambda_m\}$ is the corresponding set of eigenvalues.

Many mathematical subtleties have been ignored in the preceding discussion. Orders of expectation and integration have been freely interchanged without stating requirements on the functions involved to ensure validity. For our purposes, it suffices simply to state that the autocorrelation function $\Gamma_U(t_2, t_1)$ should be a continuous function of its arguments. For a more detailed mathematical discussion, the reader should consult Ref. 3-3.

REFERENCES

3-1 J. Dugundji, *IRE Trans. Info. Th.*, **IR-4**, 53 (1958).

3-2 D. Gabor, *J. Inst. Electr. Eng.*, **93**, Part III, 429 (1946).

3-3 M. Loève, *Probability Theory*, D. Van Nostrand, Princeton, NJ (1955). See also references for Chapter 2.

ADDITIONAL READING

B. R. Frieden, *Probability, Statistical Optics and Data Testing: Problem Solving Approach*, Springer Series in Information Sciences, Vol. 10, Springer-Verlag, Heidelberg, 1983.

PROBLEMS

3-1 Let the random process $U(t)$ be defined by

$$U(t) = A\cos(2\pi\nu t - \Phi),$$

where ν is a known constant, Φ is uniformly distributed on $(-\pi, \pi)$, and the probability density function of A is given by

$$p_A(a) = \tfrac{1}{2}\delta(a-1) + \tfrac{1}{2}\delta(a-2),$$

and A and Φ are statistically independent.

(a) Calculate $\langle u^2(t)\rangle$ for a sample function with amplitude 1 and a sample function with amplitude 2.

(b) Calculate $\overline{u^2}$.

(c) Show that

$$\overline{u^2} = \tfrac{1}{2}\langle u^2\rangle_1 + \tfrac{1}{2}\langle u^2\rangle_2,$$

where $\langle u^2 \rangle_1$ and $\langle u^2 \rangle_2$ represent the results of part a for amplitudes of 1 and 2, respectively.

3-2 Consider the random process $U(t) = A$, where A is a random variable uniformly distributed on $(-1, 1)$.

(a) Sketch some sample functions of this process.
(b) Find the time autocorrelation function of $U(t)$.
(c) Find the statistical autocorrelation function of $U(t)$.
(d) Is $U(t)$ wide-sense stationary? Is it strictly stationary?
(e) Is $U(t)$ an ergodic random process? Explain.

3-3 • An ergodic random process with autocorrelation function $\Gamma_U(\tau) = (N_0/2)\delta(\tau)$ is applied to the input of a linear, time-invariant filter with impulse response $h(t)$. The output $V(t)$ is multiplied by a delayed version of $U(t)$, forming a new random process $Z(t)$, as indicated below in Figure 3-3p. Show that the impulse response of the filter can be determined from measurements of $\langle z(t) \rangle$ as a function of delay Δ.

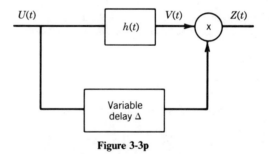

Figure 3-3p

3-4 Consider the random process

$$Z(t) = U\cos \pi t,$$

where U is a random variable with probability density function

$$p_U(u) = \frac{1}{\sqrt{2\pi}} \exp\left(\frac{-u^2}{2}\right).$$

(a) What is the probability density function of the random variable $Z(0)$?

PROBLEMS

(b) What is the joint density function of $Z(0)$ and $Z(1)$?

(c) Is this random process strictly stationary, wide-sense stationary, or ergodic?

3-5 Find the statistical autocorrelation function of the random process

$$U(t) = a_1\cos(2\pi\nu_1 t - \Phi_1) + a_2\cos(2\pi\nu_2 t - \Phi_2),$$

where a_1, a_2, ν_1, ν_2 are known constants whereas Φ_1 and Φ_2 are independent random variables uniformly distributed on $(-\pi, \pi)$. What is the power spectral density of $U(t)$?

3-6 A certain random process $U(t)$ takes on equally probable values $+1$ or 0 with changes occurring randomly in time. The probability that n changes occur in time τ is known to be

$$P_N(n) = \frac{1}{1+\alpha\tau}\left(\frac{\alpha\tau}{1+\alpha\tau}\right)^n \qquad n = 0, 1, 2, \ldots,$$

where the mean number of changes is $\bar{n} = \alpha\tau$. Find and sketch the autocorrelation function of this random process.

Hint: $\sum_{k=0}^{\infty} r^k = \frac{1}{1-r}$ when $|r| < 1$.

3-7 A certain random process $U(t)$ consists of a sum of rectangular pulses of the form $p(t - t_k) = \text{rect}((t - t_k)/b)$, occurring with mean rate $\bar{n}$. The times of occurrence are random, with the number of pulses emitted in a T-second interval being Poisson distributed with mean $\bar{n}T$. This random input is applied to a nonlinear device with input–output characteristic

$$z = \begin{cases} 1 & u > 0 \\ 0 & u = 0. \end{cases}$$

Find

(a) $\bar{z}$.

(b) $\Gamma_Z(\tau)$.

3-8 Assuming that $U(t)$ is wide-sense stationary, with mean $\bar{u}$ and variance σ^2, which of the following functions represent possible structure functions for $U(t)$?

(a) $D_U(\tau) = 2\sigma^2[1 - e^{-\alpha|\tau|}]$

(b) $D_U(\tau) = 2\sigma^2[1 - \alpha|\tau|\cos\alpha\tau]$
(c) $D_u(\tau) = 2\sigma^2[1 - \sin\alpha\tau]$
(d) $D_U(\tau) = 2\sigma^2[1 - \cos\alpha\tau]$
(e) $D_U(\tau) = 2\sigma^2[1 - \text{rect}\,\alpha\tau]$

3-9 Prove that the Hilbert transform of the Hilbert transform of $u(t)$ is $-u(t)$, up to a possible additive constant.

3-10 Parseval's theorem, in generalized form, states that for any two Fourier transformable functions $\mathbf{f}(t)$ and $\mathbf{g}(t)$ with transforms $\mathscr{F}(\nu)$ and $\mathscr{G}(\nu)$,

$$\int_{-\infty}^{\infty} \mathbf{f}(t)\mathbf{g}^*(t)\,dt = \int_{-\infty}^{\infty} \mathscr{F}(\nu)\mathscr{G}^*(\nu)\,d\nu.$$

Show that if $\mathbf{u}(t)$ and $\mathbf{v}(t)$ are analytic signals,

$$\int_{-\infty}^{\infty} \mathbf{u}(t)\mathbf{v}(t)\,dt = 0.$$

3-11 Given that the autocorrelation function of an analytic signal $\mathbf{u}(t)$ is $\Gamma_U(\tau)$, show that the autocorrelation function of $(d/dt)\mathbf{u}(t)$ is $-(\partial^2/\partial\tau^2)\Gamma_U(\tau)$.
Hint: Use frequency domain reasoning.

3-12 Find the analytic signal representation for the function

$$u(t) = \text{rect}\,t.$$

3-13 (a) Show that for an analytic signal representation of a real-valued narrowband random process, the autocorrelation function of the complex process $\mathbf{U}(t)$ (assumed wide-sense stationary) can be written in the form

$$\Gamma_U(\tau) = \mathbf{g}(\tau)e^{-j2\pi\nu_0\tau}$$

where $\mathbf{g}(\tau)$ is a slowly varying function of τ by comparison with the complex carrier.

(b) Show further that when $\mathbf{U}(t)$ has a power spectral density that is even about the center frequency ν_0, $\mathbf{g}(\tau)$ is entirely real valued.

3-14 Let $\mathbf{V}(t)$ be a linearly filtered complex-valued random process with sample functions given by

$$\mathbf{v}(t) = \int_{-\infty}^{\infty} \mathbf{h}(t-\tau)\mathbf{u}(\tau)\,d\tau,$$

where $\mathbf{U}(t)$ is a complex-valued input process and $\mathbf{h}(t)$ is the impulse response of a time-invariant linear filter.

(a) Show that, for a wide-sense-stationary input process,

$$\Gamma_V(\tau) = \mathbf{H}(\tau) * \Gamma_U(\tau),$$

where

$$\mathbf{H}(\tau) \triangleq \int_{-\infty}^{\infty} \mathbf{h}(\xi + \tau)\mathbf{h}^*(\xi)\, d\xi.$$

(b) Show that the mean-square value $|\mathbf{v}|^2$ of the output is given by

$$\overline{|\mathbf{v}|^2} = \int_{-\infty}^{\infty} \mathbf{H}(-\tau)\Gamma_U(\tau)\, d\tau.$$

3-15 Find the power spectral density of a doubly stochastic Poisson impulse process having a rate process described by

$$\Lambda(t) = \lambda_0[1 + \cos(2\pi\nu_0 t + \Phi)],$$

where Φ is a random variable uniformly distributed on $(-\pi, \pi)$ and λ_0 and ν_0 are constants.

4

Some First-Order Properties of Light Waves

Discussions of the statistical characteristics of optical radiation properly include consideration of first-order properties (i.e., at a single time instant), second-order properties (two time instants), and higher-order properties (three or more time instants). In this chapter we restrict attention to first-order properties of light waves. The discussion begins with a nonstatistical topic, the propagation of light waves under various restrictions on optical bandwidth. Attention is then turned to the first-order statistics of the amplitude and intensity of polarized, unpolarized, and partially polarized thermal light. Finally, various statistical models for the light emitted by a laser are considered.

The discussions presented in this chapter are entirely in classical terms. The reader should be aware that, paralleling the classical theory of fluctuations of light, there exists a rigorous quantum mechanical theory (see, e.g., Ref. 4-1). The quantum mechanical treatment is not covered here, partly because of the considerable background in quantum mechanics required and partly because the classical theory (together with the semiclassical treatment of detection found in Chapter 9) appears to be adequate, from a practical point of view, in nearly all experiments of interest to the optical systems engineer.

Throughout this chapter, and indeed throughout the entire book, we deal with a *scalar* theory of light waves. The scalar quantities dealt with may be regarded as representing one polarization component of the electric or magnetic field, with the approximation that all such components can be treated independently. This approximation neglects the coupling between various components of the electric and magnetic fields imposed by Maxwell's equations. Fortunately, the experiments presented in Ref. 4-2 indicate that the scalar theory yields accurate results provided only moderate or small diffraction angles are involved.

4.1 PROPAGATION OF LIGHT WAVES

As necessary background for material to follow in later chapters, we turn attention to a nonstatistical topic, namely, the propagation of light waves. This discussion is simply a brief review and tabulation of the important results. For more detailed treatments of the problem, see, for example, Ref. 4-3, Chapter 8, or Ref. 4-4, Chapter 3.

4.1.1 Monochromatic Light

Let $u(P, t)$ represent[†] the scalar amplitude of one polarization component of the electric or magnetic field associated with a monochromatic optical disturbance. (In accord with the philosophy of the scalar theory, we treat each component independently.) Here P represents a position in space and t a point in time. The analytic signal associated with $u(P, t)$ is

$$\mathbf{u}(P,t) = \mathbf{U}(P,\nu)\exp(-j2\pi\nu t), \qquad (4.1\text{-}1)$$

where ν is the frequency of the wave and $\mathbf{U}(P, \nu)$ is its phasor amplitude.

Let this wave be incident from the left on the infinite surface Σ shown in Fig. 4-1. We wish to specify the phasor amplitude of the field at the point P_0 to the right of the surface in terms of the field on Σ. The solution to this problem can be found in most standard texts on optics (again, consult Ref. 4-3 or 4-4, e.g.). We express the solution here in a form known as the *Huygens–Fresnel principle*, which states that, provided the distance r (see Fig. 4-1) is much greater than a wavelength λ,

$$\mathbf{U}(P_0,\nu) = \frac{1}{j\lambda} \iint_\Sigma \mathbf{U}(P_1,\nu) \frac{e^{j2\pi(r/\lambda)}}{r} \chi(\theta)\, dS \qquad (4.1\text{-}2)$$

where $\lambda = c/\nu$ is the wavelength of the light (c is the velocity of light), r is the distance from P_1 to P_0, θ is the angle between the line joining P_0 to P_1 and the normal to Σ (see Fig. 4-1), and $\chi(\theta)$ is an "obliquity factor" with the properties $\chi(0) = 1$ and $0 \le \chi(\theta) \le 1$.

[†] Beginning in this chapter and continuing hereafter, we drop the notational distinction between a random process and its sample functions. Although it is useful in purely statistical discussions to represent the process by a capital letter and a sample function by a lowercase letter, this distinction is seldom necessary in the discussion of physical applications of the theory. An exception is the notation for probability density functions. Such functions are generally subscripted by a capital letter representing the random variable of concern.

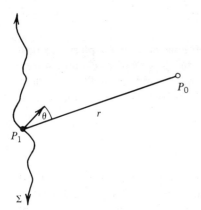

Figure 4-1. Propagation geometry.

The Huygens–Fresnel principle can be interpreted in a quasiphysical way as follows. Each point on Σ acts as a new "secondary source" of spherical waves. The strength of the secondary source at P_1 is proportional to $(j\lambda)^{-1}U(P_1,\nu)$, and that source radiates with a directional amplitude pattern $\chi(\theta)$.

The Huygens–Fresnel principle, as expressed by Eq. (4.1-2), will serve as our fundamental physical law governing the propagation of monochromatic light. In addition, as we see in the sections to follow, it can be used to find similar relations for nonmonochromatic light.

4.1.2 Nonmonochromatic Light

Let $u(P,t)$ be a nonmonochromatic wave, with an associated analytic signal $\mathbf{u}(P,t)$. Although $u(P,t)$ is in general not Fourier transformable, we can truncate it to the interval $(-T/2, T/2)$, yielding a Fourier-transformable function $u_T(P,t)$. Now $u_T(P,t)$ can be represented by an analytic signal $\mathbf{u}_T(P,t)$ that is Fourier transformable, even though its imaginary part is not truncated.

From the basic properties of analytic signals, in particular Eq. (3.8-8), we have

$$\mathbf{u}_T(P,t) = \int_0^\infty 2\mathscr{U}_T(P,\nu)e^{-j2\pi\nu t}\,d\nu, \qquad (4.1\text{-}3)$$

where $\mathscr{U}_T(P,\nu)$ is the Fourier transform of the real signal $u_T(P,t)$. Using this relationship, we now derive an expression for $\mathbf{u}(P_0,t)$ in terms of $\mathbf{u}(P_1,t)$, where P_0 and P_1 are as shown previously in Fig. 4-1.

PROPAGATION OF LIGHT WAVES

To begin, we note

$$\mathbf{u}(P_0, t) = \lim_{T \to \infty} \mathbf{u}_T(P_0, t) = \lim_{T \to \infty} \int_0^\infty 2\mathcal{U}_T(P_0, \nu) e^{-j2\pi\nu t} \, d\nu. \tag{4.1-4}$$

But from the Huygens–Fresnel principle, as expressed by Eq. (4.1-2),

$$\mathcal{U}_T(P_0, \nu) = \frac{1}{j\lambda} \iint_\Sigma \mathcal{U}_T(P_1, \nu) \frac{e^{j2\pi(r/\lambda)}}{r} \chi(\theta) \, dS. \tag{4.1-5}$$

Noting that $\lambda = c/\nu$, we use Eq. (4.1-3) and change orders of integration to write

$$\mathbf{u}_T(P_0, t) = \int_0^\infty 2\mathcal{U}_T(P_0, \nu) e^{-2\pi\nu t} \, d\nu$$

$$= \iint_\Sigma \frac{2\chi(\theta)}{2\pi cr} \left[\int_0^\infty (-j2\pi\nu) \mathcal{U}_T(P_1, \nu) e^{-j2\pi\nu[t - (r/c)]} \, d\nu \right] dS. \tag{4.1-6}$$

Differentiation of Eq. (4.1-3) with respect to t yields

$$\frac{d}{dt} \mathbf{u}_T(P_1, t) = 2 \int_0^\infty (-j2\pi\nu) \mathcal{U}_T(P_1, \nu) e^{-j2\pi\nu t} \, d\nu, \tag{4.1-7}$$

and as a consequence the bracketed quantity in (4.1-6) can be expressed in terms of a time derivative. The result is

$$\mathbf{u}_T(P_0, t) = \iint_\Sigma \frac{(d/dt)\mathbf{u}_T[P_1, t - (r/c)]}{2\pi cr} \chi(\theta) \, dS. \tag{4.1-8}$$

Finally, letting $T \to \infty$, we obtain the fundamental relationship describing the propagation of nonmonochromatic waves,

$$\mathbf{u}(P_0, t) = \iint_\Sigma \frac{(d/dt)\mathbf{u}[P_1, t - (r/c)]}{2\pi cr} \chi(\theta) \, dS. \tag{4.1-9}$$

In closing, the reader is reminded that our derivation utilized the form of the Huygens–Fresnel principle valid only when the distance r in Fig. 4-1 is always much greater than the wavelength λ, and hence a similar restriction applies to Eq. (4.1-9). This condition is well satisfied in all problems of interest to us in the future.

4.1.3 Narrowband Light

As a final relation of future interest, we derive a specialized form of Eq. (4.1-9) valid for nonmonochromatic light that is *narrowband*, that is, light with bandwidth $\Delta \nu$ much smaller than the center frequency $\bar{\nu}$.

According to Eq. (4.1-6), we can write

$$\mathbf{u}_T(P_0, t) = \iint_\Sigma \frac{2\chi(\theta)}{jcr} \int_0^\infty \nu \, \mathcal{U}_T(P_1, \nu) e^{-j2\pi\nu[t-(r/c)]} \, d\nu \, dS.$$

(4.1-10)

Now noting that $\Delta \nu \ll \bar{\nu}$, the following approximation can be made with good accuracy:

$$\mathbf{u}_T(P_0, t) \cong \iint_\Sigma \frac{\bar{\nu}}{jcr} \left\{ \int_0^\infty 2 \mathcal{U}_T(P_1, \nu) e^{-j2\pi\nu[t-(r/c)]} \, d\nu \right\} \chi(\theta) \, dS.$$

(4.1-11)

The quantity within braces is simply $\mathbf{u}_T[P_1, t - (r/c)]$. Thus, with the definition $\bar{\lambda} \triangleq c/\bar{\nu}$, and letting T grow infinitely large, we find

$$\mathbf{u}(P_0, t) \cong \iint_\Sigma \frac{1}{j\bar{\lambda}r} \mathbf{u}\left(P_1, t - \frac{r}{c}\right) \chi(\theta) \, dS. \qquad (4.1\text{-}12)$$

This relationship will serve as our fundamental law of propagation for narrowband disturbances. Again, it is strictly valid only for $r \gg \bar{\lambda}$.

This concludes our discussion of the nonstatistical propagation laws obeyed by light waves. These first-order relationships will be particularly useful to us in Chapter 5. Our attention is now turned to first-order *statistical* properties of various kinds of light waves.

4.2 POLARIZED AND UNPOLARIZED THERMAL LIGHT

A great majority of optical sources, both natural and man-made, emit light by means of *spontaneous* emission from a collection of excited atoms or molecules. Such is the case for the sun, incandescent bulbs, and gas

POLARIZED AND UNPOLARIZED THERMAL LIGHT

discharge lamps, for example. A large collection of atoms or molecules, excited to high energy states by thermal, electrical, or other means, randomly and independently drop to lower energy states, emitting light in the process. Such radiation, consisting of a large number of independent contributions, is referred to as *thermal light*.

To be contrasted with the chaotic wave emitted by a thermal source is the relatively well-ordered *stimulated* radiation emitted by a laser. Excited atoms or molecules, confined within a resonant cavity, radiate synchronously, or in unison, in a well-ordered and highly dependent fashion. Such light, which we refer to simply as *laser light*, is discussed in Section 4.4.

Both thermal light and laser light consist of waves that fluctuate randomly with time. Thus either kind of light must ultimately be treated as a random process. In this section we concentrate on the first-order statistics of the amplitude and intensity of thermal light.

4.2.1 Polarized Thermal Light

Consider the light emitted by a thermal source and passed by a polarization analyzer, with polarization direction lying, for example, along the x axis. The real-valued function $u_X(P, t)$ represents the x-component of the electric field vector, observed at point P and time t. Because of the presence of the polarization analyzer, the y-component of the field $u_Y(P, t)$ is zero. For the present, we refer to such a light wave as *polarized thermal light*, although a more general definition of polarized light emerges in later discussions (see Section 4.3).

Since the source in question is thermal, the time waveform $u_X(P, t)$ can be regarded as a sum of a great many independent contributions,

$$u_X(P, t) = \sum_{\substack{\text{all} \\ \text{atoms}}} u_i(P, t), \qquad (4.2\text{-}1)$$

where $u_i(P, t)$ is the x-component of the field contributed by the ith atom. Since the number of radiating atoms is usually enormous, we conclude, with the help of the central limit theorem, that $u_X(P, t)$ is a *Gaussian random process* for a polarized thermal source.

Often it is most convenient to work with the analytic signal representation of the polarized wave $\mathbf{u}_X(P, t)$ or alternatively with the complex envelope

$$\mathbf{A}_X(P, t) = \mathbf{u}_X(P, t) e^{j2\pi \bar{\nu} t},$$

where $\bar{\nu}$ is the center frequency of the wave. For such representations we

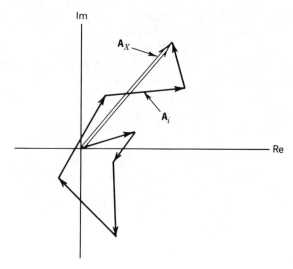

Figure 4-2. Complex envelope of polarized thermal light at a fixed time and a fixed point in space.

have

$$\mathbf{u}_X(P,t) = \sum_{\substack{\text{all} \\ \text{atoms}}} \mathbf{u}_i(P,t) \qquad (4.2\text{-}2)$$

$$\mathbf{A}_X(P,t) = \sum_{\substack{\text{all} \\ \text{atoms}}} \mathbf{A}_i(P,t), \qquad (4.2\text{-}3)$$

where $\mathbf{u}_i(P,t)$ and $\mathbf{A}_i(P,t)$ are the analytic signal and complex envelope representations, respectively, of the wave component contributed by the ith elementary radiator. When the central limit theorem is applied to both the real and imaginary parts of (4.2-2) and (4.2-3), we see that, under the assumption that the various contributions are randomly phased and independent, $\mathbf{u}_X(P,t)$ and $\mathbf{A}_X(P,t)$ are both *circular complex Gaussian random processes*.

Figure 4-2 shows the complex envelope $\mathbf{A}_X(P,t)$ at a particular point P and time instant t, consisting of a great many independent complex phasors. Since there is no relationship between the phases of the individual atomic contributions, we can reasonably model the phases of the $\mathbf{A}_i(P,t)$ as statistically independent and uniformly distributed on $(-\pi, \pi)$.† Thus

---

†If the arrival time of the radiation from a particular atom is totally unpredictable, the phase of that radiation is uniformly distributed on the primary interval.

$\mathbf{A}_X(P, t)$ has all the properties of the random phasor sum discussed in Section 2.9. In particular, its real and imaginary parts are independent, identically distributed zero-mean Gaussian random variables.

Detectors of optical radiation respond not to field strength, but rather to optical power or *intensity*. Accordingly, the statistical properties of the intensity of an optical wave are of considerable practical importance. We define the *instantaneous intensity* $I_X(P, t)$ of the polarized wave to be the squared modulus of the analytic signal representation of the field,

$$I_X(P, t) \triangleq |\mathbf{u}_X(P, t)|^2 = |\mathbf{A}_X(P, t)|^2. \qquad (4.2\text{-}4)$$

We reserve the unmodified term "intensity" for the time average, or under the assumption of ergodicity, the ensemble average of the instantaneous intensity $I_X(P, t)$,

$$I_X(P) \triangleq \langle I_X(P, t) \rangle = \bar{I}_X(P). \qquad (4.2\text{-}5)$$

The instantaneous intensity is, of course, a random process. Since $I_X(P, t)$ is the squared length of a random phasor sum, we can readily use the knowledge developed in Section 2.9 to find its first-order probability density function. For brevity, we use the notation

$$A \triangleq |\mathbf{A}_X(P, t)|, \qquad I \triangleq I_X(P, t)$$

in the discussion to follow. We know that A obeys a Rayleigh probability density function,

$$p_A(A) = \begin{cases} \dfrac{A}{\sigma^2} \exp(-A^2/2\sigma^2) & A \geq 0 \\ 0 & \text{otherwise,} \end{cases} \qquad (4.2\text{-}6)$$

where σ^2 represents the variance of the real and imaginary parts of $\mathbf{A}_X(P, t)$. The transformation

$$I = A^2, \qquad A = \sqrt{I}$$

is monotonic on $(0, \infty)$, and thus we can use Eq. (2.5-11) to write

$$p_I(I) = p_A(A = \sqrt{I}) \left| \frac{dA}{dI} \right|$$

$$= \frac{\sqrt{I}}{\sigma^2} e^{-I/2\sigma^2} \cdot \frac{1}{2\sqrt{I}}$$

$$= \begin{cases} \dfrac{1}{2\sigma^2} \exp\left(-\dfrac{I}{2\sigma^2}\right) & I \geq 0 \\ 0 & \text{otherwise.} \end{cases} \qquad (4.2\text{-}7)$$

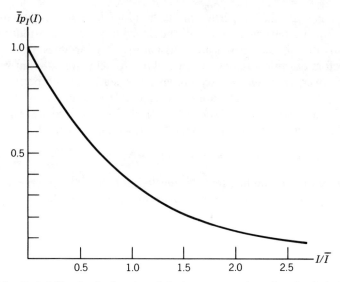

Figure 4-3. Probability density function of the instantaneous intensity of polarized thermal light.

Thus the instantaneous intensity obeys a negative exponential probability density function. This distribution has the important property that its standard deviation σ_I is equal to its mean $\bar{I}$, both of which equal $2\sigma^2$,

$$\sigma_I = \bar{I} = 2\sigma^2. \qquad (4.2\text{-}8)$$

Hence in a slightly more compact notation we have

$$p_I(I) = \begin{cases} \dfrac{1}{\bar{I}} \exp\left(-\dfrac{I}{\bar{I}}\right) & I \geq 0 \\ 0 & \text{otherwise.} \end{cases} \qquad (4.2\text{-}9)$$

This density function is plotted in Fig. 4-3.

Knowing the properties of polarized thermal light, we turn now to consideration of unpolarized thermal light.

4.2.2 Unpolarized Thermal Light

Light from a thermal source is regarded as unpolarized if two conditions are met. First, we require that the intensity of the light passed by a polarization analyzer, situated in a plane perpendicular to the direction of propagation of the wave, be independent of the rotational orientation of the analyzer. Second, we require that any two orthogonal field components $\mathbf{u}_X(P,t)$ and $\mathbf{u}_Y(P,t)$ have the property that $\langle \mathbf{u}_X(t+\tau)\mathbf{u}_Y^*(t)\rangle$ is identically zero for all rotational orientations of the $X-Y$ coordinate axes and for all delays τ. (For a further discussion of unpolarized light, see Section 4.3.) This type of light is also often referred to as "natural" light.

Since the light arises from a thermal source, the arguments of the previous section can be applied to each individual polarization component, yielding the conclusion that $\mathbf{u}_X(P,t)$ and $\mathbf{u}_Y(P,t)$ are circular complex Gaussian random processes. Furthermore, since they are uncorrelated for all relative time delays, the two processes are statistically independent.

The instantaneous intensity of the wave is defined by

$$I(P,t) \triangleq |\mathbf{u}_X(P,t)|^2 + |\mathbf{u}_Y(P,t)|^2$$
$$= |\mathbf{A}_X(P,t)|^2 + |\mathbf{A}_Y(P,t)|^2$$
$$= I_X(P,t) + I_Y(P,t). \quad (4.2\text{-}10)$$

From the previous section, $I_X(P,t)$ and $I_Y(P,t)$ each obey negative-exponential statistics. Moreover, from the definition of unpolarized light, $I_X(P,t)$ and $I_Y(P,t)$ have equal means,

$$\bar{I}_X(P) = \bar{I}_Y(P) = \tfrac{1}{2}\bar{I}(P), \quad (4.2\text{-}11)$$

and are statistically independent random processes. To find the first-order probability density function of the total instantaneous intensity, we must find the density function of the sum of two independent random variables having identical density functions

$$p_{I_X}(I_X) = \frac{2}{\bar{I}} \exp\left(-2\frac{I_X}{\bar{I}}\right)$$
$$p_{I_Y}(I_Y) = \frac{2}{\bar{I}} \exp\left(-2\frac{I_Y}{\bar{I}}\right). \quad (4.2\text{-}12)$$

With the help of Eq. (2.6-10) and Fig. 4-4, we write the required convolution as

$$p_I(I) = \begin{cases} \int_0^I \left(\frac{2}{\bar{I}}\right)^2 \exp\left(-2\frac{\xi}{\bar{I}}\right)\exp\left[-\frac{2}{\bar{I}}(I-\xi)\right]d\xi & I \geq 0 \\ 0 & \text{otherwise} \end{cases}$$

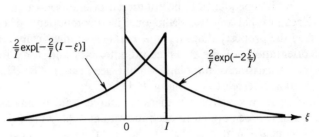

Figure 4-4. Factors in the integrand of the convolution equation.

or

$$p_I(I) = \begin{cases} \left(\dfrac{2}{\bar{I}}\right)^2 I \exp\left(-2\dfrac{I}{\bar{I}}\right) & I \geq 0 \\ 0 & \text{otherwise.} \end{cases} \quad (4.2\text{-}13)$$

This density function is plotted in Fig. 4-5.

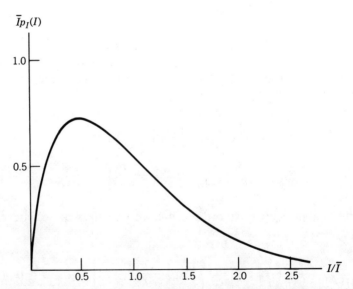

Figure 4-5. Probability density function of the instantaneous intensity of unpolarized thermal light.

Note that unpolarized thermal light has considerably less probability of having an extremely small value of instantaneous intensity than does polarized thermal light. In addition, we can readily show that the ratio of standard deviation σ_I to mean $\bar{I}$, which was unity for polarized thermal light, is reduced to a value of $\sqrt{1/2}$ for unpolarized thermal light.

4.3 PARTIALLY POLARIZED THERMAL LIGHT

Having discussed the properties of polarized and unpolarized thermal light, we are led naturally to inquire whether a more general theory exists, a theory capable of dealing with intermediate cases of *partial polarization*. Such a theory does indeed exist, and we accordingly devote effort to presenting it here. To do so requires some initial explanation of a matrix theory capable of conveniently describing partially polarized light and the transformations to which it may be subjected. For more detailed discussions of the general subject of partial polarization, the reader may wish to consult Ref. 4-3, Section 10.8, or Ref. 4-5.

4.3.1 Passage of Narrowband Light Through Polarization-Sensitive Instruments

We consider now a mathematical formalism for describing the effects of various optical instruments on the polarization of transmitted light. A convenient formalism was first developed by R. C. Jones (Ref. 4-6) for monochromatic waves. The same formalism can be used for narrowband light, provided the bandwidth of the light is so narrow that the instrument in question affects all spectral components identically (Ref. 4-7).

Let $\mathbf{u}_X(t)$ and $\mathbf{u}_Y(t)$ represent the X- and Y-components of the electric or magnetic field at a particular point P in space. The state of this field is represented by a two-element column matrix $\underline{U}$,

$$\underline{U} = \begin{bmatrix} \mathbf{u}_X(t) \\ \mathbf{u}_Y(t) \end{bmatrix}. \qquad (4.3\text{-}1)$$

Suppose that the light is now passed through an optical instrument that may contain polarization sensitive elements (polarizers, retardation plates, etc.) and consider the field leaving the instrument at a point P' that is the geometric projection of P through the instrument. The state of this field is represented by a matrix $\underline{U}'$ similar to (4.3-1), but with elements $\mathbf{u}'_X(t)$ and $\mathbf{u}'_Y(t)$. Now if the instrument in question contains only linear elements, as is most often the case, the matrix $\underline{U}'$ can be expressed in terms of $\underline{U}$ by the simple matrix formula

$$\underline{U}' = \begin{bmatrix} \mathbf{u}'_X(t) \\ \mathbf{u}'_Y(t) \end{bmatrix} = \begin{bmatrix} l_{11} & l_{12} \\ l_{21} & l_{22} \end{bmatrix} \begin{bmatrix} \mathbf{u}_X(t) \\ \mathbf{u}_Y(t) \end{bmatrix} = \underline{L}\underline{U} \qquad (4.3\text{-}2)$$

where $\underline{L}$ is the 2×2 *polarization matrix* representing the effects of the instrument.

The matrix representations of some very simple types of physical operations will be useful to us in future discussions. First, and perhaps simplest, we consider the effect of a rotation of the X–Y coordinate system. This simple operation can be regarded as an "instrument" that transforms the original field components $\mathbf{u}_X(t)$ and $\mathbf{u}_Y(t)$ into new components $\mathbf{u}'_X(t)$ and $\mathbf{u}'_Y(t)$ according to the matrix operator

$$\underline{L} = \begin{bmatrix} \cos\theta & \sin\theta \\ -\sin\theta & \cos\theta \end{bmatrix} \qquad (4.3\text{-}3)$$

where θ is the rotation angle illustrated in Fig. 4-6.

A second important type of simple instrument is a retardation plate which, by means of a birefringent material, introduces a relative delay between the two polarization components. If the velocities of propagation of the X- and Y-polarization components are v_X and v_Y, a plate of thickness d introduces a time delay

$$\tau_d = d\left(\frac{1}{v_X} - \frac{1}{v_Y}\right) \qquad (4.3\text{-}4)$$

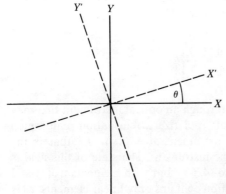

Figure 4-6. Old (X, Y) and new (X', Y') coordinate systems after rotation by angle θ.

of the X-component relative to the Y-component. In accord with the narrowband condition, we require that τ_d be much less than $1/\Delta\nu$. In this case the retardation plate can be represented by a matrix (written in symmetrical form for simplicity)

$$\underline{L} = \begin{bmatrix} e^{j(\delta/2)} & 0 \\ 0 & e^{-j(\delta/2)} \end{bmatrix} \qquad (4.3\text{-}5)$$

where

$$\delta = \frac{2\pi dc}{\lambda}\left(\frac{1}{v_X} - \frac{1}{v_Y}\right) \qquad (4.3\text{-}6)$$

is the phase delay of the X-component relative to the Y-component. We note in passing that both the rotation matrix (4.3-3) and the retardation matrix (4.3-5) are *unitary* matrices; that is, they have the property that $\underline{L}\underline{L}^\dagger = \underline{\mathscr{I}}$, where $\underline{L}^\dagger$ is the hermitian conjugate of $\underline{L}$ and $\underline{\mathscr{I}}$ is the identity matrix

$$\underline{L}^\dagger = \begin{bmatrix} l_{11}^* & l_{21}^* \\ l_{12}^* & l_{22}^* \end{bmatrix}, \qquad \underline{\mathscr{I}} = \begin{bmatrix} 1 & 0 \\ 0 & 1 \end{bmatrix}. \qquad (4.3\text{-}7)$$

As a final example, we mention without proof (see Problem 4-12) that the matrix representation of a polarization analyzer, oriented at angle α to the X axis, is

$$\underline{L}(\alpha) = \begin{bmatrix} \cos^2\alpha & \sin\alpha\cos\alpha \\ \sin\alpha\cos\alpha & \sin^2\alpha \end{bmatrix}. \qquad (4.3\text{-}8)$$

Thus each type of polarization instrument has its own matrix representation. Furthermore, if light is passed through a series of such instruments, their combined effect can be represented by a single matrix which is a product of the individual matrices involved. Thus if light passes through instruments with matrices $\underline{L}_1, \underline{L}_2, \ldots, \underline{L}_N$, we have

$$\underline{U}' = \underline{L}_N \cdots \underline{L}_2 \underline{L}_1 \underline{U}, \qquad (4.3\text{-}9)$$

and the total effect is equivalent to a single instrument with matrix

$$\underline{L} = \underline{L}_N \cdots \underline{L}_2 \underline{L}_1, \qquad (4.3\text{-}10)$$

where the usual rules of matrix multiplication are to be observed.

4.3.2 The Coherency Matrix

We now consider the problem of describing the state of polarization of a wave. In general, the direction of the electric vector may fluctuate with time in a complicated deterministic or random manner. A useful description is supplied by the so-called coherency matrix introduced by Wiener (Ref. 4-8) and Wolf (Ref. 4-7).

Consider the 2 × 2 matrix defined by

$$\underline{\mathbf{J}} \triangleq \langle \underline{\mathbf{U}}\underline{\mathbf{U}}^\dagger \rangle \qquad (4.3\text{-}11)$$

where the infinite time average $\langle \cdot \rangle$ is to be applied to each of the elements of the product matrix. Equivalently, $\underline{\mathbf{J}}$ may be expressed as

$$\underline{\mathbf{J}} = \begin{bmatrix} \mathbf{J}_{xx} & \mathbf{J}_{xy} \\ \mathbf{J}_{yx} & \mathbf{J}_{yy} \end{bmatrix}, \qquad (4.3\text{-}12)$$

where

$$\mathbf{J}_{xx} \triangleq \langle \mathbf{u}_X(t)\mathbf{u}_X^*(t) \rangle \qquad \mathbf{J}_{yx} \triangleq \langle \mathbf{u}_Y(t)\mathbf{u}_X^*(t) \rangle$$

$$\mathbf{J}_{xy} \triangleq \langle \mathbf{u}_X(t)\mathbf{u}_Y^*(t) \rangle \qquad \mathbf{J}_{yy} \triangleq \langle \mathbf{u}_Y(t)\mathbf{u}_Y^*(t) \rangle. \qquad (4.3\text{-}13)$$

The matrix $\underline{\mathbf{J}}$ so defined is called the *coherency matrix* of the wave. The elements on the main diagonal of $\underline{\mathbf{J}}$ are clearly the average intensities of the X- and Y-polarization components. The off-diagonal elements are the cross-correlations of the two polarization components.

From a purely mathematical point of view, we can identify some fundamental properties of the coherency matrix. First, from (4.3-13) it is clear that $\mathbf{J}_{xx}$ and $\mathbf{J}_{yy}$ are always nonnegative and real. Second, the element $\mathbf{J}_{yx}$ is equal to the complex conjugate of the element $\mathbf{J}_{xy}$. Thus $\underline{\mathbf{J}}$ is an *hermitian* matrix and can be written in the form

$$\underline{\mathbf{J}} = \begin{bmatrix} \mathbf{J}_{xx} & \mathbf{J}_{xy} \\ \mathbf{J}_{xy}^* & \mathbf{J}_{yy} \end{bmatrix}. \qquad (4.3\text{-}14)$$

Furthermore, by a direct application of Schwarz's inequality to the definition of $\mathbf{J}_{xy}$, we can show that

$$|\mathbf{J}_{xy}| \leq [\mathbf{J}_{xx}\mathbf{J}_{yy}]^{1/2}, \qquad (4.3\text{-}15)$$

PARTIALLY POLARIZED THERMAL LIGHT

and hence the determinant of $\underline{\mathbf{J}}$ is nonnegative

$$\det[\underline{\mathbf{J}}] = \mathbf{J}_{xx}\mathbf{J}_{yy} - |\mathbf{J}_{xy}|^2 \geq 0. \qquad (4.3\text{-}16)$$

Equivalently, we say that $\underline{\mathbf{J}}$ is nonnegative definite. Finally, the matrix $\underline{\mathbf{J}}$ has the important property that its trace is equal to the average intensity of the wave,

$$\text{tr}[\underline{\mathbf{J}}] = \mathbf{J}_{xx} + \mathbf{J}_{yy} = \bar{I}. \qquad (4.3\text{-}17)$$

When an optical wave passes through a polarization instrument, its coherency matrix is in general modified. Let $\underline{\mathbf{J}}'$ represent the coherency matrix at the output of the instrument and $\underline{\mathbf{J}}$ the coherency matrix at the input. How are $\underline{\mathbf{J}}'$ and $\underline{\mathbf{J}}$ related? The answer is easily found for narrowband light by substituting (4.3-2), which describes the transformation of the wave components, into the definition (4.3-11) of the coherency matrix. The result is

$$\underline{\mathbf{J}}' = \underline{\mathbf{L}}\,\underline{\mathbf{J}}\,\underline{\mathbf{L}}^\dagger. \qquad (4.3\text{-}18)$$

where we have used the fact that $(\underline{\mathbf{L}}\underline{\mathbf{U}})^\dagger = \underline{\mathbf{U}}^\dagger\underline{\mathbf{L}}^\dagger$.

Specific forms of the coherency matrix under various conditions of polarization can readily be deduced simply from the definitions of its elements. Some obvious examples are:

$$\text{Linear polarization in the } X \text{ direction} \qquad \underline{\mathbf{J}} = \bar{I}\begin{bmatrix} 1 & 0 \\ 0 & 0 \end{bmatrix} \qquad (4.3\text{-}19)$$

$$\text{Linear polarization in the } Y \text{ direction} \qquad \underline{\mathbf{J}} = \bar{I}\begin{bmatrix} 0 & 0 \\ 0 & 1 \end{bmatrix} \qquad (4.3\text{-}20)$$

$$\text{Linear polarization at } +45° \text{ to the } X \text{ axis} \qquad \underline{\mathbf{J}} = \frac{\bar{I}}{2}\begin{bmatrix} 1 & 1 \\ 1 & 1 \end{bmatrix}. \qquad (4.3\text{-}21)$$

Less obvious is the case of circularly polarized light. A wave is circularly polarized if the average intensity passed by a polarization analyzer is independent of the angular orientation of the analyzer *and* if the direction of the electric vector rotates with uniform angular velocity and period $1/\bar{\nu}$. The circular polarization is said to be in the *right-hand sense* if the direction of the vector rotates with time in a clockwise sense when the wave is viewed head-on (i.e., looking toward the source). The polarization is circular in the *left-hand sense* if the rotation is counterclockwise.

For right-hand circular polarization, the analytic signals $\mathbf{u}_X(t)$ and $\mathbf{u}_Y(t)$ take the form

$$\mathbf{u}_X(t) = \mathbf{A}(t)e^{-j2\pi\bar{\nu}t}$$

$$\mathbf{u}_Y(t) = \mathbf{A}(t)e^{-j[2\pi\bar{\nu}t+(\pi/2)]}, \quad (4.3\text{-}22)$$

where $\mathbf{A}(t)$ is a slowly varying complex envelope. Note that in a time interval $\Delta t \ll 1/\Delta\nu$, $\mathbf{A}(t)$ is approximately constant and the electric vector simply undergoes a rapid rotation of direction. The coherency matrix for this kind of light is readily found by substituting (4.3-22) in the definitions (4.3-13), with the result

$$\text{Right-hand circular polarization} \quad \mathbf{J} = \frac{\bar{I}}{2}\begin{bmatrix} 1 & j \\ -j & 1 \end{bmatrix}. \quad (4.3\text{-}23)$$

For left-hand circular polarization, the corresponding relationships are

$$\mathbf{u}_X(t) = \mathbf{A}(t)e^{-j2\pi\bar{\nu}t}$$

$$\mathbf{u}_Y(t) = \mathbf{A}(t)e^{-j[2\pi\bar{\nu}t-(\pi/2)]} \quad (4.3\text{-}24)$$

and

$$\text{Left-hand circular polarization} \quad \mathbf{J} = \frac{\bar{I}}{2}\begin{bmatrix} 1 & -j \\ j & 1 \end{bmatrix}. \quad (4.3\text{-}25)$$

Note in particular that for both types of circular polarization, the average intensities of the two polarization components are equal, but in addition the two components are *perfectly correlated*, as they have a correlation coefficient with unity magnitude,

$$|\mu_{xy}| \triangleq \frac{|\mathbf{J}_{xy}|}{[\mathbf{J}_{xx}\mathbf{J}_{yy}]^{1/2}} = 1. \quad (4.3\text{-}26)$$

Next, the important case of "natural" light is considered. By this term we mean that the light has two important properties. First, like circularly polarized light, natural light has equal average intensity in all directions; that is, if the wave is passed through a polarization analyzer, the average transmitted intensity is independent of the angular orientation of the analyzer. Unlike the case of circular polarization, however, natural light is characterized by a direction of polarization that fluctuates randomly with time, all directions being equally likely. The analytic signals representing the two polarization components of natural light can be written in the form

$$\mathbf{u}_X(t) = \mathbf{A}(t)\cos\theta(t)e^{-j2\pi\bar{\nu}t}$$
$$\mathbf{u}_Y(t) = \mathbf{A}(t)\sin\theta(t)e^{-j2\pi\bar{\nu}t}, \quad (4.3\text{-}27)$$

where $\mathbf{A}(t)$ is a slowly varying complex envelope describing the phasor amplitude of the electric vector at time t, and $\theta(t)$ is the slowly varying angle of polarization with respect to the X axis. If the angle θ is uniformly distributed on $(-\pi, \pi)$, the coherency matrix is readily found to be

$$\underline{\mathbf{J}} = \frac{\bar{I}}{2}\begin{bmatrix} 1 & 0 \\ 0 & 1 \end{bmatrix} = \frac{\bar{I}}{2}\underline{\mathscr{I}} \quad (4.3\text{-}28)$$

where again $\underline{\mathscr{I}}$ is the identity matrix.

It is a simple matter to show (see Problem 4-3) that if light with a coherency matrix given by (4.3-28) is passed through any instrument described by a *unitary* polarization matrix (e.g., a coordinate rotation or a retardation plate), the coherency matrix remains in the form (4.3-28). If the coherency matrix has this form, therefore, it is impossible to reintroduce correlation between the X and Y field components by means of an instrument with a unitary polarization matrix.

In closing this basic discussion of the coherency matrix, we point out that the elements of this matrix have the virtue that they are *measurable* quantities. Clearly, $\mathbf{J}_{xx}$ and $\mathbf{J}_{yy}$, which represent the average intensities of the X- and Y-polarization components, can be directly measured with the aid of a polarization analyzer, oriented sequentially in the X and Y directions. To measure the complex-valued element $\mathbf{J}_{xy}$, two measurements are required. If a polarization analyzer is set at $+45°$ to the X axis, the transmitted intensity is (see Problem 4-4)

$$\bar{I}_1 = \tfrac{1}{2}\left[\mathbf{J}_{xx} + \mathbf{J}_{yy} + \mathbf{J}_{xy} + \mathbf{J}^*_{xy}\right]$$
$$= \tfrac{1}{2}\left[\mathbf{J}_{xx} + \mathbf{J}_{yy}\right] + \operatorname{Re}\{\mathbf{J}_{xy}\}. \quad (4.3\text{-}29)$$

Since $\mathbf{J}_{xx}$ and $\mathbf{J}_{yy}$ are known, the real part of $\mathbf{J}_{xy}$ is thus determined. Now if a quarter-wave plate is introduced to retard the Y-component with respect to the X-component by $\pi/2$ radians, followed by a polarization analyzer oriented again at $45°$ to the X axis, the transmitted intensity is (see Problem 4-5)

$$\bar{I}_2 = \tfrac{1}{2}\left[\mathbf{J}_{xx} + \mathbf{J}_{yy} - j\left(\mathbf{J}_{xy} - \mathbf{J}^*_{xy}\right)\right]$$
$$= \tfrac{1}{2}\left[\mathbf{J}_{xx} + \mathbf{J}_{yy}\right] + \operatorname{Im}\{\mathbf{J}_{xy}\}, \quad (4.3\text{-}30)$$

thus allowing the imaginary part of $\mathbf{J}_{xy}$ to be determined. Since $\mathbf{J}_{yx} = \mathbf{J}_{xy}^*$, the entire coherency matrix has thus been established.

4.3.3 The Degree of Polarization

It would be highly desirable, both aesthetically and from a practical point of view, to find a single parameter that will characterize the degree to which a wave can be said to be polarized. For the case of a linearly polarized wave, this parameter should have its maximum value (unity for convenience), for such a wave is fully polarized by any reasonable definition. For circularly polarized light, the parameter should again have its maximum value, for such light can be made linearly polarized, without loss of energy, by means of a quarter-wave retardation plate. For the case of natural light, the parameter should have value zero, for the polarization direction is totally random and unpredictable in this case.

A parameter that measures the degree of statistical dependence between the two polarization components would be ideally suited for our purpose. In general, however, such a parameter would require full knowledge of the joint statistics of $\mathbf{u}_X(t)$ and $\mathbf{u}_Y(t)$. For simplicity, a more limited measure of polarization is adopted, one that depends only on the correlation parameters $\mathbf{J}_{xx}$, $\mathbf{J}_{yy}$, and $\mathbf{J}_{xy}$ of the coherency matrix. Such a definition is quite adequate in most applications, particularly if the light is thermal in origin. However, it is not difficult to find an example of a light wave that has a coherency matrix identical with that of natural light and yet has a fully deterministic and predictable behavior of its polarization direction (see Problem 4-6). Recognizing these possible pitfalls, we consider the definition of a *degree of polarization* $\mathscr{P}$ based on the properties of the coherency matrix.

What are the key differences between the coherency matrices of light that we would logically call fully polarized (e.g., linearly or circularly polarized) and light that we would logically call unpolarized (e.g., natural light)? The differences are not merely the presence or absence of off-diagonal elements, for such elements are zero in both Eqs. (4.3-19) and (4.3-28), yet the former corresponds to fully polarized light and the latter to unpolarized light.

Some help is afforded by the following physical observations. For light polarized at 45° to the X axis, it is possible to diagonalize the coherency matrix by means of a simple coordinate rotation, changing (4.3-21) to (4.3-19), for example. Similarly, for the case of circularly polarized light, a quarter-wave plate followed by a coordinate rotation of 45° results in light linearly polarized along the X axis and thus diagonalizes the coherency matrix. In both cases a lossless polarization transformation has eliminated the off-diagonal elements. Perhaps, then, the key difference between polarized and unpolarized light lies in the form of the coherency matrix *after diagonalization*.

Further support for this idea is afforded by some very general results from matrix theory. It is possible to show that for every hermitian matrix $\underline{J}$, there exists a *unitary* matrix transformation $\underline{P}$ such that

$$\underline{P}\underline{J}\underline{P}^\dagger = \begin{bmatrix} \lambda_1 & 0 \\ 0 & \lambda_2 \end{bmatrix}, \quad (4.3\text{-}31)$$

where λ_1 and λ_2 are the (real-valued) eigenvalues of $\underline{J}$ (Ref. 4-9). Furthermore, any coherency matrix $\underline{J}$ can be shown to be nonnegative definite; therefore, λ_1 and λ_2 are nonnegative. If these results are interpreted physically, for every wave there exists a lossless polarization instrument that will eliminate all correlation between the X- and Y-polarization components. The required instrument (i.e., the required $\underline{P}$) depends on the initial coherency matrix $\underline{J}$, but can always be realized with a combination of a coordinate rotation and a retardation plate (Ref. 4-10).

If λ_1 and λ_2 are identical (as for natural light), clearly the degree of polarization (however we define it) must be zero. If either λ_1 or λ_2 is zero (as for light polarized linearly along the X or Y axes), the degree of polarization must clearly be unity. To arrive at a logical definition of the degree of polarization, we note that the diagonalized coherency matrix can always be rewritten in the following way:

$$\begin{bmatrix} \lambda_1 & 0 \\ 0 & \lambda_2 \end{bmatrix} = \begin{bmatrix} \lambda_2 & 0 \\ 0 & \lambda_2 \end{bmatrix} + \begin{bmatrix} \lambda_1 - \lambda_2 & 0 \\ 0 & 0 \end{bmatrix}, \quad (4.3\text{-}32)$$

where we have assumed, without loss of generality, that $\lambda_1 \geq \lambda_2$. The first matrix on the right is recognized as representing unpolarized light of average intensity $2\lambda_2$, whereas the second matrix represents linearly polarized light of intensity $\lambda_1 - \lambda_2$. Thus light with arbitrary polarization properties can be represented as a sum of polarized and unpolarized components. We define the *degree of polarization* of the wave as the ratio of the intensity of the polarized component to the total intensity,

$$\mathscr{P} \triangleq \frac{\lambda_1 - \lambda_2}{\lambda_1 + \lambda_2}. \quad (4.3\text{-}33)$$

Thus a general definition has been arrived at.

The degree of polarization can be expressed more explicitly in terms of the elements of the original coherency matrix, if desired. To do so, we note that the eigenvalues λ_1 and λ_2 are, by definition, solutions to the equation

$$\det[\mathbf{J} - \lambda \mathbf{\mathcal{I}}] = 0. \tag{4.3-34}$$

Straightforward solution of the resulting quadratic equation in λ yields

$$\lambda_{1,2} = \frac{1}{2}\mathrm{tr}[\mathbf{J}]\left[1 \pm \sqrt{1 - 4\frac{\det[\mathbf{J}]}{(\mathrm{tr}[\mathbf{J}])^2}}\right]. \tag{4.3-35}$$

Thus the degree of polarization can be written as

$$\mathcal{P} = \sqrt{1 - 4\frac{\det[\mathbf{J}]}{(\mathrm{tr}[\mathbf{J}])^2}}. \tag{4.3-36}$$

It is not difficult to show that any unitary transformation of the coherency matrix does not affect the trace of that matrix. As a consequence, we can always regard the intensity of a partially polarized wave as being the sum of the intensities λ_1 and λ_2 of two *uncorrelated* field components. The average intensities of these components are expressible in terms of the degree of polarization $\mathcal{P}$ as follows

$$\lambda_1 = \tfrac{1}{2}\bar{I}(1 + \mathcal{P})$$

$$\lambda_2 = \tfrac{1}{2}\bar{I}(1 - \mathcal{P}), \tag{4.3-37}$$

where we have simply noted that $\mathrm{tr}[\mathbf{J}] = \bar{I}$ and substituted (4.3-36) in (4.3-35). If the light is thermal in origin, lack of correlation implies statistical independence of both the field components and the corresponding intensities.

The preceding discussion of partially polarized light is not an exhaustive one, for many interesting subjects have been omitted. We mention in particular the Stokes parameters and Mueller matrices, neither of which have been treated here. However, we limit our discussion to those aspects that will be useful to us in later material, and hence the reader is referred to Refs. 4-3 (Chapter 10), 4-5, and 4-11 for more complete discussions.

4.3.4 First-Order Statistics of the Instantaneous Intensity

We close this discussion of partially polarized light with a derivation of the probability density function of the instantaneous intensity of thermal light with an arbitrary degree of polarization $\mathcal{P}$. As we have seen in the previous

PARTIALLY POLARIZED THERMAL LIGHT

section, it is always possible to express the instantaneous intensity of a partially polarized wave as the sum of two uncorrelated intensity components

$$I(P,t) = I_1(P,t) + I_2(P,t). \quad (4.3\text{-}38)$$

Furthermore, if the light is thermal in origin, the intensity components are also statistically independent, as a result of the independence of the underlying complex Gaussian field components. The average intensities of these two components are, from Eq. (4.3-37),

$$\bar{I}_1 = \tfrac{1}{2}(1 + \mathscr{P})\bar{I}$$

$$\bar{I}_2 = \tfrac{1}{2}(1 - \mathscr{P})\bar{I}, \quad (4.3\text{-}39)$$

where $\bar{I}$ is the total average intensity.

Since I_1 and I_2 are squared moduli of circular complex Gaussian fields, each obeys negative-exponential statistics; that is,

$$p_{I_1}(I_1) = \frac{2}{(1+\mathscr{P})\bar{I}} \exp\left\{-\frac{2I_1}{(1+\mathscr{P})\bar{I}}\right\}$$

$$p_{I_2}(I_2) = \frac{2}{(1-\mathscr{P})\bar{I}} \exp\left\{-\frac{2I_2}{(1-\mathscr{P})\bar{I}}\right\} \quad (4.3\text{-}40)$$

for $I_1 \geq 0$ and $I_2 \geq 0$. The probability density function for the total intensity I is most easily found by first calculating the characteristic function $\mathbf{M}_I(\omega)$. Using the independence of I_1 and I_2, we can express the characteristic function as a product of two characteristic functions (cf. Problem 4-2):

$$\mathbf{M}_I(\omega) = \left[\frac{1}{1 - j\frac{\omega}{2}(1+\mathscr{P})\bar{I}}\right]\left[\frac{1}{1 - j\frac{\omega}{2}(1-\mathscr{P})\bar{I}}\right]$$

$$= \frac{(1+\mathscr{P})/2\mathscr{P}}{1 - j\frac{\omega}{2}(1+\mathscr{P})\bar{I}} - \frac{(1-\mathscr{P})/2\mathscr{P}}{1 - j\frac{\omega}{2}(1-\mathscr{P})\bar{I}}, \quad (4.3\text{-}41)$$

where a partial fraction expansion has been used in the last step. A Fourier inversion now yields a density function of the form

$$p_I(I) = \frac{1}{\mathscr{P}\bar{I}}\left\{\exp\left[-\frac{2I}{(1+\mathscr{P})\bar{I}}\right] - \exp\left[-\frac{2I}{(1-\mathscr{P})\bar{I}}\right]\right\}. \quad (4.3\text{-}42)$$

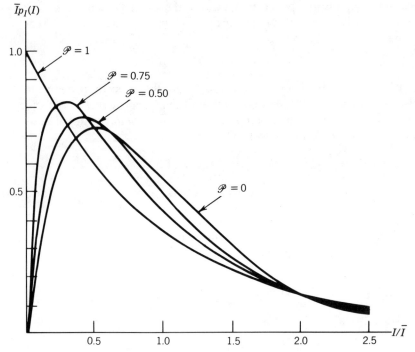

Figure 4-7. Probability density function of the instantaneous intensity of a thermal source with degree of polarization $\mathscr{P}$.

This density function is plotted in Fig. 4-7 for several values of $\mathscr{P}$. The results are seen to agree with Figs. 4-3 and 4-5 for the cases $\mathscr{P} = 1$ and $\mathscr{P} = 0$, respectively.

Finally, for a partially polarized thermal source it can readily be shown (see Problem 4-7), that the standard deviation σ_I of the instantaneous intensity is given by

$$\sigma_I = \sqrt{\frac{1 + \mathscr{P}^2}{2}}\, \bar{I}. \qquad (4.3\text{-}43)$$

4.4 LASER LIGHT

Having examined the first-order properties of thermal light, which is the type of light most often encountered in practice, we now turn attention to the more difficult problem of modeling the first-order properties of light generated by a laser oscillator. The problem is made difficult not only by the complicated physics that describes the operation of even the simplest kind of laser, but also by the vast multitude of types of laser that exist. No one model could be hoped to accurately describe the statistical properties of laser light in all possible cases. The best that can be done is to present several models that describe certain idealized properties of laser light.

By way of background, we briefly describe in an intuitive way the principle of laser action. A laser consists of a collection of atoms or molecules (the "active medium") excited by an energy source (the "pump") and contained within a resonant cavity that provides feedback. Spontaneous emission from the active medium is reflected from the end mirrors of the cavity and passes again through the active medium, where it is reinforced by additional stimulated emission. Stimulated emission contributions from different passes through the active medium will add constructively only for certain discrete frequencies or modes.

Whether a given mode breaks into oscillation depends on whether the gain of the active medium exceeds the various inherent losses for that particular mode frequency. We say loosely that a given mode is at "threshold" when the gain just equals the losses. The gain can be increased by increasing the power of the pump. When oscillation develops, however, nonlinearities of the process introduce a saturation of the gain, preventing further increase of gain with increased pump power. Nonetheless, as we shall see, the statistical properties of the emitted radiation are influenced by the degree to which the pump exceeds threshold. In addition, as the pump power increases, generally speaking, more modes of the cavity reach threshold, and the output contains several oscillating lines at different frequencies.

Our initial considerations are restricted to the case of single-mode laser oscillation. Later we focus on the more common but more difficult case of multimode oscillation.

4.4.1 Single-Mode Oscillation

The most highly idealized model of laser light is a purely monochromatic oscillator of known amplitude S, known frequency ν_0, and fixed but unknown absolute phase ϕ. The real-valued representation of such a signal, assumed linearly polarized, is

$$u(t) = S\cos[2\pi\nu_0 t - \phi]. \tag{4.4-1}$$

To incorporate the fact that we never know the absolute phase of the oscillation, ϕ must be regarded as a random variable, uniformly distributed on $(-\pi, \pi)$. The result is a random process representation that is both stationary and ergodic.

The first-order statistics of the instantaneous amplitude can be most easily found by calculating its characteristic function. Since the process is stationary, we can set $t = 0$, in which case

$$\mathbf{M}_U(\omega) = E[\exp(j\omega S \cos\phi)]$$

$$= \frac{1}{2\pi} \int_{-\pi}^{\pi} \exp(j\omega S \cos\phi) \, d\phi = J_0(\omega S), \quad (4.4\text{-}2)$$

where J_0 is a Bessel function of the first kind, zero order. Fourier inversion of this function yields (Ref. 4-12, p. 366) a probability density function

$$p_U(u) = \begin{cases} \left[\pi\sqrt{S^2 - u^2}\right]^{-1} & |u| \leq S \\ 0 & \text{otherwise,} \end{cases} \quad (4.4\text{-}3)$$

which is plotted in Fig. 4.8a.

As for the intensity of the signal $u(t)$, we have

$$I = \left| S \exp[-j(2\pi\nu_0 t - \phi)] \right|^2 = S^2.$$

Thus the probability density function for I can be written

$$p_I(I) = \delta(I - S^2), \quad (4.4\text{-}4)$$

which is shown in Fig. 4-8b.

A first step toward a more realistic model is taken by incorporating the fact that no real oscillation has a perfectly constant phase. Rather, to a

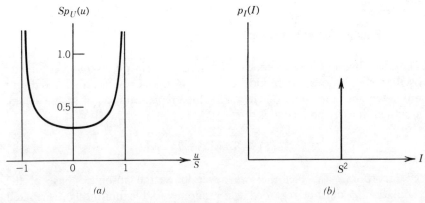

Figure 4-8. Probability density functions $p_U(u)$ of the amplitude and $p_I(I)$ of the intensity of a perfectly monochromatic wave of unknown phase.

LASER LIGHT

degree that depends on the type of laser and the precautions taken for stability, the phase undergoes random fluctuations with time. Thus we modify Eq. (4.4-1) to read

$$u(t) = S\cos[2\pi\nu_0 t - \theta(t)], \qquad (4.4\text{-}5)$$

where $\theta(t)$ represents the temporal fluctuations of the phase.

The randomly varying phase component $\theta(t)$ can arise from a variety of sources, including acoustically coupled vibrations of the end mirrors of the laser cavity and noise inherent in the output of any noise-driven nonlinear oscillator. In all cases the phase fluctuations can be interpreted as arising from a random fluctuation of the frequency of the oscillation.

To make these ideas more precise, let the total phase of the oscillating mode be represented by $\psi(t)$,

$$\psi(t) = 2\pi\nu_0 t - \theta(t). \qquad (4.4\text{-}6)$$

The *instantaneous frequency* of the oscillation can then be defined by

$$\nu_i(t) \triangleq \frac{1}{2\pi}\frac{d}{dt}\psi(t) = \nu_0 - \frac{1}{2\pi}\frac{d\theta}{dt} \qquad (4.4\text{-}7)$$

and is seen to consist of a mean ν_0 minus a randomly fluctuating component

$$\nu_R(t) \triangleq \frac{1}{2\pi}\frac{d\theta(t)}{dt}. \qquad (4.4\text{-}8)$$

In most cases of interest, the physical process causing frequency fluctuations can be regarded as generating a zero mean, stationary fluctuation $\nu_R(t)$ of the instantaneous frequency. It follows that

$$\theta(t) = 2\pi \int_{-\infty}^{t} \nu_R(\xi)\, d\xi \qquad (4.4\text{-}9)$$

is a *nonstationary* random process, although the following argument shows it to be stationary in first increments. The structure function of $\theta(t)$ is independent of the time origin, as demonstrated by

$$D_\theta(t_2, t_1) = \overline{[\theta(t_2) - \theta(t_1)]^2}$$

$$= 4\pi^2 \overline{\left\{ \int_{-\infty}^{\infty} \text{rect}\left[\frac{\xi - \left(\frac{t_1 + t_2}{2}\right)}{t_2 - t_1}\right] \nu_R(\xi)\, d\xi \right\}^2}$$

$$= 8\pi^2 \tau \int_0^\tau \left(1 - \frac{\eta}{\tau}\right) \Gamma_\nu(\eta)\, d\eta, \qquad (4.4\text{-}10)$$

where Γ_ν is the autocorrelation function of $\nu_R(t)$, $\tau = t_2 - t_1$, and manipulations similar to those used prior to Eq. (3.4-9) have been carried out. If the delay τ is much longer than the correlation time of $\nu_R(t)$, the structure function becomes

$$D_\theta(\tau) \cong 8\pi^2 \tau \int_0^\infty \Gamma_\nu(\eta)\, d\eta, \tag{4.4-11}$$

or, in words, the mean square phase difference is linearly proportional to the time separation τ. Such a property is also characteristic of a diffusion process and of Brownian motion of a free particle.

As for the probability density functions of the amplitude and intensity of the wave with constant strength and randomly varying phase, they are identical with those of Fig. 4-8, for the phase is again uniformly distributed on the interval $(-\pi, \pi)$ and the intensity remains constant.

A final step in sophistication of the model is to allow the amplitude of the mode to fluctuate randomly in time, as invariably happens in practice to some degree. A solution to the linearized Van der Pol oscillator equation (Ref. 4-13) describing a CW laser oscillator operating well above threshold shows that the emitted wave has a time structure of the form

$$u(t) = S\cos[2\pi\nu_0 t - \theta(t)] + u_n(t), \tag{4.4-12}$$

where S and ν_0 are regarded as known constants, $\theta(t)$ is a randomly time-varying phase of the diffusion-type discussed above, and $u_n(t)$ is a weak stationary noise process, with a spectrum centered at ν_0 and a relatively narrow bandwidth ($\Delta\nu \ll \nu_0$). The strength of the noise component diminishes as the laser operates further and further above threshold.

It can be argued from a physical viewpoint that the first term of (4.4-12) represents the result of stimulated emission, whereas the second term represents a small residual amount of spontaneous emission. In this case it is reasonable to ascribe Gaussian statistics to $u_n(t)$ and to assume that it is independent of $\theta(t)$. At a fixed time t, the first term has a probability density function given by (4.4-3), whereas the second term has a Gaussian density function. For operation well above threshold, the Gaussian function has a standard deviation σ that is much less than S. Therefore, the convolution of the two density functions yields a slightly smoothed version of Fig. 4.8a for the density function of amplitude.

As for the intensity of the mode, we note that it is the squared length of a strong constant-amplitude, random-phase phasor $\mathbf{S}$ plus a weak circular complex Gaussian phasor $\mathbf{A}_n$ representing the complex envelope of the Gaussian noise term. The probability density function of I can be found by noting that

$$I = |\mathbf{S} + \mathbf{A}_n|^2 \cong |\mathbf{S}|^2 + 2\,\text{Re}\{\mathbf{S}^*\mathbf{A}_n\}. \quad (4.4\text{-}13)$$

Now

$$\mathbf{S} = Se^{j\theta}, \qquad \mathbf{A}_n = A_n e^{j\phi_n}$$

where A_n, θ, and ϕ_n are independent and θ and ϕ_n are uniformly distributed on $(-\pi, \pi)$. The real part of $2\mathbf{S}^*\mathbf{A}_n$ is a Gaussian random variable,[†] with zero mean and variance

$$\sigma_I^2 = 4S^2\overline{A_n^2}\overline{\cos^2(\theta - \phi_n)} = 4I_S\bar{I}_N \cdot \tfrac{1}{2}$$

$$= 2I_S\bar{I}_N. \quad (4.4\text{-}14)$$

We conclude that the intensity I obeys (approximately) a Gaussian density function

$$p_I(I) \cong \frac{1}{\sqrt{4\pi I_S \bar{I}_N}} \exp\left\{-\frac{(I - I_S)^2}{4I_S\bar{I}_N}\right\} \quad (4.4\text{-}15)$$

valid for $I_S \gg \bar{I}_N$.

An alternative solution for the probability density function of the intensity of a laser operating above or below threshold has been found by Risken (Ref. 4-14), who solved a nonlinear Fokker–Planck equation to obtain the probability density function directly. The result is a density function of the form

$$p_I(I) = \begin{cases} \dfrac{2}{\pi I_0}\dfrac{1}{1 + \text{erf}\,w}\exp\left\{-\left(\dfrac{I}{\sqrt{\pi}\,I_0} - w\right)^2\right\} & I \geq 0 \\ 0 & \text{otherwise,} \end{cases}$$

$$(4.4\text{-}16)$$

where I_0 is the average intensity at threshold; w is a parameter that varies from large negative values well below threshold, to zero at threshold, to

---

[†] Note that $\text{Re}\{S^*A_n\} = SA_n\cos(\phi_n - \theta)$. Since ϕ_n is uniformly distributed and A_n is Rayleigh distributed, the resulting product obeys Gaussian statistics.

large positive values well above threshold; and erf w is a standard error function,

$$\operatorname{erf} w = \frac{2}{\sqrt{\pi}} \int_0^w \exp(-x^2)\, dx, \qquad \operatorname{erf}(-w) = -\operatorname{erf} w. \qquad (4.4\text{-}17)$$

The average intensity of the laser output is related to the average intensity at threshold by

$$\bar{I} = I_0 \left[\sqrt{\pi}\, w + \frac{e^{-w^2}}{1 + \operatorname{erf} w} \right]. \qquad (4.4\text{-}18)$$

When $w \ll 0$, the laser is well below threshold, and $p_I(I)$ is approximately negative exponential, as for thermal light,

$$p_I(I) \cong \begin{cases} \dfrac{2|w|}{\sqrt{\pi}\, I_0} \exp\left\{ -\dfrac{2|w|}{\sqrt{\pi}\, I_0} I \right\} & I \geq 0 \\ 0 & I < 0. \end{cases} \qquad (4.4\text{-}19)$$

When $w = 0$, the laser is just at threshold, and $p_I(I)$ has the shape of half of a Gaussian curve,

$$p_I(I) = \begin{cases} \dfrac{2}{\pi I_0} \exp\left\{ -\dfrac{I^2}{\pi I_0^2} \right\} & I \geq 0 \\ 0 & I < 0. \end{cases} \qquad (4.4\text{-}20)$$

Finally, in the most common case of a laser far above threshold, $w \gg 0$, and $p_I(I)$ has the form of a Gaussian density with mean $\bar{I} = w\sqrt{\pi}\, I_0$,

$$p_I(I) \cong \begin{cases} \dfrac{1}{\pi I_0} \exp\left\{ -\left(\dfrac{I - w\sqrt{\pi}\, I_0}{\sqrt{\pi}\, I_0} \right)^2 \right\} & I \geq 0 \\ 0 & I < 0. \end{cases} \qquad (4.4\text{-}21)$$

Recall that the previous approximation (4.4-15) predicted a similar result, which suggests the association

$$\begin{aligned} I_S &\cong w\sqrt{\pi}\, I_0 \\ \bar{I}_N &\cong \frac{\sqrt{\pi}\, I_0}{4w} \end{aligned} \qquad (w \gg 0). \qquad (4.4\text{-}22)$$

LASER LIGHT

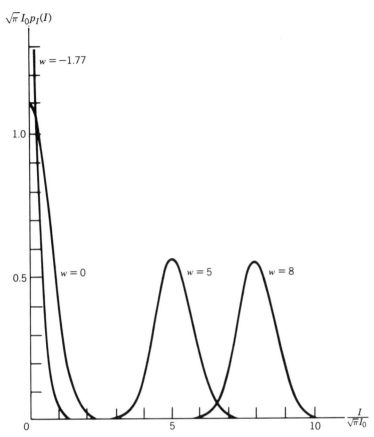

Figure 4-9. Risken's solution for the probability density function of the intensity of a laser oscillator (single mode).

The general probability density function of Eq. (4.4-16) is plotted for several values of w in Fig. 4-9.

In our future discussions it is often convenient to assume that the laser is operating so far above threshold that fluctuations of intensity are insignificant. Thus the randomly phase-modulated cosine of Eq. (4.4-5) is most commonly used to represent the light from a single-mode laser.

4.4.2 Multimode Laser Light

Whereas single-mode oscillation can be achieved with some lasers if special precautions are taken, lasers are more commonly found to oscillate in a multitude of transverse and/or longitudinal modes. Assuming that the laser is oscillating well above threshold, a reasonable model for the steady-state output is

$$u(t) = \sum_{i=1}^{N} S_i \cos[2\pi \nu_i t - \theta_i(t)], \qquad (4.4\text{-}23)$$

where N is the number of modes, S_i and ν_i are the amplitude and center frequency of the ith mode, and $\theta_i(t)$ is a time varying phase associated with that mode.

The most commonly used model for multimode laser light assumes that the modes oscillate *independently*, with no appreciable degree of phase locking. Such a model must be used with great caution, however. If the phase fluctuations are caused by vibrations of the end mirrors of the laser, then clearly the fluctuations of the various modes will be statistically dependent. Furthermore, even if the phase fluctuations arise as an integral part of the oscillation mechanism, the laser is fundamentally a nonlinear device, and significant mode coupling can occur as a result of these nonlinearities. For example, if a frequency component generated by nonlinear intermodulation between two modes happens to coincide with the frequency of a third mode, some degree of phase locking can occur. Such effects are particularly strong for a laser operating well above threshold, where the nonlinearities are most significant. (For a review of techniques for intentionally introducing mode locking in lasers, see, for example, Ref. 4-15).

Recognizing that the model is not valid under many conditions, we nonetheless investigate the properties of light emitted by a laser oscillating in several *independent* modes. A reasonable approximation to this condition can be obtained for a gas laser oscillating just above threshold, although strictly speaking, a spontaneous emission Gaussian noise term should be added to the model (4.4-23) in this case. (However, it should be noted that just above threshold the laser may well oscillate in only one or two modes).

The characteristic function of the amplitude of a single mode is given, according to (4.4-2), by

$$\mathbf{M}_i(\omega) = J_0(\omega S_i). \qquad (4.4\text{-}24)$$

For N independent modes, the characteristic function is

$$\mathbf{M}_U(\omega) = \prod_{i=1}^{N} J_0(\omega S_i), \qquad (4.4\text{-}25)$$

and if all modes have equal amplitudes $\sqrt{\bar{I}/N}$, the result is

$$\mathbf{M}_U(\omega) = J_0^N\left(\omega\sqrt{\frac{\bar{I}}{N}}\right). \tag{4.4-26}$$

To obtain the probability density function, the characteristic function must be Fourier transformed.

Hodara (Ref. 4-16) and Mandel (Ref. 4-17) have shown that for two equal strength modes, the density function for amplitude is

$$p_U(u) = \begin{cases} \dfrac{1}{\pi^2}\sqrt{\dfrac{2}{\bar{I}}}\, K\!\left(\sqrt{1-\dfrac{u^2}{2\bar{I}}}\right) & |u| < \sqrt{2\bar{I}} \\ 0 & \text{otherwise,} \end{cases} \tag{4.4-27}$$

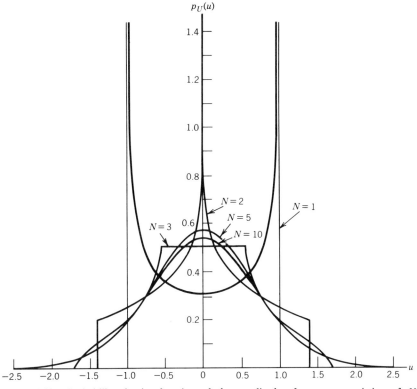

Figure 4-10. Probability density function of the amplitude of a wave consisting of N equal-strength independent modes. The total intensity $\bar{I}$ is held constant and equal to unity. A Gaussian curve is indistinguishable from the $N = 10$ curve on this plot.

where $K(\cdot)$ is a complete elliptic integral of the first kind. For our purposes, we simply subject Eq. (4.4-26) to a digital Fourier transformation and plot the curves of $p_U(u)$ for various values of N in Fig. 4-10.

As more and more independent modes are added, the probability density function is seen to approach Gaussian form, as expected in accordance with the central limit theorem. For N as small as 5, there is little visible difference between the true density function and a Gaussian function. From the point of view of classical, first-order statistics, there is little difference between multimode laser light ($N \geq 5$) and thermal light, provided the major assumption of no phase locking is satisfied.

As for the probability density function of the intensity of multimode laser light, the problem is even more difficult than for amplitude. We consider first the case of two independent modes, with intensities $k\bar{I}$ and $(1-k)\bar{I}$. Reference to Fig. 4-11 and the law of cosines shows that the total instantaneous intensity can be expressed as

$$I = k\bar{I} + (1-k)\bar{I} + 2\sqrt{k(1-k)}\,\bar{I}\cos\psi = \bar{I}\left[1 + 2\sqrt{k(1-k)}\,\cos\psi\right], \tag{4.4-28}$$

where

$$\psi = 2\pi(\nu_2 - \nu_1)t - \theta_2(t) + \theta_1(t). \tag{4.4-29}$$

As a result of the uniform distributions of θ_2 and θ_1 and their assumed statistical independence, ψ is uniformly distributed on $(-\pi, \pi)$. Thus the

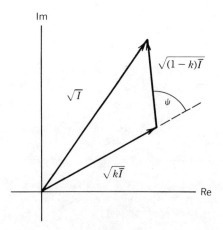

Figure 4-11. Phasor diagram to aid in the computation of I.

characteristic function of the instantaneous intensity is

$$\mathbf{M}_I(\omega) = \frac{1}{2\pi} \int_{-\pi}^{\pi} \exp\{j\omega\bar{I}[1 + 2\sqrt{k(1-k)}\cos\psi]\}\,d\psi$$

$$= \exp(j\omega\bar{I})\,J_0\!\left(2\omega\bar{I}\sqrt{k(1-k)}\right). \qquad (4.4\text{-}30)$$

Fourier inversion yields a probability density function

$$p_I(I) = \begin{cases} \left[\pi\sqrt{\left(2\bar{I}\sqrt{k(1-k)}\right)^2 - (I-\bar{I})^2}\right]^{-1} & \text{for} \quad \begin{array}{c} \bar{I}\left[1 - 2\sqrt{k(1-k)}\right] \\ < I < \\ \bar{I}\left[1 + 2\sqrt{k(1-k)}\right] \end{array} \\ 0 & \text{otherwise.} \end{cases}$$

(4.4-31)

This density function is shown plotted in Fig. 4-12 for various values of k.

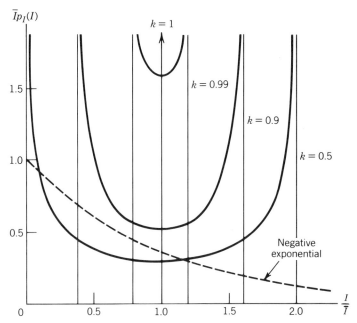

Figure 4-12. Probability density function of the instantaneous intensity of two independent modes with fraction k of the total intensity in one mode and $(1-k)$ in the other.

Also shown dotted is the negative exponential distribution associated with the intensity fluctuations of thermal light. This curve is approached as the number of independent modes is increased.

Although the density function for I with more than two modes present is not readily found, it is possible to calculate the standard deviation of the intensity with N equal strength independent modes present and to compare it with the standard deviation of I for thermal light. The reader is asked to verify in Problem 4-11 that the ratio of standard deviation to mean intensity with N equal strength independent modes satisfies the equation

$$\frac{\sigma_I}{\bar{I}} = \sqrt{1 - \frac{1}{N}}. \qquad (4.4\text{-}32)$$

This dependence on N is illustrated in Fig. 4-13. Note that as N increases, the ratio $\sigma_I/\bar{I}$ approaches the value unity characteristic of polarized thermal light. When more than five independent modes are present, the ratio is within 10% of the value appropriate for thermal light. Thus we again see that the approach towards "pseudothermal" light is very rapid with increasing number of modes.

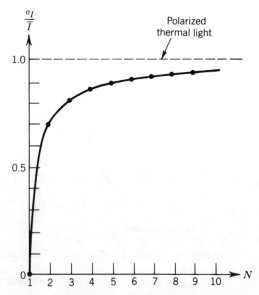

Figure 4-13. Ratio of standard deviation σ_I to mean $\bar{I}$ for the intensity of light emitted by a laser oscillating in N independent, equal-strength modes.

LASER LIGHT

Finally, we emphasize again that the resemblance of multimode laser light to thermal light is true only when the various oscillating modes are uncoupled. In practice, the situations in which this assumption is satisfied are probably rather limited.

4.4.3 Pseudothermal Light Produced by Passing Laser Light Through a Moving Diffuser

A light wave having first-order classical statistical properties indistinguishable from polarized thermal light can be produced by passing laser light (single mode or multimode) through a moving diffuser. Such light differs from thermal light primarily through the much greater energy it possesses per temporal fluctuation interval (or "correlation time"); this point is treated in more detail in Chapter 9.

Figure 4-14 illustrates the experimental arrangement for producing pseudothermal light of this type. A laser illuminates a diffuser, such as ground glass. On a very fine spatial scale, the diffuser introduces extremely complex and irregular deformations of the incident wavefront, with phase changes generally many times 2π radians. At a distant point P_0 the light may be regarded as consisting of many independent contributions from different "correlation areas" on the diffuser, where the diffuser is regarded as one particular realization drawn from an ensemble of possible diffusers. These contributions are randomly phased, and hence the complex field observed may be regarded as resulting from a random phasor sum. The field thus obeys complex Gaussian statistics, and the intensity obeys negative exponential statistics over an ensemble of microscopically dissimilar diffusers, as it has been the assumed that negligible depolarization of the light has occurred.

If the diffuser is now moved continuously, the field and intensity fluctuate with time, taking on many independent realizations of the underlying

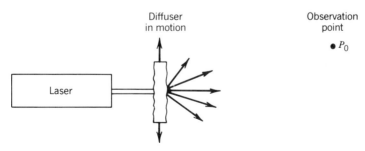

Figure 4-14. Production of pseudothermal light from laser light by means of a moving diffuser.

statistical distribution. Thus the light intensity fluctuates randomly in time, obeying negative exponential statistics as for polarized thermal light, but with a far narrower bandwidth than true thermal light. For a further detailed discussion of the relationship between this kind of light and ordinary thermal light, see Ref. 4-18.

REFERENCES

4-1 R. J. Glauber, "Photon Statistics," in *Laser Handbook*, Vol. 1 (F. T. Arecchi and E. O. Schulz-Dubois, editors), North-Holland Publishing Company, Amsterdam, pp. 1–43 (1972).

4-2 S. Silver, *J. Opt. Soc. Am.*, **52**, 131 (1962).

4-3 M. Born and E. Wolf, *Principles of Optics*, 2nd rev. ed., MacMillan Company, New York (1964).

4-4 J. W. Goodman, *Introduction to Fourier Optics*, McGraw-Hill Book Company, New York (1968).

4-5 M. V. Klein, *Optics*, John Wiley & Sons, New York (1970).

4-6 R. C. Jones, *J. Opt. Soc. Am.*, **31**, 488 (1941); **31**, 500 (1941); **32**, 486 (1942); **37**, 107 (1947); **37**, 110 (1947); **38**, 671 (1948).

4-7 E. Wolf, *Nuovo Cimento*, **13**, 1165 (1959).

4-8 N. Wiener, *J. Math. Phys.* (*MIT*), **7**, 109 (1927–1928).

4-9 D. C. Murdoch, *Linear Algebra for Undergraduates*, John Wiley & Sons, New York (1957).

4-10 G. B. Parrent and P. Roman, *Nuovo Cimento*, **15**, 370 (1960).

4-11 E. L. O'Neill, *Introduction to Statistical Optics*, Addison-Wesley, Reading, MA (1963).

4-12 R. M. Bracewell, *The Fourier Transform and its Applications*, McGraw-Hill Book Company, New York (1965).

4-13 J. A. Armstrong and A. W. Smith, "Experimental Studies of Intensity Fluctuations in Lasers," *Progress in Optics*, Vol. VI (E. Wolf, editor), North-Holland Publishing Company, Amsterdam, pp. 211–257 (1967).

4-14 H. Risken, "Statistical Properties of Laser Light," *Progress in Optics*, Vol. VIII (E. Wolf, editor), North-Holland Publishing Company, Amsterdam, pp. 239–294 (1970).

4-15 L. Allen and D. G. C. Jones, "Mode Locking in Gas Lasers," *Progress in Optics*, Vol. IX (E. Wolf, editor), North-Holland Publishing Company, Amsterdam, pp. 181–233 (1971).

4-16 H. Hodara, *IEEE Wescon Proceedings*, paper 17.4 (1964).

4-17 L. Mandel, *Phys. Rev.*, **138**, B753 (1965).

4-18 W. Martienssen and E. Spiller, *Am. J. Phys.*, **32**, 919 (1964).

ADDITIONAL READING

B. Daino, P. Spano, M. Tamburrini, and S. Piazzolla, "Phase Noise and Spectral Line Shape in Semiconductor Lasers," *IEEE J. Quant. Electron.*, **QE-19**, 266–270 (1983).

PROBLEMS

4-1 Starting with Eq. (4.1-10), show that if $\Delta\nu \ll \bar{\nu}$ and $r \ll c/\Delta\nu$ for all P_1, then

$$\mathbf{u}(P_0, t) \cong \iint_\Sigma \frac{e^{j2\pi(r/\bar{\lambda})}}{j\bar{\lambda}r} \mathbf{u}(P_1, t)\chi(\theta)\, dS$$

can be used to describe the propagation of $\mathbf{u}(P, t)$.

4-2 Show that the characteristic function of the intensity of polarized thermal light is given by

$$\mathbf{M}_I(\omega) = \frac{1}{1 - j\omega\bar{I}}.$$

4-3 Show that the coherency matrix of natural light is unaffected by any unitary polarization transformation.

4-4 By finding the trace of the transformed coherency matrix, show that the intensity transmitted by a polarization analyzer set at $+45°$ to the X axis can be expressed as

$$I = \tfrac{1}{2}\left[\mathbf{J}_{xx} + \mathbf{J}_{yy}\right] + \operatorname{Re}\{\mathbf{J}_{xy}\},$$

where $\mathbf{J}_{xx}$, $\mathbf{J}_{yy}$, and $\mathbf{J}_{xy}$ are elements of the coherency matrix of the *incident* light.

4-5 By finding the trace of the transformed coherency matrix, show that the intensity transmitted by a quarter-wave plate followed by a polarization analyzer set at $+45°$ to the X axis can be expressed as

$$I = \tfrac{1}{2}\left[\mathbf{J}_{xx} + \mathbf{J}_{yy}\right] + \operatorname{Im}\{\mathbf{J}_{xy}\}$$

where $\mathbf{J}_{xx}$, $\mathbf{J}_{yy}$, and $\mathbf{J}_{xy}$ are again elements of the coherency matrix of the *incident* light, and it has been assumed that the quarter-wave plate delays $\mathbf{u}_Y$ with respect to $\mathbf{u}_X$ by $90°$.

4-6 Consider a light wave that has X- and Y-polarization components of its electric field at point P given by

$$\mathbf{u}_X(t) = \exp\left[-j2\pi\left(\bar{\nu} - \frac{\Delta\nu}{2}\right)t\right]$$

$$\mathbf{u}_Y(t) = \exp\left[-j2\pi\left(\bar{\nu} + \frac{\Delta\nu}{2}\right)t\right]$$

(a) Show that at time t the electric vector makes an angle

$$\theta(t) = \tan^{-1}\left\{\frac{\cos 2\pi\left(\bar{\nu} + \dfrac{\Delta\nu}{2}\right)t}{\cos 2\pi\left(\bar{\nu} - \dfrac{\Delta\nu}{2}\right)t}\right\}$$

with respect to the X axis, and thus the polarization direction is entirely deterministic.

(b) Show that such light has a coherency matrix that is identical with that of natural light, for which the polarization direction is entirely random.

4-7 Show that the standard deviation σ_I of the instantaneous intensity of partially polarized thermal light is

$$\sigma_I = \sqrt{\frac{1 + \mathscr{P}^2}{2}}\,\bar{I}$$

as asserted in Eq. (4.3-43).

4-8 Consider the analytic signal representation of a monochromatic signal

$$\mathbf{u}(t) = S\exp[-j(2\pi\nu_0 t - \phi)],$$

where S and ν_0 are known constants, whereas ϕ is a random variable uniformly distributed on $(-\pi, \pi)$. Let

$$u^{(r)}(t) = \text{Re}\{\mathbf{u}(t)\} = S\cos[2\pi\nu_0 t - \phi]$$

$$u^{(i)}(t) = \text{Im}\{\mathbf{u}(t)\} = -S\sin[2\pi\nu_0 t - \phi]$$

(a) Show that the conditional density function of $u^{(i)}$, given $u^{(r)}$, is

$$p_{i|r}(u^{(i)}|u^{(r)}) = \sqrt{S^2 - (u^{(r)})^2}\,\delta\!\left[(u^{(i)})^2 + (u^{(r)})^2 - S^2\right]$$

(b) Show that the joint density function $p(u^{(r)}, u^{(i)})$ is given by

$$p(u^{(r)}, u^{(i)}) = \frac{1}{\pi}\delta\!\left[(u^{(r)})^2 + (u^{(i)})^2 - S^2\right]$$

PROBLEMS

(c) Show that $u^{(i)}$ obeys the same probability density function as $u^{(r)}$, that is,

$$p_i(u^{(i)}) = \frac{1}{\pi\sqrt{S^2 - (u^{(i)})^2}} \quad (|u^{(i)}| < S)$$

(d) Show that, whereas $E[u^{(r)}u^{(i)}] = 0$, $u^{(r)}$ and $u^{(i)}$ are not independent.

Hint:

$$\delta[f(x)] = \sum_{\substack{\text{all} \\ \text{roots} \\ x_n \text{ of} \\ f(x)}} \frac{\delta(x - x_n)}{\left|\frac{df}{dx}\right|_{x=x_n}}$$

(see Ref. 2-4, pp. 37 and 38).

4-9 Present an argument demonstrating that thermal light remains thermal light after propagation to a distant observation point, but that laser light may or may not retain the form

$$\mathbf{u}(t) = S\exp\{-j[2\pi\nu_0 t - \theta(t)]\}.$$

4-10 Consider a single-mode laser emitting light described by the analytic signal

$$\mathbf{u}(t) = \exp\{-j[2\pi\nu_0 t - \theta(t)]\}.$$

(a) Assuming that $\Delta\theta(t)$ is an ergodic random process, show that the autocorrelation function of $\mathbf{u}(t)$ is given by

$$\Gamma_U(t_2, t_1) = e^{-j2\pi\nu_0\tau} \mathbf{M}_{\Delta\theta}(1),$$

where $\mathbf{M}_{\Delta\theta}(\omega)$ is the characteristic function of the phase difference $\Delta\theta = \theta(t_2) - \theta(t_1)$.

(b) Show that for a zero mean Gaussian $\theta(t)$, arising from a stationary instantaneous frequency process,

$$\Gamma_U(\tau) = e^{-j2\pi\nu_0\tau} e^{-(1/2)D_\theta(\tau)},$$

where $D_\theta(\tau)$ is the structure function of the phase process $\theta(t)$.

4-11 Let the field emitted by a laser oscillating in N equal-strength but independent modes be represented by

$$\mathbf{u}(t) = \sum_{k=1}^{N} \exp\{-j(2\pi\bar{\nu}_k t - \phi_k)\},$$

where the ϕ_k are uniformly distributed on $(-\pi, \pi)$ and are statistically independent. Find an expression for the ratio of the standard deviation of intensity σ_I to the mean intensity $\bar{I}$, expressing the result as a function of N.

4-12 Show that the Jones matrix of a polarization analyzer set at angle $+\alpha$ to the x axis is given by

$$\underline{\mathbf{L}}(\alpha) = \begin{bmatrix} \cos^2\alpha & \sin\alpha\cos\alpha \\ \sin\alpha\cos\alpha & \sin^2\alpha \end{bmatrix}$$

Is this matrix unitary?

4-13 Show that the second moment $\overline{I^2}$ of the intensity of a wave is *not* equal to the fourth moment $\overline{[u^{(r)}]^4}$ of the real amplitude of that wave, the difference being due to a low-pass filtering operation (and a scaling by a factor of 2) that are implicit in the definition of intensity.

5

Coherence of Optical Waves

The statistical properties of light play an important role in determining the outcome of most optical experiments. In many cases of practical importance, however, a satisfactory description of the experiment can be formed with far less than a complete statistical model. Most commonly, a description in terms of certain second-order averages known as *coherence functions* is entirely adequate for predicting experimental outcomes. Attention is focused in this chapter on the properties of such second-order averages.

The origins of the modern concept of coherence can be found in the scientific literature of the late nineteenth and early twentieth centuries. Particularly noteworthy early contributions were made by E. Verdet (Ref. 5-1), M. vonLaue (Ref. 5-2), M. Berek (Ref. 5-3), P. H. van Cittert (Ref. 5-4), F. Zernike (Ref. 5-5), and others. In more recent times, developments of major importance are found in the work of H. H. Hopkins (Ref. 5-6), A. Blanc-Lapierre and P. Dumontet (Ref. 5-7), and E. Wolf (Ref. 5-8). These few references are far from a complete list of important advances, but fortunately the interested reader can easily trace the historical evolution of these ideas with the help of two volumes of reprints of original papers, together with an extensive bibliography, available under the editorship of L. Mandel and E. Wolf (Ref. 5-9).

Before proceeding with detailed discussions, it is perhaps worth briefly mentioning the distinction between two types of coherence, *temporal* coherence and *spatial* coherence. When considering temporal coherence, we are concerned with the ability of a light beam to interfere with a delayed (but not spatially shifted) version of itself. We refer to such division of a light beam as *amplitude splitting*. On the other hand, when considering spatial coherence we are concerned with the ability of a light beam to interfere with a spatially shifted (but not delayed) version of itself. We refer to this type of division of light as *wavefront splitting*. Clearly, the ideas can be generalized to allow both temporal and spatial shifting, which will lead us to the concept of the mutual coherence function. The type of coherence that is needed in any particular case depends on the particular experiment we are attempting to understand on an analytical basis. These ideas are developed in greater detail in the sections that follow.

For alternative discussions of much of the material covered here, the reader can consult Refs. 5-10 through 5-14.

5.1 TEMPORAL COHERENCE

Let $\mathbf{u}(P, t)$ be the complex scalar representation of an optical disturbance at point P in space and instant t in time. Associated with $\mathbf{u}(P, t)$ is a complex envelope $\mathbf{A}(P, t)$. Since $\mathbf{u}(P, t)$ has a finite bandwidth $\Delta \nu$, we expect the amplitude and phase of $\mathbf{A}(P, t)$ to be changing at a rate determined by $\Delta \nu$. If a finite time duration τ is of interest, we expect $\mathbf{A}(P, t)$ to remain relatively constant during the interval τ provided $\tau \ll 1/\Delta \nu$. In other words, the time functions $\mathbf{A}(P, t)$ and $\mathbf{A}(P, t + \tau)$ are highly correlated, or *coherent*, provided τ is much less than the "coherence time" $\tau_c \cong 1/\Delta \nu$.

The concept of temporal coherence can be given a more precise definition and description by considering the interference of light waves in an interferometer first introduced by Michelson (Ref. 5-15).

5.1.1 The Michelson Interferometer

Consider the interferometer illustrated in Fig. 5-1. Light from a point source S is collimated (i.e., the rays are made parallel) by the lens L_1 and falls on the beam splitter (a partially reflecting mirror) BS. A portion of the incident light is reflected and passes to the moveable mirror M_1. This light is reflected from M_1, is again incident on the beam splitter, and a portion is again transmitted, this time to the lens L_2, which brings the rays to a focus on detector D.

Simultaneously, a portion of the original light from S is transmitted by the beam splitter, passes through the compensating plate C, is incident on and reflected from the fixed mirror M, and again passes through the compensating plate. A portion of this light is reflected from the beam splitter and finally is focused on the detector D by lens L_2. Thus the intensity of the light incident on the detector is determined by interference of the light from the two arms of the interferometer.

The compensator C serves the purpose of assuring that the light in both arms of the interferometer travels the same distance *in glass*, thus guaranteeing that both beams have suffered the same dispersion in passage from the source S to the detector D.

If the mirror M_1 is moved from the position required for equal pathlengths in the two arms of the interferometer, a relative time delay is

TEMPORAL COHERENCE

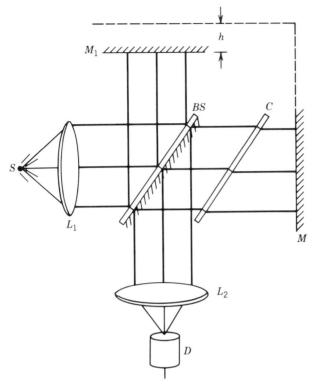

Figure 5-1. The Michelson interferometer, including the point source S, the lenses L_1 and L_2, mirrors M_1 and M, beam splitter BS, compensator C, and detector D.

introduced between the two interfering beams. As the mirror moves, the light falling on the detector passes from a state of constructive interference to a state of destructive interference and back to constructive interference, with a mirror movement of $\bar{\lambda}/2$ (a pathlength difference of $\bar{\lambda}$) between bright fringes. Superimposed on this rapid oscillation of intensity is a gradually tapering envelope of fringe modulation, caused by the finite bandwidth of the source and the gradual decorrelation of the complex envelope of the light as the pathlength difference increases. A typical pattern of interference is shown in Fig. 5-2, with intensity plotted against mirror displacement h from the position of equal pathlengths. Such a display of intensity vs. pathlength difference is referred to as an *interferogram*.

The general behavior of the interferogram can be explained in simple physical terms. The extended spectrum of the source can be regarded as consisting of many monochromatic components. Each such component generates a perfectly periodic contribution to the interferogram, but with a

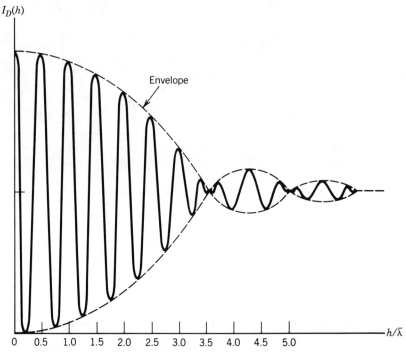

Figure 5-2. Intensity incident on the detector D versus normalized mirror displacement $h/\bar{\lambda}$, where $\bar{\lambda}$ is the mean wavelength. The envelope of the fringe pattern is drawn dotted.

period depending on its particular optical frequency. At zero pathlength difference ($h = 0$), all such components add in phase, producing a large central peak in the interferogram. As the mirror is displaced from the zero-delay position, each monochromatic fringe suffers a phase shift that depends on its particular temporal frequency. The result is a partially destructive addition of the elementary fringes and a consequent drop in the fringe depth on the interferogram. When the relative delay grows large enough, the addition of elementary fringes is nearly totally destructive, and the interferogram remains at its constant average value.

It is evident from the preceding discussions that the drop in the fringe depth of the interferogram can be explained in either of two equivalent ways, in terms of a "dephasing" of elementary fringes or in terms of a loss of correlation due to the finite pathlength delay. The role of the autocorrelation function of the light beam will become more evident in the simple analysis that follows.

5.1.2 Mathematical Description of the Experiment

The response of the detector D is governed by the *intensity* of the optical wave falling on its surface. For virtually all applications involving true thermal light, the detector may be assumed to average over a time duration that is infinitely long. (Effects of finite averaging time, which can be important with pseudothermal light, are treated in Section 6.2.) Taking account of the relative time delay $2h/c$ suffered by the light in the arm with the moveable mirror, the intensity incident on the detector can be written as

$$I_D(h) = \left\langle \left| K_1 \mathbf{u}(t) + K_2 \mathbf{u}\left(t + \frac{2h}{c}\right) \right|^2 \right\rangle, \qquad (5.1\text{-}1)$$

where K_1 and K_2 are real numbers determined by the losses in the two paths and $\mathbf{u}(t)$ is the analytic signal representation of the light emitted by the source. Expanding this expression, we find

$$I_D(h) = K_1^2 \langle |\mathbf{u}(t)|^2 \rangle + K_2^2 \left\langle \left| \mathbf{u}\left(t + \frac{2h}{c}\right) \right|^2 \right\rangle$$

$$+ K_1 K_2 \left\langle \mathbf{u}\left(t + \frac{2h}{c}\right) \mathbf{u}^*(t) \right\rangle$$

$$+ K_1 K_2 \left\langle \mathbf{u}^*\left(t + \frac{2h}{c}\right) \mathbf{u}(t) \right\rangle. \qquad (5.1\text{-}2)$$

Thus the important role played by the autocorrelation function of the light wave in determining the observed intensity becomes evident.

Because of the fundamental role played by the time averages in (5.1-2), special symbols are adopted for them. In particular we use the notation

$$I_0 \triangleq \langle |\mathbf{u}(t)|^2 \rangle = \left\langle \left| \mathbf{u}\left(t + \frac{2h}{c}\right) \right|^2 \right\rangle \qquad (5.1\text{-}3)$$

and

$$\Gamma(\tau) = \langle \mathbf{u}(t + \tau) \mathbf{u}^*(t) \rangle. \qquad (5.1\text{-}4)$$

The function $\Gamma(\tau)$, which is the autocorrelation function of the analytic signal $\mathbf{u}(t)$, is known as the *self coherence function* of the optical disturbance. In this abbreviated notation we write the detected intensity as

$$I_D = (K_1^2 + K_2^2)I_0 + K_1 K_2 \Gamma\left(\frac{2h}{c}\right) + K_1 K_2 \Gamma^*\left(\frac{2h}{c}\right)$$

$$= (K_1^2 + K_2^2)I_0 + 2K_1 K_2 \operatorname{Re}\left\{\Gamma\left(\frac{2h}{c}\right)\right\}. \qquad (5.1\text{-}5)$$

In many cases it is convenient to work with a normalized version of the self coherence function, rather than the self coherence function itself. Noting that $I_0 = \Gamma(0)$, we choose to normalize by this quantity, yielding

$$\gamma(\tau) = \frac{\Gamma(\tau)}{\Gamma(0)}, \qquad (5.1\text{-}6)$$

which is known as the *complex degree of coherence* of the light. We note for future reference the important properties

$$\gamma(0) = 1 \quad \text{and} \quad |\gamma(\tau)| \leq 1 \qquad (5.1\text{-}7)$$

[cf. Eq. (3.4-5)]. In terms of this quantity, the detector intensity is given by

$$I_D(h) = (K_1^2 + K_2^2)I_0 \left[1 + \frac{2K_1 K_2}{K_1^2 + K_2^2}\operatorname{Re}\left\{\gamma\left(\frac{2h}{c}\right)\right\}\right]. \qquad (5.1\text{-}8)$$

With the goal of reaching an analytic expression that clearly describes an interferogram of the type depicted in Fig. 5-2, we express the complex degree of coherence in the following general form:

$$\gamma(\tau) = \gamma(\tau)\exp\{-j[2\pi\bar{\nu}\tau - \alpha(\tau)]\}, \qquad (5.1\text{-}9)$$

where $\gamma(\tau) = |\gamma(\tau)|$, $\bar{\nu}$ is the center frequency of the light and $\alpha(\tau) \triangleq \arg\{\gamma(\tau)\} + 2\pi\bar{\nu}\tau$. Using this expression, assuming equal losses in the two arms of the interferometer ($K_1 = K_2 = K$), and noting that $\bar{\nu}/c = 1/\bar{\lambda}$, we can express the interferogram in the form

$$I_D(h) = 2K^2 I_0\left\{1 + \gamma\left(\frac{2h}{c}\right)\cos\left[2\pi\left(\frac{2h}{\bar{\lambda}}\right) - \alpha\left(\frac{2h}{c}\right)\right]\right\}. \qquad (5.1\text{-}10)$$

The expression (5.1-10) can now be compared with Fig. 5-2, which was asserted to be typical of the structure of the interferogram. In the vicinity of zero relative pathlength difference ($h \cong 0$), we have $\gamma(2h/c) \cong 1$ and

$\alpha(2h/c) \cong 0$ from Eq. (5.1-7). Thus near the origin, the interferogram consists of a fully modulated cosine, with intensity varying from $4K^2 I_0$ to zero about a mean level $2K^2 I_0$. As the pathlength difference h is increased, the amplitude modulation $\gamma(2h/c)$ falls from unity towards zero, and in addition the fringes may suffer a phase modulation $\alpha(2h/c)$, depending on the nature of the spectrum of the light.

The depth of the fringes observed in the vicinity of any pathlength difference h can be described in precise terms using the concept of *fringe visibility* first introduced by Michelson. The visibility of a sinusoidal fringe pattern is defined by

$$\mathcal{V} \triangleq \frac{I_{\max} - I_{\min}}{I_{\max} + I_{\min}} \qquad (5.1\text{-}11)$$

where $I_{\max}$ and $I_{\min}$ are the intensities at the maximum and minimum of the fringe. In the near vicinity of mirror displacement h, the interferogram in Eq. (5.1-10) can be seen to have a visibility

$$\mathcal{V}(h) = \left| \gamma\!\left(\frac{2h}{c}\right) \right| = \gamma\!\left(\frac{2h}{c}\right) \qquad (5.1\text{-}12)$$

when losses in the two arms are equal. The reader can readily show that for unequal losses, the visibility is

$$\mathcal{V}(h) = \frac{2 K_1 K_2}{K_1^2 + K_2^2} \gamma\!\left(\frac{2h}{c}\right). \qquad (5.1\text{-}13)$$

As the pathlength difference $2h$ grows large, the visibility of the fringes drops, and we say that the relative coherence of the two beams has diminished. When the visibility has fallen to approximately zero, we say that the pathlength difference has exceeded the *coherence length* of the light, or equivalently, that the relative time delay has exceeded the *coherence time*.

Clearly, then, *the concept of temporal coherence has to do with the ability of two relatively delayed light beams to form fringes*. Note that all the preceding definitions have utilized time averages. If the random processes of concern are ergodic, ensemble averages could be used instead. In addition, there are some cases in which we must deal with nonergodic wavefields and for which we use exclusively ensemble averages (see Section 7.5.2). In the next section we explore in more detail the relation of the interferogram to the power spectral density of the light beam.

5.1.3 Relationship of the Interferogram to the Power Spectral Density of the Light Beam

As we have seen, the character of the interferogram obtained from a Michelson interferometer is determined by the self coherence function $\Gamma(\tau)$, or equivalently by the complex degree of coherence $\gamma(\tau)$, of the light emitted by the source. In addition, we know from Section 3.4 that, for a stationary random process, an intimate relationship exists between these correlation functions and the power spectral density of the source. In particular, from Eq. (3.8-34) we have

$$\Gamma(\tau) = \int_0^\infty 4\mathscr{G}^{(r,r)}(\nu)e^{-j2\pi\nu\tau}\,d\nu, \qquad (5.1\text{-}14)$$

where $\mathscr{G}^{(r,r)}(\nu)$ is the power spectral density of the real-valued optical disturbance $u^{(r)}(t)$. Equivalently, we can express the complex degree of coherence $\gamma(\tau)$ in terms of $\mathscr{G}^{(r,r)}(\nu)$ by

$$\gamma(\tau) = \frac{\int_0^\infty 4\mathscr{G}^{(r,r)}(\nu)e^{-j2\pi\nu\tau}\,d\nu}{\int_0^\infty 4\mathscr{G}^{(r,r)}(\nu)\,d\nu} = \int_0^\infty \hat{\mathscr{G}}(\nu)e^{-j2\pi\nu\tau}\,d\nu, \qquad (5.1\text{-}15)$$

where $\hat{\mathscr{G}}(\nu)$ is a normalized power spectral density,

$$\hat{\mathscr{G}}(\nu) = \begin{cases} \dfrac{\mathscr{G}^{(r,r)}(\nu)}{\int_0^\infty \mathscr{G}^{(r,r)}(\nu)\,d\nu} & \text{for } \nu > 0 \\ 0 & \text{otherwise.} \end{cases} \qquad (5.1\text{-}16)$$

We note that the normalized power spectral density has unit area,

$$\int_0^\infty \hat{\mathscr{G}}(\nu)\,d\nu = 1. \qquad (5.1\text{-}17)$$

If we know the preceding relationship between $\gamma(\tau)$ and $\hat{\mathscr{G}}(\nu)$, we can readily predict the form of the interferograms obtained with light having different shapes of power spectral density. Some specific examples are now considered. For a *low-pressure* gas discharge lamp, the shape of the power spectrum of a single line is determined primarily by the Doppler shifts of the light emitted from moving radiators that suffer infrequent collisions. In this case the spectral line is known to have approximately a *Gaussian* shape

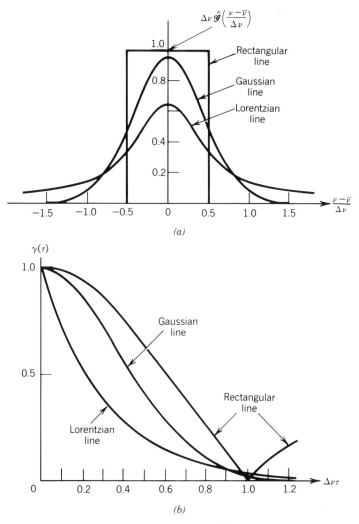

Figure 5-3. (a) Normalized power spectral density $\hat{\mathcal{G}}(\nu)$ and (b) envelope $\gamma(\tau)$ of the complex degree of coherence for three line shapes.

(Ref. 5-16),

$$\hat{\mathcal{G}}(\nu) \cong \frac{2\sqrt{\ln 2}}{\sqrt{\pi}\,\Delta\nu} \exp\left[-\left(2\sqrt{\ln 2}\,\frac{\nu - \bar{\nu}}{\Delta\nu}\right)^2\right], \qquad (5.1\text{-}18)$$

where the normalization is chosen to satisfy (5.1-17), and $\Delta\nu$ is the half-power bandwidth. This spectrum is shown in Fig. 5-3a. By a simple inverse

Fourier transformation we obtain the corresponding complex degree of coherence,

$$\gamma(\tau) = \exp\left[-\left(\frac{\pi\Delta\nu\tau}{2\sqrt{\ln 2}}\right)^2\right]\exp(-j2\pi\bar{\nu}\tau). \qquad (5.1\text{-}19)$$

Note that the phase $\alpha(\tau)$ is zero in this case, so the interferogram contains fringes of constant phase, but with visibility decreasing in accord with the modulus of $\gamma(\tau)$,

$$\gamma(\tau) = \exp\left[-\left(\frac{\pi\Delta\nu\tau}{2\sqrt{\ln 2}}\right)^2\right] \qquad (5.1\text{-}20)$$

as shown in Fig. 5-3b.

For a *high-pressure* gas discharge lamp, the spectral shape is determined primarily by the relatively frequent collisions of radiating atoms or molecules. The spectral line in this case can be shown to have a *Lorentzian* shape (Ref. 5-16),

$$\hat{\mathcal{G}}(\nu) \cong \frac{2(\pi\Delta\nu)^{-1}}{1 + \left(2\dfrac{\nu - \bar{\nu}}{\Delta\nu}\right)^2}, \qquad (5.1\text{-}21)$$

where again $\bar{\nu}$ is the center frequency of the line and $\Delta\nu$ is its half-power bandwidth (see Fig. 5-3a). The corresponding complex degree of coherence is readily shown to be

$$\gamma(\tau) = \exp[-\pi\Delta\nu|\tau|]\exp[-j2\pi\bar{\nu}\tau]. \qquad (5.1\text{-}22)$$

Again the interferogram observed with a Michelson interferometer will exhibit fringes of constant phase, but with an envelope decreasing as

$$\gamma(\tau) = \exp[-\pi\Delta\nu|\tau|]. \qquad (5.1\text{-}23)$$

This envelope is shown in Fig. 5-3b as a function of the parameter $\Delta\nu\tau$.

Occasionally in theoretical calculations it is convenient to assume a rectangular power spectral density

$$\hat{\mathcal{G}}(\nu) = \frac{1}{\Delta\nu}\text{rect}\left(\frac{\nu - \bar{\nu}}{\Delta\nu}\right). \qquad (5\text{-}1.24)$$

A simple Fourier transformation shows that the corresponding complex degree of coherence is

$$\gamma(\tau) = \text{sinc}(\Delta\nu\tau)\exp(-j2\pi\bar{\nu}\tau), \tag{5.1-25}$$

where $\text{sinc}\, x \triangleq \sin\pi x/\pi x$. In this case the envelope of the interference pattern is given by

$$\gamma(\tau) = |\text{sinc}\,\Delta\nu\tau|, \tag{5.1-26}$$

and the phase function $\alpha(\tau)$ is not zero for all τ. Rather, $\alpha(\tau)$ jumps between 0 and π radians as we pass from lobe to lobe of the sinc function,

$$\alpha(\tau) = \begin{cases} 0 & 2n < |\Delta\nu\tau| < 2n+1 \\ \pi & 2n+1 < |\Delta\nu\tau| < 2n+2 \end{cases} \quad n = 0, 1, 2, \ldots . \tag{5.1-27}$$

Both the power spectral density $\hat{\mathcal{G}}(\nu)$ and the envelope $\gamma(\tau)$ are shown in Fig. 5-3.

All the preceding examples yield interferograms that are even functions of delay h. This is a universal property of such interferograms and is simply an indication that it does not matter which of the two beams is delayed with respect to the other.

In addition, in all the examples the complex degree of coherence has been expressible as a product of $\exp(-j2\pi\bar{\nu}\tau)$ and a *real-valued* factor. This property is a result of our choice of line shapes that are *even* functions of $(\nu - \bar{\nu})$, (i.e., symmetrical about $\bar{\nu}$). More generally, the choice of an asymmetrical line profile will yield a $\gamma(\tau)$ that is the product of $\exp(-j2\pi\bar{\nu}\tau)$ and a complex-valued function. Thus the phase function $\alpha(\tau)$ can take on more general values than just 0 or π.

In many applications it is desirable to have a precise and definite meaning for the term "coherence time." Such a definition can be made in terms of the complex degree of coherence, but there are a multitude of definitions in terms of $\gamma(\tau)$ that can be imagined [see Ref. 5-17, Chapter 8, for a discussion of various possible measures of the "width" of a function such as $\gamma(\tau)$]. However, in future discussions there is one definition that arises most naturally and most frequently. Accordingly, following Mandel (Ref. 5-18), we define the *coherence time* τ_c of the disturbance $\mathbf{u}(t)$ by

$$\tau_c \triangleq \int_{-\infty}^{\infty} |\gamma(\tau)|^2 \, d\tau. \tag{5.1-28}$$

If this is to be a meaningful definition, it is necessary that τ_c have a value that is the same order of magnitude as $1/\Delta\nu$. That such is indeed the case can be found by substituting Eqs. (5.1-19), (5.1-22), and (5.1-25) into (5.1-28) and performing the required integration in each case. The results are as follows:

$$\tau_c = \sqrt{\frac{2\ln 2}{\pi}} \cdot \frac{1}{\Delta\nu} = \frac{0.664}{\Delta\nu} \qquad \text{Gaussian line}$$

$$\tau_c = \frac{1}{\pi\Delta\nu} = \frac{0.318}{\Delta\nu} \qquad \text{Lorentzian line} \qquad (5.1\text{-}29)$$

$$\tau_c = 1/\Delta\nu. \qquad \text{rectangular line}$$

Thus the order of magnitude does indeed agree with our intuition, and hence the specific definition of (5.1-28) will be used in the future. (See Problem 5-2 for calculation of some typical values of τ_c for some specific sources.)

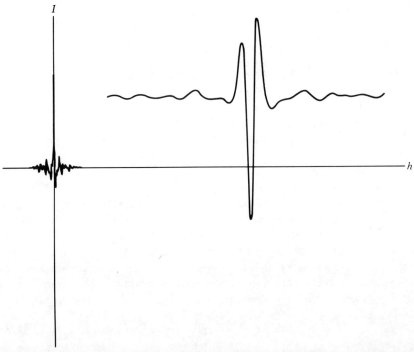

Figure 5-4. Typical midinfrared interferogram plotted with two different horizontal scales. The vertical axis represents detected intensity, and the horizontal axis represents optical path difference. The maximum optical path difference is 0.125 centimeters. (Courtesy of Peter R. Griffiths, University of California, Riverside and the American Association for the Advancement of Science. Reprinted from P. R. Griffiths, *Science*, vol. 222, pp. 297–302, 21 October 1983. Copyright 1983 by the American Association for the Advancement of Science.)

TEMPORAL COHERENCE 169

5.1.4 Fourier Spectroscopy

We have seen that the character of the interferogram observed with a Michelson interferometer can be completely determined if the power spectral density of the light is known. This intimate relationship between the interferogram and the power spectrum can be utilized for a very practical purpose. Namely, by measurement of the interferogram it is possible to determine the unknown power spectral density of the incident light. This principle forms the basis of the important field known as *Fourier spectroscopy* (for reviews of this field, see Refs. 5-19 and 5-20).

The general steps involved in obtaining a spectrum by Fourier spectroscopy are as follows. First, the interferogram must be measured. The move-

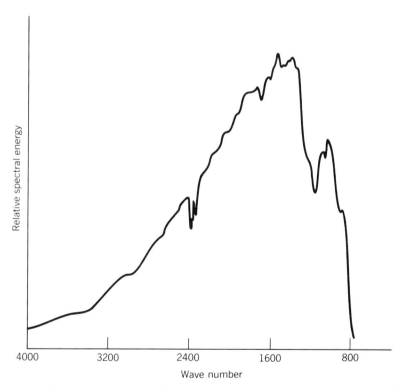

Figure 5-5. The Fourier transform of Figure 5-4, representing the spectrum of the source. The vertical axis represents power spectral density, and the horizonal axis represents optical wavenumber $(2\pi/\lambda)$ in inverse centimeters. The resolution achieved is 8 centimeters^{-1}. (Courtesy of Peter R. Griffiths, University of California, Riverside and the American Association for the Advancement of Science. Reprinted from P. R. Griffiths, *Science*, vol. 222, pp. 297–302, 21 October 1983. Copyright 1983 by the American Association for the Advancement of Science.)

able mirror travels, usually under interferometric control, from the position of zero pathlength difference into a region of large pathlength difference. The intensity of the light is measured as a function of time during this process, and the resulting interferogram is digitized. A digital Fourier transformation, usually using fast Fourier transform techniques (Ref. 5-21) yields a spectrum. A typical interferogram is shown in Fig. 5-4, and the power spectral density obtained from this interferogram is shown in Fig. 5-5.

Fourier spectroscopy has been found to offer distinct advantages over more direct methods (e.g., grating spectroscopy) in certain cases. First, there is an advantage in terms of light flux utilization (throughput), which we do not dwell on here (see Ref. 5-19). Of more direct interest to us, it was first shown by Fellgett (Ref. 5-22) that Fourier spectrometers can have an advantage over more conventional spectrometers in terms of the signal-to-noise ratio achieved in the measured spectrum. This advantage holds when the chief source of noise is additive detection noise and in general does not hold when photon noise is the limiting factor. As a consequence, Fourier spectroscopy has found considerable application in the infrared, often eliminating the need for detector refrigeration.

5.2 SPATIAL COHERENCE

In discussing temporal coherence, we noted that every real source has a finite bandwidth; therefore, for sufficiently large time delays τ, the analytic signals $\mathbf{u}(P, t)$ and $\mathbf{u}(P, t + \tau)$ become decorrelated. To concentrate on temporal coherence, we assumed that the source emitting the radiation was a perfect point source. In practice, of course, any real source must have a finite physical size, and as a consequence it is necessary to take this finite size into account. To do so leads us to the realm of *spatial coherence*. In this case we consider the two analytic signals $\mathbf{u}(P_1, t)$ and $\mathbf{u}(P_2, t)$ observed at two space points P_1 and P_2, ideally with zero relative time delay. When $P_1 = P_2$, the two waveforms are, of course, perfectly correlated. As P_1 and P_2 are moved apart, however, some degree of loss of correlation can in general be expected. We accordingly say that the wave emitted by the source has a limited spatial coherence. These ideas can be put on firmer ground by considering the interference of light in the classic experiment of Thomas Young (Ref. 5-23).

5.2.1 Young's Experiment

Consider the experiment illustrated in Fig. 5-6. A spatially extended source S illuminates an opaque screen in which two tiny pinholes have been

SPATIAL COHERENCE

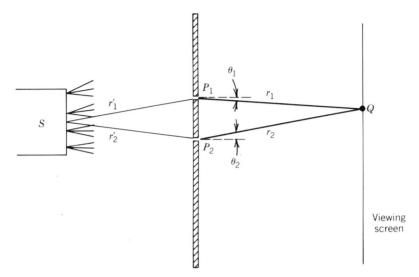

Figure 5-6. Young's interference experiment.

pierced at points P_1 and P_2. At some distance behind the opaque screen a viewing screen is placed, and the pattern of interference of the light from the two pinholes can be observed on this screen.

Light passing through the pinholes travels to the viewing screen, suffering time delays r_1/c and r_2/c, respectively, in the process. If the delay difference $(r_2 - r_1)/c$ is much less than the coherence time τ_c of the light from the source, fringes of interference can be expected, with a depth of modulation (visibility) that depends on the degree of correlation between the light waves incident on the two pinholes. Thus the cross-correlation $\langle \mathbf{u}(P_1, t + \tau)\mathbf{u}^*(P_2, t) \rangle$ can be expected to play an important role in determining the visibility of the observed fringes.

As with the case of the Michelson interferometer, there is another equivalent viewpoint that lends further insight into the character of the observed fringes. If the light is approximately monochromatic and originates from a single point source, sinusoidal fringes of high contrast are observed on the viewing screen. Now if a second point source, of the same wavelength as the first, but radiating independently, is added, a second fringe pattern is generated. The period of this fringe pattern is the same as that of the first, but the position of zero pathlength difference is shifted with respect to the corresponding position for the first fringe (see Fig. 5-7).

If the pinhole separation is small, the fringes are very coarse, and the shift of one fringe with respect to the other is a negligible fraction of a period. If the pinhole separation is large, however, the fringe period is small,

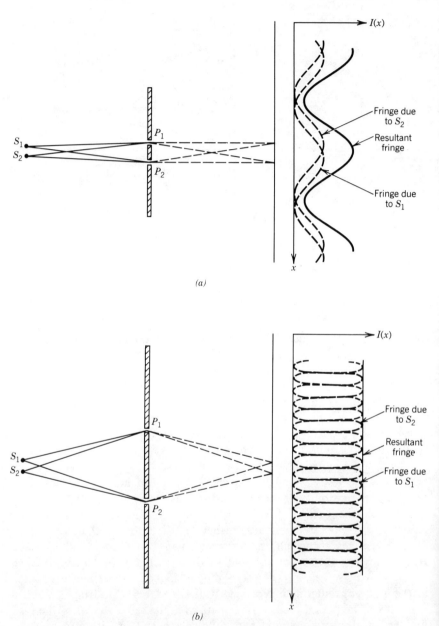

Figure 5-7. Physical explanation for the loss of fringe visibility at large pinhole spacings: (a) small pinhole separation; (b) large pinhole separation.

SPATIAL COHERENCE

and the fringe is shifted by a significant fraction of its period (perhaps even many periods). The two fringes may then partially cancel, with a loss of visibility resulting. If the source is an extended collection of many independent radiators, destructive cancellation of fringes can result in nearly a total loss of visibility for large pinhole spacings. This physical explanation is illustrated in Fig. 5-7.

To place the concepts discussed above on firmer ground and to discover the assumptions that may be buried in our intuitive discussion, we turn to a simple mathematical analysis of Young's experiment.

5.2.2 Mathematical Description of Young's Experiment

With reference again to Fig. 5-6, we wish to calculate mathematically the intensity of the light reaching point Q. As we have done previously, we again assume that the averaging time is effectively infinite, a valid assumption for true thermal light. The desired intensity is accordingly expressed as

$$I(Q) = \langle \mathbf{u}^*(Q,t)\mathbf{u}(Q,t) \rangle. \tag{5.2-1}$$

To proceed further, it is necessary to express $\mathbf{u}(Q,t)$ in more detail, presumably in terms of the analytic signals $\mathbf{u}(P_1,t)$ and $\mathbf{u}(P_2,t)$ reaching pinholes P_1 and P_2. At this point the assumption is usually made that $\mathbf{u}(Q,t)$ can be expressed as a weighted superposition of $\mathbf{u}(P_1,t)$ and $\mathbf{u}(P_2,t)$, each suitably delayed,

$$\mathbf{u}(Q,t) = \mathbf{K}_1 \mathbf{u}\left(P_1, t - \frac{r_1}{c}\right) + \mathbf{K}_2 \mathbf{u}\left(P_2, t - \frac{r_2}{c}\right), \tag{5.2-2}$$

where $\mathbf{K}_1$ and $\mathbf{K}_2$ are (possibly complex-valued) constants. With reference to Section 4.1.3, it becomes clear that such an expression is indeed possible, *provided the light is narrowband and the pinholes are not too large.* In particular, with the help of Eq. (4.1-12), we write

$$\mathbf{K}_1 \cong \iint_{\substack{\text{pinhole} \\ P_1}} \frac{\chi(\theta_1)}{j\bar{\lambda} r_1} dS_1, \qquad \mathbf{K}_2 \cong \iint_{\substack{\text{pinhole} \\ P_2}} \frac{\chi(\theta_2)}{j\bar{\lambda} r_2} dS_2, \tag{5.2-3}$$

where θ_1, θ_2, r_1, and r_2 are indicated in Fig. 5-6. (For consideration of the case of broadband light, the reader may consult Problem 5-4). In writing (5.2-3), it has been implicitly assumed that the pinholes are so small that the incident fields are constant over their spatial extent. For circular pinholes of diameter δ and a source with maximum linear dimension D, a sufficient condition to assure accuracy of this assumption is that

$$\delta \ll \frac{\bar{\lambda} z}{D}, \qquad (5.2\text{-}4)$$

where z is the normal distance from the source to the pinhole plane.

Using Eqs. (5.2-2) and (5.2-1), the intensity of the light at Q is readily shown to be

$$I(Q) = |\mathbf{K}_1|^2 \left\langle \left| \mathbf{u}\left(P_1, t - \frac{r_1}{c}\right) \right|^2 \right\rangle + |\mathbf{K}_2|^2 \left\langle \left| \mathbf{u}\left(P_2, t - \frac{r_2}{c}\right) \right|^2 \right\rangle$$
$$+ \mathbf{K}_1 \mathbf{K}_2^* \left\langle \mathbf{u}\left(P_1, t - \frac{r_1}{c}\right) \mathbf{u}^*\left(P_2, t - \frac{r_2}{c}\right) \right\rangle$$
$$+ \mathbf{K}_1^* \mathbf{K}_2 \left\langle \mathbf{u}^*\left(P_1, t - \frac{r_1}{c}\right) \mathbf{u}\left(P_2, t - \frac{r_2}{c}\right) \right\rangle. \qquad (5.2\text{-}5)$$

For convenience we again adopt some special symbols for quantities that are of particular importance. For a stationary optical source, we define

$$I^{(1)}(Q) \triangleq |\mathbf{K}_1|^2 \left\langle \left| \mathbf{u}\left(P_1, t - \frac{r_1}{c}\right) \right|^2 \right\rangle$$
$$I^{(2)}(Q) \triangleq |\mathbf{K}_2|^2 \left\langle \left| \mathbf{u}\left(P_2, t - \frac{r_2}{c}\right) \right|^2 \right\rangle \qquad (5.2\text{-}6)$$

representing, respectively, the intensities produced at Q by light from pinholes P_1 and P_2 *individually*. In addition, to account for interference effects, we introduce the definition

$$\Gamma_{12}(\tau) \triangleq \left\langle \mathbf{u}(P_1, t + \tau) \mathbf{u}^*(P_2, t) \right\rangle, \qquad (5.2\text{-}7)$$

representing the cross-correlation function of the light reaching pinholes P_1 and P_2. This function, first introduced by Wolf (Ref. 5-8), is called the *mutual coherence function* of the light and plays a fundamental role in the theory of partial coherence.

In terms of the above quantities, the intensity of Q can now be expressed in shorter form:

$$I(Q) = I^{(1)}(Q) + I^{(2)}(Q) + \mathbf{K}_1 \mathbf{K}_2^* \Gamma_{12}\left(\frac{r_2 - r_1}{c}\right)$$
$$+ \mathbf{K}_1^* \mathbf{K}_2 \Gamma_{21}\left(\frac{r_1 - r_2}{c}\right). \qquad (5.2\text{-}8)$$

SPATIAL COHERENCE

Now $\Gamma_{12}(\tau)$ can be readily shown to have the property that $\Gamma_{21}(-\tau) = \Gamma_{12}^*(\tau)$. Furthermore, since both $\mathbf{K}_1$ and $\mathbf{K}_2$ are purely imaginary numbers (Eq. 5.2-3), we see that $\mathbf{K}_1\mathbf{K}_2^* = \mathbf{K}_1^*\mathbf{K}_2 = K_1K_2$, where $K_1 = |\mathbf{K}_1|$ and $K_2 = |\mathbf{K}_2|$. Thus the expression for the intensity at Q becomes

$$I(Q) = I^{(1)}(Q) + I^{(2)}(Q) + K_1K_2\Gamma_{12}\left(\frac{r_2 - r_1}{c}\right)$$

$$+ K_1K_2\Gamma_{12}^*\left(\frac{r_2 - r_1}{c}\right),$$

or equivalently

$$I(Q) = I^{(1)}(Q) + I^{(2)}(Q) + 2K_1K_2\operatorname{Re}\left\{\Gamma_{12}\left(\frac{r_2 - r_1}{c}\right)\right\}. \quad (5.2\text{-}9)$$

A further simplification results if we introduce a normalization of the coherence function, as was done in discussing the Michelson interferometer. In this case we have, from Schwarz's inequality,

$$|\Gamma_{12}(\tau)| \leq [\Gamma_{11}(0)\Gamma_{22}(0)]^{1/2}, \quad (5.2\text{-}10)$$

where $\Gamma_{11}(\tau)$ and $\Gamma_{22}(\tau)$ are the self-coherence functions of the light at pinholes P_1 and P_2. Note that $\Gamma_{11}(0)$ and $\Gamma_{22}(0)$ represent the intensities of the light incident on the two pinholes. The inequality (5.2-10) leads us to define a normalized mutual coherence function in the form

$$\gamma_{12}(\tau) \triangleq \frac{\Gamma_{12}(\tau)}{[\Gamma_{11}(0)\Gamma_{22}(0)]^{1/2}}, \quad (5.2\text{-}11)$$

which is called the *complex degree of coherence*. [Strictly speaking, $\gamma_{12}(\tau)$ should perhaps be called the complex degree of *mutual* coherence and $\gamma(\tau)$ of Section 5.1 should be called the complex degree of *self* coherence, but this distinction is seldom worth making.] From the inequality (5.2-10) we can readily see that

$$0 \leq |\gamma_{12}(\tau)| \leq 1. \quad (5.2\text{-}12)$$

Noting further that

$$I^{(1)}(Q) = K_1^2\Gamma_{11}(0)$$

$$I^{(2)}(Q) = K_2^2\Gamma_{22}(0), \quad (5.2\text{-}13)$$

we can immediately rewrite the expression (5.2-9) for $I(Q)$ in the more convenient form

$$I(Q) = I^{(1)}(Q) + I^{(2)}(Q) + 2\sqrt{I^{(1)}(Q)I^{(2)}(Q)} \, \text{Re}\left\{\gamma_{12}\left(\frac{r_2 - r_1}{c}\right)\right\}.$$

(5.2-14)

To make further progress toward discovering the basic nature of the fringe patterns, we note that the complex degree of coherence, which is a normalized cross-correlation function of two random processes with center frequencies $\bar{\nu}$, can always be written in the form

$$\gamma_{12}(\tau) = \gamma_{12}(\tau)\exp\{-j[2\pi\bar{\nu}\tau - \alpha_{12}(\tau)]\}.$$

(5.2-15)

Substituting this expression in (5.2-14), we find

$$I(Q) = I^{(1)}(Q) + I^{(2)}(Q)$$
$$+ 2\sqrt{I^{(1)}(Q)I^{(2)}(Q)} \, \gamma_{12}\left(\frac{r_2 - r_1}{c}\right)\cos\left[2\pi\bar{\nu}\left(\frac{r_2 - r_1}{c}\right) - \alpha_{12}\left(\frac{r_2 - r_1}{c}\right)\right].$$

(5.2-16)

Although we are not yet in a position to specify precisely the geometric character of the interference pattern, we can draw some general conclusions at this point. The first two terms of Eq. (5.2-16) represent the intensities contributed by the pinholes individually. For pinholes of finite size, $I^{(1)}(Q)$ and $I^{(2)}(Q)$ will vary in the observation plane in accord with the diffraction patterns of the pinhole apertures, but for the present we assume that the pinholes are so small that these intensities are constant across the observation region. Riding on this constant bias we find a fringe pattern, with a period determined by $\bar{\nu}$ and other geometric factors, and having a slowly varying amplitude and phase modulation. In the vicinity of zero pathlength difference ($r_2 - r_1 = 0$), the fringes have a classical visibility

$$\mathscr{V} = \frac{2\sqrt{I^{(1)}I^{(2)}}}{I^{(1)} + I^{(2)}}\gamma_{12}(0).$$

(5.2-17)

Since $\gamma_{12}(0)$ represents the cross-correlation coefficient of the (underlying) waveforms $\mathbf{u}(P_1, t)$ and $\mathbf{u}(P_2, t)$, we conclude that $\gamma_{12}(0)$ [or $\mathscr{V}$ when $I^{(1)} = I^{(2)}$] is a measure of the coherence of the two optical vibrations. A description of how $\gamma_{12}(0)$ changes with changing distance between P_1 and

SPATIAL COHERENCE

P_2 is accordingly a description of the *spatial coherence* of the light striking the pinhole plane.

Note that in the general form of the Young's experiment discussed so far, both temporal and spatial coherence effects play a role. The envelope of the fringe pattern at zero pathlength difference is an indication of *spatial coherence* effects, whereas the tapering and eventual vanishing of the fringe envelope at large pathlength differences is an indication of *temporal coherence* effects. Ultimately we shall separate these two effects, but first we take up some geometric considerations that will allow us to specify the character of the fringe pattern in even greater detail.

5.2.3 Some Geometric Considerations

To specify more precisely the geometrical structure of the fringes, it is necessary to relate the delay difference $(r_2 - r_1)/c$ to various geometric factors, including the spacing of the pinholes, the distance to the observation plane, and the coordinates of the observation point Q. Such a relationship can be found with the help of Fig. 5-8. Let pinhole P_1 have transverse coordinates (ξ_1, η_1) and pinhole P_2 have transverse coordinates (ξ_2, η_2), both in the plane of the opaque screen. The viewing screen is assumed parallel to and a distance z_2 from the opaque screen. The coordinates of the observation point Q on the viewing screen are represented by (x, y).

The distances r_1 and r_2 are given exactly by the expressions

$$r_1 = \sqrt{z_2^2 + (\xi_1 - x)^2 + (\eta_1 - y)^2}$$
$$r_2 = \sqrt{z_2^2 + (\xi_2 - x)^2 + (\eta_2 - y)^2}.$$
(5.2-18)

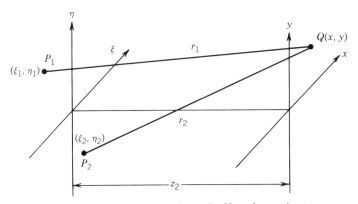

Figure 5-8. Interference geometry for Young's experiment.

To obtain a simple result, we make the usual *paraxial* approximations, valid when the pinholes and the observation point are close to the optical axis. In particular, we assume that

$$z_2 \gg \sqrt{x^2 + y^2}, \qquad z_2 \gg \sqrt{\xi_1^2 + \eta_1^2}, \qquad z_2 \gg \sqrt{\xi_2^2 + \eta_2^2}. \quad (5.2\text{-}19)$$

With these approximations we obtain

$$r_1 = z_2 \sqrt{1 + \frac{(\xi_1 - x)^2}{z_2^2} + \frac{(\eta_1 - y)^2}{z_2^2}}$$

$$\cong z_2 + \frac{(\xi_1 - x)^2}{2z_2} + \frac{(\eta_1 - y)^2}{2z_2} \quad (5.2\text{-}20)$$

and similarly

$$r_2 \cong z_2 + \frac{(\xi_2 - x)^2}{2z_2} + \frac{(\eta_2 - y)^2}{2z_2}. \quad (5.2\text{-}21)$$

Using these results, the pathlength difference takes the form

$$r_2 - r_1 \cong \frac{(\xi_2 - x)^2 + (\eta_2 - y)^2 - (\xi_1 - x)^2 - (\eta_1 - y)^2}{2z_2} \quad (5.2\text{-}22)$$

or equivalently

$$r_2 - r_1 \cong \frac{1}{2z_2}\left[(\xi_2^2 + \eta_2^2) - (\xi_1^2 + \eta_1^2) + 2(\xi_1 - \xi_2)x + 2(\eta_1 - \eta_2)y\right]. \quad (5.2\text{-}23)$$

Finally, we define the symbols

$$\rho_2 \triangleq \sqrt{\xi_2^2 + \eta_2^2}, \qquad \rho_1 = \sqrt{\xi_1^2 + \eta_1^2} \quad (5.2\text{-}24)$$

representing the distance of the pinholes from the optical axis, and

$$\Delta\xi = \xi_2 - \xi_1, \qquad \Delta\eta = \eta_2 - \eta_1 \quad (5.2\text{-}25)$$

representing the ξ and η spacings of the two pinholes. Thus the pathlength difference is expressed as

$$r_2 - r_1 \cong \frac{1}{2z_2}\left[\rho_2^2 - \rho_1^2 - 2\Delta\xi x - 2\Delta\eta y\right]. \quad (5.2\text{-}26)$$

Returning to the general expression (5.2-16) for the intensity distribution in the observation plane, Eq. (5.2-26) can be used to discover the exact form of the fringe pattern in the (x, y) plane. Referring to Fig. 5-9a, which is

SPATIAL COHERENCE

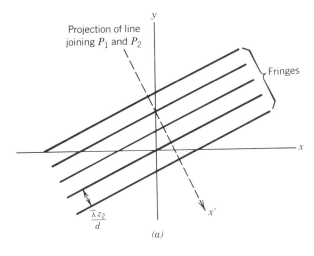

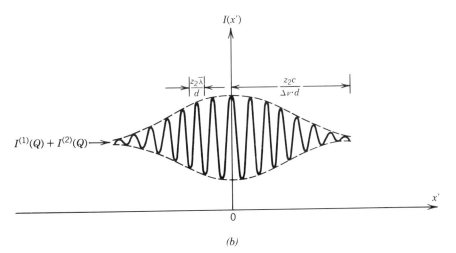

Figure 5-9. Geometric properties of the fringes.

drawn for the case of α_{12} constant, we find the fringe crests and nulls to run normal to the line joining P_1 and P_2 with a spatial fringe period given by

$$L = \frac{\bar{\lambda} z_2}{d}, \qquad (5.2\text{-}27)$$

where $\bar{\lambda} = c/\bar{\nu}$ and $d = \sqrt{(\Delta\xi)^2 + (\Delta\eta)^2}$ is the distance between the two pinholes.

Figure 5-9b shows a typical profile of the fringe along the x' axis, which passes through P_1 and P_2, with the assumptions that $I^{(1)}(Q)$ and $I^{(2)}(Q)$

are constant over the region shown, while α_{12} and $\pi(\rho_2^2 - \rho_1^2)/\bar{\lambda}z_2$ are identically zero. We note several properties of these fringes. The fringe envelope is centered at a point that corresponds to zero relative pathlength difference, which is taken to be the origin of the x' axis. The fringe period is given by (5.2-27) and the half-width of the fringe packet along the x' axis is

$$\Delta l \cong \frac{z_2 c}{\Delta \nu d}. \tag{5.2-28}$$

The total number of fringes appearing under the tapering envelope is

$$N \cong 2\frac{\Delta l}{L} = 2\frac{\bar{\nu}}{\Delta \nu}. \tag{5.2-29}$$

From the preceding discussions, it is clear that the results of a Young's interference experiment are dependent on both temporal and spatial coherence effects. Since we wish to concentrate on spatial coherence effects for the moment, it is necessary to impose further restrictions on the light that make temporal coherence effects negligible.

5.2.4 Interference Under Quasimonochromatic Conditions

To express the field incident at the observation point Q as a simple weighted sum of the (properly delayed) fields incident on the pinholes, it was necessary to assume that the light is *narrowband*. We now add a second assumption. Namely, we assume that the coherence length of the light is much greater than the maximum pathlength difference encountered in passage from the source to the interference region of interest. Stated mathematically, we require that for all source points and all points in the observation region of interest,

$$\Delta \nu \ll \bar{\nu} \quad \text{and} \quad \frac{(r_2 + r_2') - (r_1 + r_1')}{c} \ll \tau_c, \tag{5.2-30}$$

where the various distances involved are shown in Fig. 5-6. Such light is said to satisfy the *quasimonochromatic* conditions.

The addition of the second assumption in the preceding paragraph results in the assurance that the fringe contrast will be constant over the observation region of interest. Utilizing this fact, considerable simplifications in the forms of the mutual coherence function and the complex degree of coherence are possible. These functions can now be rewritten as

$$\Gamma_{12}(\tau) \cong \mathbf{J}_{12} e^{-j2\pi\bar{\nu}\tau}$$
$$\gamma_{12}(\tau) \cong \mu_{12} e^{-j2\pi\bar{\nu}\tau}, \tag{5.2-31}$$

SPATIAL COHERENCE

where

$$\mathbf{J}_{12} \triangleq \mathbf{\Gamma}_{12}(0) = \langle \mathbf{u}(P_1, t)\mathbf{u}^*(P_2, t) \rangle = \langle \mathbf{A}(P_1, t)\mathbf{A}^*(P_2, t) \rangle \quad (5.2\text{-}32)$$

is called the *mutual intensity* of the light at pinholes P_1 and P_2 and

$$\boldsymbol{\mu}_{12} \triangleq \boldsymbol{\gamma}_{12}(0) = \frac{\mathbf{J}_{12}}{[I(P_1)I(P_2)]^{1/2}} \quad (5.2\text{-}33)$$

is called the *complex coherence factor* of the light. In effect, $\mathbf{J}_{12}$ may be regarded as a *phasor amplitude* of a spatial sinusoidal fringe, whereas $\boldsymbol{\mu}_{12}$ is simply a normalized version of $\mathbf{J}_{12}$ having the property

$$0 \le |\boldsymbol{\mu}_{12}| \le 1. \quad (5.2\text{-}34)$$

Note that in writing (5.2-32) and what follows it has been tacitly assumed that $(r_2 - r_1)/c \ll \tau_c$ and $(r_2' - r_1')/c \ll \tau_c$, a slightly more restrictive condition than stated in (5.2-30).

The character of the fringe pattern can be stated more explicitly by substituting the expressions in (5.2-31) into Eqs. (5.2-9) and (5.2-14). Under paraxial conditions $[r_2 - r_1$ given by (5.2-26)] and tiny pinholes $[I^{(1)}(Q) = I^{(1)}, I^{(2)}(Q) = I^{(2)}, I^{(1)}$ and $I^{(2)}$ constants] the interference pattern in the (x, y) plane can be expressed by

$$I(x, y) = I^{(1)} + I^{(2)} + 2K_1 K_2 J_{12} \cos\left[\frac{2\pi}{\lambda z_2}(\Delta\xi x + \Delta\eta y) + \phi_{12}\right]$$

(5.2-35)

or

$$I(x, y) = I^{(1)} + I^{(2)} + 2\sqrt{I^{(1)}I^{(2)}}\,\mu_{12}\cos\left[\frac{2\pi}{\lambda z_2}(\Delta\xi x + \Delta\eta y) + \phi_{12}\right],$$

(5.2-36)

where $J_{12} = |\mathbf{J}_{12}|$, $\mu_{12} = |\boldsymbol{\mu}_{12}|$, and

$$\phi_{12} = \arg\{\mathbf{J}_{12}\} - \frac{\pi}{\lambda z_2}(\rho_2^2 - \rho_1^2) = \alpha_{12}(0) - \frac{\pi}{\lambda z_2}(\rho_2^2 - \rho_1^2).$$

(5.2-37)

Under the quasimonochromatic conditions, and assuming $I^{(1)}$ and $I^{(2)}$ are constant, the observed interference pattern has constant visibility and constant phase across the observation region. The visibility $\mathscr{V}$ may be expressed in terms of the modulus μ_{12} of the complex coherence factor by

$$\mathscr{V} = \frac{2\sqrt{I^{(1)}I^{(2)}}}{I^{(1)} + I^{(2)}} \mu_{12} \quad (I^{(1)} \neq I^{(2)}),$$
$$\mathscr{V} = \mu_{12} \quad (I^{(1)} = I^{(2)}). \tag{5.2-38}$$

When $\mu_{12} = 0$, the fringes vanish, and the two light waves are said to be mutually *incoherent*. When $\mu_{12} = 1$, the two waves are perfectly correlated,

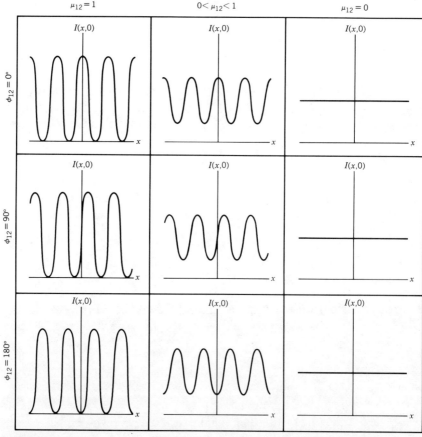

Figure 5-10. Fringe patterns obtained for various values of the complex coherence factor $(I^{(2)} = I^{(2)})$.

SPATIAL COHERENCE

and the two waves are called mutually *coherent*. For an intermediate value of μ_{12}, the two waves are *partially coherent*.

Figure 5-10 shows the character of the fringe patterns observed under various conditions on μ_{12} and ϕ_{12} and under the assumption that $I^{(1)} = I^{(2)}$. Note that the fringe position corresponding to $\phi_{12} = 0$ is arbitrary, but once selected it should be retained without change for all fringes.

A number of new quantities have been defined in this and previous sections. As an aid to the reader, we summarize the names and definitions of these quantities in Table 5-1.

5.2.5 Effects of Finite Pinhole Size

The pinholes utilized in the Young's interference experiment have, until now, been assumed to be so small that the centers of their diffraction patterns cover the entire observation region. Under quasimonochromatic

Table 5-1 Names and Definitions of Various Measures of Coherence

Symbol	Definition	Name	Temporal or Spatial Coherence
$\Gamma_{11}(\tau)$	$\langle u(P_1, t+\tau) u^*(P_1, t) \rangle$ [Note $\Gamma_{11}(0) = I(P_1)$]	Self coherence function	Temporal
$\gamma_{11}(\tau)$	$\dfrac{\Gamma_{11}(\tau)}{\Gamma_{11}(0)}$	Complex degree of (self) coherence	Temporal
$\Gamma_{12}(\tau)$	$\langle u(P_1, t+\tau) u^*(P_2, t) \rangle$	Mutual coherence function	Spatial and temporal
$\gamma_{12}(\tau)$	$\dfrac{\Gamma_{12}(\tau)}{[\Gamma_{11}(0)\Gamma_{22}(0)]^{1/2}}$	Complex degree of coherence	Spatial and temporal
J_{12}	$\langle u(P_1, t) u^*(P_2, t) \rangle$ $= \Gamma_{12}(0)$	Mutual intensity	Spatial quasimonochromatic
μ_{12}	$\dfrac{J_{12}}{[J_{11} J_{22}]^{1/2}} = \gamma_{12}(0)$	Complex coherence factor	Spatial quasimonochromatic

conditions, the result is a fringe of constant-amplitude riding on a bias level that is constant over the field of interest. The disadvantage of using such small pinholes is, of course, that little light reaches the observation plane; therefore, we must know in more detail the effects of enlarging the size of the pinholes.

Assuming that the pinholes are still sufficiently small to produce Fraunhofer diffraction patterns (rather than Fresnel diffraction patterns) in the observation plane, we can readily specify the distribution of intensity produced by each pinhole. For circular pinholes of diameter δ, we find that the intensities $I^{(1)}(Q)$ and $I^{(2)}(Q)$ produced by the pinholes individually are Airy patterns (see Ref. 5-24, pp. 63, 64):

$$I^{(1)}(Q) = \left(\frac{A}{\bar{\lambda} z_2}\right)^2 I(P_1) \left[2 \frac{J_1\left(\frac{\pi \delta}{\bar{\lambda} z_2} \sqrt{\left(x - \frac{z_1 + z_2}{z_1}\xi_1\right)^2 + \left(y - \frac{z_1 + z_2}{z_1}\eta_1\right)^2}\right)}{\frac{\pi \delta}{\bar{\lambda} z_2} \sqrt{\left(x - \frac{z_1 + z_2}{z_1}\xi_1\right)^2 + \left(y - \frac{z_1 + z_2}{z_1}\eta_1\right)^2}} \right]^2$$

$$I^{(2)}(Q) = \left(\frac{A}{\bar{\lambda} z_2}\right)^2 I(P_2) \left[2 \frac{J_1\left(\frac{\pi \delta}{\bar{\lambda} z_2} \sqrt{\left(x - \frac{z_1 + z_2}{z_1}\xi_2\right)^2 + \left(y - \frac{z_1 + z_2}{z_1}\eta_2\right)^2}\right)}{\frac{\pi \delta}{\bar{\lambda} z_2} \sqrt{\left(x - \frac{z_1 + z_2}{z_1}\xi_2\right)^2 + \left(y - \frac{z_1 + z_2}{z_1}\eta_2\right)^2}} \right]^2,$$

(5.2-39)

where z_1 and z_2 are shown in Fig. 5-11a, $A = \pi(\delta/2)^2$ is the area of a pinhole, whereas $I(P_1)$ and $I(P_2)$ are the intensities incident on the pinholes. In writing these expressions, it has been assumed that the source size is sufficiently small so as not to "smooth" these diffraction patterns. The required condition is

$$D \ll \frac{\bar{\lambda} z_1}{\delta}, \qquad (5.2\text{-}40)$$

where D is the maximum linear dimension of the source.

Figure 5-11b illustrates the overlapping diffraction patterns. Each pattern has a width of $2.44\bar{\lambda} z_2/\delta$ between first zeros, and the centers of the patterns are separated by distance

$$d' = \frac{z_1 + z_2}{z_1} d, \qquad (5.2\text{-}41)$$

SPATIAL COHERENCE

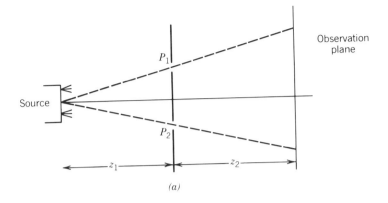

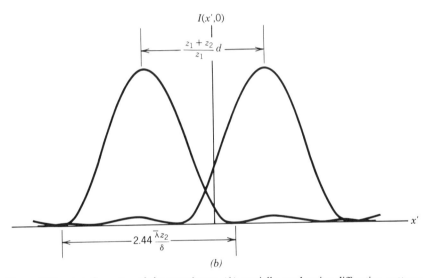

Figure 5-11. (a) Geometry of the experiment, (b) partially overlapping diffraction patterns.

where d is the separation of the pinholes. Thus we can expect nearly complete overlap of the two diffraction patterns if

$$d' \ll \frac{2.44\bar{\lambda}z_2}{\delta} \quad \text{or} \quad d \ll \frac{2.44\bar{\lambda}z_1 z_2}{(z_1 + z_2)\delta}. \quad (5.2\text{-}42)$$

If the pinholes are too far apart, the intensities $I^{(1)}(Q)$ and $I^{(2)}(Q)$ will not be equal, even if $I(P_1)$ and $I(P_2)$ are equal. Furthermore, the fringe visibility $\mathscr{V}$ will not be constant and will not equal the modulus μ_{12} of the complex coherence factor. Although μ_{12} can be recovered from the measured visibility and measured diffraction patterns by means of

$$\mu_{12} = \frac{I^{(1)}(Q) + I^{(2)}(Q)}{2\sqrt{I^{(1)}(Q)I^{(2)}(Q)}} \mathscr{V}, \tag{5.2-43}$$

the correction factor depends on which portion of the interference pattern is used for the visibility measurement, and further it will change if the pinhole separation is modified.

These difficulties can be alleviated if the interference measurements are made with a slightly different optical system illustrated in Fig. 5-12. In this case the source is placed in the front focal plane of a positive lens, the observation screen in the back focal plane of a second lens, and the pinhole screen between the two lenses. For circular pinholes with diameter δ, equal intensities $I(P_1) = I(P_2) = I$, and quasimonochromatic light, the interference pattern becomes

$$I(x, y) = 2\left(\frac{A}{\overline{\lambda}f}\right)^2 I \left[2 \frac{J_1\left(\frac{\pi\delta}{\overline{\lambda}f}\sqrt{x^2 + y^2}\right)}{\frac{\pi\delta}{\overline{\lambda}f}\sqrt{x^2 + y^2}}\right]^2$$

$$\times \left\{1 + \mu_{12}\cos\left[\frac{2\pi}{\overline{\lambda}f}(\Delta\xi x + \Delta\eta y) + \alpha_{12}\right]\right\}. \tag{5.2-44}$$

Note that, in addition to causing complete overlap of the two diffraction patterns, this optical system has the effect of canceling the phase factor

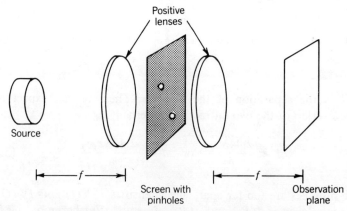

Figure 5-12. Optical system for interference experiment.

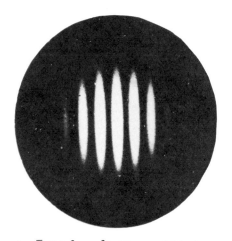

5-13. Photograph of interference pattern. (Courtesy of B. J. Thompson and the Optical Society of America. From B. J. Thompson and E. Wolf *J. Opt. Soc*; vol. 47, p. 899, 1957.)

$(\pi/\bar{\lambda}f)(\rho_2^2 - \rho_1^2)$. Figure 5-13 shows a photograph of an interference pattern obtained with such a system. The visibility of the fringes is the same at all points in the interference pattern.

5.3 CROSS-SPECTRAL PURITY

Many problems in coherence theory are simplified if the light of concern has the property that the complex degree of coherence can be factored into a product of a component depending only on spatial coordinates and a component depending only on time delay. Such a coherence function is said to be *reducible*. This property will be seen to be expressible as a different but entirely equivalent relationship in the spectral domain, where it is referred to as *cross-spectral purity*, a concept first introduced by Mandel (Ref. 5-25). By way of background, it is first helpful to consider a general problem: when two different light beams, each having the same normalized power spectral density $\hat{\mathscr{G}}(\nu)$, are superimposed, what is the shape of the power spectral density of the resultant beam? In the subsection to follow, we answer this question. Attention is then turned to the concept of cross-spectral purity and the conditions under which it can hold. The final subsection deals with an example of a light beam that is *not* cross-spectrally pure.

5.3.1 Power Spectrum of the Superposition of Two Light Beams

Consider two narrowband, statistically stationary light beams represented by analytic signals $\mathbf{u}(P_1, t)$ and $\mathbf{u}(P_2, t)$. These waves may be regarded as arising from two pinholes at points P_1 and P_2 in a Young's interference experiment. The two waves are superimposed after suffering delays τ_1 and τ_2, yielding a resultant light wave

$$\mathbf{u}(Q,t) = \mathbf{K}_1\mathbf{u}(P_1, t - \tau_1) + \mathbf{K}_2\mathbf{u}(P_2, t - \tau_2) \quad (5.3\text{-}1)$$

at a fixed point Q.

Suppose that the power spectral densities of $\mathbf{u}_1(t)$ and $\mathbf{u}_2(t)$ have identical shapes. Stated in mathematical terms, we require that their normalized power spectra [cf. Eq. (5.1-16)] be equal,

$$\hat{\mathcal{G}}_1(\nu) = \hat{\mathcal{G}}_2(\nu) \triangleq \hat{\mathcal{G}}(\nu). \quad (5.3\text{-}2)$$

Our goal is to find the relationship between the normalized spectrum at Q and the normalized spectra of the component beams.

Consider first the (self) coherence function of the light at Q. We have

$$\Gamma_Q(\tau) = \langle \mathbf{u}(Q, t+\tau)\mathbf{u}^*(Q,t) \rangle$$

$$= K_1^2 \Gamma_{11}(\tau) + K_2^2 \Gamma_{22}(\tau) + K_1 K_2 \Gamma_{12}(\tau_2 - \tau_1 + \tau)$$

$$+ K_1 K_2 \Gamma_{21}(\tau_1 - \tau_2 + \tau), \quad (5.3\text{-}3)$$

where $K_1 = |\mathbf{K}_1|$, $K_2 = |\mathbf{K}_2|$, and

$$\Gamma_{ij}(\tau) = \langle \mathbf{u}_i(t+\tau)\mathbf{u}_j^*(t) \rangle. \quad (5.3\text{-}4)$$

Recalling that $\Gamma_{21}(\tau) = \Gamma_{12}^*(-\tau)$, we can write $\Gamma_Q(\tau)$ as

$$\Gamma_Q(\tau) = K_1^2 \Gamma_{11}(\tau) + K_2^2 \Gamma_{22}(\tau) + K_1 K_2 \Gamma_{12}(\tau_2 - \tau_1 + \tau)$$

$$+ K_1 K_2 \Gamma_{12}^*(\tau_2 - \tau_1 + \tau). \quad (5.3\text{-}5)$$

Normalizing by $\Gamma_Q(0)$, and noting that, since the normalized spectra of the two beams are equal, their complex degrees of self coherence must likewise be identical, we obtain the complex degree of coherence at Q,

$$\gamma_Q(\tau) = \frac{\gamma_{11}(\tau) + A\,\text{Re}\{\gamma_{12}(\tau_2 - \tau_1 + \tau)\}}{1 + A\,\text{Re}\{\gamma_{12}(\tau_2 - \tau_1)\}}, \quad (5.3\text{-}6)$$

where the constant A is given by

$$A = \frac{2\sqrt{I^{(1)}I^{(2)}}}{I^{(1)} + I^{(2)}}. \quad (5.3\text{-}7)$$

CROSS-SPECTRAL PURITY

Moving to the spectral domain, a Fourier transformation of Eq. (5.3-6) yields the normalized power spectrum at Q,

$$\hat{\mathcal{G}}_Q(\nu) = \frac{\hat{\mathcal{G}}(\nu) + A\,\mathrm{Re}\{\hat{\mathcal{G}}_{12}(\nu)\exp[j2\pi\nu(\tau_1 - \tau_2)]\}}{1 + A\,\mathrm{Re}\{\gamma_{12}(\tau_2 - \tau_1)\}}. \quad (5.3\text{-}8)$$

Note that the denominator of (5.3-6) did not depend on τ and thus was not transformed. In addition, we have used the fact that

$$\mathcal{F}\{\mathrm{Re}\{\gamma_{12}(\tau)\}\} = \mathcal{F}\{2\gamma_{12}^{(r,r)}(\tau)\} = \mathrm{Re}\{\hat{\mathcal{G}}_{12}(\nu)\}. \quad (5.3\text{-}9)$$

[See Eq. (3.8-37) and recall that $\hat{\mathcal{G}}_{12}(\nu)$ is itself the one-sided spectrum of an analytic signal.]

The result (5.3-8) provides us with an explicit expression for the spectrum of the combined light beam at Q. We now compare this spectrum with the spectra of the original light beams.

5.3.2 Cross-Spectral Purity and Reducibility

With the result of Eq. (5.3-8) in hand, we can now investigate the conditions under which the normalized spectrum $\hat{\mathcal{G}}_Q(\nu)$ of the superimposed light waves is equal to the normalized spectrum $\hat{\mathcal{G}}(\nu)$ of the component beams. When these two spectra are equal, the light is said to be *cross-spectrally pure*, a term borrowed from the field of genetics and meant to imply that the two progenitors (the original beams that were superimposed) have produced a progeny (the new light beam) that has the same properties as the progenitors, at least as far as the shape of the power spectral density is concerned. Consider the difference of the spectra in question,

$$\hat{\mathcal{G}}_Q(\nu) - \hat{\mathcal{G}}(\nu) = \frac{A\,\mathrm{Re}\{\hat{\mathcal{G}}_{12}(\nu)e^{-j2\pi(\tau_2-\tau_1)\nu} - \gamma_{12}(\tau_2 - \tau_1)\hat{\mathcal{G}}(\nu)\}}{1 + A\,\mathrm{Re}\{\gamma_{12}(\tau_2 - \tau_1)\}}. \quad (5.3\text{-}10)$$

For this expression to be zero for all $I^{(1)}$ and $I^{(2)}$, and independent of $\tau_2 - \tau_1$, we must have

$$\hat{\mathcal{G}}_{12}(\nu)e^{-j2\pi(\tau_2-\tau_1)\nu} - \gamma_{12}(\tau_2 - \tau_1)\hat{\mathcal{G}}(\nu) = 0. \quad (5.3\text{-}11)$$

One way this requirement can be satisfied is for the light at P_1 and P_2 to be completely uncorrelated for all $\tau_2 - \tau_1$. Then

$$\hat{\mathcal{G}}_{12}(\nu) = 0, \qquad \gamma_{12}(\tau_2 - \tau_1) = 0. \quad (5.3\text{-}12)$$

However, we seek less restrictive conditions.

It soon becomes evident that condition (5.3-11) cannot be satisfied in the most general case, for the left-hand term oscillates indefinitely with $\tau_2 - \tau_1$, while the right-hand term eventually drops to zero. Therefore, we can expect that some restrictions must be placed on the delay difference if we are to achieve approximate equality.

To this end, let the delay difference be written

$$\tau_2 - \tau_1 = \tau_0 + \Delta\tau, \qquad (5.3\text{-}13)$$

where for the moment τ_0 is arbitrary, but $\Delta\tau$ is restricted to be much less than $1/\Delta\nu$, $\Delta\nu$ being the bandwidth of the light. With this restriction, is it now possible to satisfy the required equation?

For the small range of $\Delta\tau$ allowed, we can show that

$$\gamma_{12}(\tau_0 + \Delta\tau) \cong \gamma_{12}(\tau_0)\exp\{-j2\pi\bar{\nu}\Delta\tau\}, \qquad (5.3\text{-}14)$$

where $\bar{\nu}$ represents the center frequency of the cross spectrum $\hat{\mathcal{G}}_{12}(\nu)$. The steps involved in proving this assertion are:

(1) Note

$$\gamma_{12}(\tau_0 + \Delta\tau) = \int_0^\infty \hat{\mathcal{G}}_{12}(\nu)e^{-j2\pi\nu(\tau_0 + \Delta\tau)}\,d\nu. \qquad (5.3\text{-}15)$$

(2) Let $\nu = \bar{\nu} + \delta\nu$ $(-\Delta\nu/2 < \delta\nu < \Delta\nu/2)$.
(3) Approximate $\exp\{-j2\pi\delta\nu\Delta\tau\} \cong 1$.

From these three steps, Eq. (5.3-14) follows. Substitution of the approximate expression for $\gamma_{12}(\tau_0 + \Delta\tau)$ in Eq. (5.3-11) yields the following equation that must be satisfied if the normalized spectrum of the superimposed light beams is to equal that of the component beams:

$$\hat{\mathcal{G}}_{12}(\nu)e^{-j2\pi\tau_0\nu} = \gamma_{12}(\tau_0)\hat{\mathcal{G}}(\nu). \qquad (5.3\text{-}16)$$

Some comments are in order before continuing. The reader may question whether Eq. (5.3-16) can be expected to hold in general, for it appears that the left-hand side will be oscillatory in ν, as a result of the exponential, whereas the right-hand side will be nonoscillatory for any smooth $\hat{\mathcal{G}}(\nu)$. This objection is in general valid but can be overcome *if we choose the delay τ_0 correctly*. In fact, if the delays suffered by the light on its travel from the source *to* the two pinholes differ by more than $1/\Delta\nu$, the cross spectrum $\hat{\mathcal{G}}_{12}(\nu)$ will itself be oscillatory (see discussion that follows). However, if the delay difference τ_0 after the pinholes is chosen to cancel the delays suffered by the light on the way to the pinholes, the exponential term on the right of Eq. (5.3-16) will exactly cancel the oscillatory behavior of $\hat{\mathscr{G}}_{12}(\nu)$ itself. To achieve the equality desired in Eq. (5.3-16), therefore, it is necessary that the delay τ_0 be chosen so that the total delays, from source to observation plane, are equalized. This requirement is equivalent to one that chooses τ_0 to maximize $\gamma_{12}(\tau_0)$.

The preceding assertion that the cross-spectral density $\hat{\mathscr{G}}_{12}(\nu)$ can itself be oscillatory is best illustrated by considering two light beams at points P_1 and P_2 that are identical except for a relative delay $\bar{\tau}$. We suppose that one beam has been advanced by $\bar{\tau}/2$ and the other has been retarded by the same amount. The two beams have the same power spectrum $\hat{\mathscr{G}}(\nu)$. The delays in time can be represented equivalently by means of transfer functions in the frequency domain. The appropriate transfer functions are

$$\mathbf{H}_1(\nu) = \exp\left[+j2\pi\nu\frac{\bar{\tau}}{2}\right] \quad \left(\text{beam retarded by } \frac{\bar{\tau}}{2}\right)$$
$$\mathbf{H}_2(\nu) = \exp\left[-j2\pi\nu\frac{\bar{\tau}}{2}\right] \quad \left(\text{beam advanced by } \frac{\bar{\tau}}{2}\right). \tag{5.3-17}$$

Now using the expression (3.5-8) for the cross-spectral density of two linearly filtered random processes, we obtain a normalized cross-spectral density

$$\hat{\mathscr{G}}_{12}(\nu) = \hat{\mathscr{G}}(\nu)\exp\{j2\pi\nu\bar{\tau}\}, \tag{5.3-18}$$

which has an oscillatory component. Thus we see that when there are relative delays present as the light travels to the pinholes, the cross-spectral density of the light will have an oscillatory component and that proper choice of the delay τ_0 in Eq. (5.3-16) will cancel this oscillatory component.

We note in passing that it can readily be shown that the superposition of two light beams that are identical except for a relative delay $\bar{\tau}$ results in a new normalized spectrum $\hat{\mathscr{G}}'(\nu)$ of the form

$$\hat{\mathscr{G}}'(\nu) = \tfrac{1}{2}\hat{\mathscr{G}}(\nu)[1 + \cos 2\pi\nu\bar{\tau}]. \tag{5.3-19}$$

Such a spectrum is illustrated in Fig. 5-14. It should be clear that if the delay $\bar{\tau}$ satisfies $\bar{\tau} > 1/\Delta\nu$, the new spectrum will exhibit fringes and will thus be different from the original spectra of the component beams. [Spectral fringes of this type were used by W. P. Alford and A. Gold to measure the velocity of light (Ref. 5-26).] This basic phenomenon of spectral fringing must be avoided by proper choice of τ_0 if cross-spectral purity of the light is to be achieved.

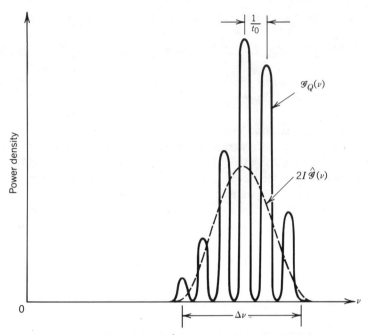

Figure 5-14. Power spectra $\mathcal{G}_Q(\nu)$ and $2I\hat{\mathcal{G}}(\nu)$, showing departure from cross-spectral purity.

Having chosen the delay term τ_0 properly, and restricting the delay term $\Delta\tau$ to be much smaller than $1/\Delta\nu$, Eq. (5.3-16) can be satisfied. It is helpful to examine the form of the same equation, but expressed in the time domain, rather than the frequency domain. An inverse Fourier transformation of the equation yields

$$\gamma_{12}(\tau + \tau_0) = \gamma_{12}(\tau_0)\gamma(\tau). \tag{5.3-20}$$

Any complex degree of coherence with the preceding property is said to be *reducible*, and we see that, within the approximations and restrictions made above, reducibility is entirely equivalent to cross-spectral purity. Note that the reducible property of the complex degree of coherence is the property we initially set out to explore. Namely we were seeking an understanding of when the complex degree of coherence factors into a product of a spatial part and a temporal part. Since τ_0 is a constant, Eq. (5.3-20) is precisely the factorization property we were seeking.

A bit of physical interpretation of the spectral representation (5.3-16) might be helpful at this point. The left-hand side of the equation can be regarded as expressing the cross-correlation between the spectral components that are in the vicinity of frequency ν for each of the two beams, but with one beam delayed with respect to the other by τ_0. The right-hand side expresses that correlation as being proportional to $\gamma_{12}(\tau_0)$. The factor $\hat{\mathcal{G}}(\nu)$ is simply a normalization that represents the relative amount of power present at frequency ν. Equation (5.3-15) can thus be interpreted as stating that, for two beams to be cross spectrally pure, *all frequency components of one beam must have the same normalized cross-correlation with the corresponding frequency components of the other beam.*

Since the delay τ_0 has been chosen to maximize $\gamma_{12}(\tau_0)$, it is clear that the quasimonochromatic conditions are satisfied, and we could equally well express the reducibility result (5.3-20) in terms of the complex coherence factor μ_{12},

$$\gamma_{12}(\tau + \tau_0) = \mu_{12}\gamma(\tau). \tag{5.3-21}$$

In closing this section, we summarize by stating that factorization of the complex degree of coherence yields great simplifications in many problems for which both temporal and spatial coherence play an important role. Such factorization is possible if the light is cross-spectrally pure. Often cross-spectral purity is simply assumed without any real justification other than the simplification that results. Such an assumption may or may not be valid in any particular case: For example, if the light arises from a source that radiates with an angularly dependent optical spectrum, cross-spectral purity generally will not hold. An example of such a source is considered in Section 5.3.3.

5.3.3 Laser Light Scattered by a Moving Diffuser

An example of light that is not cross-spectrally pure is afforded by considering the wave transmitted by a moving diffuser (such as ground glass) when illuminated by ideal laser light. The geometry is illustrated in Fig. 5-15. The *CW* laser provides plane wave illumination by essentially monochromatic light. The diffuser is moving with constant linear velocity v in the vertical direction. An opaque screen pierced by two tiny pinholes P_1 and P_2 is placed immediately adjacent to the diffuser, allowing us to perform a Young's interference experiment on the light transmitted by the diffuser. Our goal is to determine whether the complex degree of coherence $\Gamma_{12}(\tau)$ of the light transmitted by a moving diffuser can be expressed in product form

$$\gamma_{12}(\tau) = \mu_{12}\gamma(\tau), \tag{5.3-22}$$

that is, to discover whether the transmitted light is cross-spectrally pure.

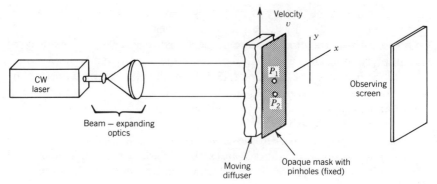

Figure 5-15. Measurement of the mutual coherence function of light transmitted by a moving diffuser.

The diffuser may be represented in terms of its amplitude transmittance $t_A(x, y)$. For simplicity we assume that the pinholes are oriented vertically along the y axis, and hence the x dependence of t_A is dropped.[†] Under unit intensity normally incident plane wave illumination, the diffuser produces an optical field with amplitude distribution

$$\mathbf{u}(y; t) = \mathbf{t}_A(y - vt)\exp(-j2\pi\bar{\nu}t), \qquad (5.3\text{-}23)$$

where $\bar{\nu}$ is the frequency of the incident laser light.

Considering pinholes located at positions y_1 and y_2, the mutual coherence function of interest is

$$\Gamma_{12}(\tau) = \Gamma(y_1, y_2; \tau) = \langle \mathbf{t}_A(y_1 - vt - v\tau)\mathbf{t}_A^*(y_2 - vt)\rangle e^{-j2\pi\bar{\nu}\tau}.$$

$$(5.3\text{-}24)$$

Neglecting any small component of absorption by the diffuser, we have

$$\langle |\mathbf{t}_A(y - vt)|^2 \rangle = 1 \qquad (5.3\text{-}25)$$

and hence $\Gamma(y_1, y_2; \tau) = \gamma(y_1, y_2; \tau)$.

The statistical fluctuations of the transmitted fields arise from the statistical structure of the diffuser. (The detailed spatial structure of the diffuser is

---

[†] Implicit here is the assumption that the pinholes are much smaller than the finest structure of t_A.

unknown a priori.) We make the reasonable assumption that the random process $t_A(y)$ is spatially ergodic[†] (and hence stationary) in y and has a statistical autocorrelation function

$$\gamma_t(\Delta y) \triangleq \overline{t_A(y + \Delta y)t_A^*(y)}. \tag{5.3-26}$$

In terms of this quantity, the complex degree of coherence $\gamma(y_1, y_2; \tau)$ of the transmitted field can now be expressed as

$$\gamma(y_1, y_2; \tau) = \gamma_t(\Delta y - v\tau)e^{-j2\pi\bar{\nu}\tau}, \tag{5.3-27}$$

where $\Delta y = y_1 - y_2$ is the pinhole separation. This complex degree of coherence is in general *not* separable into a product of space and time factors as required for cross-spectral purity. For example, when the correlation function $\gamma_t(\Delta y)$ of the diffuser has Gaussian form

$$\gamma_t(\Delta y) = \exp\left[-a(\Delta y)^2\right], \tag{5.3-28}$$

the complex degree of coherence is readily seen to be

$$\gamma(y_1, y_2; \tau) = e^{-a(\Delta y)^2} e^{-a(v\tau)^2} e^{2av\tau \Delta y} e^{-j2\pi\bar{\nu}\tau} \tag{5.3-29}$$

As an interesting exercise, the reader is asked to prove (see Problem 5-8) that if the same laser light is passed through two closely spaced diffusers, moving in exactly opposite directions with equal speeds, the transmitted light *is* cross-spectrally pure when the correlation function $\gamma_t(\Delta y)$ has Gaussian form.

5.4 PROPAGATION OF MUTUAL COHERENCE

The detailed structure of an optical wave undergoes changes as the wave propagates through space. In a similar fashion, the detailed structure of the mutual coherence function undergoes changes, and in this sense the mutual coherence function is said to propagate. In both cases the underlying physical reason for propagation rests on the wave equation obeyed by the light waves themselves. In this section we first derive some basic propagation laws obeyed by mutual coherence and later show that the mutual coherence function obeys a pair of scalar wave equations.

[†] By spatially ergodic, we mean that all space averages equal corresponding ensemble averages.

5.4.1 Solution Based on the Huygens–Fresnel Principle

The simplest method for discovering the propagation laws obeyed by mutual coherence is to begin with the Huygens–Fresnel principle, as presented previously in Section 4.1; knowing that the complex fields satisfy such equations, we can easily derive the corresponding relations for mutual coherence.

The general problem of interest is illustrated in Fig. 5-16. A light wave with arbitrary coherence properties propagates from left to right. Knowing the mutual coherence function $\Gamma(P_1, P_2; \tau)$ on the surface Σ_1, we wish to find the mutual coherence function $\Gamma(Q_1, Q_2; \tau)$ on surface Σ_2. Stated in more physical terms, our goal is to predict the results of Young's interference experiments with pinholes Q_1 and Q_2 when we know the results of Young's interference experiments with all possible pinholes P_1 and P_2.

Our analysis centers on the case of *narrowband* light, discussed in Section 4.1.3. Results for broadband light are also presented later in the section. We begin by noting that the mutual coherence function on Σ_2 is by definition

$$\Gamma(Q_1, Q_2; \tau) = \langle \mathbf{u}(Q_1, t + \tau) \mathbf{u}^*(Q_2, t) \rangle. \qquad (5.4\text{-}1)$$

The fields on Σ_2 can be related to the fields on Σ_1 with the help of Eq. (4.1-12), valid for narrowband light. In particular, we have

$$\mathbf{u}(Q_1, t + \tau) = \iint_{\Sigma_1} \frac{1}{j\bar{\lambda} r_1} \mathbf{u}\left(P_1, t + \tau - \frac{r_1}{c}\right) \chi(\theta_1)\, dS_1,$$

$$\mathbf{u}^*(Q_2, t) = \iint_{\Sigma_1} \frac{(-1)}{j\bar{\lambda} r_2} \mathbf{u}^*\left(P_2, t - \frac{r_2}{c}\right) \chi(\theta_2)\, dS_2. \qquad (5.4\text{-}2)$$

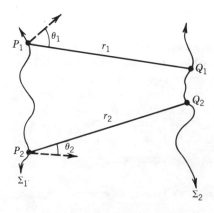

Figure 5-16. Geometry for propagation of mutual coherence, where θ_1 and θ_2 represent, respectively, the angle between $P_1 Q_1$ and the surface normal at P_1 and the corresponding angle for $P_2 Q_2$.

PROPAGATION OF MUTUAL COHERENCE

Substituting (5.4-2) in (5.4-1) and interchanging orders of integration and averaging, we find

$$\Gamma(Q_1,Q_2;\tau) = \iint_{\Sigma_1} \iint_{\Sigma_1} \frac{\left\langle \mathbf{u}\!\left(P_1, t+\tau-\frac{r_1}{c}\right)\mathbf{u}^*\!\left(P_2, t-\frac{r_2}{c}\right)\right\rangle}{(\bar{\lambda})^2 r_1 r_2}$$

$$\times \chi(\theta_1)\chi(\theta_2)\,dS_1\,dS_2. \qquad (5.4\text{-}3)$$

The time average in the integrand can be expressed in terms of the mutual coherence function on Σ_1, yielding the basic propagation law for mutual coherence (under the narrowband assumption)

$$\Gamma(Q_1,Q_2;\tau) = \iint_{\Sigma_1} \iint_{\Sigma_1} \Gamma\!\left(P_1, P_2; \tau + \frac{r_2 - r_1}{c}\right) \frac{\chi(\theta_1)}{\bar{\lambda} r_1} \frac{\chi(\theta_2)}{\bar{\lambda} r_2}\, dS_1\, dS_2.$$

$$(5.4\text{-}4)$$

The reader can readily show (see Problem 5-9), starting with Eq. (4.1-9), that for broadband light, the corresponding relationship is

$$\Gamma(Q_1,Q_2;\tau) = -\iint_{\Sigma_1}\iint_{\Sigma_1} \frac{\partial^2}{\partial \tau^2}\Gamma\!\left(P_1,P_2;\tau+\frac{r_2-r_1}{c}\right)$$

$$\times \frac{\chi(\theta_1)}{2\pi c r_1}\frac{\chi(\theta_2)}{2\pi c r_2}\,dS_1\,dS_2. \qquad (5.4\text{-}5)$$

Returning to the case of narrowband light, we now invoke the second quasimonochromatic condition, namely, that the maximum difference of pathlengths is much smaller than the coherence length of the light. With this assumption we can find the corresponding propagation laws for mutual intensity. When the quasimonochromatic conditions are satisfied, we find the mutual intensity on Σ_2 by noting that

$$\mathbf{J}(Q_1,Q_2) = \Gamma(Q_1,Q_2;0). \qquad (5.4\text{-}6)$$

Using (5.4-4) with $\tau = 0$, and further noting [cf. Eq. (5.2-31)] that

$$\Gamma\!\left(P_1,P_2;\frac{r_2-r_1}{c}\right) = \mathbf{J}(P_1,P_2)\exp\!\left[-j\frac{2\pi}{\bar{\lambda}}(r_2-r_1)\right], \qquad (5.4\text{-}7)$$

we have

$$J(Q_1, Q_2) = \iint_{\Sigma_1} \iint_{\Sigma_1} J(P_1, P_2) \exp\left[-j\frac{2\pi}{\bar{\lambda}}(r_2 - r_1)\right] \frac{\chi(\theta_1)}{\bar{\lambda} r_1} \frac{\chi(\theta_2)}{\bar{\lambda} r_2} dS_1 dS_2,$$

(5.4-8)

which represents the basic propagation law for mutual intensity.

The *intensity* distribution on the surface Σ_2 can readily be found by letting $Q_1 \to Q_2$ in (5.4-8). Thus

$$I(Q) = \iint_{\Sigma_1} \iint_{\Sigma_1} J(P_1, P_2) \exp\left[-j\frac{2\pi(r_2' - r_1')}{\bar{\lambda}}\right] \frac{\chi(\theta_1')}{\bar{\lambda} r_1'} \frac{\chi(\theta_2')}{\bar{\lambda} r_2'} dS_1 dS_2,$$

(5.4-9)

where the quantities r_1', r_2', θ_1', and θ_2' differ from r_1, r_2, θ_1, and θ_2 in Fig. 5-16 because Q_1 and Q_2 have merged. The new geometry is illustrated in Fig. 5-17.

Thus the basic propagation laws for mutual coherence and mutual intensity have been derived. The reader is reminded that, because the results were derived from the Huygens–Fresnel principle, the assumptions imposed in deriving that principle are also implicit here. In particular, the distances r_1 and r_2 (or r_1' and r_2') must be much larger than a wavelength, a condition satisfied in all applications of interest to us here.

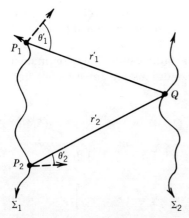

Figure 5-17. Geometry for calculation of the intensity on Σ_2.

5.4.2 Wave Equations Governing Propagation of Mutual Coherence

The basic propagation laws for mutual coherence have been found from the Huygens-Fresnel principle; however, it is of some general interest to examine the propagation problem on a more fundamental level. In this section we begin with the scalar wave equation governing the propagation of fields and show that the mutual coherence function obeys a *pair* of wave equations, a fact first discovered by E. Wolf.

In free space, the real wave disturbance $u^{(r)}(P,t)$ obeys the partial differential equation

$$\nabla^2 u^{(r)}(P,t) - \frac{1}{c^2}\frac{\partial^2}{\partial t^2}u^{(r)}(P,t) = 0, \quad (5.4\text{-}10)$$

where $\nabla^2 = \partial^2/\partial x^2 + \partial^2/\partial y^2 + \partial^2/\partial z^2$ is the Laplacian operator. Now if both sides of this equation are Hilbert transformed, after an interchange of orders of operators it follows that

$$\nabla^2 u^{(i)}(P,t) - \frac{1}{c^2}\frac{\partial^2}{\partial t^2}u^{(i)}(P,t) = 0 \quad (5.4\text{-}11)$$

where $u^{(i)}(P,t)$ is the Hilbert transform of $u^{(r)}(P,t)$. We conclude that both the real and imaginary parts of the analytic signal $\mathbf{u}(P,t)$ obey the wave equation and thus that

$$\nabla^2 \mathbf{u}(P,t) - \frac{1}{c^2}\frac{\partial^2}{\partial t^2}\mathbf{u}(P,t) = 0. \quad (5.4\text{-}12)$$

Now by definition, the mutual coherence function is given by $\Gamma_{12}(\tau) = \langle \mathbf{u}_1(t+\tau)\mathbf{u}_2^*(t)\rangle$, where $\mathbf{u}_1(t) \triangleq \mathbf{u}(P_1,t)$ and $\mathbf{u}_2(t) \triangleq \mathbf{u}(P_2,t)$. Let the operator ∇_1^2 be defined by

$$\nabla_1^2 \triangleq \frac{\partial^2}{\partial x_1^2} + \frac{\partial^2}{\partial y_1^2} + \frac{\partial^2}{\partial z_1^2}, \quad (5.4\text{-}13)$$

where P_1 has coordinates (x_1, y_1, z_1). We apply the operator ∇_1^2 directly to the definition of $\Gamma_{12}(\tau)$, yielding

$$\nabla_1^2 \Gamma_{12}(\tau) = \nabla_1^2 \langle \mathbf{u}_1(t+\tau)\mathbf{u}_2^*(t)\rangle$$
$$= \langle \nabla_1^2 \mathbf{u}_1(t+\tau)\mathbf{u}_2^*(t)\rangle. \quad (5.4\text{-}14)$$

But since

$$\nabla_1^2 \mathbf{u}_1(t+\tau) = \frac{1}{c^2} \frac{\partial^2 \mathbf{u}_1(t+\tau)}{\partial(t+\tau)^2}, \qquad (5.4\text{-}15)$$

we see†

$$\nabla_1^2 \Gamma_{12}(\tau) = \left\langle \frac{1}{c^2} \frac{\partial^2 \mathbf{u}_1(t+\tau)}{\partial(t+\tau)^2} \mathbf{u}_2^*(t) \right\rangle = \left\langle \frac{1}{c^2} \frac{\partial^2 \mathbf{u}_1(t+\tau)}{\partial \tau^2} \mathbf{u}_2^*(t) \right\rangle$$

$$= \frac{1}{c^2} \frac{\partial^2}{\partial \tau^2} \langle \mathbf{u}_1(t+\tau) \mathbf{u}_2^*(t) \rangle. \qquad (5.4\text{-}16)$$

The time-averaged quantity is simply the mutual coherence function, and hence

$$\nabla_1^2 \Gamma_{12}(\tau) = \frac{1}{c^2} \frac{\partial^2}{\partial \tau^2} \Gamma_{12}(\tau). \qquad (5.4\text{-}17)$$

In a similar fashion, the operator $\nabla_2^2 = \partial^2/\partial x_2^2 + \partial^2/\partial y_2^2 + \partial^2/\partial z_2^2$ can be applied to the definition of $\Gamma_{12}(\tau)$, yielding a second equation

$$\nabla_2^2 \Gamma_{12}(\tau) = \frac{1}{c^2} \frac{\partial^2}{\partial \tau^2} \Gamma_{12}(\tau), \qquad (5.4\text{-}18)$$

which $\Gamma_{12}(\tau)$ must also satisfy. Thus $\Gamma_{12}(\tau)$ propagates in accord with a *pair* of wave equations. The relationships derived in Section 5.4.1 are in fact certain specialized solutions of this pair of equations. For a discussion of rigorous general solutions to this pair of equations, the reader may wish to consult Ref. 5-10, Section 10-7.

As an exercise (see Problem 5-10), the reader is asked to verify that the mutual intensity $\mathbf{J}_{12}$ propagates in accord with a pair of Helmholtz equations,

$$\nabla_1^2 \mathbf{J}_{12} + (\bar{k})^2 \mathbf{J}_{12} = 0$$

$$\nabla_2^2 \mathbf{J}_{12} + (\bar{k})^2 \mathbf{J}_{12} = 0, \qquad (5.4\text{-}19)$$

where $\bar{k} = 2\pi/\bar{\lambda}$.

---

†In manipulating the right-hand side of this equation, we have used the fact that

$$\frac{\partial^2 u_1(t+\tau)}{\partial(t+\tau)^2} = \frac{\partial^2 u_1(t+\tau)}{\partial \tau^2},$$

as can readily be proved from the fundamental definition of a derivative.

PROPAGATION OF MUTUAL COHERENCE

5.4.3 Propagation of Cross-Spectral Density

Our previous discussion and our treatments to follow rest heavily on the laws of propagation of the mutual coherence function and mutual intensity. It is also possible to treat these same problems in terms of propagation of cross-spectral density, that is, the Fourier transform of the mutual coherence function. Here we briefly discuss the relationship of such solutions to those we rely on here.

From the basic definition of cross-spectral density [cf. Eq. (3.5-5)], the mutual coherence function can be expressed as an inverse Fourier transform of the cross-spectral density function,

$$\Gamma_{12}(\tau) = \int_0^\infty \mathcal{G}_{12}(\nu) e^{-j2\pi\nu\tau} d\nu, \qquad (5.4\text{-}20)$$

where it has been noted that $\mathcal{G}_{12}$ is zero for negative frequencies. Knowing the propagation equations obeyed by mutual intensity [Eqs. (5.4-17) and (5.4-18)], we can apply these laws to Eq. (5.4-20) and deduce the corresponding laws for cross-spectral density. Interchange of orders of differentiation and integration allows the new equations to be written

$$\int_0^\infty \left[\nabla_1^2 - \frac{1}{c^2} \frac{\partial^2}{\partial \tau^2} \right] \mathcal{G}_{12}(\nu) e^{-j2\pi\nu\tau} d\nu = 0$$

$$\int_0^\infty \left[\nabla_2^2 - \frac{1}{c^2} \frac{\partial^2}{\partial \tau^2} \right] \mathcal{G}_{12}(\nu) e^{-j2\pi\nu\tau} d\nu = 0. \qquad (5.4\text{-}21)$$

For these equations to hold for all delays τ and all cross-spectral densities, the integrands of the integrals on the left must vanish. Applying the τ derivatives to the exponentials, which contain the only dependence on that variable, we obtain a pair of Helmholtz equations that must be satisfied by the cross-spectral density,

$$\nabla_1^2 \mathcal{G}_{12}(\nu) + \left(\frac{2\pi\nu}{c} \right)^2 \mathcal{G}_{12}(\nu) = 0$$

$$\nabla_2^2 \mathcal{G}_{12}(\nu) + \left(\frac{2\pi\nu}{c} \right)^2 \mathcal{G}_{12}(\nu) = 0. \qquad (5.4\text{-}22)$$

The main significance of this result is appreciated by examining Eq. (5.4-19), which presents the Helmholtz equations satisfied by the mutual intensity $\mathbf{J}_{12}$. Remembering that $\bar{k} = 2\pi\bar{\nu}/c$, we see that cross-spectral density and mutual intensity obey the same set of Helmholtz equations. The only difference is the appearance of frequency ν in Eqs. (5.4-22) wherever the center frequency $\bar{\nu}$ would appear in (5.4-19). This observation leads us to the following general conclusion:

Cross-spectral densities obey the same propagation laws as do mutual intensities. To find the solution for cross-spectral density, the corresponding result for mutual intensity can be used, subject only to the requirement that the parameter $\bar{\nu}$ must be replaced by ν.

In attacking coherence problems in the frequency domain using cross-spectral density, it is sometimes useful to introduce the definition of yet another coherence quantity, known as (Ref. 5-27) the *complex degree of spectral coherence* and defined as

$$\mu_{12}(\nu) = \frac{\mathscr{G}_{12}(\nu)}{\left[\mathscr{G}_{11}(\nu)\mathscr{G}_{22}(\nu)\right]^{1/2}}, \qquad (5.4\text{-}23)$$

where $\mathscr{G}_{11}(\nu)$ and $\mathscr{G}_{22}(\nu)$ are the power spectral densities of the light at points P_1 and P_2, respectively. The complex degree of spectral coherence can be shown to satisfy the inequality

$$0 \le |\mu_{12}(\nu)| \le 1. \qquad (5.4\text{-}24)$$

The reader interested in studying proofs of these relationships is referred to Ref. 5-27. We have chosen to use mutual intensities in our analyses, rather than cross-spectral densities, primarily because $\mathbf{J}_{12}$ directly describes the amplitude and phase of a spatial fringe, whereas cross-spectral density is one step further removed from the physics of the problems of concern.

5.5 LIMITING FORMS OF THE MUTUAL COHERENCE FUNCTION

In this section we consider certain limiting conditions of coherence that are important idealizations in practical calculations. In particular, the concepts of a *coherent* wavefield and an *incoherent* wavefield are defined.

5.5.1 A Coherent Field

In terms of the definitions of coherence already introduced, we are led naturally to say that the optical waveforms observed at points P_1 and P_2,

LIMITING FORMS OF THE MUTUAL COHERENCE FUNCTION

subject to relative time delay τ, are *fully coherent* provided

$$|\gamma_{12}(\tau)| = 1. \tag{5.5-1}$$

Although this condition defines perfect coherence for particular points (P_1, P_2) and a particular time delay τ, we might inquire whether there is a more general definition that will allow the entire wavefield to be referred to as fully coherent.

One possible definition is to call a wavefield fully coherent provided

$$|\gamma_{12}(\tau)| = 1 \quad \text{for } all \ (P_1, P_2) \text{ and } all \ \tau. \tag{5.5-2}$$

Such a definition is overly restrictive, however, since no real experiment involves simultaneously all values of delay τ. Furthermore, it can be shown to be satisfied only for monochromatic waves, leading us to seek a weaker and more widely applicable definition.

A less stringent condition was introduced by Mandel and Wolf (Ref. 5-28). According to this definition, a wavefield is called fully coherent if, for every pair of points (P_1, P_2), there exists a delay τ_{12} [a function of (P_1, P_2)] such that $|\gamma_{12}(\tau_{12})| = 1$. Stated mathematically, we require that

$$\max_{\tau} |\gamma_{12}(\tau)| = 1 \quad \text{for every pair } (P_1, P_2). \tag{5.5-3}$$

If the field is cross-spectrally pure, an equivalent definition is easily seen to be

$$|\mu_{12}| = 1 \quad \text{for every pair } (P_1, P_2). \tag{5.5-4}$$

Some physical insight into the concept of perfect coherence can be obtained by expressing the condition $|\gamma_{12}(\tau_{12})| = 1$ in terms of the complex envelopes of the two wave disturbances. From the definition of the complex degree of coherence we have

$$|\gamma_{12}(\tau_{12})| = \frac{|\langle \mathbf{u}(P_1, t + \tau_{12})\mathbf{u}^*(P_2, t)\rangle|}{[\langle |\mathbf{u}(P_1, t + \tau_{12})|^2\rangle \langle |\mathbf{u}(P_2, t)|^2\rangle]^{1/2}}. \tag{5.5-5}$$

Now using the fact that

$$\mathbf{u}(P, t) = \mathbf{A}(P, t)e^{-j2\pi\bar{\nu}t} \tag{5.5-6}$$

we can equivalently write

$$|\gamma_{12}(\tau_{12})| = \frac{|\langle \mathbf{A}(P_1, t + \tau_{12})\mathbf{A}^*(P_2, t)\rangle|}{[\langle |\mathbf{A}(P_1, t + \tau_{12})|^2\rangle \langle |\mathbf{A}(P_2, t)|^2\rangle]^{1/2}}. \tag{5.5-7}$$

It is now useful to apply Schwarz's inequality, which states that

$$\left| \int f(t) g^*(t) \, dt \right| \leq \left[\int |f(t)|^2 \, dt \int |g(t)|^2 \, dt \right]^{1/2}, \qquad (5.5\text{-}8)$$

with equality if and only if

$$g(t) = k f(t), \qquad (5.5\text{-}9)$$

where k is a complex constant.

Applying (5.5-8) and (5.5-9) to (5.5-7), we see that $|\gamma_{12}(\tau_{12})| = 1$ if and only if

$$A(P_2, t) = k_{12} A(P_1, t + \tau_{12}), \qquad (5.5\text{-}10)$$

where k_{12} is a complex constant that in general depends on the points P_1 and P_2.

Stating the above result in words, a wavefield is called perfectly coherent if and only if, for every pair of points P_1 and P_2, there exists a time delay τ_{12} such that the complex envelopes of the two waveforms, relatively delayed by the required τ_{12}, differ by only a time-independent complex constant.

When the quasimonochromatic conditions are imposed on the wavefield of concern, the situation simplifies somewhat. In any one experiment it is likely that a multitude of different pinhole spacings will be involved. If we insist that the quasimonochromatic conditions be satisfied, by implication we mean they should simultaneously be satisfied for all pinhole pairs involved in the experiment, thus implying that for *all* points (P_1, P_2) the *same* delay τ_{12} should be required to eliminate temporal coherence effects. Furthermore, if we let pinhole P_1 approach pinhole P_2, thus including negligibly small (or zero) spacings in our experiment, it is clear that the unique delay τ_{12} required to maximize $|\Gamma_{12}(\tau)|$ must in fact be identically zero. From Eq. (5.5-10), the complex envelopes at P_1 and P_2 are now related by

$$A(P_2, t) = k_{12} A(P_1, t), \qquad (5.5\text{-}11)$$

where again k_{12} depends on the particular points (P_1, P_2). Thus the complex envelopes at all points vary in unison, differing from each other only by time-invariant amplitude and phase factors.

A useful special form for the mutual intensity, valid in the fully coherent, quasimonochromatic case, can be found by expressing the complex envelopes $A(P_1, t)$ and $A(P_2, t)$ in terms of the complex envelope $A(P_0, t)$ at a prechosen reference point P_0. We define time-invariant phasor amplitudes $A(P_1)$ and $A(P_2)$ in terms of the complex envelope at P_0 as follows:

$$A(P_1, t) = A(P_1) \frac{A(P_0, t)}{[I(P_0)]^{1/2}}$$

$$A(P_2, t) = A(P_2) \frac{A(P_0, t)}{[I(P_0)]^{1/2}}. \qquad (5.5\text{-}12)$$

The mutual intensity is now calculated to be

$$J_{12} = \langle A(P_1, t) A^*(P_2, t) \rangle = A(P_1) A^*(P_2). \qquad (5.5\text{-}13)$$

Alternatively, the complex coherence factor can be expressed in the form

$$\mu_{12} = \exp\{ j[\phi(P_1) - \phi(P_2)] \}, \qquad (5.5\text{-}14)$$

where

$$\phi(P_1) = \arg\{A(P_1)\}, \qquad \phi(P_2) = \arg\{A(P_2)\}. \qquad (5.5\text{-}15)$$

For fully coherent, quasimonochromatic radiation, the fringe pattern generated by a Young's interference experiment takes the form

$$I(Q) = I^{(1)}(Q) + I^{(2)}(Q)$$
$$+ 2\sqrt{I^{(1)}(Q) I^{(2)}(Q)} \cos\left[\frac{2\pi(r_2 - r_1)}{\lambda} + \phi(P_2) - \phi(P_1) \right]$$

$$(5.5\text{-}16)$$

for every pinhole pair (P_1, P_2). When the intensity of the wave is uniform, the visibility of the fringes is always unity, but the phase of the fringe pattern will change as P_1 and P_2 are changed.

5.5.2 An Incoherent Field

For a fully coherent field, the fluctuations of the complex envelopes of the wave at P_1 and P_2 are perfectly correlated, provided the appropriate delay τ_{12} is introduced. The logical opposite of a fully coherent field is called an *incoherent* field. Thus we might reasonably define a field to be incoherent if

$$|\Gamma_{12}(\tau)| = 0 \quad \text{for all } P_1 \neq P_2 \text{ and for all } \tau. \qquad (5.5\text{-}17)$$

Although such a definition is indeed the logical opposite of a fully coherent field, it is not a very useful definition in practice. The reason can readily be seen by substituting $\Gamma[P_1, P_2; \tau + (r_2 - r_1)/c] = 0$ in Eq. (5.4-4). If this is first integrated over the surface Σ_1, the integrand is zero everywhere except when $P_1 = P_2$, where it has a finite value. The result of the integration is thus precisely zero, and we have the result

$$\Gamma(Q_1, Q_2; \tau) \equiv 0. \tag{5.5-18}$$

Letting $\tau = 0$ and $Q_2 = Q_1$, we see that (5.5-18) implies that $I(Q_1) = I(Q_2) = 0$. Thus if the wavefield on Σ_1 is incoherent in the sense defined above, it does not propagate to Σ_2!

The physical explanation of the above seemingly nonphysical result lies in the *evanescent-wave* phenomenon. A wavefield incoherent in the sense of (5.5-17) has infinitesimally fine spatial structure. However, spatial structure finer than a wavelength corresponds to nonpropagating evanescent waves (see, e.g., Ref. 5-24, pp. 50 and 51). Hence the perfectly incoherent surface does not radiate.

When the evanescent-wave phenomenon is taken fully into account, it can be shown that for a propagating wave, coherence must exist over a linear dimension of at least a wavelength. For quasimonochromatic light, the mutual intensity most closely approximating incoherence, yet still corresponding to a propagating wave, is found to be (Ref. 5-11, pp. 57–60)

$$\mathbf{J}(P_1, P_2) = \sqrt{I(P_1)I(P_2)} \left[2 \frac{J_1\left(\bar{k}\sqrt{(x_1 - x_2)^2 + (y_1 - y_2)^2}\right)}{\bar{k}\sqrt{(x_1 - x_2)^2 + (y_1 - y_2)^2}} \right], \tag{5.5-19}$$

where P_1 and P_2 are assumed to lie in a plane and have coordinates (x_1, y_1) and (x_2, y_2), respectively; $J_1(x)$ is a Bessel function of the first kind, order 1; and $\bar{k} = 2\pi/\bar{\lambda}$.

In practical computations, the form (5.5-19) is rather cumbersome to use. If a wave with such a mutual intensity passes through an optical system that has resolution in the (x, y) plane that is much coarser than $\bar{\lambda}$, the exact shape of $\mathbf{J}(P_1, P_2)$ is not of consequence. In this case the mutual intensity corresponding to incoherence can be approximated by

$$\mathbf{J}(P_1, P_2) = \kappa I(P_1)\delta(x_1 - x_2, y_1 - y_2), \tag{5.5-20}$$

where $\delta(\cdot, \cdot)$ represents a two-dimensional Dirac delta function. The constant κ should be chosen to assure that the volume of the function $\mathbf{J}(P_1, P_2)$ in (5.5-20) is the same as in (5.5-19). The required value is

$$\kappa = \frac{(\bar{\lambda})^2}{\pi}. \tag{5.5-21}$$

If coherence extends over more than a wavelength, but the optical system that follows still cannot resolve a coherence area, the δ-function representation for $\mathbf{J}(P_1, P_2)$ is still valid, although the appropriate value of κ is no longer $\bar{\lambda}^2/\pi$. Since the constant κ ultimately affects the intensity level but *not* spatial structure, it is often replaced by unity for simplicity. Since this constant has dimensions of squared length [cf. Eq. (5.5-21)], however, we retain it in our future mathematical expressions to assure dimensional consistency.

5.6 THE VAN CITTERT–ZERNIKE THEOREM

In nearly all optical problems involving light that does not originate from a laser, the original optical source consists of an extended collection of independent radiators. Such a source can reasonably be modeled as incoherent in the sense defined in the preceding section, provided only that the optical elements through which the light passes are incapable of resolving the individual radiating elements on the source. Accordingly, it is of some special interest to know precisely how mutual intensity propagates away from an incoherent source. The character of the mutual intensity function produced by an incoherent source is fully described by the Van Cittert–Zernike theorem, which is undoubtedly one of the most important theorems of modern optics. As the name implies, the theorem was first demonstrated in papers by Van Cittert (Ref. 5-4) and Zernike (Ref. 5-5).

5.6.1 Mathematical Derivation

Restricting our attention to quasimonochromatic light, we have previously shown that mutual intensity propagates according to the law

$$\mathbf{J}(Q_1, Q_2) = \iint_\Sigma \iint_\Sigma \mathbf{J}(P_1, P_2) \exp\left[-j\frac{2\pi}{\bar{\lambda}}(r_2 - r_1)\right] \frac{\chi(\theta_1)}{\bar{\lambda} r_1} \frac{\chi(\theta_2)}{\bar{\lambda} r_2} dS_1 dS_2 \tag{5.6-1}$$

regardless of the initial state of coherence represented by $\mathbf{J}(P_1, P_2)$. For the special case of an *incoherent* source, we further have [Eq. (5.5-20)] that

$$\mathbf{J}(P_1, P_2) = \kappa I(P_1)\delta(|P_1 - P_2|). \qquad (5.6\text{-}2)$$

Simple substitution and use of the "sifting" or "sampling" property of the δ function yields a mutual intensity

$$\mathbf{J}(Q_1, Q_2) = \frac{\kappa}{(\bar{\lambda})^2} \iint_\Sigma I(P_1)\exp\left[-j\frac{2\pi(r_2 - r_1)}{\bar{\lambda}}\right] \frac{\chi(\theta_1)}{r_1} \frac{\chi(\theta_2)}{r_2} dS, \qquad (5.6\text{-}3)$$

where the required geometrical factors are illustrated in Fig. 5-18.

To simplify this expression further, we make certain assumptions and approximations as follows:

(1) The extents of the source and observation region are much less than the distance z separating them. Thus

$$\frac{1}{r_1} \cdot \frac{1}{r_2} \cong \frac{1}{z^2}. \qquad (5.6\text{-}4)$$

(2) Only small angles are involved. Thus

$$\chi(\theta_1) \cong \chi(\theta_2) \cong 1. \qquad (5.6\text{-}5)$$

The mutual intensity in the observation region now takes the form

$$\mathbf{J}(Q_1, Q_2) = \frac{\kappa}{(\bar{\lambda}z)^2} \iint_\Sigma I(P_1)\exp\left[-j\frac{2\pi}{\bar{\lambda}}(r_2 - r_1)\right] dS. \qquad (5.6\text{-}6)$$

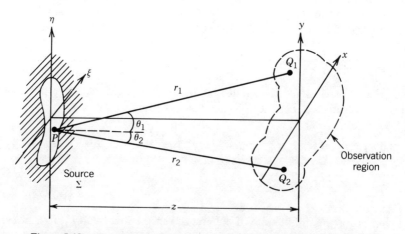

Figure 5-18. Geometry for derivation of the Van Cittert–Zernike theorem.

THE VAN CITTERT-ZERNIKE THEOREM

At this point we specifically adopt the planar geometry shown in Fig. 5-18; that is, the source and observation regions are assumed to lie in parallel planes separated by distance z. Furthermore, in accord with our previous assumptions (5.6-4) and (5.6-5), we introduce the "paraxial" approximations

$$r_2 = \sqrt{z^2 + (x_2 - \xi)^2 + (y_2 - \eta)^2} \cong z + \frac{(x_2 - \xi)^2 + (y_2 - \eta)^2}{2z}$$

$$r_1 = \sqrt{z^2 + (x_1 - \xi)^2 + (y_1 - \eta)^2} \cong z + \frac{(x_1 - \xi)^2 + (y_1 - \eta)^2}{2z}.$$

(5.6-7)

Finally, we adopt the definitions $\Delta x = x_2 - x_1$, $\Delta y = y_2 - y_1$, and the convention that $I(\xi, \eta)$ equals zero when (ξ, η) lies outside the finite source region Σ. The final form of the Van Cittert–Zernike theorem then follows:

$$\mathbf{J}(x_1, y_1; x_2, y_2) = \frac{\kappa e^{-j\psi}}{(\bar{\lambda}z)^2} \iint_{-\infty}^{\infty} I(\xi, \eta) \exp\left[j\frac{2\pi}{\bar{\lambda}z}(\Delta x \xi + \Delta y \eta)\right] d\xi \, d\eta.$$

(5.6-8)

In this expression, the phase factor ψ is given by

$$\psi = \frac{\pi}{\bar{\lambda}z}\left[(x_2^2 + y_2^2) - (x_1^2 + y_1^2)\right] = \frac{\pi}{\bar{\lambda}z}(\rho_2^2 - \rho_1^2) \quad (5.6\text{-}9)$$

where ρ_2 and ρ_1 represent, respectively, the distances of the points (x_2, y_2) and (x_1, y_1) from the optical axis.

It is often more convenient to express this theorem in normalized form, writing the complex coherence factor as

$$\mu(x_1, y_1; x_2, y_2) = \frac{e^{-j\psi} \iint_{-\infty}^{\infty} I(\xi, \eta) \exp\left[j\frac{2\pi}{\bar{\lambda}z}(\Delta x \xi + \Delta y \eta)\right] d\xi \, d\eta}{\iint_{-\infty}^{\infty} I(\xi, \eta) \, d\xi \, d\eta},$$

(5.6-10)

thus eliminating the awkward scaling factors. In most practical applications involving incoherent sources, to a good approximation $I(x_1, y_1) \cong I(x_2, y_2)$, and hence $|\mu(x_1, y_1; x_2, y_2)|$ also represents the classical visibility of the fringes that would be produced in Young's experiment.

5.6.2 Discussion

The Van Cittert–Zernike theorem, stated mathematically in Eq. (5.6-8), can be expressed in words as follows: aside from the factor $\exp(-j\psi)$ and scaling constants, the mutual intensity $\mathbf{J}(x_1, y_1; x_2, y_2)$ can be found by a two-dimensional Fourier transformation of the intensity distribution $I(\xi, \eta)$ across the source. This relationship can be likened to the relationship between the field across a coherently illuminated aperture and the field observed in the Fraunhofer diffraction pattern of that aperture, although the physical quantities involved are entirely different. In this analogy, we regard the intensity distribution $I(\xi, \eta)$ as analogous to the field across the aperture and $\mathbf{J}(x_1, y_1; x_2, y_2)$ as analogous to the field in the Fraunhofer diffraction pattern. The relationship (5.6-8) is the same as the corresponding Fraunhofer diffraction formula. We emphasize again that this analogy is only a mathematical one, however, for the physical situations described by the same equation are entirely different, as are the physical quantities involved. We further note that, as implied by Eq. (5.6-7), the Fourier transform relationship between $\mathbf{J}(x_1, y_1; x_2, y_2)$ and $I(\xi, \eta)$ holds over a wider range of distances than would the analogous Fraunhofer diffraction equation, for the paraxial approximation is valid in regions of both Fresnel and Fraunhofer diffraction (see Ref. 5-24, Chapter 4).

Noting that the modulus of the complex coherence factor $|\mu|$ depends only on the difference of coordinates $(\Delta x, \Delta y)$ in the (x, y) plane, it is possible to define the *coherence area* A_c of the light in a manner entirely analogous to the definition (5.1-28) of coherence time τ_c. For our purposes, the coherence area is defined by

$$A_c \triangleq \iint_{-\infty}^{\infty} |\mu(\Delta x, \Delta y)|^2 \, d\Delta x \, d\Delta y. \tag{5.6-11}$$

The reader may wish to prove, with the help of Problem 5-15, that for a uniformly bright incoherent source of area A_S and any shape, the coherence area A_c at distance z from the source is

$$A_c = \frac{(\bar{\lambda} z)^2}{A_S} \cong \frac{(\bar{\lambda})^2}{\Omega_S}, \tag{5.6-12}$$

where Ω_S is the solid angle subtended by the source from the origin of the observation region.

THE VAN CITTERT–ZERNIKE THEOREM

Returning to the general expression (5.6-10) for μ, we consider conditions under which the factor $\exp(-j\psi)$ can be dropped in the expression for the complex coherence factor. Since

$$\psi = \frac{\pi}{\lambda z}(\rho_2^2 - \rho_1^2) \tag{5.6-13}$$

three different conditions can be identified:

1. If the distance z is so large that $z \gg 2[(\rho_2^2 - \rho_1^2)/\bar{\lambda}]$, then $\psi \ll \pi/2$ and $\exp(-j\psi) \cong 1$.
2. If the measurement points Q_1 and Q_2 are intentionally maintained at equal distances from the optical axis (although their spacing may be changed in both size and direction), ψ is identically zero.
3. If, rather than lying in a plane, the pinholes lie on a reference sphere of radius z, centered on the source, the phase factors vanish.

In such cases the phase factor ψ can, of course, be dropped.

Finally, we remind the reader that the mathematical result relating μ_{12} to the source intensity distribution can be understood qualitatively by consideration of a simple Young's experiment with the extended source. Just as a point source will create interference fringes of perfect visibility, each point on an incoherent source will create a separate fringe of high visibility. If the source size is too large, these elementary fringe patterns add with significantly different spatial phases, and the contrast of the overall fringe pattern is reduced. The mathematical statement of the Van Cittert–Zernike theorem is simply a precise statement of this relationship between the intensity distribution across the source and resulting fringe contrast for given locations of the pinholes.

5.6.3 An Example

As an example of the use of the Van Cittert–Zernike theorem, the complex coherence factor μ_{12} of the light produced by a uniformly bright, incoherent, quasimonochromatic circular disk of radius a will be calculated. The intensity distribution of the source is thus assumed to be

$$I(\xi, \eta) = I_0 \operatorname{circ} \frac{\sqrt{\xi^2 + \eta^2}}{a}, \tag{5.6-14}$$

where

$$\operatorname{circ} w \triangleq \begin{cases} 1 & w < 1 \\ 1/2 & w = 1 \\ 0 & w > 1. \end{cases} \tag{5.6-15}$$

To find $\mathbf{J}(x_1, y_1; x_2, y_2)$, we must Fourier transform this distribution. We note first that (Ref. 5-24, pp. 15 and 16)

$$\mathscr{F}\left\{\operatorname{circ}\frac{\sqrt{\xi^2 + \eta^2}}{a}\right\} = a^2 \frac{J_1\left(2\pi a \sqrt{\nu_X^2 + \nu_Y^2}\right)}{a\sqrt{\nu_X^2 + \nu_Y^2}}, \qquad (5.6\text{-}16)$$

where $\mathscr{F}\{\cdot\}$ is a two-dimensional Fourier transform operator,

$$\mathscr{F}\{\mathbf{g}(\xi,\eta)\} \triangleq \iint_{-\infty}^{\infty} \mathbf{g}(\xi,\eta) e^{j2\pi(\xi\nu_X + \eta\nu_Y)} d\xi\, d\eta, \qquad (5.6\text{-}17)$$

and $J_1(\cdot)$ is a Bessel function of the first kind, order 1. Furthermore, in accord with the scaling factors in the exponent of Eq. (5.6-8), we must substitute

$$\nu_X = \frac{\Delta x}{\bar{\lambda} z}, \qquad \nu_Y = \frac{\Delta y}{\bar{\lambda} z}. \qquad (5.6\text{-}18)$$

The result is

$$\mathbf{J}(x_1, y_1; x_2, y_2) = \frac{\pi a^2 I_0 \kappa}{(\bar{\lambda} z)^2} e^{-j\psi} \left[2 \frac{J_1\left(\frac{2\pi a}{\bar{\lambda} z}\sqrt{(\Delta x)^2 + (\Delta y)^2}\right)}{\frac{2\pi a}{\bar{\lambda} z}\sqrt{(\Delta x)^2 + (\Delta y)^2}} \right]$$

$$(5.6\text{-}19)$$

for the mutual intensity function, and the corresponding complex coherence factor is

$$\boldsymbol{\mu}(x_1, y_1; x_2, y_2) = e^{-j\psi} \left[2 \frac{J_1\left(\frac{2\pi a}{\bar{\lambda} z}\sqrt{(\Delta x)^2 + (\Delta y)^2}\right)}{\frac{2\pi a}{\bar{\lambda} z}\sqrt{(\Delta x)^2 + (\Delta y)^2}} \right]. \qquad (5.6\text{-}20)$$

Note that the first factor $e^{-j\psi}$ depends on both (x_1, y_1) and (x_2, y_2), whereas the second factor depends only on the *spacing* of the two points, $s = \sqrt{(\Delta x)^2 + (\Delta y)^2}$. Thus the modulus $|\boldsymbol{\mu}_{12}|$ depends only on Δx and Δy and is shown in Fig. 5-19. The first zero of $J_1(2\pi a \rho)$ appears at $\rho = 0.610/a$, and hence the first zero of $|\boldsymbol{\mu}_{12}|$ occurs at spacing

$$s_0 = 0.610 \frac{\bar{\lambda} z}{a}. \qquad (5.6\text{-}21)$$

THE VAN CITTERT-ZERNIKE THEOREM

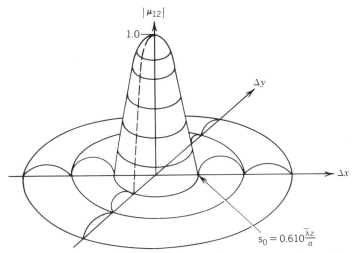

Figure 5-19. Modulus $|\mu_{12}|$ of the complex coherence factor versus coordinate differences Δx and Δy in the (x, y) plane.

Remembering our small-angle approximation, the angular diameter θ of the source, viewed from the origin of the (x, y) plane, is $\theta \cong 2a/z$. Thus the spacing for the first zero of $|\mu_{12}|$ can also be expressed as

$$s_0 = 1.22 \frac{\bar{\lambda}}{\theta}. \tag{5.6-22}$$

The coherence area of the light emitted by the source can be found with the help of the results of Problem 5-15. For a circular incoherent source of radius a, the coherence area at distance z is

$$A_c = \frac{\bar{\lambda}^2 z^2}{A_S} = \frac{\bar{\lambda}^2 z^2}{\pi a^2}. \tag{5.6-23}$$

Recognizing that only small angles are involved in our analysis, we note that the solid angle subtended by the source, viewed from the origin of the (x, y) plane, is

$$\Omega_S \cong \frac{A_S}{z^2}. \tag{5.6-24}$$

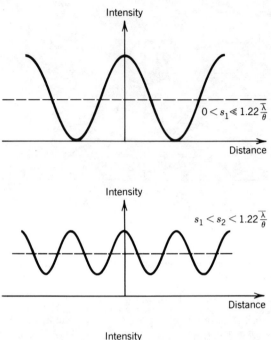
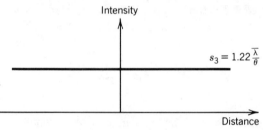
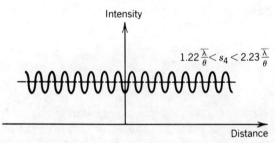

Figure 5-20. Fringe patterns produced by a circular incoherent source for various spacings s_1, s_2, s_3, s_4 of pinholes.

THE VAN CITTERT-ZERNIKE THEOREM

We see that the coherence area can be expressed as

$$A_c \cong \frac{\bar{\lambda}^2}{\Omega_S}, \qquad (5.6\text{-}25)$$

as was previously stated in Eq. (5.6-12).

Suppose that the points (x_1, y_1) and (x_2, y_2) correspond to pinholes in an opaque screen and that fringes of interference are observed some distance behind the screen. Our knowledge of the character of μ_{12} allows us to predict the character of the fringes obtained at each possible spacing s of

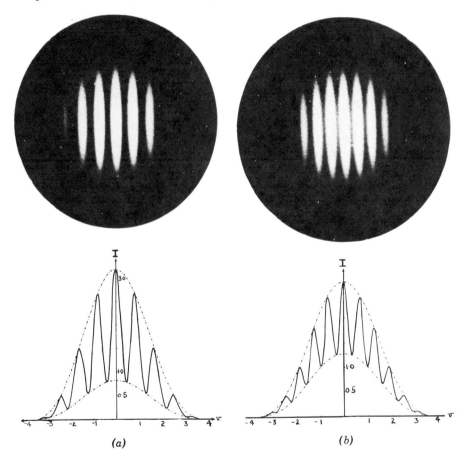

Figure 5-21. Photographs of the fringes obtained from a circular incoherent source with various spacings of pinholes. The spacing of the pinholes progressively increases in parts *a–g*. (Courtesy of B. J. Thompson and E. Wolf, *J. Opt. Soc. Am.*, vol. 47, pp. 898, 899, 1957.)

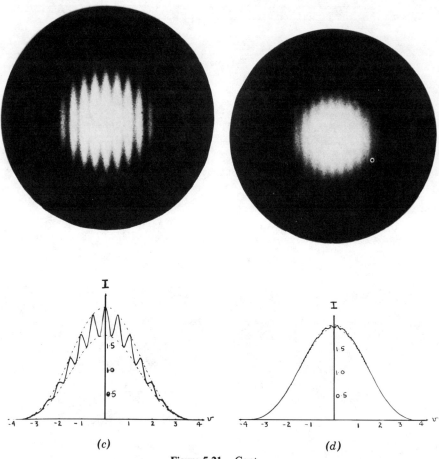

Figure 5-21. Cont.

the pinholes. Suppose, for simplicity, that (x_1, y_1) and (x_2, y_2) are always equally distant from the optical axis, thus assuring that $\psi = 0$. The predicted fringe patterns obtained for various spacings are shown in Fig. 5-20. Note the increasing spatial frequency of the fringes as s is increased, the vanishing of fringe contrast at spacing s_0, and the reversal of phase of the fringes when the spacing s corresponds to the first negative lobe of the Bessel function. Photographs of interference patterns obtained at various spacings are shown in Figure 5-21, where the finite size of the fringe patterns is due to the finite width of the diffraction patterns of the pinholes used.

Throughout our discussions it has been assumed that the circular source is centered on the optical axis. If the source is offset from this position by

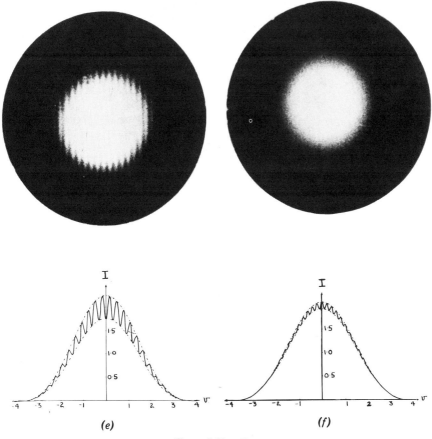

Figure 5-21. Cont.

displacements $\Delta\xi$ and $\Delta\eta$ in the (ξ, η) plane, the shift theorem of Fourier analysis (Ref. 5-24, p. 9) implies that the new complex coherence factor $\boldsymbol{\mu}'_{12}$ can be expressed in terms of the old complex coherence factor (source centered on axis) $\boldsymbol{\mu}_{12}$ by

$$\boldsymbol{\mu}'_{12} = \boldsymbol{\mu}_{12} \exp\left[j\frac{2\pi}{\lambda z}(\Delta\xi \Delta x + \Delta\eta \Delta y)\right]. \tag{5.6-26}$$

Thus the modulus $|\boldsymbol{\mu}_{12}|$ of the complex coherence factor is unaffected by a translation of the source, but the *phase* of the fringes is changed in proportion to the source translation increments $(\Delta\xi, \Delta\eta)$ and in proportion to the pinhole spacings $(\Delta x, \Delta y)$.

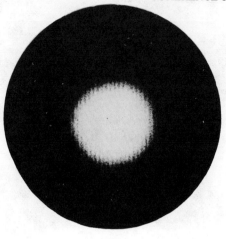

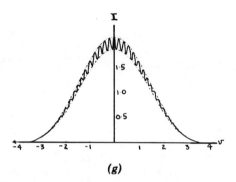

(g)

Figure 5-21. Cont.

5.6.4 A Generalized Van Cittert–Zernike Theorem[†]

In deriving the Van Cittert–Zernike theorem, the δ-function form of the source mutual intensity function was used to represent an incoherent source. We consider now a more general form of the Van Cittert–Zernike theorem that applies to a restricted class of partially coherent sources, including an incoherent source in the previous sense as a special case. The effect of a

---

[†]A rudimentary form of the result to be derived here was first demonstrated in Ref. 5-29. A more elegant and more complete result has been derived in Ref. 5-30. For a review of work in this area relating coherence theory and radiometry, see Ref. 5-31.

small but nonzero coherence area on the source will be evident from these results.

The mutual intensity function of the source is assumed to be of the form

$$\mathbf{J}(\xi_1, \eta_1; \xi_2, \eta_2) = [I(\xi_1, \eta_1) I(\xi_2, \eta_2)]^{1/2} \mu(\Delta\xi, \Delta\eta). \quad (5.6\text{-}27)$$

Implicit in this form is the assumption that the complex coherence factor μ depends only on coordinate differences $(\Delta\xi, \Delta\eta)$ in the (ξ, η) plane. Such is often the case in practice (see, e.g., Section 7.2). A radiator having a mutual intensity in the form of Eq. (5.6-27) has come to be known as a "quasi-homogeneous" source.

As a further approximation we assume that the source size is much larger than a coherence area A_c on the source and that any spatial structure associated with the source intensity distribution is coarse compared with A_c. These assumptions allow us to approximate the mutual intensity function of the source by

$$\mathbf{J}(\xi_1, \eta_1; \xi_2, \eta_2) \cong I(\bar{\xi}, \bar{\eta}) \mu(\Delta\xi, \Delta\eta), \quad (5.6\text{-}28)$$

where[†]

$$\Delta\xi = \xi_2 - \xi_1, \quad \bar{\xi} = \frac{\xi_1 + \xi_2}{2}$$

$$\Delta\eta = \eta_2 - \eta_1, \quad \bar{\eta} = \frac{\eta_1 + \eta_2}{2}. \quad (5.6\text{-}29)$$

This approximate form can now be substituted in the relationship

$$\mathbf{J}(x_1, y_1; x_2, y_2) = \frac{1}{(\bar{\lambda}z)^2} \iiiint_{-\infty}^{\infty} \mathbf{J}(\xi_1, \eta_1; \xi_2, \eta_2)$$

$$\times \exp\left[-j \frac{2\pi}{\bar{\lambda}} (r_2 - r_1)\right] d\xi_1 d\eta_1 d\xi_2 d\eta_2 \quad (5.6\text{-}30)$$

which is the general propagation law (5.4-8) for mutual intensity, taken under paraxial conditions. Under such conditions, the difference $r_2 - r_1$

[†]Equivalently, $\xi_1 = \bar{\xi} - \Delta\xi/2$, $\xi_2 = \bar{\xi} + \Delta\xi/2$, $\eta_1 = \bar{\eta} - \Delta\eta/2$, $\eta_2 = \bar{\eta} + \Delta\eta/2$.

takes the approximate form

$$r_2 - r_1 \cong \frac{1}{2z}[(x_2^2 + y_2^2) - (x_1^2 + y_1^2) + (\xi_2^2 + \eta_2^2) - (\xi_1^2 + \eta_1^2)$$
$$- 2(x_2\xi_2 + y_2\eta_2) + 2(x_1\xi_1 + y_1\eta_1)]$$
$$= \frac{1}{2z}[(x_2 + x_1)(x_2 - x_1) + (y_2 + y_1)(y_2 - y_1)$$
$$+ (\xi_2 + \xi_1)(\xi_2 - \xi_1) + (\eta_2 + \eta_1)(\eta_2 - \eta_1)$$
$$- 2(x_2\xi_2 - x_1\xi_1) - 2(y_2\eta_2 - y_1\eta_1)]. \quad (5.6\text{-}31)$$

Now we use the previous definitions for $\bar{\xi}$, $\bar{\eta}$, $\Delta\xi$, and $\Delta\eta$, and additionally define

$$\bar{x} = \frac{x_1 + x_2}{2}, \qquad \Delta x = x_2 - x_1$$
$$\bar{y} = \frac{y_1 + y_2}{2}, \qquad \Delta y = y_2 - y_1. \quad (5.6\text{-}32)$$

Substituting these definitions in expression (5.6-31), we obtain

$$r_2 - r_1 \cong \frac{1}{z}[\bar{x}\Delta x + \bar{y}\Delta y + \bar{\xi}\Delta\xi + \bar{\eta}\Delta\eta$$
$$- \Delta x\bar{\xi} - \bar{x}\Delta\xi - \Delta y\bar{\eta} - \bar{y}\Delta\eta]. \quad (5.6\text{-}33)$$

At this point we find it convenient to impose the following assumption (which is discussed in more detail shortly):

$$z > 4\frac{\bar{\xi}\Delta\xi}{\lambda} \quad \text{and} \quad z > 4\frac{\bar{\eta}\Delta\eta}{\lambda} \quad (5.6\text{-}34)$$

for all $\Delta\xi$, $\Delta\eta$, $\bar{\xi}$, and $\bar{\eta}$ of interest in the experiment. This assumption allows us to drop the third and fourth terms of (5.6-33). Now when the modified Eqs. (5.6-33) and (5.6-28) are substituted into the integral (5.6-30), taking account of the change of variables of integration, we obtain

$$\mathbf{J}(x_1, y_1; x_2, y_2) = \frac{e^{-j\psi}}{(\bar{\lambda}z)^2} \iint_{-\infty}^{\infty} I(\bar{\xi}, \bar{\eta}) \exp\left[j\frac{2\pi}{\bar{\lambda}z}(\Delta x\bar{\xi} + \Delta y\bar{\eta})\right] d\bar{\xi}\,d\bar{\eta}$$
$$\times \iint_{-\infty}^{\infty} \mu(\Delta\xi, \Delta\eta) \exp\left[j\frac{2\pi}{\bar{\lambda}z}(\bar{x}\Delta\xi + \bar{y}\Delta\eta)\right] d\Delta\xi\,d\Delta\eta,$$
$$(5.6\text{-}35)$$

where $-\psi$ is given by $(2\pi/\bar{\lambda})(\bar{x}\Delta x + \bar{y}\Delta y)$, which is equivalent to our previous definition (5.6-9).

To afford easy comparison with the previous form of the Van Cittert–Zernike theorem, we adopt a special symbol for the last double integral,

$$\kappa(\bar{x}, \bar{y}) = \iint_{-\infty}^{\infty} \mu(\Delta\xi, \Delta\eta) \exp\left[j\frac{2\pi}{\bar{\lambda}z}(\bar{x}\Delta\xi + \bar{y}\Delta\eta)\right] d\Delta\xi\, d\Delta\eta, \quad (5.6\text{-}36)$$

in which case the mutual intensity is expressed by

$$\mathbf{J}(x_1, y_1; x_2, y_2) = \frac{\kappa(\bar{x}, \bar{y})e^{-j\psi}}{(\bar{\lambda}z)^2} \iint_{-\infty}^{\infty} I(\bar{\xi}, \bar{\eta}) \exp\left[j\frac{2\pi}{\bar{\lambda}z}(\Delta x\bar{\xi} + \Delta y\bar{\eta})\right] d\bar{\xi}\, d\bar{\eta}.$$

$$(5.6\text{-}37)$$

Thus the constant κ of the previous Van Cittert–Zernike theorem has become a function of coordinates $(\bar{x}, \bar{y})$. As a consequence, the modulus $|\mu|$ of the complex coherence factor is no longer a function only of coordinate differences $(\Delta x, \Delta y)$.

Our physical interpretation of the generalized Van Cittert–Zernike theorem is as follows. Since $\mu(\Delta\xi, \Delta\eta)$ is much narrower in the $(\Delta\xi, \Delta\eta)$ plane than $I(\bar{\xi}, \bar{\eta})$ is in the $(\bar{\xi}, \bar{\eta})$ plane, the factor $\kappa(\bar{x}, \bar{y})$ will be broad in $(\bar{x}, \bar{y})$ whereas the integral will be narrow in $(\Delta x, \Delta y)$, a consequence of the reciprocal width relations of Fourier transform pairs (Ref. 5-17, pp. 148–163). We interpret the integral factor as representing the correlation of the light as a function of the separation of two exploratory points (x_1, y_1) and (x_2, y_2), whereas the factor $\kappa(\bar{x}, \bar{y})$ represents a coarse variation of average intensity in the (x, y) plane. Exactly as in the case of incoherent light, it is the *source size* that determines the coherence area of the observed wave, but in addition the *source coherence area* influences the distribution of average intensity over the (x, y) plane.

We close this section with some comments on the conditions (5.6-34) that were used to obtain the generalized result. If D represents the maximum linear dimension of the source and d_c represents the maximum linear dimension of a coherence area of the source, the required condition will be satisfied provided

$$z > 2\frac{Dd_c}{\bar{\lambda}}, \quad (5.6\text{-}38)$$

where it has been noted that $I(\bar{\xi}, \bar{\eta})$ drops to zero when $\sqrt{\bar{\xi}^2 + \bar{\eta}^2} > D/2$ and $\mu(\Delta\xi, \Delta\eta)$ drops to zero when $\sqrt{\Delta\xi^2 + \Delta\eta^2} > d_c$. Equation (5.6-38) may be interpreted as requiring that the observation distance z be at least as large as the geometric mean of the far-field distances for apertures of diameter D and d_c. As a particular example, we suppose that $D = 10^{-2}$ m, $d_c = 10^{-5}$ m, and $\bar{\lambda} = 5 \times 10^{-7}$ m, in which case it is required that z satisfy $z > 0.4$ m. The reader may wish to verify (see Problem 5-16) that, when a positive lens with focal length f is placed between the source and the observation plane and when the observation plane is the rear focal plane of that lens, the restrictions (5.6-34) are no longer necessary, and thus the generalized Van Cittert–Zernike theorem holds under a wider set of circumstances than directly treated here.

5.7 DIFFRACTION OF PARTIALLY COHERENT LIGHT BY AN APERTURE

Suppose that a quasimonochromatic wave is incident on an aperture in an opaque screen, as illustrated in Fig. 5-22. In general, this wave may be partially coherent. We wish to calculate the intensity distribution $I(x, y)$ observed across a parallel plane at distance z beyond the aperture.

5.7.1 Effect of a Thin Transmitting Structure on Mutual Intensity

The diffracting aperture shown in Fig 5-22 may be represented by an amplitude transmittance function

$$\mathbf{t}_A(\xi, \eta) = \begin{cases} 1 & (\xi, \eta) \text{ in } \Sigma \\ 0 & \text{otherwise.} \end{cases} \quad (5.7\text{-}1)$$

More generally, the aperture may contain absorbing and/or phase shifting structures that are characterized by an arbitrary complex-valued amplitude transmittance function within the aperture,[†] subject only to the constraint that $0 \le |\mathbf{t}_A| \le 1$. (For a more complete discussion of amplitude transmittance, see Section 7.1.1.)

Knowing that the mutual intensity function of the incident light is $\mathbf{J}_i(\xi_1, \eta_1; \xi_2, \eta_2)$, we inquire as to the form of the mutual intensity function $\mathbf{J}_t(\xi_1, \eta_1; \xi_2, \eta_2)$ of the transmitted light. If the complex envelope of the

---

[†] We assume that this amplitude transmittance function is independent of wavelength within the narrow bandwidth of the incident light.

DIFFRACTION OF PARTIALLY COHERENT LIGHT BY AN APERTURE

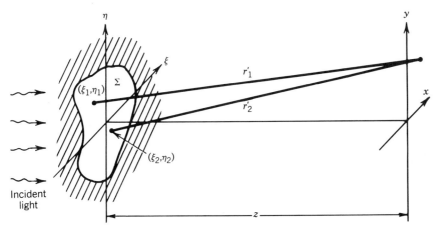

Figure 5-22. Geometry for diffraction calculation.

incident light is $\mathbf{A}_i(\xi, \eta; t)$ and the complex envelope of the transmitted light is $\mathbf{A}_t(\xi, \eta; t)$, the two can be related through the amplitude transmittance

$$\mathbf{A}_t(\xi, \eta; t) = \mathbf{t}_A(\xi, \eta) \mathbf{A}_i(\xi, \eta; t - \tau_0), \qquad (5.7\text{-}2)$$

where τ_0 is the average time delay associated with the structure. The mutual intensity of the transmitted light is thus

$$\mathbf{J}_t(\xi_1, \eta_1; \xi_2, \eta_2) = \langle \mathbf{A}_t(\xi_1, \eta_1; t) \mathbf{A}_t^*(\xi_2, \eta_2; t) \rangle$$

$$= \mathbf{t}_A(\xi_1, \eta_1) \mathbf{t}_A^*(\xi_2, \eta_2)$$

$$\times \langle \mathbf{A}_i(\xi_1, \eta_1; t - \tau_0) \mathbf{A}_i^*(\xi_2, \eta_2; t - \tau_0) \rangle. \quad (5.7\text{-}3)$$

Hence the general relationship between incident and transmitted mutual intensity is

$$\mathbf{J}_t(\xi_1, \eta_1; \xi_2, \eta_2) = \mathbf{t}_A(\xi_1, \eta_1) \mathbf{t}_A^*(\xi_2, \eta_2) \mathbf{J}_i(\xi_1, \eta_1; \xi_2, \eta_2). \quad (5.7\text{-}4)$$

5.7.2 Calculation of the Observed Intensity Pattern

To calculate the observed intensity pattern, we begin with Eq. (5.4-9), which we simplify by assuming that both the diffracting aperture and the observation region are much smaller than z. The relationship is then given by

$$I(x, y) \cong \frac{1}{(\bar{\lambda}z)^2} \iiiint_{-\infty}^{\infty} \mathbf{J}_t(\xi_1, \eta_1; \xi_2, \eta_2)$$

$$\times \exp\left[-j\frac{2\pi}{\bar{\lambda}}(r_2' - r_1')\right] d\xi_1 \, d\eta_1 \, d\xi_2 \, d\eta_2. \quad (5.7\text{-}5)$$

Representing the amplitude transmittance function t_A of the aperture by a "pupil" function $P(\xi, \eta)$, which for generality we allow to be complex-valued, we substitute (5.7-4) in (5.7-5) to obtain

$$I(x, y) \cong \frac{1}{(\bar{\lambda}z)^2} \iiiint_{-\infty}^{\infty} P(\xi_1, \eta_1) P^*(\xi_2, \eta_2) \mathbf{J}_i(\xi_1, \eta_1; \xi_2, \eta_2)$$

$$\times \exp\left[-j\frac{2\pi}{\bar{\lambda}}(r_2' - r_1')\right] d\xi_1 \, d\eta_1 \, d\xi_2 \, d\eta_2. \quad (5.7\text{-}6)$$

For simplicity of analysis, we assume that the mutual intensity function can be expressed in the form

$$\mathbf{J}_i(\xi_1, \eta_1; \xi_2, \eta_2) = I_0 \mathbf{\mu}_i(\Delta\xi, \Delta\eta). \quad (5.7\text{-}7)$$

Such is the case in many practical cases of interest.[†] For example, it is valid if the light arrives at the aperture from an incoherent source by way of a Köhler condenser system (see Section 7.2.1). In addition, we make the usual paraxial approximation,

$$r_2' - r_1' \cong \frac{1}{2z}\left[(\xi_2^2 + \eta_2^2) - (\xi_1^2 + \eta_1^2) - 2(x\Delta\xi + y\Delta\eta)\right]$$

$$= \frac{1}{z}[\bar{\xi}\Delta\xi + \bar{\eta}\Delta\eta - x\Delta\xi - y\Delta\eta], \quad (5.7\text{-}8)$$

where the definitions of Eq. (5.6-29) for $\bar{\xi}$ and $\bar{\eta}$ have been introduced.

[†] More generally, we could write

$$\mathbf{J}_i(\xi_1, \eta_1; \xi_2, \eta_2) = \mathbf{A}(\xi_1, \eta_1)\mathbf{A}^*(\xi_2, \eta_2)\mathbf{\mu}_i(\Delta\xi, \Delta\eta).$$

The factors $\mathbf{A}(\xi_1, \eta_1)$ and $\mathbf{A}^*(\xi_2, \eta_2)$ may be regarded as being incorporated in $P(\xi_1, \eta_1)$ and $P^*(\xi_2, \eta_2)$.

DIFFRACTION OF PARTIALLY COHERENT LIGHT BY AN APERTURE

Using this expression and Eq. (5.7-7) for $\mathbf{J}_i$, we obtain

$$I(x, y) = \frac{I_0}{(\bar{\lambda}z)^2} \iiiint_{-\infty}^{\infty} \mathbf{P}\left(\bar{\xi} - \frac{\Delta\xi}{2}, \bar{\eta} - \frac{\Delta\eta}{2}\right) \mathbf{P}^*\left(\bar{\xi} + \frac{\Delta\xi}{2}, \bar{\eta} + \frac{\Delta\eta}{2}\right)$$

$$\times \mu_i(\Delta\xi, \Delta\eta) \exp\left\{-j\frac{2\pi}{\bar{\lambda}z}(\bar{\xi}\Delta\xi + \bar{\eta}\Delta\eta)\right\}$$

$$\times \exp\left\{j\frac{2\pi}{\bar{\lambda}z}(x\Delta\xi + y\Delta\eta)\right\} d\bar{\xi}\, d\bar{\eta}\, d\Delta\xi\, d\Delta\eta. \quad (5.7\text{-}9)$$

Now for convenience we again make the assumption (5.6-34), or in words, we assume that z is at least as large as the geometric mean of the far-field distances for the aperture size and the coherence area size. This assumption allows us to drop the first exponential factor in (5.7-9), yielding the simpler expression

$$I(x, y) \cong \frac{I_0}{(\bar{\lambda}z)^2} \iint_{-\infty}^{\infty} \mathscr{P}(\Delta\xi, \Delta\eta) \mu_i(\Delta\xi, \Delta\eta)$$

$$\times \exp\left[j\frac{2\pi}{\bar{\lambda}z}(x\Delta\xi + y\Delta\eta)\right] d\Delta\xi\, d\Delta\eta, \quad (5.7\text{-}10)$$

where $\mathscr{P}$ is the autocorrelation function of the complex pupil function $\mathbf{P}$,

$$\mathscr{P}(\Delta\xi, \Delta\eta) \triangleq \iint_{-\infty}^{\infty} \mathbf{P}\left(\bar{\xi} - \frac{\Delta\xi}{2}, \bar{\eta} - \frac{\Delta\eta}{2}\right) \mathbf{P}^*\left(\bar{\xi} + \frac{\Delta\xi}{2}, \bar{\eta} + \frac{\Delta\eta}{2}\right) d\bar{\xi}\, d\bar{\eta}.$$

$$(5.7\text{-}11)$$

Thus the intensity distribution $I(x, y)$ in the diffraction pattern can be found from a two-dimensional Fourier transform of the product of the functions $\mathscr{P}$ and μ_i. This result is sometimes referred to as *Schell's theorem* (see Refs. 5-32 and 5-33). We attempt to provide some physical feeling for this result in the section that follows, but before doing so, some further discussion of condition (5.6-34) in the present context is in order.

First, it is a straightforward matter to show that the necessity to impose this condition vanishes when a positive lens of focal length $f = z$ is placed in contact with the aperture plane. Given that such a lens is *not* present, it should be noted that these conditions on z may be more difficult to satisfy in the present problem than in the problem presented in Section 5.6.4, for there it was explicitly stated that the coherence area was much smaller than the source size, whereas no analogous assumption has been made here. If D represents the maximum linear dimension of the aperture and d_c the maximum linear dimension of a coherence cell on the aperture, the required conditions will be met if

$$z > \begin{cases} \dfrac{2D^2}{\bar{\lambda}} & \text{if } d_c > D \\ \dfrac{2Dd_c}{\bar{\lambda}} & \text{if } d_c < D. \end{cases} \qquad (5.7\text{-}12)$$

Note that the condition $z > 2D^2/\bar{\lambda}$ is identical with the far-field or Fraunhofer condition and must be imposed if the aperture illumination approaches full coherence.

5.7.3 Discussion

The result we have referred to as Schell's theorem (Eq. 5.7-10) provides a general means for calculating the diffraction pattern generated by partially coherent illumination of an aperture. The physical implications of this result are best understood by considering some limiting cases.

First, suppose that the aperture is illuminated by a single uniform normally incident plane wave (such illumination is, of course, fully coherent). The complex coherence factor μ_i is in this case unity for all arguments, and the expression for the observed diffraction pattern becomes

$$I(x, y) \cong \frac{I_0}{(\bar{\lambda}z)^2} \iint_{-\infty}^{\infty} \mathscr{P}(\Delta\xi, \Delta\eta)\exp\left[j\frac{2\pi}{\bar{\lambda}z}(\Delta\xi x + \Delta\eta y)\right] d\Delta\xi\, d\Delta\eta,$$

$$(5.7\text{-}13)$$

where $\mathscr{P}$ is the autocorrelation function of the pupil function $\mathbf{P}$. Some consideration of the autocorrelation theorem of Fourier analysis (Ref. 5-24, p. 10) shows that this result is entirely equivalent to the more usual Fraunhofer diffraction formula applied to complex fields,

$$I(x, y) = \frac{I_0}{(\bar{\lambda}z)^2} \left| \iint_{-\infty}^{\infty} \mathbf{P}(\xi, \eta)\exp\left[j\frac{2\pi}{\bar{\lambda}z}(\xi x + \eta y)\right] d\xi\, d\eta \right|^2.$$

$$(5.7\text{-}14)$$

Consider next the opposite extreme, namely, illumination with a coherence area much smaller than the aperture size. In this case the function $\mathscr{P}$ has approximately its maximum value A (the area of the aperture) over the entire range of $(\Delta\xi, \Delta\eta)$ for which μ_i is nonzero. Hence

$$I(x,y) \cong \frac{I_0 A}{(\bar{\lambda}z)^2} \iint_{-\infty}^{\infty} \mu_i(\Delta\xi, \Delta\eta) \exp\left[j\frac{2\pi}{\bar{\lambda}z}(\Delta\xi x + \Delta\eta y)\right] d\Delta\xi\, d\Delta\eta.$$

(5.7-15)

Thus the shape of the observed intensity pattern is determined primarily by the complex coherence factor μ_i and is not really influenced by the aperture shape, provided $D \gg d_c$.

In intermediate cases, where both $\mathscr{P}$ and μ_i play a role in determining the shape of $I(x, y)$, some insight can be gained by noting that, since $I(x, y)$ depends on a Fourier transform of the *product* $\mathscr{P}\mu_i$, the shape of $I(x, y)$ will be determined by a *convolution* of the transforms of $\mathscr{P}$ and μ_i

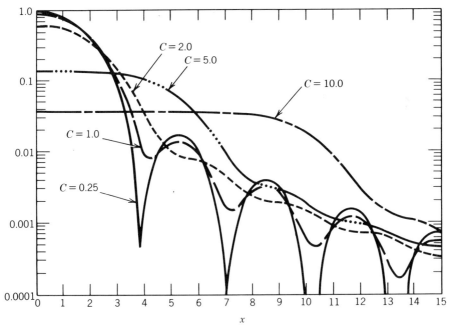

Figure 5-23. Diffraction pattern of a circular aperture for various states of transverse coherence. The parameter C represents the ratio of the area of the circular aperture to a coherence area. A circular incoherent source was assumed. The variable x has been normalized. (Ref. 5-34). (Courtesy of B. J. Thompson and the Optical Society of America.)

individually. The net result is, in general, a "smoothing" of the diffraction pattern as the coherence area is gradually reduced. Figure 5-23 (from Ref. 5-34) illustrates the gradual smoothing of the $[2(J_1(\rho)/\rho]^2$ pattern (corresponding to diffraction of coherent light by a circular aperture) as the coherence area is gradually reduced.

In closing, it should be noted that an entirely equivalent result could be obtained by an alternate approach to the problem, provided the partially coherent illumination at the aperture is produced by an incoherent source. Each point on the source may be regarded as generating a fully coherent illumination of the aperture and a corresponding diffraction pattern, but the center of each diffraction pattern depends on the location of the corresponding source point. Since the source is incoherent, all diffraction patterns are added on an intensity basis, yielding a new diffraction pattern that has been partially smoothed by the finite source size.

The second method of calculation has the advantage of conceptual simplicity but the disadvantage of being not quite as general as the first method. In particular, it is possible that the source itself may be partially coherent, in which case Schell's theorem can still be used, but the second method cannot unless it can be modified by first finding an "equivalent" incoherent source that produces the same complex coherence factor as the true partially coherent source.

REFERENCES

5-1 E. Verdet, *Ann. Scientif. l'Ecole Normale Supérieure* **2**, 291 (1865).

5-2 M. Von Laue, *Ann. Physik.*, **20**, 365 (1906); **23**, 1 (1907); **23**, 795 (1907); **30**, 225 (1909); **31**, 547 (1910); **47**, 853 (1915); **48**, 668 (1915).

5-3 M. Berek, *Z. Physik*, **36**, 675 (1926); **36** 824 (1926); **37**, 387 (1926); **40**, 420 (1927).

5-4 P. H. Van Cittert, *Physica*, **1**, 201 (1934); **6**, 1129 (1939).

5-5 F. Zernike, *Physica*, **5**, 785 (1938); *Proc. Phys. Soc.*, **61**, 158 (1948).

5-6 H. H. Hopkins, *Proc. Roy. Soc.*, **A208**, 263 (1951).

5-7 A. Blanc-Lapierre and P. Dumontet, *Compt. Rend.* (*Paris*), **238**, 1005 (1954).

5-8 E. Wolf, *Nature*, **172**, 535 (1953); *Proc. Roy. Soc.*, **A225**, 96 (1954); *Nuovo Cimento*, **12**, 884 (1954).

5-9 *Selected Papers on Coherence and Fluctuations on Light*, Vols. 1 and 2 (L. Mandel and E. Wolf, editors), Dover Publications, New York (1970).

5-10 M. Born and E. Wolf, *Principles of Optics*, 2nd rev. ed., MacMillan Company, New York, Chapter 10 (1964).

5-11 M. J. Beran and G. B. Parrent, *Theory of Partial Coherence*, Prentice-Hall, Englewood Cliffs, NJ (1964).

5-12 L. Mandel and E. Wolf, *Rev. Mod. Phys.*, **37**, 231 (1965).

5-13 J. Peřina, *Coherence of Light*, Van Nostrand Reinhold Company, London (1972).

5-14 M. Françon, *Optical Interferometry*, Academic Press, New York (1966).
5-15 A. A. Michelson, *Phil. Mag.* [5], **31**, 338 (1891); **34**, 280 (1892).
5-16 A. C. G. Mitchell and M. W. Zemansky, *Resonance Radiation and Excited Atoms*, 2nd printing, University Press, London, Chapter 3 (1961).
5-17 R. Bracewell, *The Fourier Transform and its Applications*, McGraw-Hill Book Company, New York (1965).
5-18 L. Mandel, *Proc. Phys. Soc.* (*London*), **74**, 223 (1959).
5-19 G. A. Vanasse and H. Sakai, "Fourier Spectroscopy," in *Progress in Optics*, Vol. 6, North Holland Publishing Company, Amsterdam, pp. 261–327 (1967).
5-20 L. Mertz, *Transformations in Optics*, John Wiley & Sons, New York (1965).
5-21 J. W. Cooley and J. W. Tukey, *Math. Computation*, **19**, 296 (1965).
5-22 P. Fellgett, Thesis, University of Cambridge (1951).
5-23 T. Young, *Phil. Trans. Roy. Soc.* (London) [xcii], **12**, 387 (1802).
5-24 J. W. Goodman, *Introduction to Fourier Optics*, McGraw-Hill Book Company, New York, (1968).
5-25 L. Mandel, *J. Opt. Soc. Am.*, **51**, 1342 (1961).
5-26 W. P. Alford and A. Gold, *Am. J. Phys.*, **56**, 481 (1958).
5-27 L. Mandel and E. Wolf, *J. Opt. Soc. Am.*, **66**, 529 (1976).
5-28 L. Mandel and E. Wolf, *J. Opt. Soc. Am.*, **51**, 815 (1961).
5-29 J. W. Goodman, *Proc. IEEE*, **53**, 1688 (1965).
5-30 W. H. Carter and E. Wolf, *J. Opt. Soc. Am.*, **67**, 785 (1977).
5-31 E. Wolf, *J. Opt. Soc. Am.*, **68**, 6 (1978).
5-32 A. C. Schell, Ph.D. Thesis, Massachusetts Institute of Technology (1961).
5-33 K. Singh and H. S. Dillon, *J. Opt. Soc. Am.*, **59**, 395 (1969).
5-34 R. A. Shore, B. J. Thompson, and R. E. Whitney, *J. Opt. Soc. Am.*, **56**, 733 (1966).

ADDITIONAL READING

Arvind. S. Marathay, *Elements of Optical Coherence Theory*, Wiley-Interscience, John Wiley and Sons, New York (1982).

John B. Shumaker, "Introduction to Coherence in Radiometry," in *Self-Study Manual on Optical Radiation Measurements*, Part I, Concepts, Chapter 10 (Fred E. Nicodemus, editor), Technical Note 910-6, National Bureau of Standards, Washington, DC (1983).

Robert J. Bell, *Introductory Fourier Transform Spectroscopy*, Academic Press, New York (1972).

W. H. Steel, *Interferometry*, 2nd Edition, Cambridge University Press, Cambridge (1983).

PROBLEMS

5-1 An idealized model of the (normalized) power spectral density of a gas laser oscillating in N equal-intensity axial modes is

$$\hat{\mathscr{G}}(\nu) = \frac{1}{N} \sum_{n=-(N-1)/2}^{(N-1)/2} \delta(\nu - \bar{\nu} + n\Delta\nu),$$

where $\Delta \nu$ is the mode spacing (equal to 2 × cavity length/velocity of light for axial modes), $\bar{\nu}$ is the frequency of the central mode, and N has been assumed odd for simplicity.

(a) Show that the corresponding envelope of the complex degree of coherence is

$$\gamma(\tau) = \left| \frac{\sin(N\pi\Delta\nu\tau)}{N\sin(\pi\Delta\nu\tau)} \right|$$

(b) Plot γ vs. $\Delta\nu\tau$ for $N = 3$ and $0 \leq \tau \leq 1/\Delta\nu$.

5-2 The gas mixture in a helium–neon laser (end mirrors removed) emits light at 633 nm with a Doppler-broadened spectral width of about 1.5×10^9 Hz. Calculate the coherence time τ_c and the coherence length $l_c = c\tau_c$ (c = velocity of light) for this light. Repeat for the 488 nm line of the argon ion laser, which has a Doppler-broadened line width of about 7.5×10^9 Hz.

5.3 (Lloyd's mirror.) A point source of light is placed at distance s above a perfectly reflecting mirror. At distance d away, interference fringes are observed on a screen, as shown in Fig. 5-3p: The complex degree

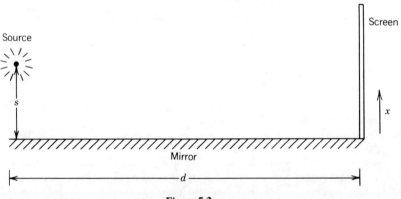

Figure 5-3p.

of coherence of the light is

$$\gamma(\tau) = e^{-\pi\Delta\nu|\tau|} e^{-j2\pi\bar{\nu}\tau}.$$

Adopting the assumptions $s \ll d$ and $x \ll d$, and taking account of a sign change of the field on reflection (polarization assumed parallel

PROBLEMS

to the mirror), find:

(a) The spatial frequency of the fringe.
(b) The classical visibility of the fringes as a function of x, assuming equal strength interfering beams.

5-4 Consider the Young's interference experiment performed with *broadband* light.

(a) Show that the field incident on the observing screen can be expressed as

$$\mathbf{u}(Q,t) = \tilde{K}_1 \frac{d}{dt}\mathbf{u}\left(P_1, t - \frac{r_1}{c}\right) + \tilde{K}_2 \frac{d}{dt}\mathbf{u}\left(P_2, t - \frac{r_2}{c}\right),$$

where

$$\tilde{K}_i \equiv \iint_{\substack{i\text{th} \\ \text{pinhole}}} \frac{\chi(\theta_i)}{2\pi c r_i} dS_i \cong \frac{\chi(\theta_i) A_i}{2\pi c r_i} \qquad i = 1, 2,$$

where A_i is the area of the ith pinhole.

(b) Using the result of part (a), show that the intensity of the light striking the screen can be expressed as

$$I(Q) = I^{(1)}(Q) + I^{(2)}(Q) - 2\tilde{K}_1 \tilde{K}_2 \text{Re}\left\{\frac{\partial^2}{\partial \tau^2}\Gamma_{12}\left(\frac{r_2 - r_1}{c}\right)\right\}$$

where

$$I^{(i)}(Q) = \tilde{K}_i^2 \left\langle \left|\frac{d}{dt}\mathbf{u}\left(P_i, t - \frac{r_i}{c}\right)\right|^2\right\rangle, \qquad i = 1, 2.$$

(c) Show that the preceding expression reduces to that obtained in Eq. (5.2-9) when the light is narrowband.

5-5 As shown in Fig. 5-5p, a positive lens with focal length f is placed in contact with the pinhole screen in a Young's interference experiment. For quasimonochromatic light, the effect of the lens can be modeled by an amplitude transmittance factor

$$\mathbf{t}_A(\rho) = \exp\left[-j\frac{\pi}{\lambda f}\rho^2\right]$$

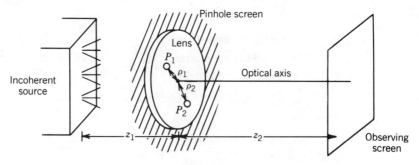

Figure 5-5p.

under paraxial conditions. Assuming a spatially incoherent source, find the relationship between z_1, z_2, and f that guarantees that the spatial phase of the fringe pattern observed depends only on the vector separation of the two pinholes and not on their absolute locations with respect to the optical axis.

5-6 Consider a Michelson interferometer that is used in a Fourier spectroscopy experiment. To obtain high resolution in the computed spectrum, it is necessary that the interferogram be measured out to large pathlength differences, where the interferogram has fallen to very small values.

(a) Show that under such conditions, the spectrum of light falling on the *detector* is significantly different than the spectrum of light entering the interferometer.

(b) If the spectrum of the light entering the interferometer is

$$\hat{\mathcal{G}}(\nu) = \frac{1}{\Delta\nu}\text{rect}\frac{\nu - \bar{\nu}}{\Delta\nu},$$

calculate the spectrum of the light falling on the detector.

5-7 In the Young's interference experiment illustrated in Fig. 5-7p, the power spectral density $\hat{\mathcal{G}}(\nu)$ of the light is measured at point Q by means of a grating spectrometer. The wave at the P_1, P_2 plane is known to be cross-spectrally pure, that is,

$$\gamma_{12}(\tau) = \mu_{12}\gamma(\tau).$$

Show that under the condition $r_2 - r_1/c \gg \tau_c$, when no interference

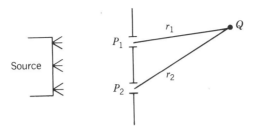

Figure 5-7p.

fringes are observed, μ_{12} can be measured by examining fringes that exist across the spectrum $\hat{\mathscr{G}}_Q(\nu)$ of the light at Q. Specify how both the modulus and phase of μ_{12} can be determined.

5-8 A monochromatic plane wave falls normally on a "sandwich" of two diffusers. The diffusers are moving in opposite directions with equal speeds, as shown in Fig. 5-8p. The amplitude transmittance $\mathbf{t}_A(x, y)$

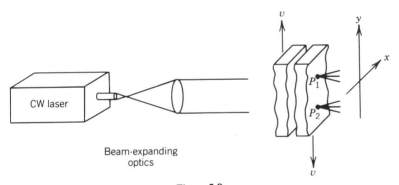

Figure 5-8p.

of the diffuser pair is assumed to be expressible as

$$\mathbf{t}_A(x, y) = \mathbf{t}_1(x, y - vt)\mathbf{t}_2(x, y + vt),$$

where $\mathbf{t}_1$ and $\mathbf{t}_2$ may be assumed to be drawn from statistically independent ensembles (since knowledge of one tells nothing about the other). Show that, if the diffusers each have a Gaussian-shaped autocorrelation function

$$\gamma_t(\Delta x, \Delta y) = \exp\left\{-a\left[(\Delta x)^2 + (\Delta y)^2\right]\right\},$$

the transmitted light is cross-spectrally pure.

5-9 Starting with Eq. (4.1-9), show that for broadband light the mutual coherence function $\Gamma(Q_1, Q_2; \tau)$ on surface Σ_2 of Fig. 5-16 can be expressed as

$$\Gamma(Q_1, Q_2; \tau) = -\iint_{\Sigma_1} \iint_{\Sigma_1} \frac{\partial^2}{\partial \tau^2} \Gamma\left(P_1, P_2; \tau + \frac{r_2 - r_1}{c}\right)$$

$$\times \frac{\chi(\theta_1)}{2\pi c r_1} \frac{\chi(\theta_2)}{2\pi c r_2} dS_1 dS_2.$$

5-10 Show that under the quasimonochromatic conditions, mutual intensity $\mathbf{J}_{12}$ obeys a pair of Helmholtz equations

$$\nabla_1^2 \mathbf{J}_{12} + (\bar{k})^2 \mathbf{J}_{12} = 0$$

$$\nabla_2^2 \mathbf{J}_{12} + (\bar{k})^2 \mathbf{J}_{12} = 0,$$

where $\bar{k} = 2\pi/\bar{\lambda}$ and

$$\nabla_1^2 \triangleq \frac{\partial^2}{\partial x_1^2} + \frac{\partial^2}{\partial y_1^2} + \frac{\partial^2}{\partial z_1^2}, \qquad \nabla_2^2 \triangleq \frac{\partial^2}{\partial x_2^2} + \frac{\partial^2}{\partial y_2^2} + \frac{\partial^2}{\partial z_2^2}.$$

5-11 A Young's interference experiment is performed in a geometry shown in Fig. 5-11p. The pinholes have finite diameter δ and spacing s. The source has bandwidth $\Delta \nu$ and mean frequency $\bar{\nu}$, and f is the focal length of the lens. Two effects cause the fringes to be attenuated away

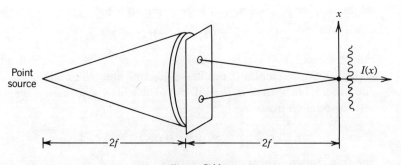

Figure 5-11p.

from the optical axis:

(a) The finite size of the pinholes.
(b) The finite bandwidth of the source.

Given δ, s and f, how small must the fractional bandwidth $\Delta\nu/\bar{\nu}$ be to assure that effect (a) dominates over effect (b)?

5-12 Prove that any monochromatic wave is perfectly coherent (in the time average sense).

5-13 The sun subtends an angle of about 32 minutes of arc (0.0093 radians) on earth. Assuming a mean wavelength of 550 nm, calculate the diameter of the coherence area of sunlight observed on earth (assume quasimonochromatic conditions).

5-14 A 1 mm pinhole is placed immediately in front of an incoherent source. The light passed by the pinhole is to be used in a diffraction experiment, for which it is desired to illuminate a distant 1mm aperture coherently. Given $\bar{\lambda} = 550$ nm, calculate the minimum distance between the pinhole source and the diffracting aperture.

5-15 Consider an incoherent source radiating with spatial intensity distribution $I(\xi, \eta)$.

(a) Using the Van Cittert-Zernike theorem and Parseval's theorem of Fourier analysis, show that the coherence area of the light (mean wavelength $\bar{\lambda}$) at distance z from the source can be expressed as

$$A_c = (\bar{\lambda}z)^2 \frac{\displaystyle\iint_{-\infty}^{\infty} I^2(\xi, \eta)\, d\xi\, d\eta}{\left[\displaystyle\iint_{-\infty}^{\infty} I(\xi, \eta)\, d\xi\, d\eta\right]^2}$$

(b) Show that if an incoherent source has an intensity distribution describable as

$$I(\xi, \eta) = I_0 P(\xi, \eta)$$

where $P(\xi, \eta)$ is a function with values 1 or 0, then

$$A_c = \frac{(\bar{\lambda}z)^2}{A_S}$$

where A_S is the area of the source.

5-16 Representing a positive lens with focal length f by an amplitude transmittance function

$$\mathbf{t}_A(\xi, \eta) = \exp\left[-j\frac{\pi}{\lambda f}(\xi^2 + \eta^2)\right]$$

and assuming that this lens is placed in contact with a partially coherent source, show that for observations in the rear focal plane of that lens, the generalized Van Cittert–Zernike theorem holds without the restrictions stated in Eq. (5.6-34).

5-17 Using a paraxial approximation, find an expression for the complex coherence factor $\mu(P_1, P_2)$ produced by a quasimonochromatic point source, with P_1 and P_2 lying in a plane, which is at distance z from the source.

6

Some Problems Involving High-Order Coherence

In Chapter 5 we dealt exclusively with problems involving second-order coherence, that is, the mutual coherence function $\Gamma_{12}(\tau)$. Such coherence functions provide only a limited description of the statistical properties of the underlying wavefields. It is quite possible for two fundamentally different types of wavefield to have indistinguishable mutual coherence functions, in which case a coherence function of order higher than 2 is required to differentiate between the two waves. In addition, we shall see that coherence functions of order higher than 2 arise quite naturally in certain physical problems.

By way of introduction, the $(n + m)$th-order coherence function of a wave $\mathbf{u}(P, t)$ is defined by

$$\Gamma_{1,2,\ldots,n+m}(t_1, t_2, \ldots, t_{n+m})$$
$$\triangleq \langle \mathbf{u}(P_1, t_1) \cdots \mathbf{u}(P_n, t_n)\mathbf{u}^*(P_{n+1}, t_{n+1}) \cdots \mathbf{u}^*(P_{n+m}, t_{n+m}) \rangle,$$

(6-1)

or, for an ergodic random process, by a corresponding ensemble average. Calculation of averages higher than second order is in general difficult mathematically, for it requires knowledge of the $(n + m)$th-order probability density function, and often the resulting integrals are very difficult to perform. Fortunately, an exception to this statement occurs for the very important case of thermal light. For such light, the moment theorem for circular complex Gaussian random variables [see Eq. (2.8-21)] can be used to write

$$\Gamma_{1,2,\ldots,2n}(t_1, t_2, \ldots, t_{2n})$$
$$= \sum_\pi \Gamma_{1p}(t_1, t_p)\Gamma_{2q}(t_2, t_q) \cdots \Gamma_{nr}(t_n, t_r),$$

(6-2)

where Σ_π denotes a summation over the $n!$ possible permutations $(p, q, \ldots, r)$ of $(1, 2, \ldots, n)$.

Use of this factorization theorem for thermal light often renders problems simple that would otherwise be extremely difficult. Factorization theorems of different forms exist for certain non-Gaussian processes as well (see, e.g., Ref. 6-1).

In the sections to follow, we present three examples of problems involving coherence of order higher than 2. First we consider some statistical properties of the time-integrated intensity of polarized thermal light. These results will be of considerable use in our later studies of photon-counting statistics (Chapter 9). Then we consider the statistical properties of mutual intensity measured with a finite averaging time. Finally, we present a fully classical analysis of the intensity interferometer.

6.1 STATISTICAL PROPERTIES OF THE INTEGRATED INTENSITY OF THERMAL OR PSEUDOTHERMAL LIGHT

In a variety of problems, including the study of photon-counting statistics, finite-time integrals of instantaneous intensity occur. In addition, an entirely analogous problem arises in considering the statistical properties of a finite space average of instantaneous intensity. Here we frame the problem in terms of time integrals, but the analysis is nearly identical for integrals over space.

Let $I(t)$ represent the instantaneous intensity of a wave observed at some specific point P. Our prime interest here is in the related quantity,

$$W(t) = \int_{t-T}^{t} I(\xi) \, d\xi \qquad (6.1\text{-}1)$$

representing the integrated value of $I(t)$ over a finite observation interval $(t - T, t)$. Note that any estimate of the average intensity of a wave must of necessity be based on a finite time average, which is nothing more than the measured value of W normalized by the averaging time T.

Throughout our discussions we shall assume that the light in question is thermal or pseudothermal in origin and that it is adequately modeled as an ergodic (and hence stationary) random process. As a consequence, the statistics of W do not depend on the particular observation time t. For mathematical convenience, we choose to let $t = T/2$, in which case Eq. (6.1-1) can be replaced by

$$W = \int_{-T/2}^{T/2} I(\xi) \, d\xi. \qquad (6.1\text{-}2)$$

PROPERTIES OF INTEGRATED INTENSITY

Our discussion is presented first with the assumption of polarized light, and then the cases of partially polarized and unpolarized light are treated.

The material to follow is divided into three parts, as we (1) derive exact expressions for the mean and variance of the integrated intensity W, (2) find an approximate expression for the first-order probability density function of W, and (3) find an exact solution for this density function.

6.1.1 Mean and Variance of the Integrated Intensity

Our initial goal is to find expressions for the mean $\overline{W}$ and variance σ_W^2 of the integrated intensity. Also of major interest is the root mean square (rms) signal-to-noise ratio,

$$\left(\frac{S}{N}\right)_{rms} = \frac{\overline{W}}{\sigma_W}, \tag{6.1-3}$$

associated with the integrated intensity, which provides us with an indication of the magnitude of the fluctuations of W relative to the mean value $\overline{W}$. For discussions of related problems, the reader may wish to consult Refs. 6-2 and 6-3.

Calculation of the mean value of W is entirely straightforward. The expected value of Eq. (6.1-2) is obtained by interchanging the orders of integration and expectation, yielding

$$\overline{W} = \int_{-T/2}^{T/2} \bar{I} \, d\xi = \bar{I}T, \tag{6.1-4}$$

which is quite independent of the state of polarization of the wave.

Calculation of the variance σ_W^2 requires a bit more effort. We have

$$\sigma_W^2 = E\left[\left(\int_{-T/2}^{T/2} I(\xi) \, d\xi\right)^2\right] - (\overline{W})^2$$

$$= \iint_{-T/2}^{T/2} \overline{I(\xi)I(\eta)} \, d\xi \, d\eta - (\overline{W})^2$$

$$= \iint_{-T/2}^{T/2} \Gamma_I(\xi - \eta) \, d\xi \, d\eta - (\overline{W})^2, \tag{6.1-5}$$

where Γ_I represents the autocorrelation function of the instantaneous intensity. Since the integrand is an even function of $(\xi - \eta)$, the double integral can be reduced to a single integral, exactly as was done in the argument leading to Eq. (3.4-9). Thus we have

$$\sigma_W^2 = T\int_{-\infty}^{\infty} \Lambda\left(\frac{\tau}{T}\right)\Gamma_I(\tau)\, d\tau - (\overline{W})^2, \qquad (6.1\text{-}6)$$

where

$$\Lambda(\tau) \triangleq \begin{cases} 1 - |\tau| & |\tau| \leq 1 \\ 0 & \text{otherwise.} \end{cases} \qquad (6.1\text{-}7)$$

At this point it is necessary to fully utilize the fact that the fields of concern arise from a *thermal* (or pseudothermal) source. The correlation function $\Gamma_I(\tau)$, which is in fact equivalent to a *fourth-order* coherence function of the underlying fields,

$$\Gamma_I(\tau) = E[\mathbf{u}(t)\mathbf{u}^*(t)\mathbf{u}(t+\tau)\mathbf{u}^*(t+\tau)], \qquad (6.1\text{-}8)$$

can be expressed in terms of the second-order coherence function of the fields in this case. Using Eq. (6-2), we have for the case of a fully polarized wave,

$$\Gamma_I(\tau) = (\bar{I})^2[1 + |\boldsymbol{\gamma}(\tau)|^2], \qquad (6.1\text{-}9)$$

where $\boldsymbol{\gamma}(\tau)$ is the complex degree of coherence of the light. Substitution of this relation in (6.1-6) yields the result

$$\sigma_W^2 = (\overline{W})^2\left[\frac{1}{T}\int_{-\infty}^{\infty} \Lambda\left(\frac{\tau}{T}\right)|\boldsymbol{\gamma}(\tau)|^2\, d\tau\right], \qquad (6.1\text{-}10)$$

for a polarized wave.

For a partially polarized wave, we know that the instantaneous intensity can be expressed in terms of two uncorrelated intensities,

$$I(t) = I_1(t) + I_2(t), \qquad (6.1\text{-}11)$$

where the mean values of $I_1(t)$ and $I_2(t)$ are

$$\bar{I}_1 = \tfrac{1}{2}\bar{I}(1 + \mathscr{P})$$

$$\bar{I}_2 = \tfrac{1}{2}\bar{I}(1 - \mathscr{P}). \qquad (6.1\text{-}12)$$

PROPERTIES OF INTEGRATED INTENSITY

Using these relations in the definition of $\Gamma_I(\tau)$, we obtain

$$\Gamma_I(\tau) = 2\bar{I}_1\bar{I}_2 + (\bar{I}_1)^2[1 + |\gamma(\tau)|^2] + (\bar{I}_2)^2[1 + |\gamma(\tau)|^2] \quad (6.1\text{-}13)$$

or

$$\Gamma_I(\tau) = (\bar{I})^2 + \tfrac{1}{2}(\bar{I})^2(1 + \mathcal{P}^2)|\gamma(\tau)|^2. \quad (6.1\text{-}14)$$

Finally, the variance σ_W^2 is given by

$$\sigma_W^2 = \frac{1 + \mathcal{P}^2}{2}(\overline{W})^2 \left[\frac{1}{T}\int_{-\infty}^{\infty}\Lambda\left(\frac{\tau}{T}\right)|\gamma(\tau)|^2\,d\tau\right] \quad (6.1\text{-}15)$$

for a partially polarized wave with degree of polarization $\mathcal{P}$.

A quantity of considerable physical interest is the rms signal-to-noise ratio of the measurement [Eq. (6.1-3)]. Using (6.1-4) and (6.1-15), we find directly that

$$\left(\frac{S}{N}\right)_{\text{rms}} = \left[\frac{2}{1 + \mathcal{P}^2}\mathcal{M}\right]^{1/2}, \quad (6.1\text{-}16)$$

where the parameter $\mathcal{M}$ is given by

$$\mathcal{M} = \left[\frac{1}{T}\int_{-\infty}^{\infty}\Lambda\left(\frac{\tau}{T}\right)|\gamma(\tau)|^2\,d\tau\right]^{-1}. \quad (6.1\text{-}17)$$

This parameter is sufficiently important, both here and in later considerations, to warrant some special discussion.

The physical meaning of the parameter $\mathcal{M}$ can best be understood by considering its limiting values. Noting that the width of the function $\Lambda(\tau/T)$ is $2T$ whereas the width of the function $|\gamma(\tau)|^2$ is roughly twice the coherence time $2\tau_c$, we can easily show that, for $T \gg \tau_c$,

$$\mathcal{M} \cong \left[\frac{1}{T}\int_{-\infty}^{\infty}|\gamma(\tau)|^2\,d\tau\right]^{-1} = \frac{T}{\tau_c} \quad (T \gg \tau_c). \quad (6.1\text{-}18)$$

In this limiting case the parameter $\mathcal{M}$ is thus the number of coherence intervals contained within the measurement time T.

For the opposite extreme of $T \ll \tau_c$, the corresponding result becomes

$$\mathcal{M} \cong \left[\frac{1}{T}\int_{-\infty}^{\infty}\Lambda\left(\frac{\tau}{T}\right)d\tau\right]^{-1} = 1 \quad (T \ll \tau_c). \quad (6.1\text{-}19)$$

This result may be interpreted as meaning that, as the measurement time shrinks, the number of coherence intervals influencing the experimental result asymptotically approaches unity. Values of $\mathcal{M}$ less than unity are not possible, for the experimental results are always influenced by the state of the fields in at least one coherence cell.

Consistent with the above arguments, in the general case of a measurement time T related arbitrarily to the coherence time τ_c, we interpret the parameter $\mathcal{M}$ as representing the number of coherence cells of the light wave that influence the experimental outcome. To specify the value of $\mathcal{M}$ in this general case, it is first necessary to know $|\gamma(\tau)|^2$, or equivalently to know the spectral distribution of the light. Analytical solutions are possible when the light has a Gaussian spectral profile (c.f. Problem 6-5) or a Lorentzian spectral profile (see Problem 6-6). The results are

Gaussian spectrum:
$$\mathcal{M} = \left\{ \frac{\tau_c}{t} \mathrm{erf}\left(\sqrt{\pi} \frac{T}{\tau_c}\right) - \frac{1}{\pi}\left(\frac{\tau_c}{T}\right)^2 \left[1 - e^{-\pi(T/\tau_c)^2}\right] \right\}^{-1},$$

(6.1-20)

where erf(x) is a standard error integral,

$$\mathrm{erf}(x) = \frac{2}{\sqrt{\pi}} \int_0^x e^{-z^2}\, dz,$$

and

Lorentzian spectrum:
$$\mathcal{M} = \left\{ \frac{\tau_c}{T} + \frac{1}{2}\left(\frac{\tau_c}{T}\right)^2 \left[e^{-2(T/\tau_c)} - 1\right] \right\}^{-1}. \quad (6.1\text{-}21)$$

For the case of a rectangular spectral profile, the corresponding result can be obtained by numerical integration (see Ref. 6-4). All three relations are illustrated in Fig. 6-1, which shows the parameter $\mathcal{M}$ plotted against T/τ (see Ref. 6-5). The dependence of $\mathcal{M}$ on the exact shape of the spectral profile is rather weak and can be ignored outside the range $0.1 < T/\tau_c < 10$

Returning to the question of the rms signal-to-noise ratio associated with measurement of W, for a polarized source we have

$$\left(\frac{S}{N}\right)_{\mathrm{rms}} = \sqrt{\mathcal{M}}. \qquad (6.1\text{-}22)$$

The dependence of $(S/N)_{\mathrm{rms}}$ on T/τ_c is also shown in Fig. 6-1. For partially polarized wave, all values should be increased by the factor $\sqrt{2/(1+\mathcal{P}^2)}$.

PROPERTIES OF INTEGRATED INTENSITY

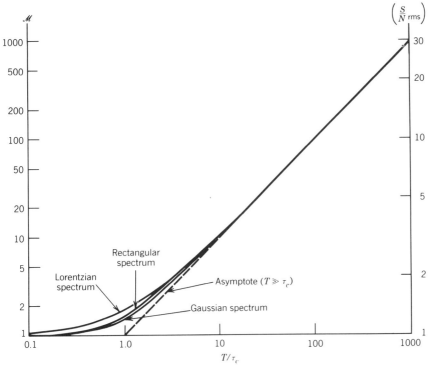

Figure 6-1. Plots of $\mathcal{M}$ versus T/τ_c, exact solutions for Gaussian, Lorentzian, and rectangular spectral profiles.

The results derived in this section can often be of practical use. For example, suppose that we wish to estimate the intensity of a polarized thermal light wave with an accuracy of 1%. Since W/T represents a finite-time average of intensity, we require measurement of W with an rms signal-to-noise ratio of 100. Referring to Fig. 6-1, we see that in this case

$$\left(\frac{S}{N}\right)_{\mathrm{rms}} \cong \sqrt{\frac{T}{\tau_c}}, \qquad (6.1\text{-}23)$$

and hence the required accuracy is achieved with $T = 10{,}000\ \tau_c$. If the mean wavelength $\bar{\lambda}$ of the source is 500 nm and its wavelength spread is as small as 0.1 nm, the coherence time $\tau_c = \bar{\lambda}^2/c\Delta\lambda$ is about 10^{-11} s. The required integration time is thus $T = 10^{-7}$ s, a duration that could easily be achieved or exceeded in most experimental situations. On the other hand, if the light is pseudothermal in origin, its bandwidth can readily be as small as 10^3 Hz, and the required integration time becomes 10 s. This condition may or may not be easily satisfied in practice, depending on the experimental constraints.[†]

6.1.2 Approximate Form for the Probability Density Function of Integrated Intensity

In some applications (e.g., see Refs. 6-4 and 6-6), knowledge of only the mean and variance of the integrated intensity is not sufficient. Rather, the entire probability density function of this quantity is desired. In this section we derive an approximate form for this density function, following the approach of Rice (Ref. 6-7) and Mandel (Ref. 6-6).

Before embarking on a derivation of these approximate results, a few remarks concerning the limiting forms of the probability density function may be helpful. First, for an integration time T that is much smaller than the coherence time τ_c of the thermal wave, the integrated intensity is, to an excellent approximation, simply the product of the instantaneous intensity and the integration time T,

$$W = \int_{-T/2}^{T/2} I(\xi)\, d\xi \cong I(0) \cdot T. \qquad (6.1\text{-}24)$$

Within a scaling factor, therefore, the probability density function of W is approximately the same as the density function of instantaneous intensity, as in Eq. (4.2-9), (4.2-13), or (4.3-42), depending on the state of polarization of the wave.

At the opposite extreme, with an integration time much longer than the coherence time, the fact that many independent fluctuations of the instantaneous intensity occur within the interval T implies, according to the central limit theorem, that the statistics of W are asymptotically Gaussian. As in all such cases involving the central limit theorem, however, care must be exercised to avoid using the "tails" of the Gaussian density function.

To find an approximate form for the density function $p_W(W)$ of integrated intensity that holds for arbitrary magnitudes of T and τ_c, we invoke a quasiphysical argument as follows. As an approximation, the smoothly fluctuating instantaneous intensity curve $I(t)$ may be replaced on the interval $(-T/2, T/2)$ by a "boxcar" function (see Fig. 6-2). The interval $(-T/2, T/2)$ is divided into m equal length subintervals. Within each

---

[†]As is shown in Chapter 9, for a true thermal light wave the fluctuations will in practice be dominated by shot noise, rather than the noise considered here. For a pseudothermal source, however, the noise considered here will often dominate.

PROPERTIES OF INTEGRATED INTENSITY

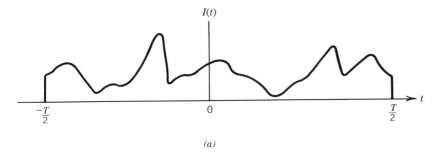

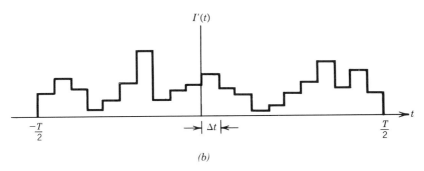

Figure 6-2. Approximation of the smoothly varying instantaneous intensity $I(t)$ by a "boxcar" function $I'(t)$.

subinterval, the approximation to $I(t)$ is constant; at the end of each subinterval, the approximate waveform jumps to a new constant value, assumed statistically independent of all preceding and following values. The probability density function of the boxcar function within any one subinterval is taken to be the same as the probability density function of the instantaneous intensity at a single time instant t [i.e., Eq. (4.2-9), (4.2-13) or (4.3-42), depending on the state of polarization].

The integrated intensity is now approximated in terms of the area under the boxcar function as follows:

$$W = \int_{-T/2}^{T/2} I(t)\, dt \cong \sum_{i=1}^{m} I_i \Delta t = \frac{T}{m} \sum_{i=1}^{m} I_i, \qquad (6.1\text{-}25)$$

where Δt is the width of one subinterval of the boxcar function and I_i is the value of the boxcar function in the ith subinterval. By hypothesis, the probability density function of each I_i is taken to be the same as the density function of the instantaneous intensity. Also by hypothesis, the various I are assumed to be statistically independent.

For the case of a polarized thermal wave, the characteristic function of I is taken (in accord with the result of Problem 4-2) to be

$$\mathbf{M}_I(\omega) = \frac{1}{1 - j\omega\bar{I}}. \qquad (6.1\text{-}26)$$

It follows directly from our hypotheses that the characteristic function of W is given approximately by

$$\mathbf{M}_W(\omega) \cong \left[\frac{1}{1 - j\omega\frac{\bar{I}T}{m}}\right]^m. \qquad (6.1\text{-}27)$$

From the table of one-dimensional Fourier transform pairs presented in Appendix A, we find that the corresponding probability density function is

$$p_W(W) \cong \begin{cases} \left(\frac{m}{\bar{I}T}\right)^m \dfrac{W^{m-1}\exp\left(-m\dfrac{W}{\bar{I}T}\right)}{\Gamma(m)} & W \geq 0 \\ 0 & \text{otherwise,} \end{cases} \qquad (6.1\text{-}28)$$

where $\Gamma(m)$ is a gamma function of argument m. This particular density function is known as a *gamma* probability density function, and accordingly the random variable W is said to be (approximately) a *gamma variate*.

Continuing with the case of a polarized wave, one problem remains: the parameters of the density function (6.1-28) must be chosen in such a way as to best match the approximate result to the true density function of W. The only two adjustable parameters available in (6.1-28) are $\bar{I}$ and m. The most common approach taken (see Refs. 6-7, 6-6, and 6-4) is to *choose the parameters $\bar{I}$ and m such that the mean and variance of the approximate density function are exactly equal to the true mean and variance of W*. The mean and variance of the gamma density function (6.1-28) can be readily shown to be

$$\bar{W} = \bar{I}T$$

$$\sigma_W^2 = \frac{(\bar{I}T)^2}{m}. \qquad (6.1\text{-}29)$$

PROPERTIES OF INTEGRATED INTENSITY

Thus the mean agrees with the true mean given by (6.1-4). For the variance of the approximate density function to agree with the true variance (6.1-10), we require that

$$m = \left[\frac{1}{T} \int_{-\infty}^{\infty} \Lambda\left(\frac{\tau}{T}\right) |\gamma(\tau)|^2 d\tau \right]^{-1} \triangleq \mathcal{M}. \qquad (6.1\text{-}30)$$

Stated in words, the number of subintervals in the boxcar function should be chosen equal to the number of coherence cells that influence the measurement of integrated intensity.

It should be noted that, in a certain sense, our quasiphysical reasoning that led to the approximate distribution (6.1-28) has broken down, for in general the parameter $\mathcal{M}$ is not an integer, whereas we implicitly assumed an integer number of subintervals in the boxcar function. At this point it is best to abandon the quasiphysical picture and simply view the gamma density function as a general approximation to the true density function, with its parameters to be chosen to benefit the accuracy of the approximation. We further note that, while choosing the parameters to match the mean and variance seems reasonable, there is no reason to assume a priori that this choice will result in the closest possible match between the true and approximate density functions at every value of W. Nonetheless, this choice is a simple one and is accordingly usually made.

The approximate density function for the integrated intensity of polarized thermal light can thus be written in final form as

$$p_W(W) \cong \begin{cases} \left(\dfrac{\mathcal{M}}{\overline{W}}\right)^{\mathcal{M}} \dfrac{W^{\mathcal{M}-1} \exp\left(-\mathcal{M}\dfrac{W}{\overline{W}}\right)}{\Gamma(\mathcal{M})} & W \geq 0 \\ 0 & \text{otherwise.} \end{cases} \qquad (6.1\text{-}31)$$

This function is plotted against W in Fig. 6-3 for various values of $\mathcal{M}$. With the help of Fig. 6-1, the value of $\mathcal{M}$ can be related to a value of T/τ_c if the spectral shape of the light is known.

When the thermal light is partially polarized, a similar approximate density function for integrated intensity can be derived. Again the integrated intensity is approximated by a boxcar function, but this time the intensity in the ith subinterval is taken to have a characteristic function

$$\mathbf{M}_I(\omega) = \left[\left(1 - j\frac{\omega}{2}(1 + \mathcal{P})\bar{I}\right)\left(1 - j\frac{\omega}{2}(1 - \mathcal{P})\bar{I}\right) \right]^{-1} \qquad (6.1\text{-}32)$$

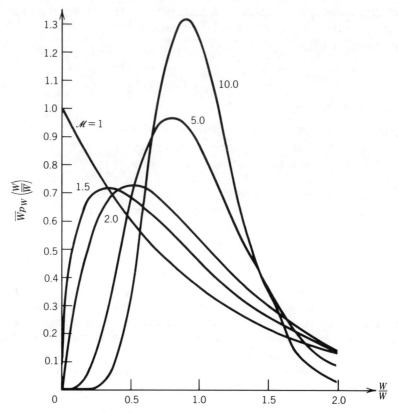

Figure 6-3. Approximate probability density function of the integrated intensity of a polarized thermal source for various values of $\mathcal{M}$.

in accord with Eq. (4.3-41). The characteristic function of the integrated intensity is then given approximately by

$$\mathbf{M}_W(\omega) \cong \left[\left(1 - j\frac{\omega}{2}(1 + \mathcal{P})\frac{\bar{I}T}{\mathcal{M}}\right)\left(1 - j\frac{\omega}{2}(1 - \mathcal{P})\frac{\bar{I}T}{\mathcal{M}}\right)\right]^{-\mathcal{M}}.$$

(6.1-33)

Two approaches to finding the probability density function of integrated intensity can be considered. One is an inversion of the characteristic function with the help of a partial fraction expansion. The other is a simple convolution of the one-dimensional probability density functions of the integrated intensities of each polarization component. The latter approach

PROPERTIES OF INTEGRATED INTENSITY

yields a probability density for the total integrated intensity given by

$$p_W(W) = \frac{\sqrt{\pi}}{\Gamma(\mathcal{M})}\left(\frac{\mathcal{M}W}{\mathcal{P}\overline{W}}\right)^{\mathcal{M}}\left[\frac{4\mathcal{P}\mathcal{M}}{(1-\mathcal{P}^2)W\overline{W}}\right]^{1/2}\exp\left(-\frac{2\mathcal{M}W}{(1-\mathcal{P}^2)\overline{W}}\right)$$

$$\times I_{\mathcal{M}-1/2}\left[\frac{2\mathcal{P}\mathcal{M}}{(1-\mathcal{P}^2)\overline{W}}\right] \quad (6.1\text{-}34)$$

for $W \geq 0$, where $\mathcal{M}$ is again given by Eq. (6.1-30) and $I_{\mathcal{M}-1/2}$ is a modified Bessel function of the first kind, order $\mathcal{M} - 1/2$.

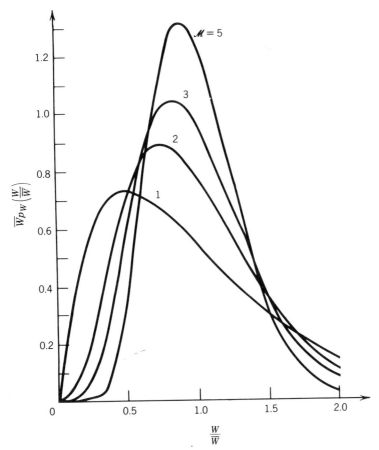

Figure 6-4. Plot of $\overline{W}p_W(W/\overline{W})$ versus $W/\overline{W}$ for various values of $\mathcal{M}$ when the light is unpolarized.

For the case of totally unpolarized light ($\mathscr{P} = 0$), the characteristic function reduces to

$$\mathbf{M}_W(\omega) \cong \left[1 - j\frac{\omega}{2}\frac{\bar{I}T}{\mathscr{M}}\right]^{-2\mathscr{M}}. \qquad (6.1\text{-}35)$$

Detailed consideration shows that Eq. (6.1-34) then reduces to

$$p_W(W) \cong \begin{cases} \left(\dfrac{2\mathscr{M}}{\overline{W}}\right)^{2\mathscr{M}} \dfrac{W^{2\mathscr{M}-1}\exp\left(-2\mathscr{M}\dfrac{W}{\overline{W}}\right)}{\Gamma(2\mathscr{M})} & W \geq 0 \\ 0 & \text{otherwise,} \end{cases} \qquad (6.1\text{-}36)$$

as can be verified by a Fourier inversion of Eq. (6.1-35). Figure 6-4 shows this probability density function for several values of $\mathscr{M}$.

6.1.3 Exact Solution for the Probability Density Function of Integrated Intensity

Whereas the approximate forms of the probability density function of integrated intensity are useful in many applications, it is also of some interest to know the exact forms of these density functions. The exact results can in fact be found for certain line shapes using the Karhunen–Loève expansion introduced in Section 3.10. For some related discussions the reader may wish to consult Refs. 6-8 through 6-10. We consider here only the case of fully polarized thermal light. The initial discussion is quite general, but our attention is ultimately limited to the case of light with a rectangular spectral profile.

In the past we have expressed the integrated intensity W in terms of the analytic signal $\mathbf{u}(t)$,

$$W = \int_{-T/2}^{T/2} \mathbf{u}(t)\mathbf{u}^*(t)\, dt; \qquad (6.1\text{-}37)$$

however, it is more convenient here to write W in terms of the complex envelope of $\mathbf{u}(t)$,

$$W = \int_{-T/2}^{T/2} \mathbf{A}(t)\mathbf{A}^*(t)\, dt \qquad (6.1\text{-}38)$$

PROPERTIES OF INTEGRATED INTENSITY

obtained by the simple substitution

$$\mathbf{u}(t) = \mathbf{A}(t)e^{-j2\pi\bar{\nu}t} \qquad (6.1\text{-}39)$$

in (6.1-37). Whereas $\mathbf{u}(t)$ has a bandpass spectrum, $\mathbf{A}(t)$ has a low-pass spectrum, as is illustrated later, in the discussion of the case of a rectangular spectral profile.

We begin by expanding the complex envelope $\mathbf{A}(t)$ on $(-T/2, T/2)$ by means of the Karhunen–Loève expansion of Eq. (3.10-1),

$$\mathbf{A}(t) = \sum_{n=0}^{\infty} \mathbf{b}_n \phi_n(t) \qquad |t| \leq \frac{T}{2}. \qquad (6.1\text{-}40)$$

Substituting this expansion in (6.1-38) and using the orthonormal properties of the functions $\phi_n(t)$ [see Eq. (3.10-2)], we obtain

$$W = \sum_{n=0}^{\infty} \sum_{m=0}^{\infty} \mathbf{b}_n \mathbf{b}_m^* \int_{-T/2}^{T/2} \phi_n(t) \phi_m^*(t)\, dt$$

$$= \sum_{n=0}^{\infty} |\mathbf{b}_n|^2. \qquad (6.1\text{-}41)$$

Thus the random variable W has been expressed exactly as an infinite sum of random variables $|\mathbf{b}_n|^2$. We turn now to considering the statistical properties of these latter random variables.

Noting from (3.10-3) that

$$\mathbf{b}_n = \int_{-T/2}^{T/2} \mathbf{A}(t)\phi_n^*(t)\, dt, \qquad (6.1\text{-}42)$$

the reader may wish to verify (see Problem 6-10) that, since the complex envelope $\mathbf{A}(t)$ of a polarized thermal light wave obeys circular complex Gaussian statistics, so do the complex coefficients $\mathbf{b}_n$. Furthermore, provided the functions $\phi_n(t)$ are solutions of the integral equation

$$\int_{-T/2}^{T/2} \Gamma_A(t_2 - t_1)\phi_n(t_2)\, dt_2 = \lambda_n \phi_n(t_1), \qquad (6.1\text{-}43)$$

the coefficients $\mathbf{b}_n$ are uncorrelated and, by virtue of their Gaussian statistics, independent. As for the coefficients $|\mathbf{b}_n|^2$, they also must be independent. Since $|\mathbf{b}_n|^2$ is the squared modulus of a circular complex Gaussian random variable, it must obey negative exponential statistics. Since by Eq.

(3.10-5) we have

$$E[|\mathbf{b}_n|^2] = \lambda_n, \quad (6.1\text{-}44)$$

the probability density function and characteristic function of $|\mathbf{b}_n|^2$ are

$$p_{|\mathbf{b}_n|^2}(|\mathbf{b}_n|^2) = \begin{cases} \dfrac{1}{\lambda_n} e^{-|\mathbf{b}_n|^2/\lambda_n} & |\mathbf{b}_n|^2 \geq 0 \\ 0 & \text{otherwise,} \end{cases} \quad (6.1\text{-}45)$$

$$\mathbf{M}_{|\mathbf{b}_n|^2}(\omega) = [1 - j\omega\lambda_n]^{-1}. \quad (6.1\text{-}46)$$

We have thus succeeded in expressing the integrated intensity W as the sum of an infinite number of statistically independent random variables, each with a known characteristic function (assuming that the eigenvalues λ_n are known). The characteristic function of W is accordingly given by

$$\mathbf{M}_W(\omega) = \prod_{n=0}^{\infty} [1 - j\omega\lambda_n]^{-1}. \quad (6.1\text{-}47)$$

Inversion of this characteristic function yields an exact probability density function of the form

$$p_W(W) = \begin{cases} \displaystyle\sum_{n=0}^{\infty} \dfrac{\lambda_n^{-1}\exp(-W/\lambda_n)}{\displaystyle\prod_{\substack{m=0 \\ m\neq n}}^{\infty}\left(1 - \dfrac{\lambda_m}{\lambda_n}\right)} & W \geq 0 \\ 0 & \text{otherwise.} \end{cases} \quad (6.1\text{-}48)$$

In order to specify numerical values of $p_W(W)$ for each W, it is necessary to assume a specific spectral profile for the optical wave. Our attention here is limited to the case of a rectangular spectrum, although the case of a Lorentzian spectrum can also be found in the literature (see Refs. 6-11 and 6-12). If the original real-valued waveform has a power spectral density

$$\mathscr{G}^{(r,r)}(\nu) = \frac{N_0}{2}\left[\operatorname{rect}\frac{\nu - \bar{\nu}}{\Delta\nu} + \operatorname{rect}\frac{\nu + \bar{\nu}}{\Delta\nu}\right], \quad (6.1\text{-}49)$$

the analytic signal $\mathbf{u}(t)$ has a power spectrum

$$\mathscr{G}(\nu) = 2N_0 \operatorname{rect}\frac{\nu - \bar{\nu}}{\Delta\nu}. \quad (6.1\text{-}50)$$

PROPERTIES OF INTEGRATED INTENSITY

It follows that the power spectral density of the complex envelope $A(t)$ is

$$\mathscr{S}(\nu) = 2N_0 \operatorname{rect} \frac{\nu}{\Delta \nu}, \tag{6.1-51}$$

and its correlation function is given by

$$\Gamma_A(\tau) = 2N_0 \frac{\sin \pi \Delta \nu \tau}{\pi \tau}. \tag{6.1-52}$$

Accordingly, the functions ϕ_n and the constants λ_n must be the eigenfunctions and eigenvalues of the integral equation

$$2N_0 \int_{-T/2}^{T/2} \frac{\sin \pi \Delta \nu (t_2 - t_1)}{\pi (t_2 - t_1)} \phi_n(t_2) \, dt_2 = \lambda_n \phi_n(t_1). \tag{6.1-53}$$

Fortunately, solutions of the integral equation

$$\int_{-T/2}^{T/2} \frac{\sin \pi \Delta \nu (t_2 - t_1)}{\pi (t_2 - t_1)} \phi_n(t_2) \, dt_2 = \tilde{\lambda}_n \phi_n(t_1) \tag{6.1-54}$$

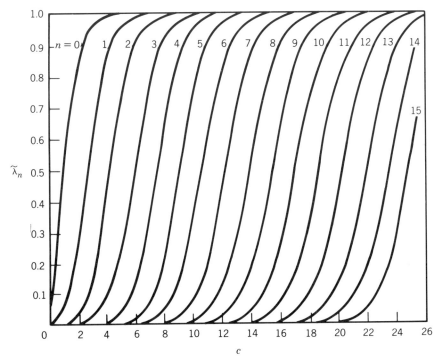

Figure 6-5. Plot of $\tilde{\lambda}_n$ versus c for various values of n. (Reprinted with permission from *The Bell System Technical Journal*. Copyright 1964, AT & T.)

have been widely studied in the literature (see Refs. 6-13 through 6-16). The eigenfunctions $\phi_n(t)$ are real valued and are known as the *prolate spheroidal functions*. The eigenvalues $\tilde{\lambda}_n$ (also real) have been tabulated and are available in both graphical and tabular form (see references cited immediately above). Both $\phi_n(t)$ and $\tilde{\lambda}_n$ depend not only on n, but also on the parameter

$$c = \frac{\pi}{2} \Delta \nu T = \frac{\pi}{2} \frac{T}{\tau_c}. \tag{6.1-55}$$

Figure 6-5 shows the values of $\tilde{\lambda}_n$ plotted against c for various values of n. Noting that $\overline{W} = \sum_{n=0}^{\infty} \tilde{\lambda}_n$, we can now plot numerical values of $\overline{W} p_W(W/\overline{W})$ against $W/\overline{W}$ according to the prescription

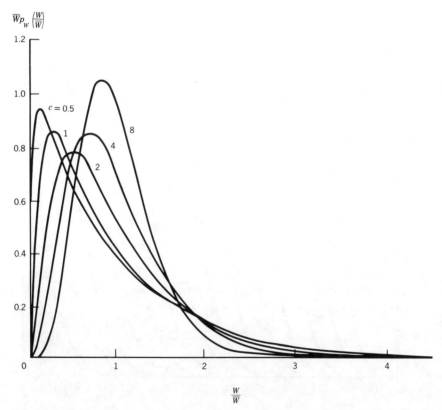

Figure 6-6. Exact probability density functions for integrated intensity; $c = 0.5, 1, 2, 4, 8$.

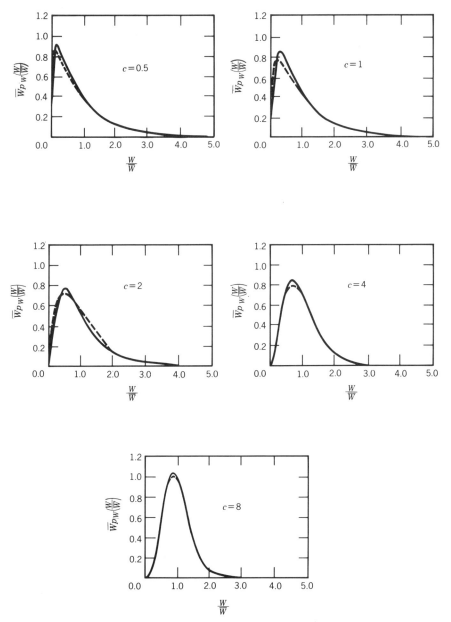

Figure 6-7. Comparison of exact and approximate probability density functions for integrated intensity. Solid lines represent exact results, dotted lines are approximate results.

$$\overline{W}p_W\left(\frac{W}{\overline{W}}\right) = \frac{\sum_{n=0}^{\infty} \frac{\overline{W}}{\lambda_n} \exp\left(-\frac{W}{\overline{W}} \cdot \frac{\overline{W}}{\lambda_n}\right)}{\prod_{\substack{m=0 \\ m \neq n}}^{\infty} \left(1 - \frac{\lambda_m}{\lambda_n}\right)}. \qquad (6.1\text{-}56)$$

This function is shown in Fig. 6-6 for $c = 0.5, 1, 2, 4$, and 8.

It is of some interest to compare the exact and approximate density functions for several values of T/τ_c. To do so conveniently, one first chooses a set of values of c for which tables are available and then converts these values of c into equivalent values of T/τ_c, using Eq. (6.1-55). In the case of the values $c = 0.5, 1, 2, 4$, and 8, we have $T/\tau_c = 0.319, 0.637, 1.273, 2.546$, and 5.093. Next, from Fig. 6-1, with use of the curve for a rectangular spectral profile, values of $\mathcal{M}$ corresponding to these values of T/τ_c are found as accurately as possible. In the present example, these values are $\mathcal{M} = 1.05, 1.22, 1.80, 3.07$, and 5.65. Finally, with use of Eq. (6.1-31) the approximate probability density functions are found and plotted along with the exact density functions. Figure 6-7 shows such plots. There is seen to be good agreement between the approximate and exact density functions for small and large T/τ_c but some noticeable discrepancies for T/τ_c close to unity.

6.2 STATISTICAL PROPERTIES OF MUTUAL INTENSITY WITH FINITE MEASUREMENT TIME

The complex-valued mutual intensity of a quasimonochromatic wave can always be interpreted physically in terms of the amplitude and spatial phase of a fringe pattern. A question of both theoretical and practical interest concerns the ultimate limits to the accuracy with which the parameters of such a fringe can be measured experimentally. Equivalently, we may inquire as to the fundamental limits to the accuracy with which the complex-valued mutual intensity can be measured.

Two fundamental limits to this accuracy can be identified. One arises from the discrete nature of the interaction of the incident waves and the measurement instrument. This limitation dominates any experiment with true thermal light and is discussed in detail in Chapter 9. The second limitation is introduced by the classical statistical fluctuations of the wavefield itself and by the (of necessity) finite duration of the measurement. This latter limitation, which is often the dominant source of errors with pseudothermal light, is the subject of interest in this section.

PROPERTIES OF MUTUAL INTENSITY

In the analysis to follow, we shall examine the statistical properties of the finite-time-averaged mutual intensity,

$$\mathbf{J}_{12}(T) = \frac{1}{T}\int_{-T/2}^{T/2} \mathbf{u}(P_1,t)\mathbf{u}^*(P_2,t)\,dt, \tag{6.2-1}$$

and in particular the dependence of those statistical properties on the duration T of the measurement. The term $\mathbf{J}_{12}(T)$ is, of course, simply an estimate of the true mutual intensity, which we represent here by $\mathbf{J}_{12}$. Clearly, as the measurement interval T increases without bound, by the definition of $\mathbf{J}_{12}$ we have

$$\lim_{T\to\infty} \mathbf{J}_{12}(T) = \mathbf{J}_{12}. \tag{6.2-2}$$

We shall assume throughout our analysis that the underlying wavefields are polarized and are of thermal or pseudothermal origin. The fields are accordingly modeled as zero-mean, ergodic, circular, complex Gaussian processes.

The statistical fluctuations of the amplitude and phase of $\mathbf{J}_{12}(T)$ are generally the quantities of ultimate interest; however, it will be convenient to first discuss the statistical properties of the real and imaginary parts of $\mathbf{J}_{12}(T)$,

$$\mathcal{R}_{12}(T) = \text{Re}\{\mathbf{J}_{12}(T)\}$$

$$\mathcal{I}_{12}(T) = \text{Im}\{\mathbf{J}_{12}(T)\}. \tag{6.2-3}$$

To facilitate the analysis, we express $\mathcal{R}_{12}(T)$ and $\mathcal{I}_{12}(T)$ in terms of $\mathbf{J}_{12}(T)$ and the underlying fields as follows:[†]

$$\mathcal{R}_{12}(T) = \frac{1}{2}[\mathbf{J}_{12}(T) + \mathbf{J}_{12}^*(T)]$$

$$= \frac{1}{2T}\int_{-T/2}^{T/2}[\mathbf{u}(P_1,t)\mathbf{u}^*(P_2,t) + \mathbf{u}^*(P_1,t)\mathbf{u}(P_2,t)]\,dt.$$

$$\tag{6.2-4}$$

[†] For an alternate method of analysis, see Ref. 6-17.

Similarly, for $\mathscr{I}_{12}(T)$ we have

$$\mathscr{I}_{12}(T) = \frac{1}{2j}\left[\mathbf{J}_{12}(T) - \mathbf{J}_{12}^*(T)\right]$$

$$= \frac{1}{2Tj}\int_{-T/2}^{T/2}\left[\mathbf{u}(P_1,t)\mathbf{u}^*(P_2,t) - \mathbf{u}^*(P_1,t)\mathbf{u}(P_2,t)\right]dt.$$

(6.2-5)

Our first task will simply be to find various moments of $\mathscr{R}_{12}(T)$ and $\mathscr{I}_{12}(T)$.

6.2.1 Moments of the Real and Imaginary Parts of $\mathbf{J}_{12}(T)$

To understand the statistical properties of $\mathbf{J}_{12}(T)$, it is first necessary to know the values of certain simple moments of the real and imaginary parts $\mathscr{R}_{12}(T)$ and $\mathscr{I}_{12}(T)$. Of particular importance are the following moments:

(1) The means $\overline{\mathscr{R}_{12}(T)}$ and $\overline{\mathscr{I}_{12}(T)}$.
(2) The variances

$$\sigma_{\mathscr{R}}^2 = \overline{\mathscr{R}_{12}^2(T)} - \left[\overline{\mathscr{R}_{12}(T)}\right]^2$$

$$\sigma_{\mathscr{I}}^2 = \overline{\mathscr{I}_{12}^2(T)} - \left[\overline{\mathscr{I}_{12}(T)}\right]^2. \qquad (6.2\text{-}6)$$

(3) The covariance

$$C_{\mathscr{R}\mathscr{I}} = \overline{\left[\mathscr{R}_{12}(T) - \overline{\mathscr{R}_{12}(T)}\right]\left[\mathscr{I}_{12}(T) - \overline{\mathscr{I}_{12}(T)}\right]}. \qquad (6.2\text{-}7)$$

The means can be calculated very quickly and easily. We simply average the expressions (6.2-4) and (6.2-5) over the statistical ensemble, yielding

$$\overline{\mathscr{R}_{12}(T)} = \frac{1}{2T}\int_{-T/2}^{T/2}\left[\overline{\mathbf{u}(P_1,t)\mathbf{u}^*(P_2,t)} + \overline{\mathbf{u}^*(P_1,t)\mathbf{u}(P_2,t)}\right]dt$$

$$\overline{\mathscr{I}_{12}(T)} = \frac{1}{2Tj}\int_{-T/2}^{T/2}\left[\overline{\mathbf{u}(P_1,t)\mathbf{u}^*(P_2,t)} - \overline{\mathbf{u}^*(P_1,t)\mathbf{u}(P_2,t)}\right]dt.$$

(6.2-8)

PROPERTIES OF MUTUAL INTENSITY

Noting that

$$\overline{\mathbf{u}(P_1,t)\mathbf{u}^*(P_2,t)} = \mathbf{J}_{12}, \quad (6.2\text{-}9)$$

we see that

$$\overline{\mathscr{R}_{12}(T)} = \frac{1}{2}[\mathbf{J}_{12} + \mathbf{J}_{12}^*] = \text{Re}\{\mathbf{J}_{12}\}$$

$$\overline{\mathscr{I}_{12}(T)} = \frac{1}{2j}[\mathbf{J}_{12} - \mathbf{J}_{12}^*] = \text{Im}\{\mathbf{J}_{12}\}. \quad (6.2\text{-}10)$$

Thus the mean values of the real and imaginary parts of the finite-time-averaged mutual intensity are equal to the real and imaginary parts of the true mutual intensity.

Calculation of the variances and covariance requires considerably more work. We illustrate only the calculation of $\sigma_{\mathscr{R}}^2$ and simply state the results for $\sigma_{\mathscr{I}}^2$ and $C_{\mathscr{R}\mathscr{I}}$. The second moment $\overline{\mathscr{R}_{12}^2(T)}$ is first calculated, following which the square of the mean $[\overline{\mathscr{R}_{12}(T)}]^2$ is subtracted. We begin with

$$\mathscr{R}_{12}^2(T) = \frac{1}{4T^2} \iint\limits_{-T/2}^{T/2} [\mathbf{u}(P_1,\xi)\mathbf{u}^*(P_2,\xi) + \mathbf{u}^*(P_1,\xi)\mathbf{u}(P_2,\xi)]$$

$$\times [\mathbf{u}(P_1,\eta)\mathbf{u}^*(P_2,\eta) + \mathbf{u}^*(P_1,\eta)\mathbf{u}(P_2,\eta)]\, d\xi\, d\eta. \quad (6.2\text{-}11)$$

Taking the average of both sides of the equation, we obtain

$$\overline{\mathscr{R}_{12}^2(T)} = \frac{1}{4T^2} \iint\limits_{-T/2}^{T/2} \left[\overline{\mathbf{u}(P_1,\xi)\mathbf{u}^*(P_2,\xi)\mathbf{u}(P_1,\eta)\mathbf{u}^*(P_2,\eta)}\right.$$

$$+ \overline{\mathbf{u}(P_1,\xi)\mathbf{u}^*(P_2,\xi)\mathbf{u}^*(P_1,\eta)\mathbf{u}(P_2,\eta)}$$

$$+ \overline{\mathbf{u}^*(P_1,\xi)\mathbf{u}(P_2,\xi)\mathbf{u}(P_1,\eta)\mathbf{u}^*(P_2,\eta)}$$

$$\left.+ \overline{\mathbf{u}^*(P_1,\xi)\mathbf{u}(P_2,\xi)\mathbf{u}^*(P_1,\eta)\mathbf{u}(P_2,\eta)}\right] d\xi\, d\eta. \quad (6.2\text{-}12)$$

Each of the fourth-order moments can be expanded by means of the complex Gaussian moment theorem, yielding,

$$\overline{\mathscr{R}_{12}^2(T)} = \frac{1}{4T^2} \iint\limits_{-T/2}^{T/2} \{[\Gamma_{12}(0)\Gamma_{12}(0) + \Gamma_{12}(\xi - \eta)\Gamma_{12}(\eta - \xi)]$$

$$+ [\Gamma_{12}(0)\Gamma_{12}^*(0) + \Gamma_{11}(\xi - \eta)\Gamma_{22}(\eta - \xi)]$$

$$+ [\Gamma_{12}^*(0)\Gamma_{12}(0) + \Gamma_{11}(\eta - \xi)\Gamma_{22}(\xi - \eta)]$$

$$+ [\Gamma_{12}^*(0)\Gamma_{12}^*(0) + \Gamma_{12}^*(\xi - \eta)\Gamma_{12}^*(\eta - \xi)]\} d\xi d\eta, \quad (6.2\text{-}13)$$

where $\Gamma_{11}(\tau)$, $\Gamma_{22}(\tau)$, and $\Gamma_{12}(\tau)$ are the self- and mutual coherence functions of $\mathbf{u}(P_1, t)$ and $\mathbf{u}(P_2, t)$.

At this point in the analysis it is helpful to make some specific assumptions about the character of the mutual coherence function $\Gamma_{12}(\tau)$. We first assume that the light is cross-spectrally pure, in which case the mutual coherence function can be written in the form

$$\Gamma_{12}(\tau) = \sqrt{I_1 I_2}\, \mu_{12} \gamma(\tau). \quad (6.2\text{-}14)$$

Second, without loss of generality, we can assume that the complex coherence factor μ_{12} is entirely real. This assumption amounts simply to the choice of a phase reference that coincides with the phase of μ_{12}. With substitution of (6.2-14) in (6.2-13), the second moment of $\mathscr{R}_{12}(T)$ becomes

$$\overline{\mathscr{R}_{12}^2(T)} = \frac{I_1 I_2}{4T^2} \iint\limits_{-T/2}^{T/2} [\mu_{12}^2 + \mu_{12}^2 \gamma(\xi - \eta)\gamma(\eta - \xi)$$

$$+ \mu_{12}^2 + \gamma(\xi - \eta)\gamma(\eta - \xi)$$

$$+ \mu_{12}^2 + \gamma(\eta - \xi)\gamma(\xi - \eta)$$

$$+ \mu_{12}^2 + \mu_{12}^2 \gamma(\xi - \eta)\gamma(\eta - \xi)] d\xi d\eta. \quad (6.2\text{-}15)$$

Noting that $\gamma(\eta - \xi) = \gamma^*(\xi - \eta)$ and collecting terms, we obtain

$$\overline{\mathscr{R}_{12}^2(T)} = I_1 I_2 \mu_{12}^2 + \frac{I_1 I_2}{2T^2}[1 + \mu_{12}^2] \iint\limits_{-T/2}^{T/2} |\gamma(\xi - \eta)|^2 d\xi d\eta.$$

$$(6.2\text{-}16)$$

The mean value of $\mathscr{R}_{12}(T)$ is simply $\sqrt{I_1 I_2}\, \mu_{12}$. Hence, subtracting the

PROPERTIES OF MUTUAL INTENSITY

square of the mean, we obtain the variance

$$\sigma_{\mathcal{R}}^2(T) = \frac{I_1 I_2}{2T^2}[1 + \mu_{12}^2] \iint_{-T/2}^{T/2} |\gamma(\xi - \eta)|^2 \, d\xi \, d\eta. \quad (6.2\text{-}17)$$

A final simplification is obtained by noting that $|\gamma|^2$ is an even function of its argument, allowing the double integral to be reduced to a single integral [cf. Eq. (6.1-6)],

$$\sigma_{\mathcal{R}}^2(T) = I_1 I_2 \frac{1 + \mu_{12}^2}{2}\left[\frac{1}{T}\int_{-\infty}^{\infty} \Lambda\left(\frac{\tau}{T}\right)|\gamma(\tau)|^2 \, d\tau\right]. \quad (6.2\text{-}18)$$

Noting that the quantity in brackets is, from Eq. (6.1-17), simply $\mathcal{M}^{-1}$, we obtain the final result,

$$\sigma_{\mathcal{R}}^2(T) = I_1 I_2 \frac{1 + \mu_{12}^2}{2\mathcal{M}}. \quad (6.2\text{-}19)$$

Proceeding in an identical fashion to calculate the variance $\sigma_{\mathcal{I}}^2$ of the imaginary part, we find

$$\sigma_{\mathcal{I}}^2(T) = I_1 I_2 \frac{1 - \mu_{12}^2}{2\mathcal{M}}. \quad (6.2\text{-}20)$$

Finally, a similar calculation shows that the covariance $C_{\mathcal{R}\mathcal{I}}$ of the real and imaginary parts is identically zero,

$$C_{\mathcal{R}\mathcal{I}} \equiv 0. \quad (6.2\text{-}21)$$

To conclude this section, we summarize the values of the various moments that have been derived here:

$$\overline{\mathcal{R}_{12}(T)} = \sqrt{I_1 I_2}\,\mu_{12}, \quad \overline{\mathcal{I}_{12}(T)} = 0$$

$$\sigma_{\mathcal{R}}^2 = I_1 I_2 \frac{1 + \mu_{12}^2}{2\mathcal{M}}, \quad \sigma_{\mathcal{I}}^2 = I_1 I_2 \frac{1 - \mu_{12}^2}{2\mathcal{M}}$$

$$C_{\mathcal{R}\mathcal{I}} \equiv 0. \quad (6.2\text{-}22)$$

These results can be made more physical by picturing the measured value of $\mathbf{J}_{12}(T)$ as consisting of a fixed phasor of length $\mathbf{J}_{12} = \sqrt{I_1 I_2}\,\mu_{12}$ along the real axis, with its tip surrounded by a "noise cloud." The measured value $\mathbf{J}_{12}(T)$ falls within the noise cloud, and thus $\mathbf{J}_{12}(T)$ differs from the true mutual intensity $\mathbf{J}_{12}$. When the measurement time T is shorter than the coherence time τ_c, we know from Fig. 6-1 that $\mathcal{M} \cong 1$. The resulting noise clouds are illustrated in Fig. 6-8 for various values of μ_{12}. Note in particular the oblong shape of the noise clouds for μ_{12} near unity, a consequence of the different values of $\sigma_{\mathcal{R}}$ and $\sigma_{\mathcal{I}}$ in this case. Of particular interest is the fact that when μ_{12} is unity, $\sigma_{\mathcal{I}} = 0$, and the noise cloud collapses onto the real axis. Thus for μ_{12} equal to unity, there will be no errors in the phase of $\mathbf{J}_{12}(T)$, regardless of the integration time! This mathematical result is simply indicative of the fact that when $\mu_{12} = 1$, the two interfering beams are perfectly coherent and have a constant phase difference independent of time. Thus the phase of the interference fringe they generate is always equal to the true phase of the mutual intensity, independent of the integration time.

When the integration time is much longer than the coherence time, the parameter $\mathcal{M}$ becomes equal to T/τ_c. The dimensions of the noise clouds shown in Fig. 6-8 are accordingly reduced by the factor $\sqrt{\tau_c/T}$. Figure 6-9 shows the results when $T = 10\tau_c$.

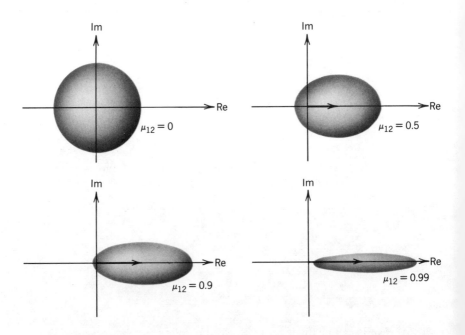

Figure 6-8. Probability clouds for $\mathbf{J}_{12}(T)$ with $T < \tau_c$ and for various values of μ_{12}.

PROPERTIES OF MUTUAL INTENSITY   263

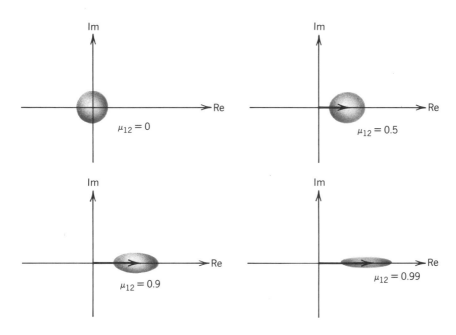

Figure 6-9. Probability clouds for $J_{12}(T)$ with $T = 10\tau_c$ and various values of μ_{12}.

6.2.2 Statistics of the Modulus and Phase of $J_{12}(T)$ for Long Integration Time and Small μ_{12}

We assume hereafter that the integration time T is much greater than the coherence time τ_c, as is usually the case in practice. We then have

$$\sigma_{\mathcal{R}}^2 \cong I_1 I_2 \frac{1+\mu_{12}^2}{2}\frac{\tau_c}{T}$$

$$\sigma_{\mathcal{I}}^2 \cong I_1 I_2 \frac{1-\mu_{12}^2}{2}\frac{\tau_c}{T}. \qquad (6.2\text{-}23)$$

Furthermore, because $T \gg \tau_c$, the quantity $J_{12}(T)$ results from integration of the quantity $u(P_1 t)u^*(P_2, t)$ over many independent fluctuation intervals. It follows directly from the central limit theorem that, for such integration times, $J_{12}(T)$ is approximately a complex Gaussian random variable. However, the complex Gaussian statistics are not circular in general (for $\sigma_{\mathcal{R}} \neq \sigma_{\mathcal{I}}$) and the mean is not zero. Because of the lack of correlation (and thus, under the Gaussian assumption, the statistical independence) of $\mathscr{R}_{12}(T)$ and $\mathscr{I}_{12}(T)$, we can write the approximate form of their joint probability density function as

$$p_{\mathscr{R}\mathscr{I}}(\mathscr{R}_{12}, \mathscr{I}_{12}) = \frac{1}{\sqrt{2\pi}\,\sigma_{\mathscr{R}}} \exp\left\{-\frac{\left(\mathscr{R}_{12} - \sqrt{I_1 I_2}\,\mu_{12}\right)^2}{2\sigma_{\mathscr{R}}^2}\right\}$$

$$\times \frac{1}{\sqrt{2\pi}\,\sigma_{\mathscr{I}}} \exp\left\{-\frac{\mathscr{I}_{12}^2}{2\sigma_{\mathscr{I}}^2}\right\}. \qquad (6.2\text{-}24)$$

The resemblance of the statistical properties of $\mathbf{J}_{12}(T)$ to the statistical properties of the sum of a constant phasor plus a random phasor sum (discussed in Section 2.9.4) may perhaps be evident to the reader. However, there is one important difference between the present case and that discussed in Section 2.9.4. In the case under consideration here, the variances of the real and imaginary parts are *not equal*, whereas in the previous case they were equal. Thus in general the statistics of the magnitude and phase of $\mathbf{J}_{12}(T)$ will *not* be the same as the statistics of the random variables A and θ in Section 2.9.4.

The statistical properties of the length of the sum of a constant phasor plus a random phasor sum with different variances along the real and imaginary axes have indeed been studied previously in the literature (see Ref. 6-18). The resulting probability density function is found to depend on two key parameters, an "asymmetry factor"

$$\mathscr{A}^2 \triangleq \frac{\sigma_{\mathscr{R}}^2}{\sigma_{\mathscr{I}}^2} = \frac{1 + \mu_{12}^2}{1 - \mu_{12}^2}, \qquad (6.2\text{-}25)$$

and a "signal-to-noise ratio"

$$K^2 \triangleq \frac{\left(\overline{\mathscr{R}}_{12}\right)^2 + \left(\overline{\mathscr{I}}_{12}\right)^2}{\sigma_{\mathscr{R}}^2 + \sigma_{\mathscr{I}}^2} = \mu_{12}^2 \frac{T}{\tau_c}. \qquad (6.2\text{-}26)$$

In this section we examine only the case of an asymmetry factor that is approximately unity, a condition satisfied for $\mu_{12} < 0.3$. In the section to follow we allow the asymmetry factor to have any value but require that the signal-to-noise ratio K^2 be much greater than unity. These two special cases are relatively easy to analyze and lend themselves to considerable physical reasoning.

PROPERTIES OF MUTUAL INTENSITY

Under the assumption that μ_{12} is less than 0.3 ($\mathscr{A}^2 \cong 1$), we have $\sigma_{\mathscr{R}}^2 \cong \sigma_{\mathscr{I}}^2$, and the joint density function of the real and imaginary parts of $\mathbf{J}_{12}(T)$ becomes

$$p_{\mathscr{R}\mathscr{I}}(\mathscr{R}_{12}, \mathscr{I}_{12}) = \frac{1}{2\pi\sigma^2} \exp\left\{-\frac{\left(\mathscr{R}_{12} - \sqrt{I_1 I_2}\,\mu_{12}\right)^2 + \mathscr{I}_{12}^2}{2\sigma^2}\right\}, \tag{6.2-27}$$

where

$$\sigma^2 = \frac{1}{2} I_1 I_2 \frac{\tau_c}{T}. \tag{6.2-28}$$

Making the identifications, $a \sim |\mathbf{J}_{12}(T)|$, $s \sim \sqrt{I_1 I_2}\,\mu_{12}$, $\theta \sim \arg\{\mathbf{J}_{12}(T)\}$, and $k^2 \sim 2K^2$ for the symbols a, s, θ, and k^2 in Section 2.9.4, we see that the probability density function of $|\mathbf{J}_{12}(T)|$ is given by Eq. (2.9-20) and the probability density function of $\arg\{\mathbf{J}_{12}(T)\}$ is given by Eq. (2.9-25).

From the probability density functions of $|\mathbf{J}_{12}(T)|$ and $\arg\{\mathbf{J}_{12}(T)\}$, which are now known for $\mu_{12} < 0.3$, it is possible to extract some information that is pertinent to the experimentalist. Our first important observation is that *the mean value of $|\mathbf{J}_{12}(T)|$ is not the same as the modulus of the true mutual intensity $|\mathbf{J}_{12}|$ when the measurement interval T is finite*. There exists a deterministic bias in any finite-time estimate $|\mathbf{J}_{12}(T)|$, which arises as a consequence of the geometric fact that the sum of a constant-length phasor and a random phasor has a length that is more likely to be longer than the constant phasor than to be shorter. This fact is illustrated in Fig. 6-10. Thus, if the value of $|\mathbf{J}_{12}(T)|$ is measured repeatedly by determining the amplitude of a fringe in a finite-time interference experiment, the arithmetic mean of those measurements will not precisely equal the fringe amplitude that would be obtained with infinite integration time, even if the finite-time measurement is repeated an infinite number of times!

The magnitude of the deterministic bias can be found quantitatively by considering the mean value $\overline{|\mathbf{J}_{12}(T)|}$. The difference between $\overline{|\mathbf{J}_{12}(T)|}$ and $|\mathbf{J}_{12}|$ is precisely the same as the difference between $\bar{a}$ and s in the context of Section 2.9.4. If we define the "fractional bias" Δ of our estimate by

$$\Delta = \frac{\overline{|\mathbf{J}_{12}(T)|} - |\mathbf{J}_{12}|}{|\mathbf{J}_{12}|} \sim \frac{\bar{a} - s}{s}, \tag{6.2-29}$$

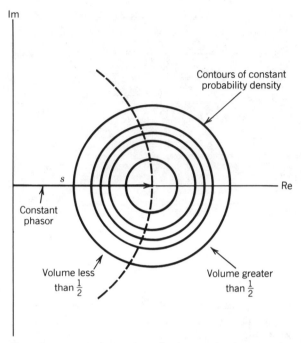

Figure 6-10. Sum of a constant phasor plus a circular complex Gaussian phasor. The volumes shown to the right and left of the dotted line are the probabilities that the resultant is longer or shorter than s, respectively.

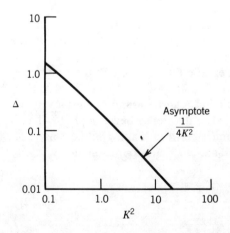

Figure 6-11. Fractional bias Δ plotted against K^2. (Reprinted with the permission of Springer-Verlag, Heidelberg.)

we obtain from Eq. (2.9-23) (with $k^2 = 2K^2$)

$$\Delta = \sqrt{\frac{\pi}{4K^2}} e^{-K^2/2} \left[(1 + K^2) I_0\left(\frac{K^2}{2}\right) + K^2 I_1\left(\frac{K^2}{2}\right) \right] - 1, \tag{6.2-30}$$

which is plotted against K^2 in Fig. 6-11. Noting that

$$K^2 = \mu_{12}^2 \frac{T}{\tau_c}, \tag{6.2-31}$$

we see that, as the integration time T grows larger and larger, the value of the fractional bias decreases monotonically.

In addition to the deterministic error represented by the bias of Fig. 6-11, there also exist random errors that result from the finite integration time T. In this case it is convenient to define an rms signal-to-noise ratio associated with the measurement of $|\mathbf{J}_{12}(T)|$ by

$$\left(\frac{S}{N}\right)_{rms} = \frac{\overline{|\mathbf{J}_{12}(T)|}}{\left[\overline{|\mathbf{J}_{12}(T)|^2} - \left[\overline{|\mathbf{J}_{12}(T)|}\right]^2\right]^{1/2}} \sim \frac{\bar{a}}{\sqrt{\overline{a^2} - (\bar{a})^2}}. \tag{6.2-32}$$

Using Eqs. (2.9-23) and (2.9-24) for $\bar{a}$ and $\overline{a^2}$, and again substituting $k^2 = 2K^2$, we find

$$\left(\frac{S}{N}\right)_{rms} = \left\{ \frac{(4/\pi)e^{K^2}(K^2 + 1)}{\left[(1 + K^2) I_0\left(\frac{K^2}{2}\right) + K^2 I_1\left(\frac{K^2}{2}\right)\right]^2} - 1 \right\}^{-1/2} \tag{6.2-33}$$

This result is shown plotted in Fig. 6-12. It represents the rms signal-to-noise ratio accurately, provided $\mu_{12} < 0.3$. Note that for $K^2 > 5$, the rms signal-to-noise ratio increases as $\sqrt{2} K$ and, therefore, in direct proportion the square root of the integration time.

Finally, we turn attention to the statistical properties of the phase of $\mathbf{J}_{12}(T)$. We can readily see from the fact that there are equal volumes under the probability contours above and below the real axis in Fig. 6-10 that the phase of $|\mathbf{J}_{12}(T)|$ is an *unbiased* estimate of the phase of $\mathbf{J}_{12}$. That is,

$$E\left[\arg\{\mathbf{J}_{12}(T)\}\right] = \arg\{\mathbf{J}_{12}\}. \tag{6.2-34}$$

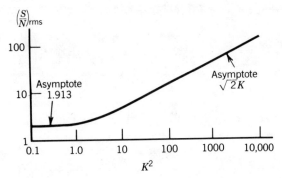

Figure 6-12. Root-mean-square signal-to-noise ratio associated with the measurement of $|J_{12}(T)|$. The asymptote 1.913 on the left represents the ratio of mean to standard deviation for a Rayleigh distribution. (Reprinted with the permission of Springer-Verlag, Heidelberg.)

This fact is also borne out by the symmetrical shapes of the probability density functions for phase shown in Fig. 2-15.

A parameter that is indicative of the experimental errors anticipated in the measurement of phase is the standard deviation of that phase σ_θ. To find σ_θ mathematically, we must find the second moment of the rather complicated probability density function for phase given by Eq. (2.9-25). The standard deviation can be found by numerical integration for each value of K^2 and is shown plotted in Fig. 6-13. For small K^2, σ_θ approaches the value 1.814 radians, the standard deviation associated with a phase uni-

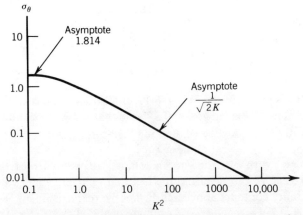

Figure 6-13. Standard deviation σ_θ of the phase of $J_{12}(T)$ as a function of K^2. The asymptote 1.814 radians represents the standard deviation of a phase uniformly distributed on $(-\pi, \pi)$. (Reprinted with the permission of Springer-Verlag, Heidelberg.)

PROPERTIES OF MUTUAL INTENSITY   269

formly distributed on $(-\pi, \pi)$. For large values of K^2, σ_θ falls as $(\sqrt{2}\,K)^{-1}$. Again the results are accurate only for $\mu_{12} < 0.3$.

The reader is reminded that all the conclusions presented in this section are valid only when the integration time T is much longer than the coherence time τ_c of the light (e.g., $T > 10\tau_c$). For shorter integration times the use of complex Gaussian statistics for $\mathbf{J}_{12}(T)$ is generally not justified.

6.2.3 Statistics of the Modulus and Phase of $\mathbf{J}_{12}(T)$ Under the Condition of High Signal-to-Noise Ratio

Although the general problem of finding the statistics of $|\mathbf{J}_{12}(T)|$ and $\arg\{\mathbf{J}_{12}(T)\}$ for arbitrary μ_{12} is not treated here, even for $T \gg \tau_c$, we can specify the statistics in one special case of interest. We again assume that $T \gg \tau_c$, allowing the use of Gaussian statistics for $\mathcal{R}_{12}(T)$ and $\mathcal{I}_{12}(T)$. However, this time we allow the asymmetry factor $\mathcal{A}^2$ to have any value between 1 and ∞ (or, equivalently, μ_{12} may have any value between 0 and 1). We require only the condition of high signal-to-noise $K^2 \gg 1$, or

$$\frac{T}{\tau_c} \gg \frac{1}{\mu_{12}^2}, \qquad (6.2\text{-}35)$$

where in practice a factor of 10 will suffice.

When $K^2 \gg 1$, the value of $\mathbf{J}_{12}(T)$ is always rather close to the ideal value $\mathbf{J}_{12}$, for the noise cloud has dimensions that are small by comparison with the length of the fixed phasor. As a consequence, consistent with the arguments presented in Section 2.9.5, with good accuracy we can regard the fluctuations of $|\mathbf{J}_{12}(T)|$ as arising primarily from the *real* part of the noise phasor and the fluctuations of $\arg\{\mathbf{J}_{12}(T)\}$ as arising primarily from the *imaginary* part of the noise phasor. Following the reasoning that led to Eq. (2.9-27), we know from the Gaussian statistics of $\mathcal{R}_{12}(T)$ that $|\mathbf{J}_{12}(T)|$ is approximately Gaussian, with probability density function

$$p_J(|\mathbf{J}_{12}(T)|) = \frac{1}{\sqrt{2\pi}\,\sigma_\mathcal{R}} \exp\left\{-\frac{\left(|\mathbf{J}_{12}(T)| - \sqrt{I_1 I_2}\,\mu_{12}\right)^2}{2\sigma_\mathcal{R}^2}\right\}, \qquad (6.2\text{-}36)$$

where again

$$\sigma_\mathcal{R}^2 = I_1 I_2 \frac{1 + \mu_{12}^2}{2} \frac{\tau_c}{T}. \qquad (6.2\text{-}37)$$

The rms signal-to-noise ratio associated with $|\mathbf{J}_{12}(T)|$ is accordingly given by

$$\left(\frac{S}{N}\right)_{rms} = \frac{\sqrt{I_1 I_2}\,\mu_{12}}{\sigma_{\mathcal{R}}} = \left[\frac{2\mu_{12}^2}{1+\mu_{12}^2}\frac{T}{\tau_c}\right]^{1/2} = \left[\frac{2K^2}{1+\mu_{12}^2}\right]^{1/2}. \quad (6.2\text{-}38)$$

We further note from the symmetry of the Gaussian density function (6.2-36) about its mean $|\mathbf{J}_{12}|$ that the deterministic bias associated with $|\mathbf{J}_{12}(T)|$ is negligible under the high signal-to-noise ratio condition.

As for the phase associated with $\mathbf{J}_{12}(T)$, we assume that its fluctuations arise primarily from $\mathcal{I}_{12}(T)$, which has variance

$$\sigma_{\mathcal{I}}^2 = I_1 I_2 \frac{1-\mu_{12}^2}{2}\frac{\tau_c}{T}. \quad (6.2\text{-}39)$$

Since $\mathcal{I}_{12}(T)$ is Gaussian with zero mean, the tangent of the angle $\arg\{\mathbf{J}_{12}(T)\}$ is also Gaussian. Further, under the high signal-to-noise ratio approximation, the tangent and the angle are approximately equal. Hence, if $\theta = \arg\{\mathbf{J}_{12}(T)\}$, we have

$$p_\theta(\theta) = \frac{1}{\sqrt{2\pi}\,\sigma_\theta}\exp\left\{-\frac{\theta^2}{2\sigma_\theta^2}\right\}, \quad (6.2\text{-}40)$$

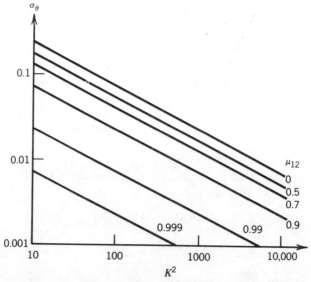

Figure 6-14. Standard deviation σ_θ of the phase of $\mathbf{J}_{12}(T)$ for various values of μ_{12} and as a function of K^2. (Reprinted with the permission of Springer-Verlag, Heidelberg.)

where
$$\sigma_\theta = \frac{\sigma_\mathscr{I}}{|\mathbf{J}_{12}|} = \left[\frac{1 - \mu_{12}^2}{K^2}\right]^{1/2} \qquad (6.2\text{-}41)$$

The approximations involved in this argument are very well satisfied for $\sigma_\theta < 0.2$ radians and for any value of K^2 greater than 10. Figure 6-14 shows σ_θ plotted against K^2 for various values of μ_{12}. From these curves we can deduce the accuracy achievable in any measurement of $\arg\{\mathbf{J}_{12}(T)\}$ for $K^2 > 10$.

The reader's attention is again called to the fact that as the complex coherence factor μ_{12} approaches unity, the errors in θ grow vanishingly small. As pointed out earlier, this result is simply evidence of the fact that the phase difference between two highly coherent beams remains almost constant.

6.3 CLASSICAL ANALYSIS OF THE INTENSITY INTERFEROMETER

The concepts of spatial and temporal coherence of light waves have been seen to arise quite naturally in the consideration of experiments that involve the interference of two light beams. Coherence effects can also be observed in a less direct (but in some respects more convenient) interferometric instrument known as the *intensity interferometer*. Such an instrument, first conceived of and demonstrated by R. Hanbury Brown and R. Q. Twiss (Refs. 6-19 through 6-23) requires the use of coherence of order higher than 2 for an understanding of its operation. A book by R. Hanbury Brown (Ref. 6-24) describes both the fascinating history of the ideas behind this interferometer and the technical developments that led to the construction of a large astronomical instrument of this kind at Narrabri, Australia.

In the material to follow, we first discuss intensity interferometry in rather qualitative terms, concentrating primarily on the basic form of the interferometer. Attention is then turned to an analysis that demonstrates how the intensity interferometer can extract information about the modulus of the complex coherence factor. Finally, a brief discussion of one component of noise associated with the interferometer output is presented. All discussion of the intensity interferometer in this chapter is presented in purely classical terms. Such an analysis is directly applicable in the radio region of the spectrum. However, the reader should bear in mind that to fully understand the capabilities and limitations of this interferometer at optical frequencies, it is essential that a detailed model be available for the process by which a light beam is converted into a photocurrent by a detector. Such considerations, which concern the discrete interaction of light and matter, are deferred to Chapter 9, where the subject of intensity interferometry is taken up again.

6.3.1 Amplitude versus Intensity Interferometry

We have seen previously (Section 5.2) that, for quasimonochromatic light, the complex coherence factor μ_{12} of the light incident at two points P_1 and P_2 in space can be measured by means of Young's interference experiment. The light waves striking P_1 and P_2 are isolated by insertion of a pair of pinholes; after passage through the pinholes these light contributions expand as spherical waves, ultimately overlapping on an observation screen or on a continuous photodetector such as photographic film. The two waves are added on an amplitude basis and then subjected to the square-law action of the intensity-sensitive detector. Associated with the detection process is a long time constant that introduces an averaging operation. The spatial distribution of time-averaged intensity was found to be a sinusoidal fringe, the visibility of which yielded information concerning the modulus of the complex coherence factor $|\mu_{12}|$ and the spatial position of which yielded information about the phase of μ_{12}.

The question now quite naturally arises as to whether it might be possible to interchange the order of some of the operations inherent in Young's interference experiment. In particular, could information about μ_{12} still be retrieved if the light waves incident at points P_1 and P_2 were directly detected at those points, the two fluctuating photocurrents brought together and forced to interact through a nonlinear electronic device, and the result of that interaction subjected to a time averaging operation? As we see in detail in the sections to follow, the answer to this question is affirmative, although it must be qualified by the statement that in general the information retrievable is not complete.

Figure 6-15 illustrates the general form of an intensity interferometer. Highly sensitive and wideband photodetectors (usually photomultiplier tubes) directly detect the light incident on points P_1 and P_2. A simple classical model of the detection process, which neglects the discrete nature of the interaction of light with the photosensitive elements (as well as other possible sources of noise), suggests that the photocurrents generated by the two photosensitive surfaces are proportional to the instantaneous light intensities incident on them. These currents are subjected to temporal smoothing by the finite response times (or limited bandwidth) of the photomultiplier structure and the electronic circuitry that follows. These unavoidable smoothing operations are represented by linear filters with impulse responses $h(t)$, assumed for simplicity to be identical in the two electrical arms of the interferometer.

Since only the fluctuations of the two photocurrents about their direct current (DC) or average values carry information regarding the correlation or coherence of the two light beams, the DC components are removed prior to bringing the two currents together. By means of filters with impulse responses $a(t)$ that average over long integration times [i.e., $a(t)$ has a much narrower bandwidth than $h(t)$], estimates of the DC photocurrents are formed and are subtracted from the two electrical signals, as shown in Fig. 6-15. These DC components will be needed later for normalization purposes, so they are presented at the output of the electrical circuitry.

The remaining fluctuating components, $\Delta i_1(t)$ and $\Delta i_2(t)$, are brought together and applied to a multiplier achieved by means of a nonlinear electrical device. The product of the two photocurrents, $\Delta i_1(t) \Delta i_2(t)$, is then subjected to a long time average, again by an electrical filter with impulse response $a(t)$. Thus the interferometer has three outputs, two representing estimates of the average photocurrents, and the third an estimate of the covariance between the two photocurrents.

A detailed discussion of the advantages and disadvantages of intensity interferometry, in comparison with more direct amplitude interferometry, is deferred to Chapter 9. We simply mention here that the intensity interferometer is far more tolerant of imperfect optical elements, imperfect path length equalization, and atmospheric "seeing" effects (cf. Chapter 8) than is the amplitude interferometer. The chief disadvantage of intensity interferometry will be found to be its comparatively poor signal-to-noise performance, which generally dictates long measurement times.

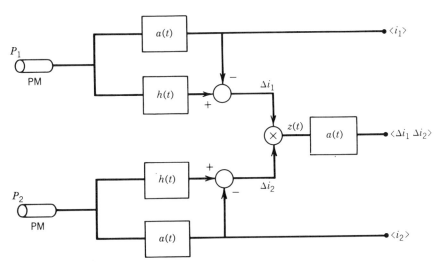

Figure 6-15. Intensity interferometer (PM = photomultiplier).

6.3.2 Ideal Output of the Intensity Interferometer

In this section our goal is to find a mathematical expression for the ideal output of the intensity interferometer. Our result will be ideal in the sense that all sources of noise are neglected. The averaging time of the filters with impulse responses $a(t)$ will be assumed to be infinitely long.

In accord with the structure illustrated in Fig. 6-15 and with the completely classical nature of our analysis, the fluctuating currents $\Delta i_1(t)$ and $\Delta i_2(t)$ are represented by

$$\Delta i_1(t) = \alpha_1 \int_{-\infty}^{\infty} h(t-\xi) I_1(\xi)\, d\xi - \langle i_1(t) \rangle$$

$$\Delta i_2(t) = \alpha_2 \int_{-\infty}^{\infty} h(t-\eta) I_2(\eta)\, d\eta - \langle i_2(t) \rangle, \qquad (6.3\text{-}1)$$

where α_1 and α_2 are constants associated with the two detectors, $I_1(t)$ and $I_2(t)$ are the instantaneous intensities incident at P_1 and P_2, and the impulse response $h(t)$ has been defined previously. We assume that the random processes $i_1(t)$ and $i_2(t)$ are ergodic, in which case the infinite time averages can be replaced by ensemble averages. Our goal is to find the expected value of the current at the output of the electronic multiplier, for it is this quantity that will yield information about the complex coherence factor.

The averaged product of $\Delta i_1(t)$ and $\Delta i_2(t)$ can be written in the form

$$\bar{z} = \overline{\Delta i_1(t)\, \Delta i_2(t)}$$

$$= \alpha_1 \alpha_2 \iint_{-\infty}^{\infty} h(t-\xi) h(t-\eta) \overline{I_1(\xi) I_2(\eta)}\, d\xi\, d\eta - \bar{i}_1 \bar{i}_2, \qquad (6.3\text{-}2)$$

where the orders of averaging and integration have been interchanged. To make further progress, we must adopt some form for the second moment of the instantaneous intensities.

We make the major assumption at this point that the light incident on the detectors is polarized thermal light. In this case, the complex Gaussian moment theorem can be utilized to demonstrate that

$$\overline{I_1(\xi) I_2(\eta)} = \overline{\mathbf{u}_1(\xi)\mathbf{u}_1^*(\xi)\mathbf{u}_2(\eta)\mathbf{u}_2^*(\eta)}$$

$$= \Gamma_{11}(0)\Gamma_{22}(0) + |\Gamma_{12}(\xi-\eta)|^2, \qquad (6.3\text{-}3)$$

where $\Gamma_{11}(\tau)$, $\Gamma_{22}(\tau)$, and $\Gamma_{12}(\tau)$ are self and mutual coherence functions of the incident fields. Substitution of (6.3-3) in (6.3-2) and recognition that

$$\alpha_1\alpha_2 \iint_{-\infty}^{\infty} h(t-\xi)h(t-\eta)\Gamma_{11}(0)\Gamma_{22}(0)\,d\xi\,d\eta = \bar{i}_1\bar{i}_2 \qquad (6.3\text{-}4)$$

yield the result

$$\bar{z} = \alpha_1\alpha_2 \iint_{-\infty}^{\infty} h(t-\xi)h(t-\eta)|\Gamma_{12}(\xi-\eta)|^2\,d\xi\,d\eta. \qquad (6.3\text{-}5)$$

A further simplification follows if we assume that the incident light is cross-spectrally pure, in which case

$$\bar{z} = \alpha_1\alpha_2\bar{i}_1\bar{i}_2|\mu_{12}|^2 \iint_{-\infty}^{\infty} h(t-\xi)h(t-\eta)|\gamma(\xi-\eta)|^2\,d\xi\,d\eta. \qquad (6.3\text{-}6)$$

At this point we have already demonstrated that the average or DC output of the multiplier is proportional to the squared modulus of the complex coherence factor. The task remains to fully evaluate the proportionality constant.

To evaluate the double integral in Eq. (6.3-6), we first make a change of variables $\zeta = \xi - \eta$, yielding (cf. Problem 3-14)

$$\iint_{-\infty}^{\infty} h(t-\xi)h(t-\eta)|\gamma(\xi-\eta)|^2\,d\xi\,d\eta = \int_{-\infty}^{\infty} H(\zeta)|\gamma(\zeta)|^2\,d\zeta, \qquad (6.3\text{-}7)$$

where

$$H(\zeta) = \int_{-\infty}^{\infty} h(\xi+\zeta)h(\xi)\,d\xi. \qquad (6.3\text{-}8)$$

Further progress is aided by some rather specific assumptions concerning the power spectral density of the light and the transfer function $\mathcal{H}(\nu)$ of the filters. For simplicity we assume that the normalized power spectral density of the light is rectangular with center frequency $\bar{\nu}$ and bandwidth $\Delta\nu$:

$$\hat{\mathcal{G}}(\nu) = \frac{1}{\Delta\nu}\text{rect}\frac{\nu-\bar{\nu}}{\Delta\nu}. \qquad (6.3\text{-}9)$$

The electrical filters we assume have transfer functions that are rectangular, centered on zero frequency, and extending to cutoff frequencies $\pm B$ Hz:

$$\mathscr{F}\{h(t)\} = \mathscr{H}(\nu) = \operatorname{rect}\frac{\nu}{2B}. \tag{6.3-10}$$

For true thermal light, the optical bandwidth $\Delta\nu$ is typically of the order of 10^{13} Hz or more, whereas the electrical bandwidth of the detectors and circuitry would seldom exceed about 10^8 Hz. Hence we are well justified in assuming that $\Delta\nu \gg B$.

To evaluate the integral of Eq. (6.3-7), we use Parseval's theorem of Fourier analysis, evaluating instead the area under the product of the Fourier transforms of $H(\tau)$ and $|\gamma(\tau)|^2$. Use of the autocorrelation theorem of Fourier analysis shows that

$$\mathscr{F}\{H(\tau)\} = |\mathscr{H}(\nu)|^2 = \operatorname{rect}\frac{\nu}{2B}$$

$$\mathscr{F}\{|\gamma(\tau)|^2\} = \int_{-\infty}^{\infty} \hat{\mathscr{G}}(\beta + \nu)\hat{\mathscr{G}}(\beta)\,d\beta$$

$$= \frac{1}{\Delta\nu}\Lambda\left(\frac{\nu}{\Delta\nu}\right) \tag{6.3-11}$$

where $\Lambda(x) = 1 - |x|$ for $|x| \leq 1$, zero otherwise. Thus the integral of interest can be rewritten

$$\int_{-\infty}^{\infty} H(\zeta)|\gamma(\zeta)|^2\,d\zeta = \frac{1}{\Delta\nu}\int_{-\infty}^{\infty} \operatorname{rect}\left(\frac{\nu}{2B}\right)\Lambda\left(\frac{\nu}{\Delta\nu}\right)d\nu$$

$$\cong \frac{2B}{\Delta\nu}, \tag{6.3-12}$$

where the approximation is valid for $B \ll \Delta\nu$.

The results of our analysis can now be summarized as follows. Subject to the assumptions we have made, the average or DC value of the multiplier output is given by

$$\bar{z} = \alpha_1\alpha_2\bar{i}_1\bar{i}_2|\mu_{12}|^2\frac{2B}{\Delta\nu}. \tag{6.3-13}$$

The final averaging filter passes this component of output unchanged. We also have available at the output separate measurements of $\bar{i}_1$ and $\bar{i}_2$, which (under the assumptions we have made) can be expressed as [cf., Eq. (6.3-4)]

$$\bar{i}_1 = \alpha_1 \bar{I}_1 \int_{-\infty}^{\infty} h(t-\xi)\,d\xi = \alpha_1 \bar{I}_1$$

$$\bar{i}_2 = \alpha_2 \bar{I}_2. \tag{6.3-14}$$

Normalization of the correlator output by these two quantities yields the result

$$\hat{\bar{z}} = \frac{\bar{z}}{\bar{i}_1 \bar{i}_2} = |\mu_{12}|^2 \frac{2B}{\Delta \nu}. \tag{6.3-15}$$

Knowing the values B and $\Delta \nu$, we can deduce the value of $|\mu_{12}|$. Note in particular that the phase of μ_{12} is simply not available at the interferometer output. We shall consider the implications of the loss of phase information in Chapter 7, when methods for image formation from interferometric data are discussed.

6.3.3 "Classical" or "Self" Noise at the Interferometer Output

There exist various sources of noise that limit the performance of the intensity interferometer. For true thermal sources in the optical region of the spectrum, the dominant source of noise is nearly always *shot noise* associated with the photodetector outputs. This type of noise is treated in detail in Chapter 9. A second type of noise, which can be dominant at radio frequencies and generally cannot be neglected for pseudothermal optical sources, is "classical" or "self" noise that arises due to the finite bandwidth of the averaging filters. The origin of this noise lies in the random fluctuations of optical waves themselves.

A complete analysis of the effects of finite averaging time should include the uncertainties associated with the estimates of all three average quantities, $\langle i_1 \rangle$, $\langle i_2 \rangle$, and $\langle \Delta i_1 \Delta i_2 \rangle$. For the purpose of simplicity, we neglect the uncertainties associated with $\langle i_1 \rangle$ and $\langle i_2 \rangle$. An assumption that these latter two quantities are known much more accurately than $\langle \Delta i_1 \Delta i_2 \rangle$ is often justified, for usually several or many different pinhole spacings must be explored, and $\langle i_1 \rangle$ and $\langle i_2 \rangle$ are thus observed over many more averaging intervals than is $\langle \Delta i_1 \Delta i_2 \rangle$ for any one spacing.

The general approach to calculating the output signal-to-noise ratio of the interferometer will be as follows. First we calculate the autocorrelation function $\Gamma_Z(\tau)$ of the signal at the output of the multiplier. We then Fourier transform this quantity to yield the power spectral density of z, following which we pass this spectrum through the averaging filter to find the output signal and self-noise powers. The ratio of these quantities yields the desired signal-to-noise ratio.

The autocorrelation function of the multiplier output is given by

$$\Gamma_Z(\tau) = \overline{\Delta i_1(t)\,\Delta i_2(t)\,\Delta i_1(t+\tau)\,\Delta i_2(t+\tau)}. \qquad (6.3\text{-}16)$$

Since $\Delta i_k(t) = i_k(t) - \overline{i_k(t)}$ ($k = 1, 2$), an evaluation of $\Gamma_Z(\tau)$ requires determination of the fourth-, third-, and second-order joint moments of the $i_k(t)$. Furthermore, since

$$i_k(t) = \alpha_k \int_{-\infty}^{\infty} h(t-\xi)|\mathbf{u}_k(\xi)|^2 \, d\xi, \qquad (6.3\text{-}17)$$

evaluation of $\Gamma_Z(\tau)$ will ultimately require use of the eighth-, sixth-, fourth-, and second-order joint moments of the fields. Such moments can be found for thermal sources by using the complex Gaussian moment theorem; however, the algebra involved in such a calculation is extremely tedious, and thus we adopt a somewhat simpler approximate approach here.

The key to simplifying the calculation lies in the assumption that the optical bandwidth $\Delta \nu$ of the incident waves far exceeds the bandwidth B of the electrical currents reaching the multiplier. Such an assumption was already made in the previous section for a different reason; it is well satisfied for true thermal sources but must be examined carefully in the case of pseudothermal sources. If indeed $\Delta \nu \gg B$, then from Eq. (6.3-17) we see that the electrical current $i_k(t)$ at any particular instant of time is the integral over many correlation intervals of the incident fields. Since the fields incident have been assumed to be complex circular Gaussian random processes (thermal light), lack of correlation implies statistical independence; each current is, in effect, the sum of a large number of statistically independent contributions, and hence by the central limit theorem, the currents $i_k(t)$ are, to a good approximation (real-valued), Gaussian random processes.

Once the $i_k(t)$ have been recognized as being approximately Gaussian, the moment theorem for real-valued Gaussian random variables [Eq. (2.7-13)] can be used to simplify the expression for $\Gamma_Z(\tau)$. With use of the definition

$$C_{jk}(\tau) = \overline{\Delta i_j(t)\,\Delta i_k(t+\tau)} \qquad (6.3\text{-}18)$$

for the covariance function of the jth and kth currents ($j = 1, 2;\ k = 1, 2$), the autocorrelation function of interest becomes

$$\Gamma_Z(\tau) = C_{12}^2(0) + C_{11}(\tau)C_{22}(\tau) + C_{12}^2(\tau), \qquad (6.3\text{-}19)$$

CLASSICAL ANALYSIS OF THE INTENSITY INTERFEROMETER 279

where we have used the fact that $C_{21}(\tau) = C_{12}(-\tau) = C_{12}(\tau)$ since the currents are real valued.

Calculation of the $C_{jk}(\tau)$ follows along lines parallel to those used earlier in the calculation of $\bar{z}$. Since $\bar{z} = C_{12}(0)$, we need only repeat that calculation with a τ inserted in the appropriate places. We obtain

$$C_{jk}(\tau) = \alpha_j \alpha_k \bar{I}_j \bar{I}_k |\mu_{jk}|^2 \int_{-\infty}^{\infty} H(\zeta + \tau) |\gamma(\zeta)|^2 \, d\zeta, \quad (6.3\text{-}20)$$

where H is again given by Eq. (6.3-8). Substituting $C_{jk}(\tau)$ into the expression for $\Gamma_Z(\tau)$, we find

$$\Gamma_Z(\tau) = \left[\alpha_1 \alpha_2 \bar{I}_1 \bar{I}_2 |\mu_{12}|^2 \int_{-\infty}^{\infty} H(\zeta) |\gamma(\zeta)|^2 \, d\zeta \right]^2$$

$$+ \left[\alpha_1 \alpha_2 \bar{I}_1 \bar{I}_2 \int_{-\infty}^{\infty} H(\zeta + \tau) |\gamma(\zeta)|^2 \, d\zeta \right]^2$$

$$+ \left[\alpha_1 \alpha_2 \bar{I}_1 \bar{I}_2 |\mu_{12}|^2 \int_{-\infty}^{\infty} H(\zeta + \tau) |\gamma(\zeta)|^2 \, d\zeta \right]^2. \quad (6.3\text{-}21)$$

The first term in this expression for $\Gamma_Z(\tau)$ represents the square of the mean of z; since it is independent of τ, it contributes a δ-function component to the power spectral density $\mathcal{G}_Z(\nu)$. The area under this δ function is the power associated with the "signal" or ideal component of the output. Using the same assumptions and approximations employed in the previous section, we express the output signal power as [cf. Eq. (6.3-13)]

$$P_S = \left[\alpha_1 \alpha_2 \bar{I}_1 \bar{I}_2 |\mu_{12}|^2 \frac{2B}{\Delta \nu} \right]^2. \quad (6.3\text{-}22)$$

To find the noise power at the output of the averaging filter, we must Fourier transform the last two terms of Eq. (6.3-21) and multiply the resulting spectral distribution by the squared modulus of the transfer function of the averaging filter. First we note that, for the spectrum of Eq. (6.3-9) and the transfer function of Eq. (6.3-10), Parseval's theorem allows us to write [cf. Eq. (6.3-12)]

$$\int_{-\infty}^{\infty} H(\zeta + \tau) |\gamma(\zeta)|^2 \, d\zeta = \frac{1}{\Delta \nu} \int_{-\infty}^{\infty} \text{rect} \frac{\nu}{2B} \Lambda\left(\frac{\nu}{\Delta \nu}\right) e^{-j2\pi\nu\tau} \, d\nu.$$

$$(6.3\text{-}23)$$

Clearly a Fourier transform of this quantity (with respect to τ) yields

$$\mathcal{F}\left\{\int_{-\infty}^{\infty} H(\zeta + \tau)|\gamma(\zeta)|^2 \, d\zeta\right\} = \left[\frac{1}{\Delta\nu}\text{rect}\frac{\nu}{2B}\right]\Lambda\left(\frac{\nu}{\Delta\nu}\right)$$

$$\cong \frac{1}{\Delta\nu}\text{rect}\frac{\nu}{2B} \qquad (B \ll \Delta\nu).$$

(6.3-24)

According to the autocorrelation theorem, the Fourier transform of the square of the integral must be given by

$$\mathcal{F}\left\{\left[\int_{-\infty}^{\infty} H(\zeta + \tau)|\gamma(\zeta)|^2 \, d\zeta\right]^2\right\} = \left(\frac{1}{\Delta\nu}\right)^2 \int_{-\infty}^{\infty} \text{rect}\left(\frac{\beta + \nu}{2B}\right)\text{rect}\frac{\beta}{2B} \, d\beta$$

$$= \frac{2B}{(\Delta\nu)^2}\Lambda\left(\frac{\nu}{2B}\right). \qquad (6.3-25)$$

Hence, from Eq. (6.3-21), the power spectral density of the noise component of the multiplier output is given by

$$[\mathcal{G}_Z(\nu)]_N = \alpha_1^2 \alpha_2^2 \bar{I}_1^2 \bar{I}_2^2 (1 + |\mu_{12}|^4) \frac{2B}{(\Delta\nu)^2}\Lambda\left(\frac{\nu}{2B}\right). \qquad (6.3-26)$$

This noise spectrum now passes through the averaging filter with transfer function assumed to be of the form

$$\mathcal{A}(\nu) = \text{rect}\frac{\nu}{2b} \qquad (b \le B). \qquad (6.3-27)$$

The noise power transmitted by the filter is found by evaluating

$$P_N = \int_{-\infty}^{\infty} |\mathcal{A}(\nu)|^2 [\mathcal{G}_Z(\nu)]_N \, d\nu, \qquad (6.3-28)$$

which, with the help of the integral identity

$$\int_{-\infty}^{\infty} \text{rect}\frac{\nu}{2b}\Lambda\left(\frac{\nu}{2B}\right) d\nu = 2b\left(1 - \frac{b}{4B}\right) \qquad (6.3-29)$$

(valid for $b \le B$), yields

$$P_N = \alpha_1^2 \alpha_2^2 \bar{I}_1^2 \bar{I}_2^2 (1 + |\mu_{12}|^4) \frac{4bB}{\Delta\nu^2}\left(1 - \frac{b}{4B}\right). \qquad (6.3-30)$$

The ratio of signal and noise powers is thus

$$\frac{P_S}{P_N} = \frac{|\mu_{12}|^4}{1 + |\mu_{12}|^4} \frac{B/b}{\left(1 - \frac{b}{4B}\right)} \tag{6.3-31}$$

under the assumptions used, namely, that $b \leq B \ll \Delta\nu$. The rms signal-to-noise ratio at the output is the square root of this quantity,

$$\left(\frac{S}{N}\right)_{\text{rms}} = \frac{|\mu_{12}|^2}{\sqrt{1 + |\mu_{12}|^4}} \sqrt{\frac{B/b}{\left(1 - \frac{b}{4B}\right)}}. \tag{6.3-32}$$

Clearly, the wider the premultiplication bandwidth B and the narrower the postmultiplication bandwidth b, the better the final signal-to-noise ratio. We see further that, under the assumption $B \ll \Delta\nu$, the optical bandwidth has no effect on the signal-to-noise ratio. When the modulus of the complex coherence factor is less than about 0.4, the first ratio in Eq. (6.3-32) is equal to $|\mu_{12}|^2$ within 1% accuracy. As $|\mu_{12}|$ approaches unity, this factor approaches $1/\sqrt{2}$.

Finally, the reader is reminded that the preceding expression for signal-to-noise ratio includes only the classical or self-noise. The signal-to-noise ratio at optical frequencies is usually dominated by photon-induced fluctuations, which are discussed in Chapter 9. Further discussion of intensity interferometry is deferred until that chapter.

REFERENCES

6-1 B. Picinbono and E. Boileau, *J. Opt. Soc. Am.*, **58**, 784 (1968).
6-2 L. Mandel, *Proc. Phys. Soc.*, **72**, 1037 (1958).
6-3 S. Lowenthal and Y. Joyeux, *J. Opt. Soc. Am.*, **61**, 847 (1971).
6-4 J. W. Goodman, *Proc. IEEE*, **53**, 1688 (1965).
6-5 J. Bures, C. Delisle, and A. Zardecki, *Can. J. Phys.*, **50**, 760 (1972).
6-6 L. Mandel, *Proc. Phys. Soc.*, **74**, 233 (1959).
6-7 S. O. Rice, *Bell Syst. Tech. J.*, **24**, 46 (1945).
6-8 M. A. Condie, "An Experimental Investigation of the Statistics of Diffusely Reflected Coherent Light," thesis for the degree Engineer, Department of Electrical Engineering, Stanford University, Stanford, California (1965).
6-9 C. L. Mehta, *J. Opt. Soc. Am.*, **58**, 1233 (1968).
6-10 C. L. Mehta, "Theory of Photoelectric Counting," in *Progress in Optics*, Vol. VIII (E. Wolf, editor), North Holland Publishing Company, Amsterdam, pp. 373–440 (1970).

6-11 D. Slepian, *Bell Syst. Tech. J.*, **37**, 163 (1958).
6-12 D. Slepian and H. O. Pollak, *Bell Syst. Tech. J.*, **40**, 43 (1961).
6-13 D. Slepian, *Bell Syst. Tech. J.*, **43**, 3009 (1964).
6-14 D. Slepian, *J. Opt. Soc. Am.*, **55**, 1110 (1965).
6-15 D. Slepian, *J. Math. Phys.*, **44**, 99 (1965).
6-16 B. Roy Frieden, "Evaluation, Design and Extrapolation Methods for Optical Signals, Based on Use of the Prolate Functions," in *Progress in Optics*, Vol. IX (E. Wolf, editor), North Holland Publishing Company, Amsterdam, pp. 311–407 (1971).
6-17 J. W. Goodman, *Appl. Phys.*, **2**, 95 (1973).
6-18 P. Beckmann and A. Spizzichino, *The Scattering of Electromagnetic Waves from Rough Surfaces*, Pergamon Press, Oxford, p. 125 (1963).
6-19 R. Hanbury Brown and R. Q. Twiss, *Phil. Mag.*, **45**, 663 (1954).
6-20 R. Hanbury Brown and R. Q. Twiss, *Proc. Roy. Soc. A*, **242**, 300 (1957).
6-21 R. Hanbury Brown and R. Q. Twiss, *Proc. Roy. Soc. A*, **243**, 291 (1957).
6-22 R. Hanbury Brown and R. Q. Twiss, *Proc. Roy. Soc. A*, **248**, 199 (1958).
6-23 R. Hanbury Brown and R. Q. Twiss, *Proc. Roy. Soc. A*, **248**, 222 (1958).
6-24 R. Hanbury Brown, *The Intensity Interferometer*, Taylor and Francis, London (1974).

PROBLEMS

6-1 Show that for quasimonochromatic, stationary thermal light, the fourth-order coherence function

$$\Gamma_{1234}(t_1, t_2, t_3, t_4) = E\left[\mathbf{u}(P_1, t_1)\mathbf{u}(P_2, t_2)\mathbf{u}^*(P_3, t_3)\mathbf{u}^*(P_4, t_4)\right]$$

can be expressed as

$$\Gamma_{1234}(t_1, t_2, t_3, t_4) = \sqrt{I_1 I_2 I_3 I_4}\left[\mu_{13}\mu_{24} + \mu_{14}\mu_{23}\right]e^{-j2\pi\nu_0(t_1+t_2-t_3-t_4)},$$

where

$$\mu_{mn} = \frac{E\left[\mathbf{u}(P_m, t)\mathbf{u}^*(P_n, t)\right]}{\sqrt{E\left[|\mathbf{u}(P_m, t)|^2\right] E\left[|\mathbf{u}(P_n, t)|^2\right]}}$$

6-2 The output of a single-mode, well-stabilized laser is passed through a spatially distributed phase modulator (e.g., a transparent acoustic cell). The field observed at point P_k at the output of the modulator is of the form

$$\mathbf{u}(P_k, t) = \sqrt{I_k}\exp\{-j[2\pi\nu_0 t - \phi(P_k, t)]\},$$

PROBLEMS 283

where ν_0 is the laser frequency, I_k is the intensity at P_k, and $\phi(P_k, t)$ is the phase modulation imparted to the wave at point P_k. The phase modulation is chosen to be a stationary, zero-mean Gaussian random process. Noting that $\Delta\phi = \phi(P_1, t) - \phi(P_2, t)$ is also a zero mean, stationary Gaussian process, show that the second order coherence function of the modulated wave is

$$\Gamma_{12}(t_1 - t_2) = \sqrt{I_1 I_2}\, e^{-j2\pi\nu_0(t_1 - t_2)} e^{-\sigma_\phi^2[1 - \gamma_\phi(P_1, P_2; t_1 - t_2)]},$$

where σ_ϕ^2 is the variance of $\phi(P, t)$ (assumed independent of P) and γ_ϕ is the normalized cross-correlation function of $\phi(P_1, t)$ and $\phi(P_2, t)$.

6-3 For the same light described in Problem 6-2, show that the fourth-order coherence function obeys the factorization theorem

$$\Gamma_{1234}(t_1, t_2, t_3, t_4) = \sqrt{I_1 I_2 I_3 I_4}\, \exp\{-j2\pi\nu_0(t_1 + t_2 - t_3 - t_4)\}$$

$$\times \left| \frac{\gamma_{23}(t_2 - t_3)\gamma_{13}(t_1 - t_3)\gamma_{24}(t_2 - t_4)\gamma_{14}(t_1 - t_4)}{\gamma_{12}(t_1 - t_2)\gamma_{34}(t_3 - t_4)} \right|.$$

6-4 The correlation time τ_c of a certain pseudothermal source is 10^{-4} seconds. A Young's interference experiment is performed under quasimonochromatic conditions with $\mu_{12} \cong 0.01$. The amplitude and phase of the resulting fringe pattern are measured with a finite integration time T.

(a) How long must the integration time T be to assure that the fractional bias Δ is less than 0.01?

(b) How long must T be to assure that the fringe amplitude is measured with an rms signal-to-noise ratio of 100?

(c) How long must T be to assure that the fringe phase θ is measured with a standard deviation σ_θ less than $2\pi/100$?

6-5 Show that for a Gaussian spectral profile, the parameter $\mathcal{M}$ of Eq. (6.1-17) is given exactly by

$$\mathcal{M} = \left\{ \frac{\tau_c}{T} \operatorname{erf}\left(\sqrt{\pi}\,\frac{T}{\tau_c}\right) - \frac{1}{\pi}\left(\frac{\tau_c}{T}\right)^2 \left[1 - e^{-\pi(T/\tau_c)^2}\right] \right\}^{-1}$$

6-6 Show that for a Lorentzian spectral profile, the parameter $\mathcal{M}$ is given by

$$\mathcal{M} = \left[\frac{\tau_c}{T} + \frac{1}{2}\left(\frac{\tau_c}{T}\right)^2 (e^{-2T/\tau_c} - 1)\right]^{-1}$$

6-7 Examination of Fig. 6-5 shows that a relatively abrupt threshold in the values of $\tilde{\lambda}_n$ occurs as n is varied. In particular, as a rough approximation,

$$\tilde{\lambda}_n \cong \begin{cases} 1 & n < n_{\text{crit}} \\ 0 & n > n_{\text{crit}}, \end{cases}$$

where $n_{\text{crit}} = 2c/\pi$. Show that this approximate distribution of $\tilde{\lambda}_n$ leads to a gamma probability density function for W. Compare the number n_{crit} with the parameter $\mathcal{M}$.

6-8 For the intensity interferometer presented in Section 6.3, suppose that the postdetection filters have impulse responses

$$h(t) = \frac{1}{T_1} \text{rect}\frac{t}{T_1},$$

whereas the averaging filter has impulse response

$$a(t) = \frac{1}{T_2} \text{rect}\frac{t}{T_2}.$$

Using assumptions similar to those introduced in sections 6.3-2 and 6.3-3, calculate

(a) $\bar{z}$.
(b) P_N.
(c) $(S/N)_{\text{rms}}$.

6-9 Find what modification of Eq. (6.3-32) occurs if the light incident on the detectors is partially polarized.

6-10 Consider the statistical properties of the coefficients b_n defined by Eq. (6.1-42),

$$b_n = \int_{-T/2}^{T/2} \mathbf{A}(t)\phi_n^*(t)\,dt,$$

where $\mathbf{A}(t)$ is a stationary, circularly complex Gaussian random process, whereas $\phi_n(t)$ is an arbitrary complex-valued weighting function.

(a) On what grounds can we argue that the real and imaginary parts of $\mathbf{b}_n$ are Gaussian random variables?

(b) Using the circularity of $\mathbf{A}(t)$, prove that

$$E[\text{Re}(\mathbf{b}_n)] = E[\text{Im}(\mathbf{b}_n)] = 0.$$

(c) Using the circularity of $\mathbf{A}(t)$, prove that

$$E\{[\text{Re}(\mathbf{b}_n)]^2\} = E\{[\text{Im}(\mathbf{b}_n)]^2\}.$$

(d) Using the circularity of $\mathbf{A}(t)$, prove that

$$E[\text{Re}(\mathbf{b}_n) \cdot \text{Im}(\mathbf{b}_n)] = 0.$$

Hint: See Eq. (2.8-20).

7

Effects of Partial Coherence on Imaging Systems

A common function of an imaging system is to provide an observer with visual information that is more detailed and/or more accurate than could be obtained with the unaided eye. In other cases, the function of the system may be simply to provide a semipermanent record (e.g., a photograph) of an object of interest. In both cases the fidelity with which the image renders information about the object is an issue of great importance.

To fully understand the quantitative relationship between an object and its image, it is not sufficient to know only the light transmitting, reflecting, or emitting properties of the object and the laws that govern the passage of light waves through the optical instrument. Rather, it is essential to know in addition the coherence properties of the light illuminating or being radiated by the object, for these properties have a profound influence on the character of the image that is ultimately observed.

The primary goal of this chapter is to develop in a logical way the relationship that exists between an object and its image, taking full account of the coherence properties of the light that illuminates or leaves the object. The second goal is to arrive at an understanding as to when an imaging system can be expected to behave as an *incoherent* system (linear in intensity), when it behaves as a *coherent* system (linear in complex amplitude), and when some intermediate form of behavior can be expected. Third, we wish to develop an understanding of certain interferometric types of imaging systems, which effectively measure the coherence of the radiation impinging on them, and from this information derive an image. (Such imaging systems are used routinely in radio astronomy.) Finally, we introduce the reader to the concept of "speckle" in coherent imaging systems and explore the means by which *ensemble-average* coherence can be useful in describing its properties.

An excellent review of the subject of image formation with partially coherent light has been published by Thompson (Ref. 7-1). The subject is also treated in various books that deal with coherence theory (Refs. 7-2

SOME PRELIMINARY CONSIDERATIONS

through 7-4). The reader may wish to consult the pioneering papers of H. H. Hopkins (Refs. 7-5 and 7-6) on this subject.

7.1 SOME PRELIMINARY CONSIDERATIONS

Before embarking on detailed analyses relating the object and image, we first present several basic coherence relationships that will be useful in future discussions. These relations concern the effects on partial coherence when light propagates through transmitting objects and through certain simple optical systems.

7.1.1 Effects of a Thin Transmitting Object on Mutual Coherence

In many optical imaging systems, the objects are transilluminated and the images are formed from the transmitted light. Here we investigate the effects of "thin" transmitting objects on mutual coherence.

With reference to Fig. 7-1a, we define a transmitting object to be "thin" if a ray of light entering the object at point (x, y) exits from the object at essentially the same transverse coordinates. Clearly, no true object can be perfectly thin in the sense used here, for a ray entering at some angle to the z axis will invariably exit at slightly different transverse coordinates. Furthermore, if the thickness of the object is not perfectly uniform, or if the

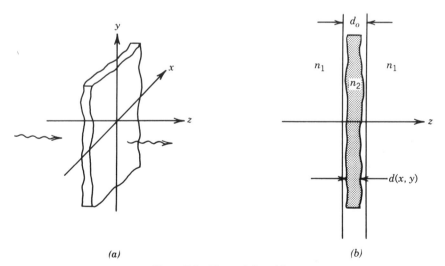

(a)   (b)

Figure 7-1. Transmitting object.

refractive index varies from point to point, refraction within the object will modify the position at which a given ray exits. Nonetheless, many objects are approximately thin in the sense used here, and the concept serves as a useful idealization.

To analyze the effects of such an object on mutual coherence, we first develop a relationship between the incident and transmitted fields. As shown in Fig. 7-1b, the object is assumed to be imbedded in a uniform nonabsorbing medium with (real-valued) refractive index n_1. The object itself is taken to have variable thickness $d(x, y)$, a variable real component of refractive index $n_2(x, y)$ (which accounts for a variable velocity of propagation in the object, through $v = c/n_2$) and a variable component of absorption that is accounted for by a real-valued multiplicative component $B(x, y)$ that reduces the amplitude of the transmitted field. For simplicity, both n_2 and B are assumed to be independent of wavelength.

We construct two planes normal to the z axis, separated by constant distance d_o, and between which lies the object of interest (Fig. 7-1b). Our goal is to specify the relationship between the field $\mathbf{u}_i(x, y; t)$ incident at the left-hand plane and the transmitted field $\mathbf{u}_t(x, y; t)$ at the right-hand plane. The delay suffered by the wave at coordinates (x, y) is given by

$$\delta = \frac{d(x, y)}{c/n_2(x, y)} + \frac{d_o - d(x, y)}{c/n_1} \tag{7.1-1}$$

or

$$\delta = \frac{[n_2(x, y) - n_1]d(x, y)}{c}, \tag{7.1-2}$$

where a constant term $d_o n_1/c$ that is independent of the object has been dropped in the second equation. Taking account of the reduction of light amplitude by the factor $B(x, y)$, we see that the incident and transmitted fields are related by

$$\mathbf{u}_t(x, y; t) = B(x, y)\mathbf{u}_i(x, y; t - \delta(x, y)), \tag{7.1-3}$$

where $\delta(x, y)$ is given by Eq. (7.1-2).

To find the effect of the transmitting object on the mutual coherence function of the light, we substitute relationship (7.1-3) into the definition of the mutual coherence function,

$$\Gamma_t(P_1, P_2; \tau) = \langle \mathbf{u}_t(P_1, t + \tau)\mathbf{u}_t^*(P_2, t) \rangle$$

$$= B(P_1)B(P_2)\langle \mathbf{u}_i(P_1, t + \tau - \delta(P_1))\mathbf{u}_i^*(P_2, t - \delta(P_2)) \rangle,$$

$$\tag{7.1-4}$$

where $P_1 \sim (x_1, y_1)$, $P_2 \sim (x_2, y_2)$. Expressing the time average in terms of the mutual coherence function of the incident light, we find the following fundamental relationship between the incident and transmitted mutual coherence functions:

$$\Gamma_t(P_1, P_2; \tau) = B(P_1)B(P_2)\Gamma_i(P_1, P_2; \tau - \delta(P_1) + \delta(P_2)).$$

(7.1-5)

When the light is narrowband, it is convenient to express the analytic signal representation of the fields in terms of a time-varying phasor,

$$\mathbf{u}_i(P, t) = \mathbf{A}_i(P, t)e^{-j2\pi\bar{\nu}t},$$

(7.1-6)

where $\bar{\nu}$ is the nominal center frequency of the disturbance. Using this representation for the fields, we express the mutual coherence function Γ_i of the incident fields in the form

$$\Gamma_i(P_1, P_2; \tau) = \langle \mathbf{A}_i(P_1, t + \tau)\mathbf{A}_i^*(P_2, t)\rangle e^{-j2\pi\bar{\nu}\tau}.$$

(7.1-7)

Using this form in (7.1-5), we obtain

$$\Gamma_t(P_1, P_2; \tau) = B(P_1)e^{j2\pi\bar{\nu}\delta(P_1)}B(P_2)e^{-j2\pi\bar{\nu}\delta(P_2)}$$

$$\times \langle \mathbf{A}_i(P_1, t + \tau - \delta(P_1) + \delta(P_2))\mathbf{A}_i^*(P_2, t)\rangle e^{-j2\pi\bar{\nu}\tau}.$$

(7.1-8)

Now if

$$|\delta(P_1) - \delta(P_2)| \ll \frac{1}{\Delta\nu} \approx \tau_c$$

(7.1-9)

for all P_1, P_2, the time average will be independent of $\delta(P_1)$ and $\delta(P_2)$. Under such a condition we find

$$\Gamma_t(P_1, P_2; \tau) = \mathbf{t}(P_1)\mathbf{t}^*(P_2)\Gamma_i(P_1, P_2; \tau),$$

(7.1-10)

where $\mathbf{t}(P)$ is the *amplitude transmittance* of the object at P, as defined by

$$\mathbf{t}(P) \triangleq B(P)e^{j2\pi\bar{\nu}\delta(P)}.$$

(7.1-11)

The relationship (7.1-10) between the incident and transmitted mutual coherence functions is widely used in the literature, but as the preceding argument shows, it is strictly valid only if all the delay differences induced by the object are much less than the coherence time of the light [cf. the quasimonochromatic conditions given in Eq. (5.2-30)].

One further simplification will be useful to us in the future. If the delay τ of importance in a given physical experiment is always much smaller than the coherence time τ_c, Eq. (7.1-10) then implies that the incident and transmitted mutual intensities are related by

$$\mathbf{J}_t(P_1, P_2) = \mathbf{t}(P_1)\mathbf{t}^*(P_2)\mathbf{J}_i(P_1, P_2). \qquad (7.1\text{-}12)$$

This relationship is used whenever the quasimonochromatic assumptions are valid.

7.1.2 Time Delays Introduced by a Thin Lens

With reference to Fig. 7-2, consider a lens composed of spherical surfaces and which may be regarded as being "thin" in the sense defined in the previous section. We wish to calculate the time delay $\delta(x, y)$ suffered by the light that strikes the lens at transverse coordinates (x, y). To obtain results that are applicable to a variety of different types of lenses, we adopt the sign convention that, as light travels from left to right, each convex surface encountered is taken to have a positive radius of curvature, whereas each concave surface is taken to have a negative radius of curvature. For the lens shown in Fig. 7-2, radius R_1 is positive, whereas radius R_2 is negative.

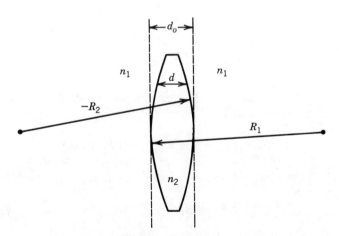

Figure 7-2. Thin lens.

SOME PRELIMINARY CONSIDERATIONS

The lens is assumed to be composed of a material that has refractive index n_2 and to be embedded in a medium with refractive index n_1; both refractive indices are taken to be independent of frequency over the frequency band of interest. The total time delay suffered by a ray passing between the two parallel planes represented by dotted lines in Fig. 7-2 is given (within the "thin" lens approximation) by

$$\delta(x, y) = \frac{n_2 d(x, y)}{c} + \frac{n_1 [d_o - d(x, y)]}{c}. \qquad (7.1\text{-}13)$$

Some simple trigonometric calculations (Ref. 7-7, pp. 78, 79) show that the thickness of the lens at coordinates (x, y) is given by

$$d(x, y) = d_o - R_1 \left(1 - \sqrt{1 - \frac{x^2 + y^2}{R_1^2}} \right)$$

$$+ R_2 \left(1 - \sqrt{1 - \frac{x^2 + y^2}{R_2^2}} \right). \qquad (7.1\text{-}14)$$

At this point it is convenient to introduce the *paraxial approximation*, representing the square roots of Eq. (7.1-14) by the first two terms of their binomial expansions. With these approximations the thickness function $d(x, y)$ becomes

$$d(x, y) = d_o - \frac{x^2 + y^2}{2} \left(\frac{1}{R_1} - \frac{1}{R_2} \right). \qquad (7.1\text{-}15)$$

Incorporating this result in Eq. (7.1-13), we obtain for the delay

$$\delta(x, y) = \frac{n_2 d_o}{c} - \frac{(n_2 - n_1)}{c} \left(\frac{1}{R_1} - \frac{1}{R_2} \right) \left(\frac{x^2 + y^2}{2} \right). \qquad (7.1\text{-}16)$$

Defining the *focal length f* of the lens by

$$\frac{1}{f} \triangleq (n_2 - n_1) \left(\frac{1}{R_1} - \frac{1}{R_2} \right), \qquad (7.1\text{-}17)$$

we find our final expression for the delay suffered by the ray at (x, y):

$$\delta(x, y) \cong \frac{n_2 d_o}{c} - \frac{x^2 + y^2}{2cf}. \qquad (7.1\text{-}18)$$

Often it is convenient to have an expression for the amplitude transmittance function that describes a thin lens. Within the approximations used above, a suitable expression is

$$\mathbf{t}_l(x, y) = \exp\{j2\pi\bar{\nu}\delta(x, y)\}$$

$$= \exp\left\{-j\frac{\pi}{\lambda f}(x^2 + y^2)\right\}, \tag{7.1-19}$$

where a factor independent of (x, y) has been dropped from the final form. According to the discussion of the previous section, this representation is strictly valid only if the bandwidth $\Delta\nu$ of the light satisfies

$$|\delta(x_2, y_2) - \delta(x_1, y_1)| \ll \frac{1}{\Delta\nu} \tag{7.1-20}$$

for all (x_1, y_1) and (x_2, y_2) of interest. For problems involving propagation of light up to the lens, passage through the lens, and further propagation beyond the lens, however, it is the difference of *total* propagation time delays that must satisfy a condition analogous to (7.1-20). When such a "total time delay" condition is satisfied, use of the amplitude transmittance expression (7.1-19) will yield a correct relationship between initial and final mutual intensities even though the time delay restriction (7.1-20) may not be valid for the lens alone.

7.1.3 Focal-Plane-to-Focal-Plane Coherence Relationships

In this section we derive the relationship that exists between the mutual intensities in the front and back focal planes of a thin positive lens. As illustrated in Fig. 7-3, these particular planes are defined as being per-

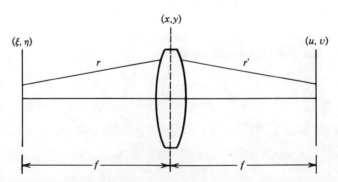

Figure 7-3. Geometry for calculation of focal-plane-to-focal-plane coherence relationships.

SOME PRELIMINARY CONSIDERATIONS

pendicular to a line passing through the centers of curvature of the two lens surfaces (i.e., the *optical axis*) and at distances f in front of and behind the lens.

The light leaving the front focal plane is assumed to be quasimonochromatic and to have a mutual intensity $\mathbf{J}_o'(\xi_1, \eta_1; \xi_2, \eta_2)$. Using Eq. (5.4-8), which describes the effect of propagation on mutual intensity, we can calculate the mutual intensity $\mathbf{J}_l(x_1, y_1; x_2, y_2)$ incident on the lens. To simplify the calculation as much as possible, we make the paraxial or small-angle assumption, in which case the following approximations may be used in Eq. (5.4-8):

$$\chi(\theta_1) \cong \chi(\theta_2) \cong 1$$

$$\frac{1}{\bar{\lambda} r_1} \cong \frac{1}{\bar{\lambda} r_2} \cong \frac{1}{\bar{\lambda} f} \qquad (7.1\text{-}21)$$

$$r_2 - r_1 \cong \frac{(x_2 - \xi_2)^2 + (y_2 - \eta_2)^2 - (x_1 - \xi_1)^2 - (y_1 - \eta_1)^2}{2f}.$$

Thus incident on the lens we find the mutual intensity to be

$$\mathbf{J}_l(x_1, y_1; x_2, y_2) = \frac{1}{(\bar{\lambda} f)^2} \exp\left\{-j\frac{\pi}{\bar{\lambda} f}\left[(x_2^2 + y_2^2) - (x_1^2 + y_1^2)\right]\right\}$$

$$\times \iiint\!\!\!\int_{-\infty}^{\infty} \mathbf{J}_o'(\xi_1, \eta_1; \xi_2, \eta_2) \exp\left\{-j\frac{\pi}{\bar{\lambda} f}\left[(\xi_2^2 + \eta_2^2) - (\xi_1^2 + \eta_1^2)\right]\right\}$$

$$\times \exp\left\{j\frac{2\pi}{\bar{\lambda} f}\left[x_2\xi_2 + y_2\eta_2 - x_1\xi_1 - y_1\eta_1\right]\right\} d\xi_1\, d\eta_1\, d\xi_2\, d\eta_2.$$

$$(7.1\text{-}22)$$

Passage of the light through the lens is accounted for by means of the lens amplitude transmittance function of Eq. (7.1-19). The mutual intensity of the light leaving the lens is thus

$$\mathbf{J}_l'(x_1, y_1; x_2, y_2) = \mathbf{J}_l(x_1, y_1; x_2, y_2) \exp\left\{j\frac{\pi}{\bar{\lambda} f}\left[(x_2^2 + y_2^2) - (x_1^2 + y_1^2)\right]\right\}.$$

$$(7.1\text{-}23)$$

Finally, the mutual intensity leaving the lens must propagate an additional distance f to the rear focal plane. Again using the propagation law in Eq. (5.4-8) under paraxial conditions, we obtain a mutual intensity in the rear focal plane given by

$$\mathbf{J}_f(u_1, v_1; u_2, v_2) = \frac{1}{(\bar{\lambda}f)^2} \exp\left\{-j\frac{\pi}{\bar{\lambda}f}\left[(u_2^2 + v_2^2) - (u_1^2 + v_1^2)\right]\right\}$$

$$\times \iiiint_{-\infty}^{\infty} \mathbf{J}_l(x_1, y_1; x_2, y_2)$$

$$\times \exp\left\{j\frac{2\pi}{\bar{\lambda}f}\left[u_2 x_2 + v_2 y_2 - u_1 x_1 - v_1 y_1\right]\right\} dx_1\, dy_1\, dx_2\, dy_2.$$

(7.1-24)

At this point the relationship between the mutual intensities in the two focal planes involves eight integrals. Fortunately, four of these integrals can be eliminated, leaving a comparatively simple relationship between the two quantities of interest. To achieve this simplification, we perform the integrations with respect to x_1, y_1, x_2, and y_2 first. Collecting all terms that depend on these variables, we find the integration of interest to be

$$\mathscr{I} = \iiiint_{-\infty}^{\infty} \exp\left\{-j\frac{\pi}{\bar{\lambda}f}\left[(x_2^2 + y_2^2) - (x_1^2 + y_1^2)\right]\right\}$$

$$\times \exp\left\{j\frac{2\pi}{\bar{\lambda}f}\left[x_2(\xi_2 + u_2) + y_2(\eta_2 + v_2)\right.\right.$$

$$\left.\left. - x_1(\xi_1 + u_1) - y_1(\eta_1 + v_1)\right]\right\} dx_1\, dy_1\, dx_2\, dy_2. \quad (7.1\text{-}25)$$

With the help of the Fourier transform relationship (see Appendix A, table A-1)

$$\int_{-\infty}^{\infty} \exp\left(-j\frac{\pi}{\bar{\lambda}f}x^2\right) e^{j2\pi\nu x}\, dx = \sqrt{-j\bar{\lambda}f}\; e^{+j\pi\bar{\lambda}f\nu^2}, \quad (7.1\text{-}26)$$

the four integrals can be evaluated, with the result

$$\mathscr{I} = (\bar{\lambda}f)^2 \exp\left\{j\frac{\pi}{\bar{\lambda}f}\left[(\xi_2 + u_2)^2 + (\eta_2 + v_2)^2 - (\xi_1 + u_1)^2 - (\eta_1 + v_1)^2\right]\right\}.$$

(7.1-27)

SOME PRELIMINARY CONSIDERATIONS

Substitution of this expression into the remaining quadruple integral yields

$$\mathbf{J}_f(u_1, v_1; u_2, v_2) = \frac{1}{(\overline{\lambda}f)^2} \iiiint_{-\infty}^{\infty} \mathbf{J}_o'(\xi_1, \eta_1; \xi_2, \eta_2)$$

$$\times \exp\left\{ j\frac{2\pi}{\overline{\lambda}f} [\xi_2 u_2 + \eta_2 v_2 - \xi_1 u_1 - \eta_1 v_1] \right\} d\xi_1 \, d\eta_1 \, d\xi_2 \, d\eta_2.$$

(7.1-28)

The preceding relationship between $\mathbf{J}_f$ and $\mathbf{J}_o'$ represents the final result of our analysis. Although this may appear to be a complicated expression at first glance, it can be stated in words quite simply: *the mutual intensities in the front and back focal planes of a thin positive lens are* (up to proportionality and scaling constants) *a four-dimensional Fourier transform pair*. The spatial frequencies of the Fourier transform operator are given explicitly by

$$\nu_1 = -\frac{u_1}{\overline{\lambda}f} \qquad \nu_3 = \frac{u_2}{\overline{\lambda}f}$$

$$\nu_2 = -\frac{v_1}{\overline{\lambda}f} \qquad \nu_4 = \frac{v_2}{\overline{\lambda}f}. \qquad (7.1\text{-}29)$$

Also of interest is the intensity of the light incident on the rear focal plane and its relationship to the mutual intensity in the front focal plane. Setting $u_1 = u_2 = u$ and $v_1 = v_2 = v$ in Eq. (7.1-28), we find

$$I_f(u, v) = \frac{1}{(\overline{\lambda}f)^2} \iiiint_{-\infty}^{\infty} \mathbf{J}_o'(\xi_1, \eta_1; \xi_2, \eta_2)$$

$$\times \exp\left\{ j\frac{2\pi}{\overline{\lambda}f} [u(\xi_2 - \xi_1) + v(\eta_2 - \eta_1)] \right\} d\xi_1 \, d\eta_1 \, d\xi_2 \, d\eta_2.$$

(7.1-30)

Finally, we examine the severity of the quasimonochromatic condition that must be satisfied if the result (7.1-28) is to hold. We require that the difference of total delays from (ξ_1, η_1) to (u_1, v_1) and (ξ_2, η_2) to (u_2, v_2) be much less than the coherence time of the light. This difference of delays is given explicitly by

$$\tau_2 - \tau_1 = \frac{r_2 - r_1}{c} + \delta_2 - \delta_1 + \frac{r_2' - r_1'}{c}, \qquad (7.1\text{-}31)$$

where r_2 and r_1 refer to distances traveled by rays from the front focal plane to the lens, r_2' and r_1' are distances traveled by these same rays from the lens to the rear focal plane, and δ_2 and δ_1 are the time delays introduced by the lens. Note that a ray through point (ξ, η) in the front focal plane will reach the point (u, v) in the rear focal plane only if it has a unique angle with respect to the optical axis. Taking these geometric factors into account, using paraxial approximations throughout, and dropping constant factors, we see that the requirement becomes

$$\left| \frac{\xi_2 u_2 + \eta_2 v_2 - \xi_1 u_1 - \eta_1 v_1}{fc} \right| \ll \tau_c. \tag{7.1-32}$$

If the regions of concern in the (ξ, η) and (u, v) planes are of dimensions $L_0 \times L_0$ and $L_f \times L_f$, respectively, the required condition will be satisfied for all points of interest provided

$$\frac{L_0 L_f}{f} \ll l_c, \tag{7.1-33}$$

where $l_c = c\tau_c$ is the coherence length of the light. If, for example, $L_0 = L_f = 5$ centimeters and $f = 1$ meter, the coherence length of the light must be considerably greater than 2.5 millimeters, a rather stringent condition.

We note in closing that the four-dimensional Fourier transform relationship between mutual intensities in the front and back focal planes is directly analogous to the two-dimensional Fourier transform relationship between the complex fields in the focal planes of a fully coherent optical system (Ref. 7-7, Section 5.2). The four-dimensional relation is more general, however, since it holds for partially coherent systems.

7.1.4 Object–Image Coherence Relations for a Single Thin Lens

In this section we find the relationship between the mutual intensities in the object and image planes of a simple thin lens. The geometry of interest is illustrated in Fig. 7-4. The procedure used for calculating the mutual intensity in the image plane will be analogous to that used in the previous section. The mutual intensity will be propagated to the lens, the amplitude transmittance function of the lens will be applied, and the transmitted mutual intensity will be propagated to the image plane. In this case the finite aperture of the lens will be taken into account explicitly by using a lens amplitude transmittance function of the form

$$\mathbf{t}_l(x, y) = \mathbf{P}(x, y)\exp\left[-j\frac{\pi}{\lambda f}(x^2 + y^2)\right], \tag{7.1-34}$$

SOME PRELIMINARY CONSIDERATIONS

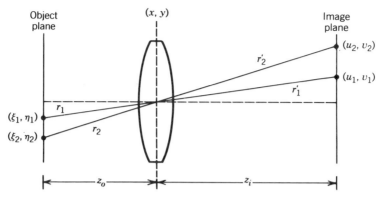

Figure 7-4. Imaging geometry.

where $\mathbf{P} = 0$ outside the lens aperture, the phase of $\mathbf{P}$ accounts for any lens aberrations that may exist, and $|\mathbf{P}|$ may vary within the aperture to account for apodizations. Most commonly, we shall assume that $|\mathbf{P}| = 1$ and $\arg\{\mathbf{P}\} = 0$ within the aperture. Paraxial approximations are again used throughout. Since a reasonably detailed analysis was carried out in the previous section, many of the details are omitted this time.

If $\mathbf{J}_o'$ again represents the mutual intensity leaving the object plane, and if z_0 represents the distance from the object plane to the lens, the mutual intensity $\mathbf{J}_l'$ leaving the lens is found to be

$$\mathbf{J}_l'(x_1, y_1; x_2, y_2) = \frac{1}{(\overline{\lambda} z_0)^2} \mathbf{P}(x_1, y_1) \mathbf{P}^*(x_2, y_2)$$

$$\times \exp\left\{-j\frac{\pi}{\lambda}\left(\frac{1}{z_0} - \frac{1}{f}\right)\left[(x_2^2 + y_2^2) - (x_1^2 + y_1^2)\right]\right\}$$

$$\times \iiiint_{-\infty}^{\infty} \mathbf{J}_o'(\xi_1, \eta_1; \xi_2, \eta_2) \exp\left\{-j\frac{\pi}{\overline{\lambda} z_0}\left[(\xi_2^2 + \eta_2^2) - (\xi_1^2 + \eta_1^2)\right]\right\}$$

$$\times \exp\left\{j\frac{2\pi}{\overline{\lambda} z_0}[x_2\xi_2 + y_2\eta_2 - x_1\xi_1 - y_1\eta_1]\right\} d\xi_1 d\eta_1 d\xi_2 d\eta_2.$$

(7.1-35)

Propagating an additional distance z_i to the image plane, we find the formidable result

$$J_i(u_1, v_1; u_2, v_2) = \frac{1}{(\bar{\lambda}z_0)^2} \frac{1}{(\bar{\lambda}z_i)^2} \exp\left\{-j\frac{\pi}{\bar{\lambda}z_i}\left[(u_2^2 + v_2^2) - (u_1^2 + v_1^2)\right]\right\}$$

$$\times \iiiint_{-\infty}^{\infty} d\xi_1 \, d\eta_1 \, d\xi_2 \, d\eta_2 \, J_o'(\xi_1, \eta_1; \xi_2, \eta_2) \exp\left\{-j\frac{\pi}{\bar{\lambda}z_0}\left[(\xi_2^2 + \eta_2^2) - (\xi_1^2 + \eta_1^2)\right]\right\}$$

$$\times \iiiint_{-\infty}^{\infty} dx_1 \, dy_1 \, dx_2 \, dy_2 \, \mathbf{P}(x_1, y_1)\mathbf{P}^*(x_2, y_2)$$

$$\times \exp\left\{-j\frac{\pi}{\bar{\lambda}}\left(\frac{1}{z_0} + \frac{1}{z_i} - \frac{1}{f}\right)\left[(x_2^2 + y_2^2) - (x_1^2 + y_1^2)\right]\right\}$$

$$\times \exp\left\{j\frac{2\pi}{\bar{\lambda}}\left[x_2\left(\frac{u_2}{z_i} + \frac{\xi_2}{z_0}\right) + y_2\left(\frac{v_2}{z_i} + \frac{\eta_2}{z_0}\right) - x_1\left(\frac{u_1}{z_i} + \frac{\xi_1}{z_0}\right) - y_1\left(\frac{v_1}{z_i} + \frac{\eta_1}{z_0}\right)\right]\right\}.$$

(7.1-36)

The task remains to simplify this result to a more usable form.

As an initial simplification, we note that, if the final plane is indeed at a proper distance to yield an image, the *lens law* must be satisfied,

$$\frac{1}{z_0} + \frac{1}{z_i} - \frac{1}{f} = 0, \qquad (7.1\text{-}37)$$

and the quadratic phase exponential in x_1^2, y_1^2, x_2^2, and y_2^2 becomes equal to unity. A further simplification results if we define the *amplitude spread function* of the system to be

$$K(u, v; \xi, \eta) = \frac{\exp\left\{j\frac{\pi}{\bar{\lambda}z_i}(u^2 + v^2)\right\}\exp\left\{j\frac{\pi}{\bar{\lambda}z_0}(\xi^2 + \eta^2)\right\}}{(\bar{\lambda}z_i)(\bar{\lambda}z_0)}$$

$$\times \iint_{-\infty}^{\infty} \mathbf{P}(x, y) \exp\left\{-j\frac{2\pi}{\bar{\lambda}z_i}\left[\left(u + \frac{z_i}{z_0}\xi\right)x + \left(v + \frac{z_i}{z_0}\eta\right)y\right]\right\} dx \, dy,$$

(7.1-38)

in which case the expression for the mutual intensity in the image plane can

SOME PRELIMINARY CONSIDERATIONS

be reduced to

$$\mathbf{J}_i(u_1,v_1;u_2,v_2) = \iiiint_{-\infty}^{\infty} \mathbf{J}_o'(\xi_1,\eta_1;\xi_2,\eta_2)\mathbf{K}(u_1,v_1;\xi_1,\eta_1)$$
$$\times \mathbf{K}^*(u_2,v_2;\xi_2,\eta_2)\,d\xi_1\,d\eta_1\,d\xi_2\,d\eta_2. \quad (7.1\text{-}39)$$

The intensity distribution in the image plane is found by setting $u_1 = u_2 = u$ and $v_1 = v_2 = v$ in the preceding result, yielding

$$I_i(u,v) = \iiiint_{-\infty}^{\infty} \mathbf{J}_o'(\xi_1,\eta_1;\xi_2,\eta_2)\mathbf{K}(u,v;\xi_1,\eta_1)\mathbf{K}^*(u,v;\xi_2,\eta_2)\,d\xi_1\,d\eta_1\,d\xi_2\,d\eta_2.$$

$$(7.1\text{-}40)$$

We now determine the restriction that must be imposed in order to assure that the quasimonochromatic conditions are satisfied for the mutual intensity calculation given above. We are aided in this task by the fact that, for an aberration-free system, the optical pathlengths traveled by all rays from a given object point to its Gaussian image point are equal (Ref. 7-8, p. 130). Therefore, it suffices for us to consider the pathlengths traveled by the central rays shown in Fig. 7-4. Within the accuracy of the thin lens approximation, the distances traveled within the lens are the same for all central rays. Hence the total pathlength difference for the rays illustrated is simply $r_2 + r_2' - r_1 - r_1'$. Using paraxial approximations for all four of the quantities involved, we find that the quasimonochromatic condition will be satisfied provided

$$\frac{(\xi_2^2 + \eta_2^2) - (\xi_1^2 + \eta_1^2)}{2z_0} + \frac{(u_2^2 + v_2^2) - (u_1^2 + v_1^2)}{2z_i} \ll l_c \quad (7.1\text{-}41)$$

for all points of interest, where l_c is again the coherence length of the light. If the object field is $L_0 \times L_0$ in size and the image field is $L_i \times L_i$, the worst-case requirement becomes

$$\frac{L_0^2}{4z_0} + \frac{L_i^2}{4z_i} \ll l_c. \quad (7.1\text{-}42)$$

If $L_0 = L_i = 2$ centimeters and $z_0 = z_i = 20$ centimeters, the coherence length must be considerably greater than 1 millimeter.

Finally, we note that, although we have found the relationship between mutual intensities in the object and image planes, we are not yet in a position to totally specify the image intensity that results from a specific object. To do so requires that we take account of the coherence properties of the object illumination, or equivalently the character of the source that illuminates the object. Such calculations are the subject of Section 7.2.

7.1.5 Relationship Between Mutual Intensities in the Exit Pupil and the Image

As the final fundamental imaging coherence relationship to be examined, we consider the dependence of the mutual intensity in the image plane on the mutual intensity in the exit pupil of a rather general imaging system.

All imaging systems, regardless of their detailed structure, contain somewhere an aperture that limits the angular extent of the pencil of rays converging toward an ideal image point. The image of this aperture stop formed by the optical elements that follow it is called the *exit pupil* of the imaging system. Similarly, the image of this aperture stop formed by the optical elements that precede it is called the *entrance pupil* of the imaging system. These concepts are illustrated in Fig. 7-5. In part (a) the limiting aperture is the lens itself, in which case the physical aperture, entrance pupil, and exit pupil all coincide. In part (b) the aperture stop occurs before the lens, in which case the entrance pupil coincides with the physical aperture, but the exit pupil is the image of that aperture, as shown by the dotted lines. Finally, in part (c), the exit pupil coincides with the physical stop while the entrance pupil is its image, represented by a dotted line. These definitions apply to systems of arbitrary complexity.

Our goal is to find the relationship between the mutual intensity distribution in the exit pupil and the mutual intensity distribution in the image. The results will be in simplest form if we express the pupil mutual intensity function on a sphere of radius z_i (the distance from the exit pupil to the image plane) centered on the origin in the image plane. Figure 7-6 illustrates the geometry. The relationship between the two mutual intensities of interest can be found if we begin with the basic relationship of Eq. (5.4-8), which we restate here as

$$\mathbf{J}_i(u_1, v_1; u_2, v_2) = \iint_{\Sigma_1} \iint_{\Sigma_1} \mathbf{J}_p'(x_1, y_1; x_2, y_2)$$

$$\times \exp\left\{-j\frac{2\pi}{\lambda}(r_2 - r_1)\right\} \frac{\chi(\theta_1)}{\lambda r_1} \frac{\chi(\theta_2)}{\lambda r_2} dx_1 dy_1 dx_2 dy_2.$$

$$(7.1\text{-}43)$$

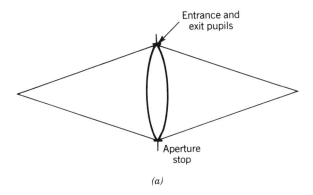

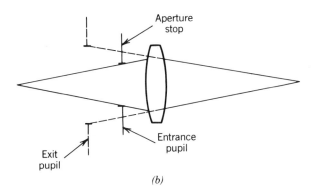

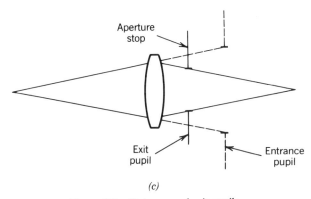

Figure 7-5. Entrance and exit pupils.

Here $\mathbf{J}'_p$ is the mutual intensity transmitted by the exit pupil and Σ_1 represents the limiting aperture of that pupil. With the usual small-angle assumption,

$$\chi(\theta_1) \cong \chi(\theta_2) \cong 1$$

$$\bar{\lambda} r_1 \cong \bar{\lambda} r_2 \cong \bar{\lambda} z_i. \qquad (7.1\text{-}44)$$

In addition, as indicated by the geometry shown in Fig. 7-6, a general distance r from point (x, y) on the exit sphere to point (u, v) in the image plane is given by

$$r = \sqrt{z_i^2 + u^2 + v^2 - 2xu - 2yv}$$

$$\cong z_i \left[1 + \frac{u^2 + v^2}{2z_i^2} - \frac{xu + yv}{z_i^2} \right]. \qquad (7.1\text{-}45)$$

Thus the relationship between the two mutual intensities takes the form

$$\mathbf{J}_i(u_1, v_1; u_2, v_2) = \frac{1}{(\bar{\lambda} z_i)^2} \exp\left\{ j \frac{\pi}{\bar{\lambda} z_i} \left[(u_1^2 + v_1^2) - (u_2^2 + v_2^2) \right] \right\}$$

$$\times \iiiint_{-\infty}^{\infty} \mathbf{J}'_p(x_1, y_1; x_2, y_2)$$

$$\times \exp\left\{ j \frac{2\pi}{\bar{\lambda} z_i} (x_2 u_2 + y_2 v_2 - x_1 u_1 - y_1 v_1) \right\} dx_1 \, dy_1 \, dx_2 \, dy_2,$$

$$(7.1\text{-}46)$$

where we have incorporated the finite limits posed by Σ_1 into the definition of $\mathbf{J}'_p$. Note that when it is the image *intensity* (rather than mutual intensity) that is of prime concern, Eq. (7.1-46) reduces to

$$I_i(u, v) = \frac{1}{(\bar{\lambda} z_i)^2} \iiiint_{-\infty}^{\infty} \mathbf{J}'_p(x_1, y_1; x_2, y_2)$$

$$\times \exp\left\{ j \frac{2\pi}{\bar{\lambda} z_i} \left[(x_2 - x_1) u + (y_2 - y_1) v \right] \right\} dx_1 \, dy_1 \, dx_2 \, dy_2.$$

$$(7.1\text{-}47)$$

METHODS FOR CALCULATING IMAGE INTENSITY

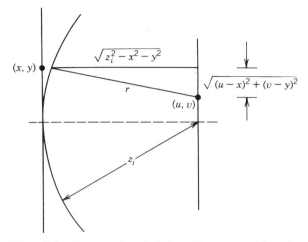

Figure 7-6. Geometry for calculation of image mutual intensity.

We shall make good use of this relationship between image plane intensity and pupil mutual intensity in our discussion of image formation from an interferometric point of view (Section 7.4). For the moment, the results (7.1-46) and (7.1-47) simply remain basic relationships that we can call upon in the future as they are needed.

7.2 METHODS FOR CALCULATING IMAGE INTENSITY

The purpose of developing a theory of image formation in partially coherent light is to allow us to calculate the intensity distribution expected in the image plane in any given experimental situation, and in so doing to develop an understanding of the individual effects of the illumination, the object, and the imaging optics. Hopefully an improved ability to interpret the results of experiments will follow from such a theoretical understanding. Accordingly, in the sections that follow, we describe several different methods of analysis, all of which can be used to predict the image intensity that will be obtained in a given experiment.

7.2.1 Integration over the Source

When the illumination of the object is derived from a quasimonochromatic, spatially incoherent source, as is often the case, there exists a method for calculating image intensity that has the special appeal of conceptual simplicity. Each point on the source is considered individually, the image intensity produced by the light from that single point is calculated, and the image intensity contributions from all such points are added, with a weighting proportional to the source intensity distribution. Simple addition of the image intensity distributions is possible as a result of the assumed incoherence of the original source.

To examine this approach in more detail, consider the geometry illustrated in Fig. 7-7. The object is located in the (ξ, η) plane and is illuminated by means of an optical system to the left of that plane. An image is formed in the (u, v) plane by the optical system on the right. We assume that, under the quasimonochromatic conditions, each optical system can be represented by an amplitude spread function (impulse response). The symbols $\mathbf{F}(\xi, \eta; \alpha, \beta)$ and $\mathbf{K}(u, v; \xi, \eta)$ are used to represent such spread functions of the illuminating and imaging systems, respectively.

A single point at coordinates (α, β) on the source emits light representable by a time-varying phasor $\mathbf{A}_S(\alpha, \beta; t)$. The illumination from this point reaches the object and passes through it, producing a time-varying phasor amplitude $\mathbf{A}'_o(\xi, \eta; \alpha, \beta; t)$ (to the right of the object) given by

$$\mathbf{A}'_o(\xi, \eta; \alpha, \beta; t) = \mathbf{F}(\xi, \eta; \alpha, \beta)\mathbf{t}_o(\xi, \eta)\mathbf{A}_S(\alpha, \beta; t - \delta_1), \quad (7.2\text{-}1)$$

where δ_1 is a time delay that depends on (ξ, η) and (α, β) and $\mathbf{t}_o(\xi, \eta)$ is the amplitude transmittance of the object, which has been assumed to be independent of the particular source point providing the illumination. Finally, the time varying phasor amplitude of the light reaching coordinates

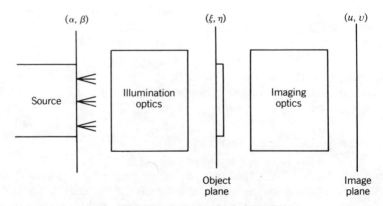

Figure 7-7. Illumination and imaging systems.

METHODS FOR CALCULATING IMAGE INTENSITY

(u, v) on the image plane from source point (α, β) is given by

$$\mathbf{A}_i(u, v; \alpha, \beta; t) = \iint_{-\infty}^{\infty} \mathbf{K}(u, v; \xi, \eta) \mathbf{t}_o(\xi, \eta)$$
$$\times \mathbf{F}(\xi, \eta; \alpha, \beta) \mathbf{A}_S(\alpha, \beta; t - \delta_1 - \delta_2) \, d\xi \, d\eta, \quad (7.2\text{-}2)$$

where δ_2 is a time delay that depends on (u, v) and (ξ, η).

The intensity of the light reaching image coordinates (u, v) from the source point at (α, β) can now be calculated to be

$$I_i(u, v; \alpha, \beta) = \langle |\mathbf{A}_i(u, v; \alpha, \beta; t)|^2 \rangle$$
$$= \iiiint_{-\infty}^{\infty} \mathbf{K}(u, v; \xi_1, \eta_1) \mathbf{K}^*(u, v; \xi_2, \eta_2) \mathbf{t}_o(\xi_1, \eta_1) \mathbf{t}_o^*(\xi_2, \eta_2)$$
$$\times \mathbf{F}(\xi_1, \eta_1; \alpha, \beta) \mathbf{F}^*(\xi_2, \eta_2; \alpha, \beta)$$
$$\times \langle \mathbf{A}_S(\alpha, \beta; t - \delta_1 - \delta_2) \mathbf{A}_S^*(\alpha, \beta; t - \delta_1' - \delta_2') \rangle \, d\xi_1 \, d\eta_1 \, d\xi_2 \, d\eta_2. \quad (7.2\text{-}3)$$

Under the quasimonochromatic assumption, the delay differences satisfy

$$|\delta_1 + \delta_2 - \delta_1' - \delta_2'| \ll \tau_c, \quad (7.2\text{-}4)$$

with the result that the time-averaged quantity in Eq. (7.2-3) reduces to $I_S(\alpha, \beta)$, the intensity of the source at (α, β). Finally we integrate the partial intensity $I_i(u, v; \alpha, \beta)$ over the source coordinates (α, β), giving the result

$$I_i(u, v) = \iint_{-\infty}^{\infty} I_S(\alpha, \beta) \iiiint_{-\infty}^{\infty} \mathbf{K}(u, v; \xi_1, \eta_1) \mathbf{K}^*(u, v; \xi_2, \eta_2)$$
$$\times \mathbf{F}(\xi_1, \eta_1; \alpha, \beta) \mathbf{F}^*(\xi_2, \eta_2; \alpha, \beta) \mathbf{t}_o(\xi_1, \eta_1) \mathbf{t}_o^*(\xi_2, \eta_2) \, d\xi_1 \, d\eta_1 \, d\xi_2 \, d\eta_2 \, d\alpha \, d\beta. \quad (7.2\text{-}5)$$

With knowledge of I_S, $\mathbf{F}$, $\mathbf{K}$ and $\mathbf{t}_o$, it is now possible to calculate the image intensity distribution.

Two different object illumination systems are often encountered in practice. First, as illustrated in Fig. 7-8a, if the incoherent source is

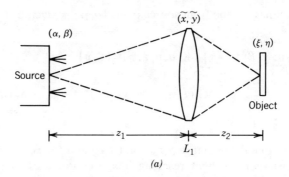

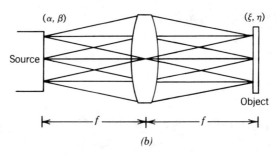

Figure 7-8. Two common illumination systems: (*a*) critical illumination; (*b*) Köhler's illumination.

relatively uniform, the illumination optics may simply image the source onto the object, perhaps with some magnification or demagnification. For the simple single-lens imaging system shown in the figure, the amplitude spread function **F** is then of the form [cf. Eq. (7.1-38)]

$$\mathbf{F}(\xi, \eta; \alpha, \beta) = \frac{\exp\left\{j\frac{\pi}{\bar{\lambda}z_1}(\alpha^2 + \beta^2)\right\}}{(\bar{\lambda}z_1)} \cdot \frac{\exp\left\{j\frac{\pi}{\bar{\lambda}z_2}(\xi^2 + \eta^2)\right\}}{(\bar{\lambda}z_2)}$$

$$\times \iint_{-\infty}^{\infty} \mathbf{P}_c(\tilde{x}, \tilde{y})\exp\left\{-j\frac{2\pi}{\bar{\lambda}z_2}[(\xi + M\alpha)\tilde{x} + (\eta + M\beta)\tilde{y}]\right\} d\tilde{x}\, d\tilde{y},$$

(7.2-6)

where $\mathbf{P}_c$ is the pupil function of the lens L_1 (perhaps as determined by an aperture stop rather than by the size of the lens itself), $M = z_2/z_1$ is the magnification of the illumination system, and $(\tilde{x}, \tilde{y})$ are coordinates in the plane of the lens L_1. When used to provide illumination in a microscope, the optical system between the source and the object is called the *condenser* system; the system illustrated in Fig. 7-8a is said to provide *critical illumination* (Ref. 7-2, Section 10.5.2).

An alternative class of illumination systems is illustrated in simple form in Fig. 7-8b. In this case the source is effectively imaged at infinite distance from the object. As a consequence, nonuniformities of the source brightness distribution are not imaged onto the object, and a highly uniform field of illumination is provided. When used in a microscope, this general class of illumination system is said to provide *Köhler's illumination* (Ref. 7-2, Section 10.5.2; Ref. 7-9). From the Fourier transforming properties of a thin positive lens, the amplitude spread function for the simple system shown in Fig. 7-8b is given by

$$\mathbf{F}(\xi, \eta; \alpha, \beta) = \frac{1}{\lambda f} \exp\left\{-j\frac{2\pi}{\lambda f}(\xi\alpha + \eta\beta)\right\} \quad (7.2\text{-}7)$$

when the finite extent of the lens aperture is neglected.

We conclude this section by noting that, whereas integration over the source is a method of calculation which has conceptual simplicity, it is not necessarily the simplest method to use in practice. For any given problem, the various possible approaches should be considered, for one approach may be distinctly easier than another, depending on the problem at hand. We now turn to considering a second method of calculation.

7.2.2 Representation of the Source by an Incident Mutual Intensity Function

A somewhat more common approach to the calculation of image intensity distributions is arrived at if the explicit integration over the source is suppressed and the effects of the source are represented by a mutual intensity function describing the illumination incident on the object. We suppose that, under the quasimonochromatic assumption, the time-varying phasor amplitude $\mathbf{A}_i(u, v; t)$ of the light arriving at image coordinates (u, v) can be represented in terms of the time-varying phasor amplitude $\mathbf{A}_o(\xi, \eta; t)$ of the light *incident* on the object at coordinates (ξ, η) by

$$\mathbf{A}_i(u, v; t) = \iint_{-\infty}^{\infty} \mathbf{K}(u, v; \xi, \eta) \mathbf{t}_o(\xi, \eta) \mathbf{A}_o(\xi, \eta; t - \delta) \, d\xi \, d\eta,$$

$$(7.2\text{-}8)$$

where **K** is the amplitude spread function of the imaging system, t_o is again the amplitude transmittance of the object, and δ is a time delay that depends on (ξ, η) and (u, v). The intensity at (u, v) is thus given by

$$I_i(u, v) = \langle |\mathbf{A}_i(u, v; t)|^2 \rangle$$

$$= \iiiint_{-\infty}^{\infty} \mathbf{K}(u, v; \xi_1, \eta_1) \mathbf{K}^*(u, v; \xi_2, \eta_2) t_o(\xi_1, \eta_1) t_o^*(\xi_2, \eta_2)$$

$$\times \langle \mathbf{A}_o(\xi_1, \eta_1; t - \delta_1) \mathbf{A}_o^*(\xi_2, \eta_2; t - \delta_2) \rangle \, d\xi_1 \, d\eta_1 \, d\xi_2 \, d\eta_2.$$

(7.2-9)

Under the quasimonochromatic assumption, $|\delta_1 - \delta_2| \ll \tau_c$, and hence

$$\langle \mathbf{A}_o(\xi_1, \eta_1; t - \delta_1) \mathbf{A}_o^*(\xi_2, \eta_2; t - \delta_2) \rangle = \mathbf{J}_o(\xi_1, \eta_1; \xi_2, \eta_2),$$

(7.2-10)

where $\mathbf{J}_o$ is the mutual intensity distribution *incident* on the object. The final expression for image intensity is now given by

$$I_i(u, v) = \iiiint_{-\infty}^{\infty} \mathbf{K}(u, v; \xi_1, \eta_1) \mathbf{K}^*(u, v; \xi_2, \eta_2) t_o(\xi_1, \eta_1) t_o^*(\xi_2, \eta_2)$$

$$\times \mathbf{J}_o(\xi_1, \eta_1; \xi_2, \eta_2) \, d\xi_1 \, d\eta_1 \, d\xi_2 \, d\eta_2.$$

(7.2-11)

Knowledge of **K**, t_o, and $\mathbf{J}_o$ allows calculation of I_i.

Now it could be argued that, whereas the expression (7.2-5) for I_i requires six integrations and (7.2-11) requires only four, the latter is not really simpler than the former, for four integrations are in general required to determine $\mathbf{J}_o$. This observation is strictly true; however, an *incoherent* source was assumed in arriving at Eq. (7.2-5), and when a similar assumption is made here, the calculation of $\mathbf{J}_o$ requires only two integrations, yielding again a total of six, as we now illustrate.

Suppose that, as illustrated in Fig. 7-9, an incoherent source is placed at an arbitrary distance z_1 in front of a simple thin positive lens, and that the object under illumination lies at distance z_2 behind the lens. The source is assumed to subtend a sufficiently large angle at the lens so that, as determined by the Van Cittert–Zernike theorem, the coherence area of the light incident on the lens is extremely small in comparison with the area of

METHODS FOR CALCULATING IMAGE INTENSITY

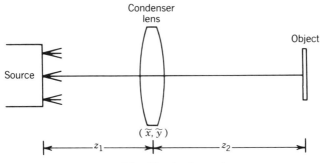

Figure 7-9. Illumination optics.

the lens. With reference to Eqs. (5.6-24) and (5.6-25), this requirement can be stated as

$$A_l A_S \gg (\bar{\lambda} z_1)^2, \qquad (7.2\text{-}12)$$

where A_l and A_S are the areas of the lens and the source, respectively. We argue in what follows that, under suitable conditions including (7.2-12), the lens aperture itself may be regarded as a source of approximately incoherent illumination, and that the illuminating mutual intensity $\mathbf{J}_o$ may be calculated rather simply by applying the Van Cittert–Zernike theorem to the lens pupil as a source.

To examine this argument critically, we first note that, by applying the Van Cittert–Zernike theorem to the incoherent source with intensity distribution $I_S(\alpha, \beta)$, the mutual intensity incident on the lens is given by [cf. Eq. (5.6-8)]

$$\mathbf{J}_I(\tilde{x}_1, \tilde{y}_1; \tilde{x}_2, \tilde{y}_2) = \frac{\kappa \exp\left\{-j\frac{\pi}{\bar{\lambda} z_1}\left[(\tilde{x}_2^2 + \tilde{y}_2^2) - (\tilde{x}_1^2 + \tilde{y}_1^2)\right]\right\}}{(\bar{\lambda} z_1)^2}$$

$$\times \iint_{-\infty}^{\infty} I_S(\alpha, \beta) \exp\left\{j\frac{2\pi}{\bar{\lambda} z_1}(\Delta\tilde{x}\alpha + \Delta\tilde{y}\beta)\right\} d\alpha\, d\beta,$$

$$(7.2\text{-}13)$$

where $\Delta\tilde{x} = \tilde{x}_2 - \tilde{x}_1$ and $\Delta\tilde{y} = \tilde{y}_2 - \tilde{y}_1$. The mutual intensity transmitted by the lens takes the form

$$\mathbf{J}_l'(\tilde{x}_1, \tilde{y}_1; \tilde{x}_2, \tilde{y}_2) = \mathbf{P}_c(\tilde{x}_1, \tilde{y}_1)\mathbf{P}_c^*(\tilde{x}_2, \tilde{y}_2)$$

$$\times \exp\left\{-j\frac{\pi}{\bar{\lambda}f}\left[(\tilde{x}_1^2 + \tilde{y}_1^2) - (\tilde{x}_2^2 + \tilde{y}_2^2)\right]\right\}\mathbf{J}_l(\tilde{x}_1, \tilde{y}_1; \tilde{x}_2, \tilde{y}_2),$$

(7.2-14)

where $\mathbf{P}_c(\tilde{x}_1, \tilde{y}_1) = P_c(\tilde{x}, \tilde{y})\exp[-jW(\tilde{x}, \tilde{y})]$ represents the complex pupil function of the condenser lens and $W(x, y)$ represents a slowly varying aberration phase describing departures of the wavefront from a perfect Gaussian reference sphere.

Now because of the very narrow width of the coherence area incident on the lens, as determined by the narrow width of the Fourier transform in Eq. (7.2-13), the transmitted mutual intensity is nonzero only for very small $\Delta \tilde{x}$ and $\Delta \tilde{y}$. Accordingly, we make the following assumptions, valid for sufficiently small $\Delta \tilde{x}$ and $\Delta \tilde{y}$ (cf. Problem 7-1):

$$\exp\left\{-j\frac{\pi}{\bar{\lambda}}\left(\frac{1}{z_1} - \frac{1}{f}\right)\left[(\tilde{x}_2^2 + \tilde{y}_2^2) - (\tilde{x}_1^2 + \tilde{y}_1^2)\right]\right\} \cong 1, \quad (7.2\text{-}15a)$$

$$\mathbf{P}_c(\tilde{x}_1, \tilde{y}_1)\mathbf{P}_c^*(\tilde{x}_2, \tilde{y}_2) \cong |\mathbf{P}_c(\tilde{x}_1, \tilde{y}_1)|^2, \quad (7.2\text{-}15b)$$

with the result that the transmitted mutual intensity is of the form

$$\mathbf{J}_l'(\tilde{x}_1, \tilde{y}_1; \tilde{x}_2, \tilde{y}_2) \cong \frac{\kappa}{(\bar{\lambda}z_1)^2}|\mathbf{P}_c(\tilde{x}_1, \tilde{y}_1)|^2 \mathcal{I}_S\left(\frac{\Delta \tilde{x}}{\bar{\lambda}z_1}, \frac{\Delta \tilde{y}}{\bar{\lambda}z_1}\right),$$

(7.2-16)

where $\mathcal{I}_S(\nu_X, \nu_Y)$ is the two-dimensional Fourier transform of the source intensity distribution $I_S(\alpha, \beta)$.

By our assumption (7.2-12), the function $\mathcal{I}_S(\Delta \tilde{x}/\bar{\lambda}z_1, \Delta \tilde{y}/\bar{\lambda}z_1)$ is an exceedingly narrow function of $(\Delta \tilde{x}, \Delta \tilde{y})$. Accordingly, we regard Eq. (7.2-16) as describing the mutual intensity of a new source (the lens pupil) that is for all practical purposes spatially incoherent and with intensity distribution proportional to $|\mathbf{P}_c(\tilde{x}_1, \tilde{y}_1)|^2$. Now we apply the Van Cittert–Zernike theorem to this new source, allowing us to specify the

METHODS FOR CALCULATING IMAGE INTENSITY

mutual intensity incident on the object as

$$\mathbf{J}_o(\xi_1, \eta_1; \xi_2, \eta_2) = \frac{\kappa' \exp\left\{-j\frac{\pi}{\bar{\lambda} z_2}\left[(\xi_2^2 + \eta_2^2) - (\xi_1^2 + \eta_1^2)\right]\right\}}{(\bar{\lambda} z_2)^2}$$

$$\times \iint_{-\infty}^{\infty} |\mathbf{P}_c(\tilde{x}_1, \tilde{y}_1)|^2 \exp\left\{j\frac{2\pi}{\bar{\lambda} z_2}(\Delta\xi \tilde{x}_1 + \Delta\eta \tilde{y}_1)\right\} d\tilde{x}_1 d\tilde{y}_1.$$

(7.2-17)

Note in particular that, subject to the approximation (7.2-15b), *the mutual intensity incident on the object is independent of any aberrations that may exist in the illumination system*, a fact first noted by Zernike (Ref. 7-10). The calculation of the mutual intensity incident on the object has been reduced to the problem of Fourier transforming the squared modulus of the lens pupil function. When the lens is not apodized (i.e., $|\mathbf{P}_c| = 0$ or 1), then $|\mathbf{P}_c|^2 = |\mathbf{P}_c|$ and it suffices to Fourier transform the aperture function itself. In addition we note that, provided the assumptions used in arriving at Eq. (7.2-17) remain valid, the mutual intensity $\mathbf{J}_0$ incident on the object is independent of the distance z_1 of the source in front of the lens.

Of course, the function $\mathcal{J}_S$ of Eq. (7.2-16) is never infinitely narrow. An application of the *generalized* Van Cittert–Zernike theorem to Eq. (7.2-16) demonstrates that our conclusion remains valid for an $\mathcal{J}_S$ of finite width provided that (see Problem 7-2)

$$\left(\frac{z_2}{z_1}\right)^2 A_S \gg A_o, \quad (7.2\text{-}18)$$

where A_o is the area of the object.

Finally, we close this section with a discussion of one other circumstance under which it is relatively easy to calculate $\mathbf{J}_o$. Suppose that the incoherent source and the object of Fig. 7-9 are each one focal length from the lens (i.e., $z_1 = z_2 = f$). In addition, suppose that the lens is substantially larger than the source and the object, so that the finite size of the lens pupil can be neglected. Representing the mutual intensity of the source by

$$\mathbf{J}_S(\alpha_1, \beta_1; \alpha_2, \beta_2) = \kappa I_S(\alpha_1, \beta_1)\delta(\alpha_1 - \alpha_2, \beta_1 - \beta_2), \quad (7.2\text{-}19)$$

we use the four-dimensional Fourier transform of Eq. (7.1-28) (with $\mathbf{J}'_o$ replaced by $\mathbf{J}_S$, and $\mathbf{J}_f$ replaced by $\mathbf{J}_o$) to calculate the mutual intensity incident on the object. The result of this simple calculation is

$$\mathbf{J}_o(\Delta\xi, \Delta\eta) = \frac{\kappa}{(\bar{\lambda}f)^2} \iint_{-\infty}^{\infty} I_S(\alpha, \beta) \exp\left\{ j\frac{2\pi}{\bar{\lambda}f}(\alpha\Delta\xi + \beta\Delta\eta) \right\} d\alpha\, d\beta,$$

(7.2-20)

where $\Delta\xi = \xi_2 - \xi_1$ and $\Delta\eta = \eta_2 - \eta_1$. Thus the mutual intensity incident on the object is a function only of the *differences* of coordinates in the object plane and can easily be found by Fourier transforming the source intensity distribution. In many applications both (7.2-20) and (7.2-17) are relatively easy operations to perform, leaving the bulk of the problem of image intensity calculation in the evaluation of the four integrals of Eq. (7.2-11). Examples are deferred to Section 7.3.

7.2.3 The Four-Dimensional Linear Systems Approach

We saw in Section 7.1.4 [in particular, Eq. (7.1-39)] that, if an imaging system is described by an amplitude spread function $\mathbf{K}(u, v; \xi, \eta)$, representing the amplitude of the field at image coordinates (u, v) that results from an object consisting of a δ-function amplitude at (ξ, η), the object and image mutual intensities are related by

$$\mathbf{J}_i(u_1, v_1; u_2, v_2) = \iiiint_{-\infty}^{\infty} \mathbf{J}_o'(\xi_1, \eta_1; \xi_2, \eta_2)$$

$$\times \mathbf{K}(u_1, v_1; \xi_1, \eta_1) \mathbf{K}^*(u_2, v_2; \xi_2, \eta_2)\, d\xi_1\, d\eta_1\, d\xi_2\, d\eta_2.$$

(7.2-21)

Such an equation may be called a four-dimensional *superposition integral* and is characteristic of a linear system. Thus it is possible to view an imaging operation as a four-dimensional linear system, with the mutual intensity transmitted by the object as the input, and the mutual intensity appearing in the image plane as the output. The quantity $\mathbf{K}(u_1, v_1; \xi_1, \eta_1)\mathbf{K}^*(u_2, v_2; \xi_2, \eta_2)$ may be regarded as the impulse response of this system, that is, the mutual intensity observed at image coordinates (u_1, v_1, u_2, v_2) in response to an object mutual intensity consisting of an impulse at $(\xi_1, \eta_1, \xi_2, \eta_2)$.

Under certain circumstances it is possible to reduce the general superposition integral (7.2-21) to a somewhat simpler *convolution integral* of the form

$$\mathbf{J}_i(u_1,v_1;u_2,v_2) = \iiiint_{-\infty}^{\infty} \mathbf{J}_o'(\xi_1,\eta_1;\xi_2,\eta_2)\mathbf{K}(u_1-\xi_1,v_1-\eta_1)$$
$$\times \mathbf{K}^*(u_2-\xi_2,v_2-\eta_2)\,d\xi_1\,d\eta_1\,d\xi_2\,d\eta_2,$$

(7.2-22)

in which the impulse response $\mathbf{K}(u_1-\xi_1,v_1-\eta_1)\mathbf{K}^*(u_2-\xi_2,v_2-\eta_2)$ depends only on the coordinate *differences* $(u_1-\xi_1)$, $(v_1-\eta_1)$, $(u_2-\xi_2)$, and $(v_2-\eta_2)$. When such is the case, the system is said to be *space invariant* or *isoplanatic*. Clearly, if the amplitude spread function $\mathbf{K}$ is space invariant in two dimensions, the impulse response of the four-dimensional system of concern here will likewise be space invariant.

The conditions under which space invariance of the amplitude spread function $\mathbf{K}$ can be assumed are nontrivial and are by no means always satisfied. For example, the spread function $\mathbf{K}$ in Eq. (7.1-38) is far from space invariant in the form presented there. In general, the following conditions must be satisfied in order to make space invariance a reasonable assumption:

(1) The object coordinates (ξ,η) must have been normalized in such a way that the magnification between the (ξ,η) and (u,v) coordinate systems is unity.
(2) The object coordinate axes must be directed in such a way that the effects of image inversion are removed from the mathematics.
(3) The amplitude spread function $\mathbf{K}$ must be free from space-variant phase factors, such as the terms $\exp\{j(\pi/\bar{\lambda}z_0)(\xi^2+\eta^2)\}$ and $\exp\{j(\pi/\bar{\lambda}z_i)(u^2+v^2)\}$ in Eq. (7.1-38).

Conditions 1 and 2 are easily satisfied by proper scaling and direction of the object coordinates. Condition 3 is more difficult to satisfy. It has been shown that these phase factors do not appear in the amplitude spread function of a two-lens telecentric imaging system (Ref. 7-11). In addition, the phase factor depending on $\xi^2+\eta^2$ can be eliminated by proper choice of the illuminating optics (Ref. 7-12). Both phase factors can be removed by placement of positive lenses with proper focal lengths against the object and the image planes. Alternatively, coherence may exist over such small areas in the object and image planes that the phase factors associated with $\mathbf{K}$ and $\mathbf{K}^*$ may cancel each other within the range of separations of interest.

If indeed conditions 1 through 3 above are satisfied, then the four-dimensional convolution of Eq. (7.2-22) may be used to represent the mapping of

$\mathbf{J}'_o$ into $\mathbf{J}_i$. In such a case we should quite naturally investigate the form of this relationship in the Fourier domain, where convolutions are represented by simple multiplications of transforms (Ref. 7-7, pp. 19 and 20). Accordingly, we define the four-dimensional Fourier spectra of the object and image mutual intensities by

$$\mathscr{J}'_o(\nu_1, \nu_2, \nu_3, \nu_4) \triangleq \mathscr{F}\{\mathbf{J}'_o\}$$

$$\mathscr{J}_i(\nu_1, \nu_2, \nu_3, \nu_4) \triangleq \mathscr{F}\{\mathbf{J}_i\}, \qquad (7.2\text{-}23)$$

where the operator $\mathscr{F}\{\ \}$ is defined by

$$\mathscr{F}\{\ \} = \iiiint_{-\infty}^{\infty}\{\ \}\exp[j2\pi(\nu_1 x_1 + \nu_2 x_2 + \nu_3 x_3 + \nu_4 x_4)]\,dx_1\,dx_2\,dx_3\,dx_4$$

$$(7.2\text{-}24)$$

and (x_1, x_2, x_3, x_4) are dummy variables of integration representing the four arguments of the mutual intensity functions taken in the order in which they are written.

In a similar fashion we define the four-dimensional *transfer function* of the space-invariant, linear system by

$$\mathscr{H}(\nu_1, \nu_2, \nu_3, \nu_4) = \mathscr{F}\{\mathbf{K}(x_1, x_2)\mathbf{K}^*(x_3, x_4)\} \qquad (7.2\text{-}25)$$

where we have noted that in the space-invariant case, each amplitude spread function depends on only two independent variables. Due to the separability of the four-dimensional spread function into two factors, $\mathscr{H}$ likewise separates into two factors, with the result

$$\mathscr{H}(\nu_1, \nu_2, \nu_3, \nu_4) = \mathscr{K}(\nu_1, \nu_2)\mathscr{K}^*(-\nu_3, -\nu_4), \qquad (7.2\text{-}26)$$

where $\mathscr{K}$ represents the two-dimensional Fourier transform of the amplitude spread function

$$\mathscr{K}(\nu_1, \nu_2) = \iint_{-\infty}^{\infty} \mathbf{K}(x_1, x_2) e^{j2\pi(\nu_1 x_1 + \nu_2 x_2)}\,dx_1\,dx_2. \qquad (7.2\text{-}27)$$

Thus the effect of the imaging system is represented in the four-dimensional frequency domain by

$$\mathcal{J}_i(\nu_1,\nu_2,\nu_3,\nu_4) = \mathcal{K}(\nu_1,\nu_2)\mathcal{K}^*(-\nu_3,-\nu_4)\mathcal{J}'_o(\nu_1,\nu_2,\nu_3,\nu_4).$$
(7.2-28)

Now since $\mathcal{K}$ is the two-dimensional Fourier transform of **K**, and since from Eq. (7.1-38) **K** is, in turn, related to the two-dimensional Fourier transform of the pupil function **P**, we can expect some rather direct relationship to exist between $\mathcal{K}$ and **P**. In fact, when conditions 1 through 3 for space invariance are satisfied, the amplitude spread function $\mathbf{K}(u - \xi, v - \eta)$ takes the form [cf. Eq. (7.1-38)]

$$\mathbf{K}(u-\xi, v-\eta) = C \iint_{-\infty}^{\infty} \mathbf{P}(x,y)$$

$$\times \exp\left\{-j2\pi\left[(u-\xi)\frac{x}{\bar{\lambda}z_i} + (v-\eta)\frac{y}{\bar{\lambda}z_i}\right]\right\} \frac{dx}{\bar{\lambda}z_i}\frac{dy}{\bar{\lambda}z_i},$$
(7.2-29)

where C is a constant. A two-dimensional Fourier transform of this equation yields, up to an unimportant scaling constant C,

$$\mathcal{K}(\nu_1,\nu_2) = \mathbf{P}(\bar{\lambda}z_i\nu_1, \bar{\lambda}z_i\nu_2) \qquad (7.2\text{-}30)$$

and the relation between the spectra of the mutual intensities [Eq. (7.2-28)] becomes

$$\mathcal{J}_i(\nu_1,\nu_2,\nu_3,\nu_4) = \mathbf{P}(\bar{\lambda}z_i\nu_1,\bar{\lambda}z_i\nu_2)\mathbf{P}^*(-\bar{\lambda}z_i\nu_3,-\bar{\lambda}z_i\nu_4)\mathcal{J}'_o(\nu_1,\nu_2,\nu_3,\nu_4).$$
(7.2-31)

Clearly, the transfer function will drop to zero when ν_1, ν_2, ν_3, or ν_4 exceed certain limits imposed by the pupil functions.

Unfortunately, the Fourier relationship (7.2-31) is not very useful in this form, for the spectrum $\mathcal{J}'_o$ depends on the properties of both the object and the illumination. Some further investigation of the character of $\mathcal{J}'_o$ is thus required in order to understand the role that each of these separate physical quantities plays.

We assume, as we have in the past, that the object is illuminated from behind (i.e., transilluminated), and that it has amplitude transmittance t_o. In addition, we assume that the mutual intensity $\mathbf{J}_o$ of the light illuminating the object depends only on the coordinate differences $\Delta\xi = \xi_2 - \xi_1$, $\Delta\eta = \eta_2 - \eta_1$, as is often the case in practice. Thus the transmitted mutual intensity is of the form

$$\mathbf{J}_o'(\xi_1,\eta_1;\xi_2,\eta_2) = \mathbf{J}_o(\Delta\xi,\Delta\eta)\mathbf{t}_o(\xi_1,\eta_1)\mathbf{t}_o^*(\xi_2,\eta_2). \quad (7.2\text{-}32)$$

The four-dimensional Fourier transform of $\mathbf{J}_o'$ takes the form

$$\mathscr{I}_o'(\nu_1,\nu_2,\nu_3,\nu_4) = \iiiint_{-\infty}^{\infty} \mathbf{J}_o(\Delta\xi,\Delta\eta)\mathbf{t}_o(\xi_1,\eta_1)\mathbf{t}_o^*(\xi_2,\eta_2)$$

$$\times \exp\{j2\pi(\nu_1\xi_1 + \nu_2\eta_1 + \nu_3\xi_2 + \nu_4\eta_2)\}\, d\xi_1\, d\eta_1\, d\xi_2\, d\eta_2.$$

$$(7.2\text{-}33)$$

With a change of variables $\xi_2 = \Delta\xi + \xi_1$, $\eta_2 = \Delta\eta + \eta_1$, this transform can be written

$$\mathscr{I}_o'(\nu_1,\nu_2,\nu_3,\nu_4) = \iint_{-\infty}^{\infty} d\xi_1\, d\eta_1\, \mathbf{t}_o(\xi_1,\eta_1) e^{j2\pi[(\nu_1+\nu_3)\xi_1 + (\nu_2+\nu_4)\eta_1]}$$

$$\times \iint_{-\infty}^{\infty} d\Delta\xi\, d\Delta\eta\, \mathbf{J}_o(\Delta\xi,\Delta\eta)\mathbf{t}_o^*(\xi_1 + \Delta\xi, \eta_1 + \Delta\eta) e^{j2\pi[\nu_3\Delta\xi + \nu_4\Delta\eta]}.$$

$$(7.2\text{-}34)$$

The last double integral is recognized as the Fourier transform of the product of two functions and as such may be evaluated as the convolution of their individual transforms. With appropriate manipulations we find that the second integral can be expressed as

$$\iint_{-\infty}^{\infty} \mathscr{J}_o(p,q)\mathscr{T}_o^*(p - \nu_3, q - \nu_4) e^{-j2\pi[\xi_1(\nu_3-p) + \eta_1(\nu_4-q)]}\, dp\, dq,$$

$$(7.2\text{-}35)$$

where $\mathscr{T}_o$ is the two-dimensional Fourier transform of $\mathbf{t}_o$. Substitution of

METHODS FOR CALCULATING IMAGE INTENSITY

this result in Eq. (7.2-34) yields the final result

$$\mathcal{J}_o'(\nu_1,\nu_2,\nu_3,\nu_4) = \iint_{-\infty}^{\infty} \mathcal{T}_o(p+\nu_1,q+\nu_2)\mathcal{T}_o^*(p-\nu_3,q-\nu_4)\mathcal{J}_o(p,q)\,dp\,dq.$$

(7.2-36)

Although it is difficult to visualize the meaning of the expression above, some help is possible from examination of Fig. 7-10. Suppose that the amplitude transmittance $\mathbf{t}_o$ of the object is a *bandlimited* function, that is, its frequency spectrum is nonzero only within a circle of radius ρ_o in the two-dimensional frequency domain. In addition, when the illumination arises from a large incoherent source, the arguments leading to Eq. (7.2-17) imply that the function $\mathcal{J}_o(p,q)$ is simply a scaled version of the squared modulus of the pupil function of the condenser (or illumination) optics,

$$\mathcal{J}_o(p,q) = C|\mathbf{P}_c(-\bar{\lambda}z_2 p, -\bar{\lambda}z_2 q)|^2,$$

(7.2-37)

where C is a constant and z_2 is the distance from the condenser lens to the object. Thus $\mathcal{J}_o$ is also nonzero over only a finite range of its arguments. As illustrated in Fig. 7-10, the spectrum $\mathcal{J}_o'$ of the transmitted mutual intensity is found by integrating the product of the three partially overlapping functions $\mathcal{J}_o(p,q)$, $\mathcal{T}_o(p+\nu_1, q+\nu_2)$, and $\mathcal{T}_o^*(p-\nu_3, q-\nu_4)$. Clearly,

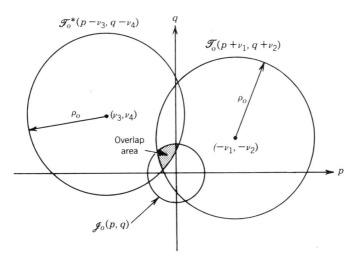

Figure 7-10. Region of overlap.

as the frequencies ν_1, ν_2, ν_3, and ν_4 grow larger, the degree of overlap decreases, and the four-dimensional spectrum $\mathscr{J}'_o$ drops in value.

By combining our previous results, in particular Eqs. (7.2-28) and (7.2-36), it is now possible to express the four-dimensional spectrum $\mathscr{J}_i$ of the image mutual intensity in terms of two-dimensional spectra of the various other quantities involved. The relationship is given by

$$\mathscr{J}_i(\nu_1, \nu_2, \nu_3, \nu_4) = \mathscr{K}(\nu_1, \nu_2)\mathscr{K}^*(-\nu_3, -\nu_4)$$

$$\times \iint_{-\infty}^{\infty} \mathscr{T}_o(p + \nu_1, q + \nu_2)\mathscr{T}_o^*(p - \nu_3, q - \nu_4)\mathscr{J}_o(p, q)\,dp\,dq.$$

(7.2-38)

The reader is reminded that $\mathscr{K}$ can be expressed in terms of the pupil function of the imaging lens, and, under usual circumstances (large incoherent source), $\mathscr{J}_0$ can be expressed in terms of the pupil function of the condenser lens.

The theory developed in the preceding parts of this section has established a framework within which it is possible to calculate the four-dimensional spectrum of the image mutual intensity. More commonly it is of interest to know the image intensity $I_i(u, v)$ or, alternatively, its two-dimensional Fourier spectrum $\mathscr{J}_i(\nu_U, \nu_V)$. Therefore, we now explore the application of this theory to the problem finding these two quantities.

The mutual intensity $\mathbf{J}_i$ can, of course, be found by taking an inverse Fourier transform of the spectrum $\mathscr{J}_i$, which in turn we can calculate by the formalism developed above. It is a straightforward exercise to show that, if we set $u_1 = u_2 = u$ and $v_1 = v_2 = v$ in this inverse transform relationship, we obtain an expression for $I_i(u, v)$ in terms of $\mathscr{J}_i(\nu_1, \nu_2, \nu_3, \nu_4)$ as follows (see Problem 7-3):

$$I_i(u, v) = \iiiint_{-\infty}^{\infty} \mathscr{J}_i(\nu_1, \nu_2, \nu_3, \nu_4)$$

$$\times \exp\{-j2\pi[u(\nu_1 + \nu_3) + v(\nu_2 + \nu_4)]\}\,d\nu_1\,d\nu_2\,d\nu_3\,d\nu_4.$$

(7.2-39)

Having found $\mathscr{J}_i$ by the methods described earlier, we can then calculate the image intensity using Eq. (7.2-39).

In some cases we may prefer to know the two-dimensional Fourier spectrum of the image intensity, a quantity defined by

$$\mathscr{S}_i(\nu_U, \nu_V) \triangleq \iint_{-\infty}^{\infty} I_i(u,v) \exp\{ j2\pi(\nu_U u + \nu_V v)\} \, du\, dv. \quad (7.2\text{-}40)$$

We could, of course, simply apply this transform to I_i (which we know how to calculate from the earlier discussion). It is possible to relate $\mathscr{S}_i$ directly to $\mathscr{J}_i$, however, thus eliminating several steps of integral transforms. Again, a straightforward exercise in manipulating Fourier integrals leads one from Eq. (7.2-39) to the relationship (see Problem 7-4)

$$\mathscr{S}_i(\nu_U, \nu_V) = \iint_{-\infty}^{\infty} \mathscr{J}_i(\nu_1, \nu_2, \nu_U - \nu_1, \nu_V - \nu_2) \, d\nu_1 d\nu_2. \quad (7.2\text{-}41)$$

Now using Eq. (7.2-38) in (7.2-41), and changing variables of integration to $z_1 = p + \nu_1$, $z_2 = q + \nu_2$, we obtain the following expression for the spectrum of the image intensity:

$$\mathscr{S}_i(\nu_U, \nu_V) = \iint_{-\infty}^{\infty} dz_1 dz_2 \, \mathscr{T}_o(z_1, z_2) \mathscr{T}_o^*(z_1 - \nu_U, z_2 - \nu_V)$$

$$\times \left[\iint_{-\infty}^{\infty} dp\, dq \, \mathscr{K}(z_1 - p, z_2 - q) \mathscr{K}^*(z_1 - p - \nu_U, z_2 - q - \nu_V) \mathscr{J}_o(p,q) \right]. \quad (7.2\text{-}42)$$

Note that the quantity in square brackets is totally independent of the object and is a complete description of the effects of the optical system from source to image plane. This quantity is often referred to as the *transmission cross-coefficient* (Ref. 7-2, p. 530; Ref. 7-1, p. 190). Its evaluation requires integration over three partially overlapping functions, much analogous to the evaluation illustrated in Fig. 7-10.

In closing this section, it must be said that the "linear systems" treatment of partially coherent imaging is not a simple theory, at least by comparison with the more common linear systems approaches to fully coherent and incoherent imaging. Nonetheless, for those who are highly familiar with two-dimensional Fourier transform theory, and can extrapolate this experience to the case of four-dimensional systems, the formalism does provide a viable approach to the analysis of partially coherent imaging systems.

7.2.4 The Incoherent and Coherent Limits

In this section we investigate the properties of the image, as predicted by the previous theories, in the limiting cases of perfect incoherence and perfect coherence of the object illumination. We find the form of I_i in these two cases using the method presented in Section 7.2.2. In addition, for illustrative purposes, we find corresponding expressions for the two-dimensional spectrum $\mathscr{I}_i$ of the image intensity using the results given in Section 7.2.3. Finally, we discuss physical criteria for determining whether a given system may be regarded as being fully incoherent or fully coherent from a practical point of view.

The case of total incoherence of the object illumination may be represented mathematically by a mutual intensity incident on the object of the form

$$\mathbf{J}_o(\Delta\xi, \Delta\eta) = \kappa I_o \delta(\Delta\xi, \Delta\eta), \tag{7.2-43}$$

where κ is a constant, I_o is a constant intensity, and δ is a two-dimensional Dirac δ function. Substituting this expression in Eq. (7.2-11), we find that, under the assumption of a space-invariant system,

$$I_i(u,v) = \kappa I_o \iint_{-\infty}^{\infty} |\mathbf{K}(u-\xi, v-\eta)|^2 |\mathbf{t}_o(\xi,\eta)|^2 \, d\xi \, d\eta. \tag{7.2-44}$$

Thus the image intensity is found to be (up to a constant multiplier) the convolution of the object intensity transmittance $|\mathbf{t}_o|^2$ with an intensity spread function $|\mathbf{K}|^2$. Clearly, an incoherent system is linear in *intensity*.

To find the Fourier spectrum of image intensity, we could simply Fourier transform the result above. For illustrative purposes, however, we prefer to find $\mathscr{I}_i$ with the help of Eq. (7.2-42). Fourier transformation of the mutual intensity $\mathbf{J}_o$ in Eq. (7.2-43) yields

$$\mathscr{I}_o(p,q) = \kappa I_o, \tag{7.2-45}$$

in which case the spectrum $\mathscr{I}_i$ becomes

$$\mathscr{I}_i(\nu_U, \nu_V) = \kappa I_o \left[\iint_{-\infty}^{\infty} \mathscr{T}_o(z_1, z_2) \mathscr{T}_o^*(z_1 - \nu_U, z_2 - \nu_V) \, dz_1 \, dz_2 \right]$$

$$\times \left[\iint_{-\infty}^{\infty} \mathscr{K}(p',q') \mathscr{K}^*(p' - \nu_U, q' - \nu_V) \, dp' \, dq' \right], \tag{7.2-46}$$

METHODS FOR CALCULATING IMAGE INTENSITY   321

where the changes of variable $p' = z_1 - p$, $q' = z_2 - q$ have been made in the second integral. The first bracketed integral represents the Fourier transform of the object intensity transmittance. The second bracketed integral represents the Fourier transform of the intensity spread function.

The transfer of frequency components of object intensity to image intensity is governed by the second bracketed expression in Eq. (7.2-46). By convention, we represent this transfer factor in normalized form,

$$\mathcal{H}(\nu_U, \nu_V) \triangleq \frac{\iint_{-\infty}^{\infty} \mathcal{K}(p', q') \mathcal{K}^*(p' - \nu_U, q' - \nu_V) \, dp' \, dq'}{\iint_{-\infty}^{\infty} |\mathcal{K}(p', q')|^2 \, dp' \, dq'},$$

(7.2-47)

which is known as the *optical transfer function*, or for brevity, the *OTF*. This quantity, originally introduced by Duffieux (Ref. 7-13), represents the complex factor applied by the imaging system to the complex exponential component of object intensity with frequency (ν_U, ν_V), relative to the factor applied to the zero-frequency component. As indicated by Eq. (7.2-30), the function $\mathcal{K}$ is proportional to the complex pupil function **P** of the imaging optics. Thus the OTF may also be expressed as a normalized autocorrelation of the complex pupil function,

$$\mathcal{H}(\nu_U, \nu_V) = \frac{\iint_{-\infty}^{\infty} \mathbf{P}(p', q') \mathbf{P}^*(p' - \bar{\lambda} z_i \nu_U, q' - \bar{\lambda} z_i \nu_V) \, dp' \, dq'}{\iint_{-\infty}^{\infty} |\mathbf{P}(p', q')|^2 \, dp' \, dq'},$$

(7.2-48)

a well-known and important result.

Turning next to the case of fully coherent illumination, we take the mutual intensity of the object illumination to be

$$\mathbf{J}_o(\Delta \xi, \Delta \eta) = I_o,$$ (7.2-49)

thereby assuming plane wave illumination normal to the object. Substituting this form in Eq. (7.2-11), and again assuming a space-invariant system, we obtain

$$I_i(u, v) = I_o \left| \iint_{-\infty}^{\infty} \mathbf{K}(u - \xi, v - \eta) \mathbf{t}_o(\xi, \eta) \, d\xi \, d\eta \right|^2.$$ (7.2-50)

If we now define the time-invariant phasor image distribution by

$$\mathbf{A}_i(u,v) \triangleq \sqrt{I_o} \iint_{-\infty}^{\infty} \mathbf{K}(u-\xi, v-\eta) \mathbf{t}_o(\xi,\eta) \, d\xi \, d\eta, \quad (7.2\text{-}51)$$

we see that the amplitude distribution $\mathbf{A}_i$ is proportional to a convolution of the amplitude spread function $\mathbf{K}$ with the amplitude transmittance $\mathbf{t}_o$ of the object. Clearly, the fully coherent system is linear in *complex amplitude*. See Problem 7-9 for a generalization of this result.

To find the Fourier spectrum of image intensity, we first note that a Fourier transform of Eq. (7.2-49) yields

$$\mathscr{I}_o(p,q) = I_o \delta(p,q). \quad (7.2\text{-}52)$$

Substitution of this form in (7.2-42) yields a spectrum of image intensity given by

$$\mathscr{I}_i(\nu_U, \nu_V) = I_o \iint_{-\infty}^{\infty} \mathscr{T}_o(z_1, z_2) \mathscr{K}(z_1, z_2) \mathscr{T}_o^*(z_1 - \nu_U, z_2 - \nu_V)$$

$$\times \mathscr{K}^*(z_1 - \nu_U, z_2 - \nu_V) \, dz_1 \, dz_2. \quad (7.2\text{-}53)$$

Consistent with the autocorrelation theorem of Fourier analysis, we regard (7.2-53) as the autocorrelation function of the spectrum

$$\mathscr{A}_i(\nu_U, \nu_V) = \sqrt{I_o} \, \mathscr{K}(\nu_U, \nu_V) \mathscr{T}_o(\nu_U, \nu_V). \quad (7.2\text{-}54)$$

Comparison with Eq. (7.2-51) demonstrates that $\mathscr{A}_i$ is the Fourier transform of the image amplitude distribution $\mathbf{A}_i$.

Aside from a multiplicative constant, the transfer function of the coherent imaging system is clearly given by

$$\mathbf{H}(\nu_U, \nu_V) = \mathscr{K}(\nu_U, \nu_V) = \mathbf{P}(\bar{\lambda} z_i \nu_U, \bar{\lambda} z_i \nu_V). \quad (7.2\text{-}55)$$

This transfer function is referred to as the *amplitude transfer function* or the *coherent transfer function*.

Finally, we turn to the question of when it is valid, from a practical point of view, to assume that an imaging system behaves essentially as predicted by the idealized incoherent or coherent theories presented previously. Clues to the answers are provided by Eq. (7.2-11), which we rewrite here for the case of a space-invariant system and an incoherent source,

$$I_i(u,v) = \iiiint_{-\infty}^{\infty} \mathbf{K}(u - \xi, v - \eta)\mathbf{K}^*(u - \xi - \Delta\xi, v - \eta - \Delta\eta)$$

$$\times \mathbf{t}_o(\xi, \eta)\mathbf{t}_o^*(\xi + \Delta\xi, \eta + \Delta\eta)\mathbf{J}_o(\Delta\xi, \Delta\eta)\,d\xi\,d\eta\,d\Delta\xi\,d\Delta\eta.$$

(7.2-56)

For the case of coherent illumination, we require that $\mathbf{J}_o(\Delta\xi, \Delta\eta)$ be essentially constant over the entire range of $(\Delta\xi, \Delta\eta)$ for which the integrand of Eq. (7.2-56) has value significantly greater than zero. Clearly, the integrand will vanish if $\Delta\xi$ or $\Delta\eta$ is greater than the width of the object, for then

$$\mathbf{t}_o(\xi, \eta)\mathbf{t}_o^*(\xi + \Delta\xi, \eta + \Delta\eta) = 0. \qquad (7.2\text{-}57)$$

In virtually all cases of interest, however, the amplitude spread function $\mathbf{K}$ has a width, referred to the (ξ, η) plane, that is much narrower than the width of the object. Hence when $\Delta\xi$ or $\Delta\eta$ exceeds the width of $\mathbf{K}$, the integrand becomes very small, as a result of

$$\mathbf{K}(u - \xi, v - \eta)\mathbf{K}^*(u - \xi - \Delta\xi, v - \eta - \Delta\eta) \approx 0. \qquad (7.2\text{-}58)$$

We conclude that the system will behave approximately as a fully coherent system provided the incoherent source of illumination is so small as to produce a coherence area on the object that considerably exceeds the area covered by the amplitude spread function, that area being referred to the object plane. Alternatively, but equivalently, we require that the angular subtence of the source, as seen from the object, must be considerably smaller than the angular subtense of the entrance pupil of the imaging optics.

For the incoherent case, we require that $\mathbf{J}_o(\Delta\xi, \Delta\eta)$ be nonzero only when $(\Delta\xi, \Delta\eta)$ are so small that

$$\mathbf{K}(u - \xi, v - \eta)\mathbf{K}^*(u - \xi - \Delta\xi, v - \eta - \Delta\eta)\mathbf{t}_o(\xi, \eta)\mathbf{t}_o^*(\xi + \Delta\xi, \eta + \Delta\eta)$$

$$\cong |\mathbf{K}(u - \xi, v - \eta)|^2|\mathbf{t}_o(\xi, \eta)|^2.$$

(7.2-59)

Clearly, a necessary condition is that the coherence area of the object illumination be smaller than *both* the area covered by the amplitude spread function and the area of the smallest structure in the object amplitude transmittance t_o. Stated in alternate but equivalent terms, the angular subtense θ_s of the incoherent source, as viewed from the object, must be considerably larger than both the angular subtense θ_p of the entrance pupil of the imaging optics and the angular subtense θ_o of the cone of angles that would be generated by the object under normally incident plane wave illumination (i.e., the subtense of the angular spectrum of the object). Thus we require

$$\theta_s > \theta_p \quad \text{and} \quad \theta_s > \theta_o. \tag{7.2-60}$$

Although the preceding argument provides *necessary* conditions for incoherent imaging, a sufficient condition requires a bit more discussion. Since the terms involving $\mathbf{K}$ and t_o are multiplied together in Eq. (7.2-59), the angular spectra of $\mathbf{K}$ and t_o must actually be *convolved* together in order to determine sufficient conditions on the angular subtenses involved. When this is done, a single necessary and sufficient condition is arrived at:

$$\theta_s \geq \theta_o + \theta_p. \tag{7.2-61}$$

Stated in physical terms, it is necessary that the angular subtense of the source, when centered about the highest diffraction angle introduced by the object, at least fill the angular subtense of the imaging optics. See Problem 7-10 for a consideration of this idea in a specific case.

In closing this section, we mention that a particularly important problem, and one that has been much treated in the literature, is the question of when a microdensitometer (an instrument used for measuring the fine-scale density structure of photographic transparencies) can be treated as an incoherent imaging system. The reader is referred to Refs. 7-14 and 7-15 for detailed discussions of this question.

7.3 SOME EXAMPLES

Several theoretical approaches to the evaluation of image intensity were outlined in Section 7.2. We turn now to the application of these methods to some specific examples.

7.3.1 The Image of Two Closely Spaced Points

Consider an object consisting of two small pinholes. We suppose the object to be transilluminated and to have an amplitude transmittance well approximated by

$$t_o(\xi, \eta) = a\delta\left(\xi - \frac{S}{2}, \eta\right) + a\delta\left(\xi + \frac{S}{2}, \eta\right), \tag{7.3-1}$$

SOME EXAMPLES

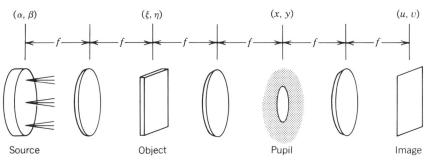

Figure 7-11. Telecentric imaging system. All lenses have the same focal length f.

where a is a constant and S represents the separation of the two transmitting points.

This object is assumed to be illuminated with partially coherent light having mutual intensity $\mathbf{J}_o(\Delta\xi, \Delta\eta)$ and to be imaged by the space-invariant system illustrated in Fig. 7-11. Substituting (7.3-1) in (7.2-56), we obtain (after some manipulation)

$$I_i(u,v) = I_1 \left[\left| \mathbf{K}\left(u - \frac{S}{2}, v\right) \right|^2 + \left| \mathbf{K}\left(u + \frac{S}{2}, v\right) \right|^2 \right.$$

$$\left. + 2\,\mathrm{Re}\left\{ \mu \mathbf{K}\left(u - \frac{S}{2}, v\right) \mathbf{K}^*\left(u + \frac{S}{2}, v\right) \right\} \right], \quad (7.3\text{-}2)$$

where

$$I_1 = a^2 \mathbf{J}_o(0,0),$$

$$\mu = \frac{a^2}{I_1} \mathbf{J}_o(S,0). \quad (7.3\text{-}3)$$

In arriving at this result we have used the fact that the source intensity distribution is real valued; therefore, its Fourier transform satisfies

$$\mathbf{J}_o(-S, 0) = \mathbf{J}_o^*(S, 0). \quad (7.3\text{-}4)$$

The amplitude spread function $\mathbf{K}$ is related to the complex pupil function $\mathbf{P}$ by

$$\mathbf{K}(u,v) = \frac{1}{\lambda f} \iint_{-\infty}^{\infty} \mathbf{P}(x, y) \exp\left\{ -j\frac{2\pi}{\lambda f}(ux + vy) \right\} dx\,dy. \quad (7.3\text{-}5)$$

If the pupil function **P** has hermitian symmetry [i.e., $\mathbf{P}(-x, -y) = \mathbf{P}^*(x, y)$], such as it does for an aberration-free, circular pupil, then $\mathbf{K}(u, v)$ is entirely real ($\mathbf{K} = K$). The image intensity then takes the form

$$I_i(u,v) = I_1 \left[K^2\left(u - \frac{S}{2}, v\right) + K^2\left(u + \frac{S}{2}, v\right) \right.$$
$$\left. + 2\mu K\left(u - \frac{S}{2}, v\right) K\left(u + \frac{S}{2}, v\right) \cos\phi \right], \quad (7.3\text{-}6)$$

where $\mu = |\mu|$ and $\phi = \arg\{\mu\}$. In the specific case of a circular pupil, the amplitude spread function takes the form

$$K(u,v) = K(\rho) = \frac{\pi r_p^2}{\lambda f} \left[2 \frac{J_1\left(\frac{2\pi r_p \rho}{\lambda f}\right)}{\frac{2\pi r_p \rho}{\lambda f}} \right] \quad (7.3\text{-}7)$$

where $\rho = \sqrt{u^2 + v^2}$ and r_p is the radius of the exit pupil.

Grimes and Thompson (Ref. 7-16) have calculated the distributions of image intensity for various separations of the two points and various complex coherence factors. In Fig. 7-12 such distributions are shown for the case of an aberration-free circular pupil, a complex coherence factor ranging from 1.0 to -1.0 in steps of 0.2, and a separation of the two points given by

$$S = 0.6366 \frac{\lambda f}{r_p}, \quad (7.3\text{-}8)$$

which is just slightly greater than the so-called Rayleigh limit of resolution ($S = 0.6098 \lambda f / r_p$).

Note that if $\mu = -1.0$, the two points are illuminated coherently but with a 180° phase difference, and the intensity at the midpoint between them always falls to zero, regardless of their separation. If the illumination of the object is provided by an incoherent source through a condenser system, then there is a particular effective source size that yields the most negative possible value of μ and hence the greatest possible dip of intensity in the image plane. The optimum effective source size depends on the separation of the two points and the intensity distribution associated with the effective source. (Some source distributions are incapable of producing negative values of μ.) These points are pursued further in Problems 7-5 and 7-6.

We note in closing that the question of when two closely spaced point sources are barely resolved is a complex one and lends itself to a variety of rather subjective answers. According to the so-called Rayleigh resolution criterion, two equally bright points are barely resolved when the first zero of

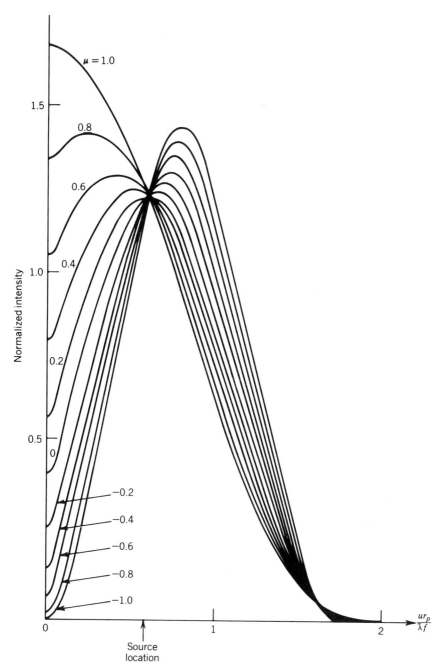

Figure 7-12. One side of the intensity distribution in the image of two point sources, with the complex degree of coherence as a parameter. The intensity distribution is symmetrical about $x = 0$, and I is normalized. The separation S of the two point sources is $0.6366\,\bar{\lambda}f/r_p$. The normalization is such that $K(0,0) = 1$

the Airy pattern of the image of one point exactly coincides with the central maximum of the Airy pattern of the image of the second point. Under such a condition the intensity at the midpoint of the image intensity distribution is 26.5% smaller than the intensity at either peak. An alternative definition is the so-called Sparrow criterion, which states that two point sources are just resolved if the second derivative of the image intensity pattern vanishes at the point midway between the Gaussian image points. In fact, the ability to resolve two point sources depends fundamentally on the signal-to-noise ratio associated with the detected image intensity pattern, and for this reason criteria that do not take account of noise are subjective. Nonetheless, such criteria may yield useful rules of thumb for engineering practice. For further discussion of these questions, the reader may wish to consult Ref. 7-1.

7.3.2 The Image of a Sinusoidal Amplitude Object

Consider next an object that has an amplitude transmittance of the form

$$\mathbf{t}_o(\xi, \eta) = \tfrac{1}{2}[1 + \cos 2\pi\nu_0 \xi]. \tag{7.3-9}$$

Such an object is often referred to as a *sinusoidal amplitude grating*. Note that the intensity transmittance associated with such an object is

$$|\mathbf{t}_o|^2 = \tfrac{1}{4} + \tfrac{1}{2}\cos 2\pi\nu_0\xi + \tfrac{1}{4}\cos^2 2\pi\nu_0\xi$$

$$= \tfrac{3}{8} + \tfrac{1}{2}\cos 2\pi\nu_0\xi + \tfrac{1}{8}\cos 4\pi\nu_0\xi. \tag{7.3-10}$$

We wish to compare the intensity distribution that appears in the image of such an object with the intensity transmittance above, taking into account the partial coherence of the object illumination (cf. Ref. 7-17).

For this particular problem, the frequency-domain method of analysis is the most convenient approach. We begin with the inverse Fourier transform of Eq. (7.2-42), writing the image intensity as

$$I_i(u, v) = \iint_{-\infty}^{\infty} d\nu_U d\nu_V \exp\{-j2\pi(u\nu_U + v\nu_V)\}$$

$$\times \iint_{-\infty}^{\infty} dz_1 dz_2 \mathcal{T}_o(z_1, z_2) \mathcal{T}_o^*(z_1 - \nu_U, z_2 - \nu_V)$$

$$\times \iint_{-\infty}^{\infty} dp\, dq\, \mathcal{K}(z_1 - p, z_2 - q) \mathcal{K}^*(z_1 - p - \nu_U, z_2 - q - \nu_V) \mathcal{J}_o(p, q).$$

$$\tag{7.3-11}$$

SOME EXAMPLES

For the object of Eq. (7.3-9), we have

$$\mathcal{T}_o(z_1, z_2) = \tfrac{1}{2}\delta(z_1, z_2) + \tfrac{1}{4}\delta(z_1 - \nu_0, z_2) + \tfrac{1}{4}\delta(z_1 + \nu_0, z_2)$$

(7.3-12)

as the spectrum of the amplitude transmittance. The δ-function expressions for $\mathcal{T}_o(z_1, z_2)$ and $\mathcal{T}_o^*(z_1 - \nu_U, z_2 - \nu_V)$ are substituted in Eq. (7.3-11), and the orders of integration are interchanged. Integrating first with respect to (ν_U, ν_V), then with respect to (z_1, z_2), and finally with respect to (p, q), after much use of the sifting property of δ functions, we obtain

$$I(u, v) = A + B\cos 2\pi\nu_0 u + C\cos 4\pi\nu_0 u, \qquad (7.3\text{-}13)$$

where for real-valued $\mathcal{K} = \mathcal{K}$ and $\mathcal{J}_o = \mathcal{J}_o$,

$$A = \tfrac{1}{4}\iint_{-\infty}^{\infty}\mathcal{J}_o(p, q)\big[\mathcal{K}^2(-p, -q) + \tfrac{1}{4}\mathcal{K}^2(\nu_0 - p, -q)$$

$$+ \tfrac{1}{4}\mathcal{K}^2(-\nu_0 - p, -q)\big]\,dp\,dq$$

$$B = \tfrac{1}{4}\iint_{-\infty}^{\infty}\mathcal{J}_o(p, q)\big[\mathcal{K}(-p, -q)\mathcal{K}(\nu_0 - p, -q)$$

$$+ \mathcal{K}(-p, -q)\mathcal{K}(-\nu_0 - p, -q)\big]\,dp\,dq$$

$$C = \tfrac{1}{8}\iint_{-\infty}^{\infty}\mathcal{J}_o(p, q)\mathcal{K}(-\nu_0 - p, -q)\mathcal{K}(\nu_0 - p, -q)\,dp\,dq.$$

(7.3-14)

Although the partially coherent imaging system is nonlinear, it is sometimes useful to consider an *apparent transfer function*, defined by

$$\mathcal{H}_A(\nu_U, \nu_V) = \frac{\text{modulation of frequency component } (\nu_U, \nu_V) \text{ at output}}{\text{modulation of frequency component } (\nu_U, \nu_V) \text{ at input}}$$

For the particular object of concern here, the input modulation at ν_0 is [from Eq. (7.3-10)] $m_i = 4/3$, whereas the output modulation is $m_o = B/A$.

Hence

$$\mathcal{H}_A(\nu_0) = \frac{3B}{4A}.$$

For the frequency component at $2\nu_0$, the corresponding result is

$$\mathcal{H}_A(2\nu_0) = \frac{3C}{A}.$$

These quantities have been calculated by Becherer and Parrent (Ref. 7-17),

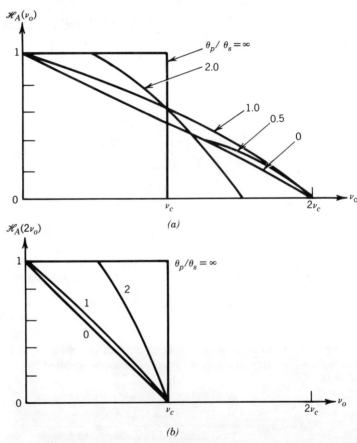

Figure 7-13. Apparent transfer functions for frequency components at (a) ν_0 and (b) $2\nu_0$, plotted against the frequency ν_0 of the amplitude grating, with the coherence of the illumination as a parameter; ν_c represents the cutoff fequency of the amplitude transfer function $\nu_c = \theta_p/\bar{\lambda}$. (Courtesy of R. J. Becherer and the Optical Society of America, Ref. 7-17.)

as a function of ν_0, for the case of a slit incoherent source and a slit pupil function. If θ_s represents the angle subtended by the source, and θ_p the angle subtended by the imaging pupil, both as viewed from the object, then $\mathcal{H}_A(\nu_0)$ and $\mathcal{H}_A(2\nu_0)$ are found to be functions of θ_p/θ_s (as well as ν_0). This dependence is an indication of the dependence of system performance on the coherence of the object illumination. Figures 7-13a and 7-13b show the apparent transfer functions for frequencies ν_0 and $2\nu_0$ for various values of θ_p/θ_s. It should be noted that $\theta_p/\theta_s \to 0$ implies the illumination is approaching total incoherence, whereas $\theta_p/\theta_s \to \infty$ implies the approach is to perfect coherence.

7.4 IMAGE FORMATION AS AN INTERFEROMETRIC PROCESS

Considerable insight into the character of images formed under various conditions of illumination can be gained by adopting the point of view that image formation is an *interferometric* process. Such an approach also suggests various novel means for gathering image data. The interferometric approach has been used by radio astronomers for many years, since the highest resolution images of radio sources must in most cases be gathered with interferometers rather than with continuous reflecting antennas (Refs. 7-18 and 7-19). The value of the interferometric viewpoint in optics was pointed out at an early date by G. L. Rogers (Ref. 7-20) for the case of a fully incoherent object.

7.4.1 An Imaging System as an Interferometer

We are familiar with the idea that, in Young's interference experiment, the light passing through two small pinholes can ultimately interfere to produce a sinusoidal fringe with a spatial frequency that is dependent on the separation of the pinholes. Now the exit pupil of an imaging system may be regarded as consisting of a multitude of (fictitious) pinholes, side by side, and the observed image intensity distribution as being built up of a multitude of sinusoidal fringes generated by all possible pairs of such pinholes.

A frequency component of image intensity with spatial frequencies (ν_U, ν_V) must arise from at least one pair of pinholes in the exit pupil with separations

$$\Delta x = \bar{\lambda} z_i \nu_U$$
$$\Delta y = \bar{\lambda} z_i \nu_V. \quad (7.4\text{-}1)$$

The amplitude and phase of the sinusoidal fringe contributed by a pair with coordinates (x_1, y_1) and (x_2, y_2) are determined by the amplitude and phase of the mutual intensity $\mathbf{J}'_p(x_1, y_1; x_2, y_2)$ transmitted by the exit pupil. Since many pinhole pairs with the separations $(\Delta x, \Delta y)$ exist in the exit pupil, the total amplitude and phase of the spectral component $\mathcal{I}_i(\nu_U, \nu_V)$ of image intensity must be calculated by adding all fringes with frequencies (ν_U, ν_V), taking proper account of both their amplitudes and their spatial phases.

A mathematical version of this conclusion can be found by beginning with Eq. (7.1-47) relating image intensity with the mutual intensity in the exit pupil. Since our main interest is in the Fourier spectrum $\mathcal{I}_i(\nu_U, \nu_V)$ of the image intensity, we Fourier transform (7.1-47) with respect to the variables (u, v). Interchanging orders of integration, we find

$$\mathcal{I}_i(\nu_U, \nu_V) = \frac{1}{(\bar{\lambda} z_i)^2} \iiiint_{-\infty}^{\infty} dx_1\, dy_1\, dx_2\, dy_2\, \mathbf{J}'_p(x_1, y_1; x_2, y_2)$$

$$\times \iint_{-\infty}^{\infty} du\, dv \exp\left\{ j2\pi \left[\left(\nu_U + \frac{x_2 - x_1}{\bar{\lambda} z_i} \right) u + \left(\nu_V + \frac{y_2 - y_1}{\bar{\lambda} z_i} \right) v \right] \right\}.$$

(7.4-2)

The last double integral is simply equal to $\delta\{\nu_U + [(x_2 - x_1)/\bar{\lambda} z_i], \nu_V + [(y_2 - y_1)/\bar{\lambda} z_i]\}$. Integrating next with respect to (x_2, y_2), and using the sifting property of the δ function, we find

$$\mathcal{I}_i(\nu_U, \nu_V) = \iint_{-\infty}^{\infty} \mathbf{J}'_p(x_1, y_1; x_1 - \bar{\lambda} z_i \nu_U, y_1 - \bar{\lambda} z_i \nu_V)\, dx_1\, dy_1.$$

(7.4-3)

Thus to find the complex value of the image spectrum at (ν_U, ν_V), we integrate (or add) all possible values of the mutual intensity, with fixed separation $(\bar{\lambda} z_i \nu_U, \bar{\lambda} z_i \nu_V)$, as the free variables (x_1, y_1) run over the pupil plane. This result is entirely equivalent to the idea of adding all Young's fringe patterns generated by pinhole pairs with spacings $(\Delta x = \bar{\lambda} z_i \nu_U, \Delta y = \bar{\lambda} z_i \nu_V)$.

Of course, in practice the exit pupil has finite physical extent. The mutual intensity $\mathbf{J}'_p$ leaving the exit pupil can be expressed in terms of the mutual intensity $\mathbf{J}_p$ incident on the exit pupil by

$$\mathbf{J}'_p(x_1, y_1; x_2, y_2) = \mathbf{P}(x_1, y_1)\mathbf{P}^*(x_2, y_2)\mathbf{J}_p(x_1, y_1; x_2, y_2), \quad (7.4\text{-}4)$$

IMAGE FORMATION AS AN INTERFEROMETRIC PROCESS   333

where the complex pupil function **P** is determined by the bounds of the exit pupil, any apodization that might exist, and phase errors or aberrations associated with the system. The finite extent of the exit pupil limits the area over which (x_1, y_1) can run for any fixed $(\bar{\lambda} z_i \nu_U, \bar{\lambda} z_i \nu_V)$, a fact that becomes more evident if (7.4-4) is substituted in (7.4-3), yielding

$$\mathscr{S}_i(\nu_U, \nu_V) = \iint_{-\infty}^{\infty} \mathbf{P}(x_1, y_1) \mathbf{P}^*(x_1 - \bar{\lambda} z_i \nu_U, y_1 - \bar{\lambda} z_i \nu_V)$$

$$\times \mathbf{J}_p(x_1, y_1; x_1 - \bar{\lambda} z_i \nu_U, y_1 - \bar{\lambda} z_i \nu_V) \, dx_1 \, dy_1. \quad (7.4\text{-}5)$$

For the case of an unobstructed, circular exit pupil of radius r_p, the region of integration in (x_1, y_1) is the shaded area in Fig. 7-14. Thus the fixed separation shown in Fig. 7-14 may be regarded as being slid within the shaded area to all possible locations that fully contain it, but with the relative orientation of the two pinholes (i.e., their *vector spacing*) remaining unchanged.

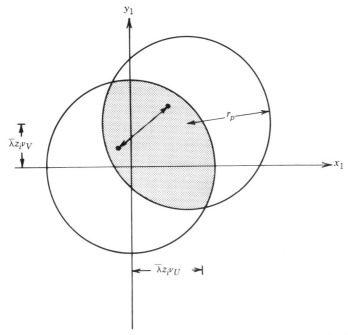

Figure 7-14. Region of integration for calculating the spectrum of image intensity [frequency (ν_U, ν_V)].

When the original object is illuminated incoherently, the calculation of $\mathcal{I}_i(\nu_U, \nu_V)$ becomes especially simple. By the Van Cittert–Zernike theorem, the mutual intensity distribution incident on a reference sphere of radius z_o in the entrance pupil (see Fig. 7-15) is a function only of the separations Δx and Δy in that pupil. (The quadratic phase factors associated with the Van Cittert–Zernike theorem vanish due to the use of this reference sphere.) The exit pupil is simply the image of the entrance pupil. Hence within the confines of the pupil, the mutual intensity $\mathbf{J}_p$ incident on the sphere of radius z_i in the exit pupil is identical (up to a possible magnification) with the mutual intensity incident on the reference sphere in the entrance pupil; therefore, $\mathbf{J}_p$ is a function only of the coordinates differences $(\Delta x, \Delta y)$ and is independent of (x_1, y_1). In this case Eq. (7.4-5) for $\mathcal{I}_i(\nu_U, \nu_V)$ becomes

$$\mathcal{I}_i(\nu_U, \nu_V) = \mathbf{J}_p(\bar{\lambda} z_i \nu_U, \bar{\lambda} z_i \nu_V)$$

$$\times \iint_{-\infty}^{\infty} \mathbf{P}(x_1, y_1)\mathbf{P}^*(x_1 - \bar{\lambda} z_i \nu_U, y_1 - \bar{\lambda} z_i \nu_V) \, dx_1 \, dy_1.$$

(7.4-6)

This result implies that, for an incoherent object and an optical system that is free from aberrations (i.e., for a pupil function that is real and nonnegative), as a pinhole pair with a fixed separation is slid around the exit pupil, the phases of Young's fringe contributions will be identical for all locations. Hence these "elementary fringes" add constructively, producing a fringe with increased amplitude. The weighting factor applied by the optical system to the frequency component at (ν_U, ν_V) is simply the autocorrelation

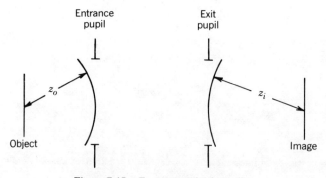

Figure 7-15. Entrance and exit pupils.

integral in Eq. (7.4-6), which (when there is no apodization) reduces to the area of overlap of two exit pupils displaced by $(\bar\lambda z_i\nu_U, \bar\lambda z_i\nu_V)$, as shown previously in Fig. 7-14. The optical transfer function of the system is this area of overlap normalized by the area of overlap when $\nu_U = \nu_V = 0$ (i.e., the area of the exit pupil).

If the object is again incoherent but the system has aberrations, then as the pinhole pair is slid around the exit pupil, the elementary fringes contributed by a single vector spacing will in general have different spatial phases for different locations of the pinhole pair. Since such fringes will not add constructively, the weighting factor applied by the optical system for that frequency is reduced in accord with the autocorrelation integral in Eq. (7.4-6). The optical transfer function in this case becomes

$$\mathcal{H}(\nu_U, \nu_V) = \frac{\displaystyle\iint_{-\infty}^{\infty} \mathbf{P}(x_1, y_1)\mathbf{P}^*(x_1 - \bar\lambda z_i\nu_U, y_1 - \bar\lambda z_i\nu_V)\, dx_1\, dy_1}{\displaystyle\iint_{-\infty}^{\infty} |\mathbf{P}(x_1, y_1)|^2\, dx_1\, dy_1}.$$

(7.4-7)

Finally, if the object illumination is partially coherent, the situation becomes more complicated. The mutual intensity $\mathbf{J}_p$ is now a function of *both* the location *and* the separation of the pinholes, and thus it no longer factors outside the integral of Eq. (7.4-5). In this case the amplitudes and phases of the elementary fringes can change as the pinhole pair is moved around the exit pupil, even if the system is free from aberrations. Hence it is not possible to identify a weighting factor associated with the system alone for any given spatial frequency (ν_U, ν_V). Nonetheless, Eq. (7.4-5) remains a very revealing result, for it does tell us explicitly how the frequency component at (ν_U, ν_V) is built up from elementary fringes, even if the amplitudes and phases of these fringes do depend in a rather complicated way on which parts of the pupil we are considering. The reader may wish to consult Ref. 7-21 for further discussion of the partially coherent case.

7.4.2 Gathering Image Information with Interferometers

For the discussions of this section, attention is restricted to fully incoherent objects. We have previously seen that for such objects and for an aberration-free optical system, a single pair of pinholes placed in the exit pupil and having vector separation $(\bar{\lambda}z_i\nu_U, \bar{\lambda}z_i\nu_V)$ will yield an image fringe having amplitude proportional to the modulus of the pupil mutual intensity function $\mathbf{J}_p$ and a spatial phase identical with the phase of $\mathbf{J}_p$. In turn, according to the Van Cittert–Zernike theorem, $\mathbf{J}_p$ is a scaled version of the two-dimensional Fourier transform of the object intensity distribution. Thus measurement of the parameters of this single fringe yields knowledge (up to a real proportionality constant) of the object spectrum at frequency (ν_U, ν_V). Different pinhole pairs with the same vector spacing yield identical fringes. Therefore, the *redundancy* of the optical system (i.e., the multitude of ways a single vector spacing is embraced by the pupil) serves to increase the signal-to-noise ratio of the measurement but does not in any other way contribute new information.

When the optical system contains aberrations, or when it is situated in an inhomogeneous medium that generates aberrations, the presence of redundancy can in fact sometimes be harmful. In this case Young's fringes with identical spatial frequency add with different spatial phases, reducing the contrast and also the accuracy with which fringe amplitude can be measured (Ref. 7-22).

In some cases it may be desired to extend the range of vector spacings observed by the system, but without building an optical system having a lens or a mirror with correspondingly large aperture. As we shall discuss in more detail, such concepts lead us into the realm of *aperture synthesis* and the use of interferometers to gather object information.

In some cases we may be satisfied to extract object information less complete than a detailed image. For an object that is known to be a uniform circular radiator, it may suffice for our purposes to determine its angular diameter. For an object known to consist of two point sources, we may be concerned primarily with their angular separation and relative intensities. In such cases, sufficient information may be provided by the modulus of the object spectrum, allowing us to ignore phase information.

The simplest kind of interferometer for use in spatial information extraction is the *Fizeau stellar interferometer* (Ref. 7-23) shown in Fig. 7-16. In astronomical measurement problems, for which this interferometer was introduced, the object lies at extremely large distances from the observer, and the image plane thus coincides with the rear focal plane of the reflecting or refracting telescope. To construct a Fizeau interferometer, a mask is placed in an image of the pupil of the telescope, effectively allowing only two small pencil beams, separated by an average spacing of $(\Delta x, \Delta y)$ on the primary collector, to interfere in the focal plane. The contrast or visibility of the fringes observed in the focal plane is determined by the modulus of the complex coherence factor of the light incident on the two effective pupil

IMAGE FORMATION AS AN INTERFEROMETRIC PROCESS   337

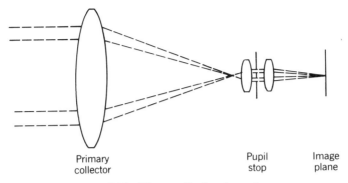

Figure 7-16. Fizeau stellar interferometer.

openings,

$$|\mu_p(\Delta x, \Delta y)| = \left|\frac{\mathbf{J}_p(\Delta x, \Delta y)}{\mathbf{J}_p(0,0)}\right|, \quad (7.4\text{-}8)$$

where $\mathbf{J}_p$ is the mutual intensity of the light incident on the aperture of the primary collector.

For a uniformly bright circular source of radius r_s at distance z, the complex coherence factor of the light incident on the telescope pupil is of the form

$$\mu_p(\Delta x, \Delta y) = 2\left[\frac{J_1\left(\frac{2\pi r_s}{\lambda z}\sqrt{(\Delta x)^2+(\Delta y)^2}\right)}{\frac{2\pi r_s}{\lambda z}\sqrt{(\Delta x)^2+(\Delta y)^2}}\right]. \quad (7.4\text{-}9)$$

Equivalently, this expression can be written in terms of the angular diameter $\theta_s = 2r_s/z$ of the source,

$$\mu_p(\Delta x, \Delta y) = 2\left[\frac{J_1\left(\frac{\pi\theta_s}{\lambda}\sqrt{(\Delta x)^2+(\Delta y)^2}\right)}{\frac{\pi\theta_s}{\lambda}\sqrt{(\Delta x)^2+(\Delta y)^2}}\right]. \quad (7.4\text{-}10)$$

Note that the fringes entirely vanish ($|\mu_p| = 0$) when the spacing $s = \sqrt{(\Delta x)^2+(\Delta y)^2}$ is such that a zero of the Bessel function J_1 occurs. The smallest spacing yielding this condition is

$$s_0 = 1.22 \frac{\bar{\lambda}}{\theta_s}. \qquad (7.4\text{-}11)$$

Thus it is possible to measure the angular diameter of the source by gradually increasing the spacing of the two openings until the fringes first vanish. The diameter of the source is then given by

$$\theta_s = 1.22 \frac{\bar{\lambda}}{s_0}. \qquad (7.4\text{-}12)$$

The reader may well wonder why the Fizeau stellar interferometer, which uses only a portion of the telescope aperture, is in any way preferred to the full telescope aperture in this task of measuring the angular diameter of a distant object. The answer lies in the effects of the random spatial and temporal fluctuations of the earth's atmosphere ("atmospheric seeing"), which are discussed in more detail in Chapter 8. For the present it suffices to say that it is easier to detect the vanishing of the contrast of a fringe in the presence of atmospheric fluctuations than it is to determine the diameter of an object from its highly blurred image.

The chief shortcoming of the Fizeau stellar interferometer lies in the fact that it can be used only to measure the diameters of relatively large sources. The maximum spacings that can be explored are limited by the physical diameter of the telescope used, and the number of stellar sources with diameters suitable for measurement by even the largest optical telescopes is extremely limited.

The limited range of spacings afforded by unaided telescopes was vastly extended by an interferometer invented by Michelson (Ref. 7-24), known as the *Michelson stellar interferometer*. As illustrated in Fig. 7-17 for the most common case of a reflecting telescope, two movable mirrors are mounted on a long rigid cross-arm. Light is directed from these two mirrors into the primary collector of the telescope, and the two pencil beams are merged in the focal plane, just as in the case of the Fizeau interferometer. Now, however, the range of spacings that can be explored is not limited by the physical extent of the telescope aperture, and diameters of much smaller sources can be measured. A 20-foot interferometer of this kind was constructed and successfully used by Michelson (Ref. 7-25) and Michelson and Pease (Ref. 7-26). A 50-ft interferometer was also built but never worked as well as had been hoped.

The problems encountered in attempting to operate a Michelson stellar interferometer are far from trivial. The entire instrument must be carefully

IMAGE FORMATION AS AN INTERFEROMETRIC PROCESS

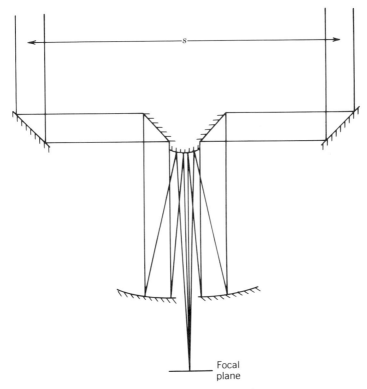

Figure 7-17. Michelson steller interferometer.

aligned, and the pathlength differences in the two arms must be maintained equal within a fraction of the coherence length of the light. In order to collect as much light as possible, the broadest possible bandwidth of light is used, which, in turn, means that the coherence length of the light is extremely short. In addition to these difficulties, the random spatial variations of the refractive index of the earth's atmosphere introduce aberrations, which in turn limit the maximum usable size of the interferometer mirrors to no more than about 10 centimeters in diameter. The random temporal variations of the atmospheric effects introduce a time-varying phase difference between the two paths, with the result that the fringe phase varies rapidly with time. It was a fortunate event that the integration time of the human eye was sufficiently short that Michelson could still detect the presence or absence of fringes, even though those fringes were moving rapidly.

More modern versions of optical stellar interferometers have been proposed and used in recent years (Refs. 7-27 through 7-31), including the intensity interferometer discussed in Chapters 6 and 9 and the stellar speckle interferometer discussed in Chapter 8.

7.4.3 The Importance of Phase Information

The simplest possible use of a Michelson stellar interferometer is for determination of the particular spacing s_0 where the fringes first vanish, and thereby determining the angular diameter of a distant source. A more ambitious undertaking would be to measure the modulus of the complex coherence factor $|\mu_p|$ for an entire range of two-dimensional spacings and hopefully to recover from this data image information more detailed than just an angular diameter. The most ambitious undertaking would be to attempt to measure both the modulus and the phase of μ_p for an entire range of two-dimensional spacings and to use these more complete data for image formation.

Clearly, if both the modulus and phase of μ_p were measured, we would then know the Fourier spectrum of the object, at least out to a limiting spatial frequency corresponding to the maximum spacing explored. An inverse Fourier transform of the measured data $\mu_p(\Delta x, \Delta y)$ would then yield the desired image, with a resolution limited by the maximum achievable spacing.

Unfortunately, in practice it is impossible to extract the true phase information from the interferometer. Although the position of a fixed fringe relative to a reference could in principle be measured, the phase of the fringe fluctuates randomly with time, as a result of both the random fluctuations of the atmosphere and the mechanical instabilities of the interferometer itself.

A more realistic task would be to attempt to measure only the modulus of μ_p on a two-dimensional array of spacings, either by a sequence of measurements or with a multielement array. The question then naturally arises as to exactly what information about the object can be derived from measurements of the modulus of its Fourier spectrum. (For a view of some early work in optics pertinent to this question, see Refs. 7-32 and 7-33.) What price do we pay for loss of phase information?

That phase information is in general extremely important for image formation is demonstrated by a simple example. With reference to Fig. 7-18, consider a one-dimensional object with a rectangular intensity profile. The corresponding complex coherence factor is a simple sinc function. Note that the negative lobes of the sinc function correspond to 180° phase reversals of the fringes produced by the interferometer, and such phase changes cannot be detected for the reasons explained earlier. Our measured data thus correspond to the modulus of the sinc function. If we treat this modulus

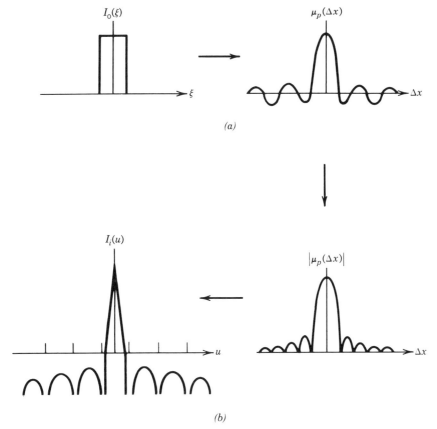

Figure 7-18. Example of consequences of losing phase information.

information as if it were the true spectrum of the object, subjecting it to an inverse Fourier transform, we obtain the "image" shown in part (b) of Fig. 7-18. Clearly, it bears little resemblance to the original object!

There do exist some cases in which the absence of phase information is of no consequence. For example, if the Fourier transform of the object is entirely real and nonnegative, the spectrum contains no phase information. An example of such an object is one with a Gaussian intensity distribution (and hence a Gaussian spectrum),

$$I_o(\xi, \eta) = I \exp\left\{-\frac{\xi^2 + \eta^2}{W^2}\right\} \qquad (7.4\text{-}13)$$

In Problem 7-7 the reader is asked to prove that any *symmetrical* one-dimensional intensity distribution can be recovered from knowledge of only the modulus of its spectrum, if proper processing of the data is performed.

A better feeling for exactly what information about the object is carried by the modulus information in the spectrum is obtained if we consider the inverse Fourier transform of $|\mu_p(\Delta x, \Delta y)|^2$, rather than just $|\mu_p|$. In this case the autocorrelation theorem of Fourier analysis implies that the recoverable "image" is of the form

$$I_i(u,v) = \iint_{-\infty}^{\infty} I_o(\xi, \eta) I_o(\xi - u, \eta - v) \, d\xi \, d\eta; \qquad (7.4\text{-}14)$$

that is, the recovered data are the autocorrelation function of the intensity distribution of the object. Such information can be useful, for example, in

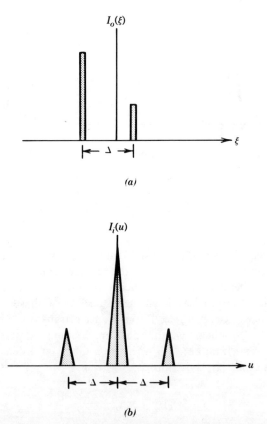

Figure 7-19. Determining the separation of two small sources from the autocorrelation function of the object: (a) object distribution; (b) autocorrelation function.

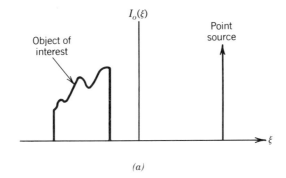

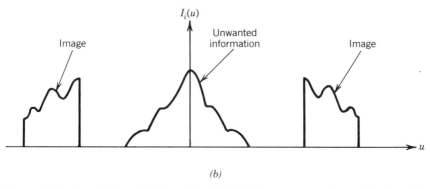

Figure 7-20. Special class of objects for which full image recovery is possible from the autocorrelation function of object intensity: (a) object; (b) autocorrelation function.

measuring the separation of two small stars, as illustrated in Fig. 7-19. The separation Δ can clearly be determined from the autocorrelation function.

There exists one special condition under which full image information can be extracted from the autocorrelation function, regardless of the symmetries of the object (Refs. 7-34 and 7-35). This condition occurs when the object of interest happens to have a point source of light near it but separated at a proper distance from it. As illustrated in Fig. 7-20, the autocorrelation function in this case contains twin images of the object, as well as unwanted information, in a form entirely analogous to that of holographic images.

7.4.4 Phase Retrieval

A tantalizing possible solution to the loss of phase information was suggested by Wolf (Ref. 7-36) in 1962. Although the context in which the suggestion was made was Fourier spectroscopy, the ideas apply equally well in the present context of spatial interferometry. For simplicity we discuss the problem using functions of one variable, but generalization to functions of two variables is possible. For discussion of the one-dimensional problem in greater detail, see Refs. 7-37 and 7-38.

We begin by assuming that the incoherent object of concern is known a priori to be spatially bounded; that is, the intensity distribution $I_o(\xi)$ describing the object is nonzero over only a finite interval on the ξ axis. Without loss of generality, we can assume that the origin has been chosen to assure that

$$I_o(\xi) = 0 \quad \text{all} \quad \xi \leq 0. \tag{7.4-15}$$

[The corresponding restriction in the two-dimensional case is to confine nonzero values of I_o to the upper right-hand quadrant of the (ξ, η) plane.] Now the complex coherence factor

$$\mu(\Delta x) = \mu_r(\Delta x) + j\mu_i(\Delta x) \tag{7.4-16}$$

is the normalized Fourier transform of I_o. From our knowledge of analytic signals, we know that the real and imaginary parts of the Fourier transform of a function satisfying (7.4-15) must be a Hilbert transform pair[†]

$$\mu_r(\Delta x) = \frac{1}{\pi} \int_{-\infty}^{\infty} \frac{\mu_i(\zeta)}{\zeta - \Delta x} \, d\zeta. \tag{7.4-17}$$

Now consider the complex coherence factor expressed as a function of a complex argument $z = \Delta x + jq$. The function $\mu(z)$ is then related to $I_o(\xi)$ by a one-sided Laplace transform

$$\mu(z) = b \int_0^\infty I_o(\xi) e^{j2\pi z \xi} \, d\xi, \tag{7.4-18}$$

where $s = -j2\pi z$ is the usual Laplace transform variable and b is a constant. A bit of reflection shows that $\mu(z)$ must of necessity be *analytic* (have no poles) in the upper half of the complex z plane, because of the single-sided nature of I_o. Hence the name "analytic signal" for such a function.[‡]

[†] When comparing (3.8-20) and (7.4-19), remember that the former deals with a function having a single-sided spectrum, while the latter deals with the spectrum of a single-sided function.
[‡] Note that if $I_o(\xi)$ is also zero for ξ larger than a certain *upper* bound, $\mu(z)$ also has no poles in the *lower* half of the complex z plane.

IMAGE FORMATION AS AN INTERFEROMETRIC PROCESS   345

Clearly, we can use relationship (7.4-17) to find the real part of μ given the imaginary part. Alternatively, we can find the imaginary part from the real part by the inverse Hilbert transform relationship

$$\mu_i(\Delta x) = -\frac{1}{\pi} \int_{-\infty}^{\infty} \frac{\mu_r(\zeta)}{\zeta - \Delta x} d\zeta. \qquad (7.4\text{-}19)$$

However, neither of these relations helps us with the task at hand, namely, determining the *phase* of μ from knowledge of its modulus.

As a step toward solving this problem, consider the result of taking the complex logarithm of the function $\mu(\Delta x)$. If

$$\mu(\Delta x) = |\mu(\Delta x)| \exp[j\alpha(\Delta x)], \qquad (7.4\text{-}20)$$

then

$$\ln[\mu(\Delta x)] = \ln|\mu(\Delta x)| + j\alpha(\Delta x). \qquad (7.4\text{-}21)$$

Now if it can be proved that if $\ln[\mu(\Delta x)]$ is an analytic signal, the phase will be recoverable from the modulus by the Hilbert transform relationship

$$\alpha(\Delta x) = -\frac{1}{\pi} \int_{-\infty}^{\infty} \frac{\ln|\mu(\zeta)|}{\zeta - \Delta x} d\zeta. \qquad (7.4\text{-}22)$$

Unfortunately, analyticity of $\mu(z)$ in the upper half plane of the complex z plane is not a sufficient condition to assure analyticity of $\ln[\mu(z)]$ in that same region. The most obvious reason for lack of analyticity is the possible existence of zeros of $\mu(z)$ in the upper half plane, which lead to singularities of $\ln[\mu(z)]$.

A careful examination of the mathematics of this problem (Ref. 7-39) demonstrates that, provided $\mu(\Delta x)$ is square integrable

$$\int_{-\infty}^{\infty} |\mu(\Delta x)|^2 \, d\Delta x < \infty \qquad (7.4\text{-}23)$$

and further provided it satisfies the "Paley–Wiener condition"

$$\int_{-\infty}^{\infty} \frac{\ln|\mu(\Delta x)|}{(\Delta x)^2 + 1} \, d\Delta x < \infty, \qquad (7.4\text{-}24)$$

the phase $\alpha(\Delta x)$ is given by

$$\alpha(\Delta x) = -\frac{1}{\pi} \int_{-\infty}^{\infty} \frac{\ln|\mu(\zeta)|}{\zeta - \Delta x} d\zeta + \sum_n \arg\left\{\frac{\Delta x - z_n}{\Delta x - z_n^*}\right\}, \quad (7.4\text{-}25)$$

where the z_n are the locations of the zeros of $\mu(z)$ in the upper half of the z plane.

In some cases the function $\mu(z)$ may have no zeros in the upper half plane, in which case the so-called minimum phase solution of (7.4-22) is valid (see Ref. 7-40). However, in general zeros will be present, and their locations will be unknown a priori. There has been considerable effort to use further physical constraints on $I_o(\xi)$ (e.g., positivity) to remove some of the uncertainties regarding the locations of the zeros, but even with such constraints, ambiguities remain in the general case.

Our discussion of this subject would be incomplete if we did not mention some important progress that has been made on the two-dimensional version of this "phase retrieval" problem in recent years. Fienup (Refs. 7-41 and 7-42) has applied an iterative technique to this problem and has found it to converge to correct solutions in the majority of cases involving functions $I_o(\xi, \eta)$ of considerable complexity. His results suggest that the ambiguities inherent in the solution set are less severe in two dimensions than in one. There exists some analytical work in support of this contention (Ref. 7-43). Nonetheless, it is possible to find two-dimensional cases in which ambiguities exist (Ref. 7-44).

The simplest version of the iterative method in question is implemented digitally and begins with the following assumptions:

(1) The modulus $|\mu(\Delta x, \Delta y)|$ of the complex function μ has been measured and hence is known.
(2) The object intensity $I_o(\xi, \eta)$ is identically zero outside a known region in the (ξ, η) plane.
(3) The object intensity $I_o(\xi, \eta)$ is a nonnegative function ($I_o(\xi, \eta) \geq 0$).

As a first guess at the phase of the function $\mu(\Delta x, \Delta y)$ a set of random phases may be used [it is assumed that $|\mu(\Delta x, \Delta y)|$ is known on a discrete set of samples in the $(\Delta x, \Delta y)$ plane]. An inverse Fourier transform of this initial guess, denoted $\mu^{(1)}(\Delta x, \Delta y)$, yields a spectrum $I_o^{(1)}(\xi, \eta)$ that will in general be nonzero outside the region of support of the true I_o and will also have some negative values. If the negative values are removed (e.g., by setting them equal to zero) and the nonzero values of $I_o^{(1)}$ lying outside the known region of support are replaced by zero, a Fourier transform of this

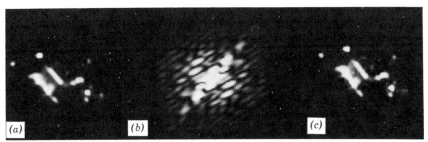

Figure 7-21. Illustration of the results of an iterative phase-retrieval algorithm (*a*) original object (a simulated spacecraft); (*b*) modulus of the Fourier spectrum of the object; (*c*) image recovered by use of an iterative algorithm. (Courtesy of J. R. Fienup and the Optical Society of America. See Ref. 7-45.)

modified intensity distribution yields a function $\tilde{\mu}^{(2)}(\Delta x, \Delta y)$ that has a new phase distribution as well as a modified modulus distribution. If we replace the modulus distribution with the original modulus information—which we know to be correct—but retain the new phase information, we have a second guess, $\mu^{(2)}(\Delta x, \Delta y)$, for the complex function μ. The process is repeated in hopes that $\lim_{n \to \infty} I_o^{(n)}(\xi, \eta) = I_o(\xi, \eta)$. Indeed, the process almost always converges, and the question then reduces to whether it has converged to the correct solution. As mentioned earlier, a correct solution is obtained in a remarkably large number of cases.

Figure 7-21 (from Ref. 7-45) shows an example of the application of an algorithm similar to the one described above. Part (*a*) of Fig. 7-21 shows the original object, a simulated image of a spacecraft. Part (*b*) shows the Fourier modulus of that image. Part (*c*) shows the image reconstructed by use of an iterative algorithm. The differences between parts (*a*) and (*c*) are difficult to discern in these reproductions and are indeed rather small.

7.5 THE SPECKLE EFFECT IN COHERENT IMAGING

When images of complex objects are formed by use of the highly coherent light produced by a laser, a very important kind of image defect soon becomes apparent. If the object is composed of surfaces that are rough on the scale of an optical wavelength (as most objects are), the image is found to have a granular appearance, with a multitude of bright and dark spots that bear no apparent relationship to the macroscopic scattering properties of the object. These chaotic and unordered patterns have come to be known

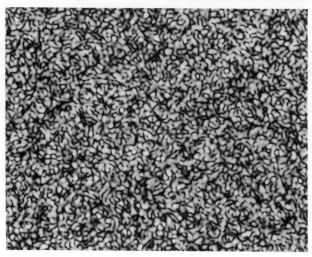

Figure 7-22. Speckle pattern in the image of a uniformly bright object. (Courtesy of J. C. Dainty and Springer-Verlag. Reprinted with the permission of Springer-Verlag, Heidelberg.)

as "speckle." Such patterns are also found in the images of transparent objects that have been illuminated by coherent light through a stationary diffuser. A typical speckle pattern appearing in the image of a uniformly reflecting surface is shown in Fig. 7-22.

Detailed analysis of the properties of speckle patterns produced by laser light began in the early 1960s; however, far earlier studies of speckle-like phenomena are found in the physics and engineering literature. Special mention should be made of the studies of "coronas" or Fraunhofer rings by Verdet (Ref. 7-46) and Lord Rayleigh (Ref. 7-47). Later, in a series of papers dealing with the scattering of light from a large number of particles, von Laue (Refs. 7-48 through 7-50) derived many of the basic properties of speckle-like phenomena.

A number of rather extensive modern references on speckle exist (Refs. 7-51 through 7-53). Ultimately, a completely rigorous understanding of speckle requires a detailed examination of the properties of electromagnetic waves after they have been reflected from or scattered by rough surfaces (Ref. 7-54). However, a good intuitive feeling for the properties of speckle can be obtained from a less rigorous consideration of the problem.

7.5.1 The Origin and First-Order Statistics of Speckle

The origin of speckle was quickly recognized by early workers in the laser field (Refs. 7-55 and 7-56). The vast majority of surfaces, whether natural or man-made, are extremely rough on the scale of an optical wavelength. Under illumination by monochromatic light, the wave reflected from such a surface consists of contributions from many different scattering points or areas. As illustrated in Fig. 7-23, the image formed at a given point in the observation plane consists of a superposition of a multitude of amplitude spread functions, each arising from a different scattering point on the surface of the object. As a consequence of the roughness of the surface, the various spread functions add with markedly different phases, resulting in a highly complex pattern of interference.

The preceding argument can also be applied to transmission objects illuminated through a diffuser. Because of the presence of the diffuser, the wavefront leaving the object has a highly corrugated and extremely complex structure. In the image of such an object we again find large fluctuations of intensity caused by the overlapping of a multitude of dephased amplitude spread functions.

Because of our lack of knowledge of the detailed microscopic structure of the complex wavefront leaving the object, it is necessary to discuss the properties of speckle in statistical terms. The statistics of concern are defined over an ensemble of objects, all with the same macroscopic properties, but differing in microscopic detail. Thus if we place a detector at a particular location in the image plane, the measured intensity cannot be predicted exactly in advance, even if the macroscopic properties of the object are known exactly. Rather, we can only predict the statistical properties of that intensity over an ensemble of rough surfaces.

Perhaps the most important statistical property of a speckle pattern is the probability density function of the intensity I observed at a point in the image. How likely are we to observe a bright peak or a dark null in the intensity? This question can be answered by noting the similarity of the problem at hand to the classical problem of the random walk (Refs. 7-57

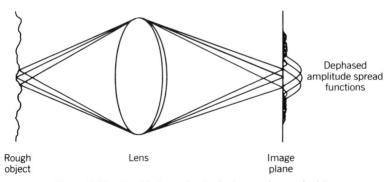

Figure 7-23. Speckle formation in the image of a rough object.

through 7-59), which was discussed in some detail in Section 2.9. The problem is also entirely analogous to that of determining the first-order statistics of the intensity of thermal light, as discussed in Section 4.2. Recalling our discussion of that section, if the phases of the individual scattered contributions from the object are approximately uniformly distributed over $(-\pi, \pi)$ (i.e., if the object is truly rough on the scale of a wavelength), the field associated with any single linear polarization component of the image must be a circular complex Gaussian random variable, and its intensity must obey negative exponential statistics

$$p_I(I) = \begin{cases} \frac{1}{\bar{I}} \exp\left(\frac{-I}{\bar{I}}\right), & I \geq 0 \\ 0 & \text{otherwise,} \end{cases} \quad (7.5\text{-}1)$$

where $\bar{I}$ is the mean intensity associated with that polarization component. If the scattered wave is partially depolarized, methods quite analogous to those used in Section 4.3.4 can be used to show that the density function for I consists of the difference of two negative exponential functions [cf. Eq. (4.3-42)]. However, we concentrate on the properties of fully polarized speckle here.

The fact that the probability density function of intensity is negative exponential implies that the fluctuations about the mean are rather pronounced. If we define the *contrast* C of a speckle pattern as the ratio of its standard deviation to its mean, for the polarized case we find

$$C = \frac{\sigma_I}{\bar{I}} = 1. \quad (7.5\text{-}2)$$

Because of this high contrast, speckle is extremely disturbing to the human observer, particularly if fine image detail is of interest, and consequently a significant loss of effective resolution results from its presence.

In closing this discussion of first-order statistics, the point should be made that the distribution of mean intensity $\bar{I}(x, y)$ in the image of a coherently illuminated rough object is identical with the image intensity that would be observed if the object were illuminated with spatially incoherent light with the same power spectral density. Incoherent illumination may be regarded as being equivalent to a rapid time sequence of spatially coherent wavefronts, with the effective phase structure of each member of the sequence being extremely complex and quite independent of the phase structure of each other member. Thus the time-integrated image intensity observed under spatially incoherent illumination is identical with the ensemble average intensity $\bar{I}(x, y)$ (assuming identical bandwidths are involved).

Hence any of the methods for analyzing the image intensity distribution for an incoherent imaging system may be used to predict the mean speckle intensity distribution in the image of a coherently illuminated rough object.

7.5.2 Ensemble Average Coherence

The light waves studied in previous sections and chapters were modeled as ergodic random processes. That is, it was assumed that time averages were identically equal to ensemble averages, and thus that the two types of averaging process could be freely interchanged. It is an important fact that when an optically rough object is illuminated by monochromatic light, the reflected waves no longer can be modeled as ergodic random processes, for time and ensemble averages are no longer equal.

The lack of ergodicity in this case is easily demonstrated by consideration of two different Young's interference experiments. First, let the light scattered from a stationary rough surface fall on a mask containing two pinholes, and observe the fringe formed on a distant observing screen. Because the light is monochromatic, it is also spatially coherent (cf. Problem 5-12), and the fringe will be found to have visibility

$$\mathscr{V} = \frac{2\sqrt{I_1 I_2}}{I_1 + I_2}, \qquad (7.5\text{-}3)$$

where I_1 and I_2 are the intensities of the light incident on the pinholes. We conclude that the modulus of the complex coherence factor $|\mu_{12}|$ must be unity, at least for the usual time-averaged definition of coherence. The extremely complex amplitude and phase distribution imparted to the wave by the rough surface has not reduced the coherence of the light, since this distribution does not change with time.

Now consider a second Young's interference experiment. In this case we shall perform ensemble averaging by successively placing objects with the same macrostructure but different microstructure (surface profile) in the illuminating beam, and time integrating on one photographic plate all the fringes generated by the succession of objects. Although any one of these component fringes has a visibility corresponding to $|\mu_{12}| = 1$, the superposition of the succession of fringes in general will not, because the phases of the component fringes will change from realization to realization. Thus the ensemble-averaged fringe will in general yield a $|\mu_{12}|$ that is quite different from unity.

Since there is a difference between ensemble averages and time averages for this type of wave, we must be careful to distinguish between time-averaged coherence and ensemble-averaged coherence. Accordingly, we shall use the ordinary symbols for coherence quantities defined by time averages and identical symbols with overbars to represent ensemble-averaged quantities. Thus we distinguish between the two mutual coherence functions $\Gamma(P_1, P_2; \tau)$ and $\overline{\Gamma}(P_1, P_2; \tau)$, the two mutual intensities $\mathbf{J}(P_1, P_2)$ and $\overline{\mathbf{J}}(P_1, P_2)$, and so on.

The wave equation governing the propagation of light is, of course, the same, whether we are ultimately interested in time-average or ensemble-average properties of the light. From this fact follows an important conclusion: *the laws governing the propagation of coherence functions are identical for time-averaged and ensemble-averaged quantities.* In other words, whereas the functional form of a mutual coherence or mutual intensity may depend on whether the average is with respect to time or with respect to the ensemble, the mathematical relationship between two coherence functions of the same type is independent of which kind of averaging is used. This fact allows us to apply all our previously acquired knowledge of the propagation of ordinary coherence functions to problems involving the propagation of ensemble-averaged coherence.

From an ensemble-averaging point of view, the mutual intensity of the light reflected or scattered from a rough surface, and observed very close to that surface, is essentially the same as the mutual intensity of an incoherent source. Over an ensemble of ideally rough surfaces, there is little relationship between the phases of the light scattered from two closely spaced surface elements, at least until the spacing becomes close to a wavelength. We state this fact mathematically by representing the mutual intensity function at the surface by

$$\overline{\mathbf{J}}(\xi_1, \eta_1; \xi_2, \eta_2) = \kappa \overline{I}(\xi_1, \eta_1)\delta(\xi_1 - \xi_2, \eta_1 - \eta_2), \quad (7.5\text{-}4)$$

where κ is a constant and $\overline{I}$ is an ensemble-averaged intensity distribution.

The mutual intensity observed on a surface some distance from the source can be calculated using the Van Cittert–Zernike theorem. By analogy with Eq. (5.6-8), therefore, the ensemble-averaged mutual intensity across a plane at distance z from the source is given by

$$\overline{\mathbf{J}}(x_1, y_1; x_2, y_2) = \frac{\kappa e^{-j\psi}}{(\overline{\lambda}z)^2} \iint_{-\infty}^{\infty} \overline{I}(\xi, \eta)\exp\left\{j\frac{2\pi}{\overline{\lambda}z}[(\Delta x\xi + \Delta y\eta)]\right\} d\xi\, d\eta,$$

$$(7.5\text{-}5)$$

where

$$\psi = \frac{\pi}{\overline{\lambda}z}\left[\left(x_2^2 + y_2^2\right) - \left(x_1^2 + y_1^2\right)\right] \quad (7.5\text{-}6)$$

THE SPECKLE EFFECT IN COHERENT IMAGING

and $\bar{I}(\xi, \eta)$ is the ensemble averaged intensity distribution across the scattering spot on the rough object.

If we are dealing with an imaging geometry as shown in Fig. 7-23, arguments similar to those used in reaching Eq. (7.2-17) can be used to predict the mutual intensity in the image. We regard the exit pupil of the imaging optics to be equivalent to a new incoherent source and apply the Van Cittert–Zernike theorem to this source. For a region of the image that has constant mean intensity, the mutual intensity takes the form

$$\bar{\mathbf{J}}_i(u_1, v_1; u_2, v_2) = \frac{\kappa}{(\bar{\lambda}z_2)^2} \exp\left\{-j\frac{\pi}{\bar{\lambda}z_2}\left[(u_2^2 + v_2^2) - (u_1^2 + v_1^2)\right]\right\}$$

$$\times \iint_{-\infty}^{\infty} |\mathbf{P}(x, y)|^2 \exp\left\{j\frac{2\pi}{\bar{\lambda}z_2}(\Delta ux + \Delta vy)\right\} dx\, dy,$$

(7.5-7)

where $\mathbf{P}$ is the complex pupil function of the imaging lens, κ is a constant defined earlier, and $\Delta u = u_2 - u_1$, $\Delta v = v_2 = v_1$.

We are now prepared to inquire as to a second basic property of a speckle pattern, namely, the distribution of scale sizes in its random spatial fluctuations. To concentrate on the speckle fluctuations as distinct from the information bearing variations of mean intensity, we suppose that the object of interest is uniformly bright. A suitable description of the speckle scale-size distribution is the spatial *power spectral density* of the speckle pattern, which we represent by $\bar{\mathcal{G}}_i(\nu_U, \nu_V)$. We calculate $\bar{\mathcal{G}}_i$ by Fourier transforming the autocorrelation function of speckle pattern

$$\bar{\mathcal{G}}_i(\nu_U, \nu_V) = \iint_{-\infty}^{\infty} \bar{\Gamma}_i(\Delta u, \Delta v) \exp\{j2\pi(\Delta u\nu_U + \Delta v\nu_V)\}\, d\Delta u\, d\Delta v,$$

(7.5-8)

where

$$\bar{\Gamma}_i(\Delta u, \Delta v) = \overline{I_i(u_1, v_1) I_i(u_1 + \Delta u, v_1 + \Delta v)}, \qquad (7.5\text{-}9)$$

and it remains to be demonstrated that $\bar{\Gamma}_i$ depends only on the coordinate differences $(\Delta u, \Delta v)$.

According to our earlier random walk arguments, the complex fields underlying the speckle pattern are circular complex Gaussian random variables. It follows from the complex Gaussian moment theorem that

$$\overline{\Gamma}_i = (\overline{I}_i)^2 [1 + |\overline{\mu}_i|^2], \qquad (7.5\text{-}10)$$

where $|\overline{\mu}_i|$ is given from (7.5-7) by

$$|\overline{\mu}_i(\Delta x, \Delta y)| = \frac{\left| \iint_{-\infty}^{\infty} |\mathbf{P}(x, y)|^2 \exp\left\{ j \frac{2\pi}{\lambda z_2} (\Delta ux + \Delta vy) \right\} dx \, dy \right|}{\iint_{-\infty}^{\infty} |\mathbf{P}(x, y)|^2 \, dx \, dy}.$$

$$(7.5\text{-}11)$$

This result demonstrates that $\overline{\Gamma}_i$ does, indeed, depend only on the coordinate differences $(\Delta u, \Delta v)$ and provides us with sufficient information to allow us to calculate the power spectral density of the speckle pattern. Using the definition

$$|\hat{\mathbf{P}}(x, y)|^2 = \frac{|\mathbf{P}(x, y)|^2}{\iint_{-\infty}^{\infty} |\mathbf{P}(x, y)|^2 \, dx \, dy} \qquad (7.5\text{-}12)$$

and substituting (7.5-10), (7.5-11), and (7.5-12) into (7.5-8), we find

$$\overline{\mathcal{G}}_i(\nu_U, \nu_V) = (\overline{I}_i)^2 \delta(\nu_U, \nu_V)$$

$$+ (\overline{I}_i)^2 \mathcal{F}\left\{ \left| \iint_{-\infty}^{\infty} |\hat{\mathbf{P}}(x, y)|^2 \exp\left\{ j \frac{2\pi}{\lambda z_2} (\Delta ux + \Delta vy) \right\} dx \, dy \right|^2 \right\},$$

$$(7.5\text{-}13)$$

where $\mathcal{F}\{\cdot\}$ is a two-dimensional Fourier transform with respect to $(\Delta u, \Delta v)$. Use of the autocorrelation theorem of Fourier analysis and the symmetry properties of the autocorrelation of a real and nonnegative function allow us to write

$$\overline{\mathcal{G}}_i(\nu_U, \nu_V) = (\overline{I}_i)^2 \bigg[\delta(\nu_U, \nu_V)$$

$$+ (\overline{\lambda z_2})^2 \iint_{-\infty}^{\infty} |\hat{\mathbf{P}}(x, y)|^2 |\hat{\mathbf{P}}(x - \overline{\lambda} z_2 \nu_U, y - \overline{\lambda} z_2 \nu_V)|^2 \, dx \, dy \bigg].$$

$$(7.5\text{-}14)$$

Aside from the uninteresting δ function at zero spatial frequency, we see that the power spectral density of a speckle pattern has the shape of the autocorrelation function of the squared modulus of a normalized pupil function. The power spectral density is independent of any aberrations that may exist in the imaging system, and in the important case of a clear, unapodized pupil (**P** = 1 or 0), the autocorrelation function of $|\hat{\mathbf{P}}|^2$ is (within a normalizing constant) equivalent to the autocorrelation function of the pupil itself.

For an imaging system with a square ($L \times L$) unapodized exit pupil, the power spectral density takes the form

$$\mathscr{G}_i(\nu_U, \nu_V) = (\bar{I}_i)^2 \left[\delta(\nu_U, \nu_V) + \left(\frac{\bar{\lambda} z_2}{L}\right)^2 \Lambda\left(\frac{\bar{\lambda} z_2}{L}\nu_U\right) \Lambda\left(\frac{\bar{\lambda} z_2}{L}\nu_V\right) \right],$$

(7.5-15)

where $\Lambda(x) = 1 - |x|$ for $|x| \leq 1$, zero otherwise. A cross section of this distribution is shown in Fig. 7-24. For a circular lens with an unobstructed pupil of diameter D, the corresponding result is

$$\mathscr{G}_i(\nu_U, \nu_V) = (\bar{I}_i)^2 \left[\delta(\nu_U, \nu_V) + 2\left(\frac{\bar{\lambda} z_2}{D}\right)^2 \right.$$

$$\left. \times \frac{2}{\pi}\left\{ \cos^{-1}\left(\frac{\bar{\lambda} z_2}{D}\nu\right) - \left(\frac{\bar{\lambda} z_2}{D}\nu\right)\left[1 - \left(\frac{\bar{\lambda} z_2}{D}\nu\right)^2\right]^{1/2}\right\} \right]$$

(7.5-16)

for $\nu \leq D/\bar{\lambda} z_2$, zero otherwise, where $\nu \triangleq \sqrt{\nu_U^2 + \nu_V^2}$.

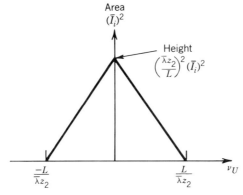

Figure 7-24. Cross section of the power spectral density of a speckle pattern resulting from an imaging system with a square exit pupil.

We conclude that in any speckle pattern, large-scale (low-frequency) fluctuations are the most populous, and no scale sizes smaller than a certain cutoff are present. The exact distribution depends on the character of the pupil function of the imaging system.

Methods for suppressing the effects of speckle in coherent imaging systems have been studied, but no general solution that eliminates speckle while maintaining perfect coherence and preserving image detail down to the diffraction limit of the imaging system has been found. Speckle suppression remains one of the most important unsolved problems of coherent imaging.

REFERENCES

7-1 B. J. Thompson, "Image Formation with Partially Coherent Light," in *Progress in Optics*, Vol. VII, (E. Wolf, editor), North Holland Publishing Company, Amsterdam (1969).

7-2 M. Born and E. Wolf, *Principles of Optics*, 2nd rev. ed., MacMillan, New York (1964).

7-3 M. J. Beran and G. B. Parrent, Jr., *Theory of Partial Coherence*, Prentice-Hall, Englewood Cliffs, NJ (1964).

7-4 J. Peřina, *Coherence of Light*, Van Nostrand Reinhold, London (1971).

7-5 H. H. Hopkins, *Proc. Roy. Soc.*, **A208**, 263–277 (1951).

7-6 H. H. Hopkins, *Proc. Roy. Soc.*, **A217**, 408–432 (1953).

7-7 J. W. Goodman, *Introduction to Fourier Optics*, McGraw-Hill Book Company, New York, (1968).

7-8 R. K. Luneburg, *Mathematical Theory of Optics*, University of California Press, Berkeley, California (1964).

7-9 R. Kingslake, *Applied Optics and Optical Engineering*, Vol. II, Academic Press, New York, p. 225, (1965).

7-10 F. Zernike, *Physica*, **5**, 785–795 (1938).

7-11 R. J. Collier, C. B. Burkhardt, and L. H. Lin, *Optical Holography*, Academic Press, New York, pp. 118–119 (1971).

7-12 D. A. Tichenor and J. W. Goodman, *J. Opt. Soc. Am.*, **62**, 293 (1972).

7-13 P. M. Duffieux, *L'Integral de Fourier et ses Applications a l'Optique*, Rennes (1946).

7-14 R. E. Swing, *J. Opt. Soc. Am.*, **62**, 199 (1972).

7-15 R. E. Kinzley, *J. Opt. Soc. Am.*, **62**, 386 (1972).

7-16 D. Grimes and B. J. Thompson, *J. Opt. Soc. Am.*, **57**, 1330 (1967).

7-17 R. J. Becherer and G. B. Parrent, *J. Opt. Soc. Am.*, **57**, 1479 (1967).

7-18 R. N. Bracewell, "Radio Astronomy Techniques," in *Encyclopedia of Physics*, Vol. 54 (S. Flügge, editor) Springer-Verlag, Berlin (1959).

7-19 Special issue on radio astronomy, *Proc. IEEE*, **61** (September 1973).

7-20 G. L. Rogers, *Proc. Phys. Soc.* (2), **81**, 323–331 (1963).

7-21 K. Dutta and J. W. Goodman, *J. Opt. Soc. Am.*, **67**, 796 (1977).

7-22 F. D. Russell and J. W. Goodman, *J. Opt. Soc. Am.*, **61**, 182 (1971).
7-23 H. Fizeau, *C. R. Acad. Sci. Paris*, **66**, 934 (1868).
7-24 A. A. Michelson, *Phil. Mag.*, (5), **30**, 1 (1890).
7-25 A. A. Michelson, *Astrophys. J.*, **51**, 257 (1920).
7-26 A. A. Michelson and F. G. Pease, *Astrophys. J.*, **53**, 249 (1921).
7-27 R. H. Miller, *Science*, **153**, 581 (1966).
7-28 D. G. Currie, S. L. Knapp, and K. M. Liewer, *Astrophys. J.*, **187**, 131 (1974).
7-29 E. S. Kulagin, *Opt. Spectrosc.*, **23**, 459 (1967).
7-30 R. Hanbury Brown, *The Intensity Interferometer*, Taylor and Francis, London (1974).
7-31 A. Labeyrie, *Astron. Astrophys.*, **6**, 85 (1970).
7-32 E. L. O'Neill and A. Walther, *Optica Acta*, **10**, 33 (1963).
7-33 A. Walther, *Optica Acta*, **10**, 41 (1963).
7-34 D. Kohler and L. Mandel, *J. Opt. Soc. Am.*, **60**, 280 (1970).
7-35 J. W. Goodman, *J. Opt. Soc. Am.*, **60**, 506 (1970).
7-36 E. Wolf, *Proc. Phys. Soc.*, **80**, 1269 (1962).
7-37 P. Roman and A. S. Marathay, *Il Nuovo Cimento* (X), **30**, 1452 (1963).
7-38 D. Dialetis and E. Wolf, *Il Nuovo Cimento* (X), **47**, 113 (1967).
7-39 H. M. Nussenzvieg, *J. Math. Phys.*, **8**, 561 (1967).
7-40 S. R. Robinson, *J. Opt. Soc. Am.*, **68**, 87 (1978).
7-41 J. R. Fienup, *Optics Lett.*, **3**, 27 (1978).
7-42 J. R. Fienup, *Opt. Eng.*, **18**, 529 (1979).
7-43 Yu M. Bruck and L. G. Sodin, *Opt. Commun.*, **30**, 304 (1979).
7-44 A. M. J. Huiser and P. van Toorn, *Optics Lett.*, **5**, 499 (1980).
7-45 J. R. Fienup, *Appl. Optics*, **21**, 2758 (1982).
7-46 E. Verdet, *Ann. Scientif. l'Ecole Normale Supérieure*, **2**, 291 (1865).
7-47 J. W. Strutt (Lord Rayleigh), *Phil. Mag.*, **10**, 73 (1880).
7-48 M. von Laue, *Sitzungsber. Akad. Wiss.* (Berlin), **44**, 1144 (1914).
7-49 M. von Laue, *Mitt. Physik. Ges.* (Zurich), **18**, 90 (1916).
7-50 M. von Laue, *Verhandl. Deut. Phys. Ges.*, **19**, 19 (1917).
7-51 J. C. Dainty, "The Statistics of Speckle Patterns," in *Progress in Optics*, Vol. XIV (E. Wolf, editor), North Holland Publishing Company, Amsterdam (1976).
7-52 J. C. Dainty (editor), *Laser Speckle and Related Phenomena* (Topics in Applied Physics, Vol. 9), Springer-Verlag, Berlin (1975).
7-53 Special issue on laser speckle, *J. Opt. Soc. Am.*, **66** (November 1976).
7-54 P. Beckmann and A. Spizzichino, *The Scattering of Electromagnetic Waves from Rough Surfaces*, Pergamon/Macmillan, New York (1963).
7-55 J. D. Rigden and E. I. Gordon, *Proc. IRE*, **50**, 2367 (1962).
7-56 B. M. Oliver, *Proc. IEEE*, **51**, 220 (1963).
7-57 K. Pearson, *A Mathematical Theory of Random Migration*, Draper's Company Research Memoirs, Biometric Series III, London (1906).
7-58 J. W. Strutt (Lord Rayleigh), *Proc. Lond. Math. Soc.*, **3**, 267 (1871).
7-59 J. W. Strutt (Lord Rayleigh), *Phil. Mag.*, **37**, 321 (1919).

ADDITIONAL READING

Cornelis van Schooneveld, Editor, *Image Formation from Coherence Functions in Astronomy*, Vol. 76 (Proceedings), D. Reidel Publishing Company, Drodrecht, Holland (1979).

J. A. Roberts, Editor, *Indirect Imaging*, Cambridge University Press, Cambridge (1984).

J. R. Fienup, T. R. Crimmins, and W. Holsztnyski, "Reconstruction of the Support of an Object from the Support of Its Autocorrelation," *J. Opt. Soc. Am.*, **72**, 610–624 (1982).

M. A. Fiddy, B. J. Brames, and J. C. Dainty, "Enforcing Irreducibility for Phase Retrieval in Two Dimensions," *Optics Lett.*, **8**, 96–98 (1983).

Feature issue on signal recovery, *J. Opt. Soc. Am.*, **73**, 1412–1526 (November 1983).

B. E. A. Saleh, "Optical bilinear transformations," *Optica Acta*, **26**, 777–799 (1979).

W. J. Tango and R. Q. Twiss, "Michelson Stellar Interferometry," in *Progress in Optics*, Vol. XVII (E. Wolf, editor), North Holland Publishing Co., Amsterdam (1980).

PROBLEMS

7-1. Given that condenser lens in Fig. 7-9 has diameter D, specify the size required of a circular source to assure that the approximation of Eq. (7.2-15a) is valid.

7-2. Use the generalized Van Cittert–Zernike theorem to prove Eq. (7.2-18).

7-3. Prove Eq. (7.2-39).

7-4. Demonstrate the correctness of Eq. (7.2-41).

7-5. In the optical system in Fig. 7-5p, a square incoherent source (L meters $\times$ L meters) lies in the source plane. The object consists of two pinholes spaced in the ξ direction by distance

$$X = \frac{\bar{\lambda} f}{D},$$

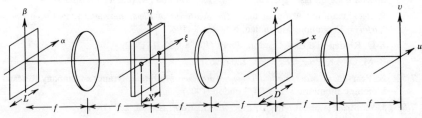

Figure 7-5*p*.

where $\bar{\lambda}$ is the mean wavelength, f is the focal length of all lenses, and D is the width of the *square* pupil plane aperture. The image intensity is observed in the (u, v) plane.

(a) What source dimension L produces the largest dip of intensity at the center of the image?

(b) For the source size found in part (*a*), what is the ratio of the intensity at the location $(u = X/2, v = 0)$ to the intensity at $(u = 0, v = 0)$?

(c) Compare the result of part (*b*) to the ratio that would be obtained if the two pinholes were illuminated with complete incoherence.

(d) Calculate the ratio of $I_i(X/2, 0)$ to $I_i(0, 0)$ when the pinholes are illuminated with perfect coherence ($\mu = 1$).

7-6. In the diagram for problem 7-5, replace the source by a thin incoherent annulus or ring, with mean radius ρ and radial width W. In addition, replace the pupil of the imaging system by a circular aperture of diameter D. The two pinholes are now separated by distance $X = 1.22\bar{\lambda}f/D$, the "Rayleigh" separation.

(a) Find the smallest radius ρ of the annular source for which the two pinholes are illuminated incoherently.

(b) Find the radius ρ of the annular source for which the central value of the intensity in the image drops to its smallest possible value relative to the peak value.

Hint: The Fourier transform of a thin uniform annulus of mean radius ρ and width W is given approximately by

$$G(\nu_X, \nu_Y) \cong 2\pi\rho W J_0\left(2\pi\rho\sqrt{\nu_X^2 + \nu_Y^2}\right),$$

where J_0 is a Bessel function of the first kind, zero order.

7-7. Prove that any symmetric one-dimensional object intensity distribution can be recovered from knowledge of only the modulus of its spectrum.

Hint: Knowledge of the modulus of the spectrum allows one to deduce the autocorrelation function of the object. Assume a space-limited object represented by a finite set of discrete samples.

7-8 It is desired to use a Michelson stellar interferometer to determine the brightness of the two components of a twin star. The individual components are known to be uniformly bright circular disks. Their angular diameters α and β and angular separation γ are all known. We also know that $\gamma \gg \alpha$, $\gamma \gg \beta$. How could we determine their relative brightness I_α/I_β from measurements of $|\mu_{12}(s)|$ with the interferometer?

7-9 Modify equations (7.2-49) and (7.2-51) to apply to a coherent imaging system in which the object is illuminated by a wave having phasor amplitude distribution $\mathbf{A}_o(\xi, \eta)$ and thus being more general than the plane-wave illumination assumed in the text.

7-10 Consider the partially coherent imaging system shown in Fig. 7-11. Note that in the absence of any object structure, the source is imaged in the pupil plane.

(a) Show that a component of object transmittance (for simplicity of infinite extent)

$$\tilde{t}_o(\xi, \eta) = \cos[2\pi\nu_0\xi]$$

generates two images of the source in the pupil plane, centered at positions

$$\bar{\xi} = \pm \bar{\lambda} f \nu_0.$$

(b) Show that, in order for these images of the source to individually fully cover the pupil, we must have

$$r_s \geq r_p + \bar{\lambda} f \nu_0$$

Note that when this is the case, the source is indistinguishable from a source of infinite extent, and thus the imaging system is incoherent. You may assume the source spectrum to be so narrow that wavelength dispersion effects can be ignored.

8

Imaging in the Presence of Randomly Inhomogeneous Media

Under ideal circumstances, the resolution achievable in an imaging experiment is limited only by our imperfect ability to make increasingly large optical elements that are free from inherent aberrations and have reasonable cost. However, these ideal conditions are seldom met in practice. Frequently the medium through which the waves must propagate while passing from the object to the imaging system is itself optically imperfect, with the result that even aberration-free optical systems may achieve actual resolutions that are far poorer than the theoretical diffraction limit.

The most important example of an imperfect optical medium in this context is the atmosphere of the Earth itself, that is, the air around us. As a consequence of the nonuniform heating of the Earth's surface by the sun, temperature-induced inhomogeneities of the refractive index of the air are ever present and can have devastating effects on the resolution achieved by large optical systems operating within such an environment.

Another common example occurs when an optical system must form images through an optical window that, because of circumstances beyond control, may be highly nonuniform in its thickness and/or refractive index.

In both of the examples just cited, the detailed structure of the optical imperfections is unknown a priori. As a consequence, it is necessary and appropriate to treat the optical distortions as random processes and to specify certain measures of the *average* performance of an optical system under such circumstances.

Two important limitations are imposed throughout this chapter. First, it is assumed throughout that the objects of interest radiate incoherently. While treatments of the problem for partially coherent objects are possible, they are generally more cumbersome than the treatment used for incoherent objects (which in many cases is already rather complicated). Furthermore, in the vast majority of problems of practical interest (e.g., in astronomy), it is highly accurate to assume that the object radiates incoherently.

A second important limitation concerns the scale size (i.e., correlation length) of the inhomogeneities present. We shall always assume that the scale sizes are much larger than the wavelength of the radiation being used. This assumption eliminates from consideration problems involving imaging through clouds or aerosols, for which the scale sizes of the inhomogeneities are comparable with or smaller than an optical wavelength and for which the refractive index changes are sharp and abrupt. This latter class of problems may be referred to as "imaging through turbid media," whereas we are concerned here with "imaging through turbulent media," for which the refractive index changes are smoother and coarser. The clear atmosphere of the Earth is the prime example of a "turbulent" medium.

The material treated in this chapter may be divided into two major topics. The first (Sections 8.1 through 8.3) concerns the effects of thin random "screens" (i.e., thin distorting structures) on the performance of optical systems. The second (Sections 8.4 through 8.9) concerns the effects of a thick inhomogeneous medium (the Earth's atmosphere) on imaging systems.

8.1 EFFECTS OF THIN RANDOM SCREENS ON IMAGE QUALITY

The effects of thin distorting layers on propagating electromagnetic waves have been treated in various places in the literature (see, e.g., Refs. 8-1 and 8-2). However, the effects of such distortions on image quality are less frequently discussed (as one exception, see Ref. 8-3). The importance of a theory that treats this subject lies not only in the understanding it yields for problems that require imaging through thin structures, but also in the physical insight it imparts to the much more complicated problem of imaging through the Earth's atmosphere.

8.1.1 Assumptions and Simplifications

In discussing the effects of thin random screens on image quality, we shall adopt the simple imaging geometry of Fig. 8-1. The object is assumed to radiate in a spatially incoherent fashion and to be describable by an intensity distribution $I_o(\xi, \eta)$. Lenses L_1 and L_2 have focal lengths f. The thin random screen is placed in the rear focal plane of L_1, which is assumed to coincide with the front focal plane of L_2. A blurred or distorted image of the object appears in the (u, v) plane and is described by the intensity distribution $I_i(u, v)$.

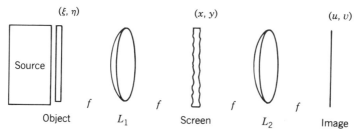

Figure 8-1. Optical system considered in random screen analysis.

We assume that the particular random screen present in the pupil of this system can be represented mathematically by a multiplicative amplitude transmittance $t_s(x, y)$. Implicit in this representation are two underlying assumptions. First, we are assuming that the screen is sufficiently thin that rays entering at coordinates (x, y) also leave at essentially the same coordinates. As a corollary, light waves emanating from all object points (ξ, η) experience the same amplitude transmittance $t_s(x, y)$. Second, we are assuming that the light is sufficiently narrowband to assure that the amplitude transmittance t_s is the same for all frequency components of the light.

It is not difficult to identify situations in which one or both of the preceding assumptions are violated. However, the simple theory that follows is more general than might appear at first glance. Once the results for a narrowband optical signal with center frequency $\bar{\nu}$ are known, results for broadband light can be obtained by integration, with $\bar{\nu}$ varying over the spectrum and with any frequency dependence of $t_s(x, y)$ explicitly included. In addition, regarding the "thin screen" assumption, whereas $t_s(x, y)$ may in fact depend on which object point (ξ, η) we consider, we shall see that it is the statistical autocorrelation function of t_s that determines image quality, and this autocorrelation function may be quite independent of (ξ, η), even though t_s is not (see Problem 8-1).

Finally, it should be mentioned that the very specific geometry assumed in Fig. 8-1 can yield results that apply to more general geometries. For example, if the screen is moved out of the common focal plane of lenses L_1 and L_2, the results of the analysis will not change. The underlying reason for this generality arises from the fact (yet to be shown) that it is the spatial *autocorrelation function* of the wave perturbations in the pupil that determines the average performance of the system. When the perturbing screen is moved away from the position shown in Fig. 8-1, the detailed structure of the field perturbations in the pupil changes, but their autocorrelation function does not (see Problem 8-2). Thus the results to be derived here apply to a broad class of imaging problems.

8.1.2 The Average Optical Transfer Function

It is customary to describe the spatial frequency response of an incoherent optical system in terms of an optical transfer function (OTF), as described in Section 7.2.4. This transfer function is represented here by $\mathcal{H}_0(\nu_U, \nu_V)$, where ν_U and ν_V are spatial frequency variables, and is given explicitly by

$$\mathcal{H}_0(\nu_U, \nu_V) = \frac{\iint_{-\infty}^{\infty} \mathbf{P}(x, y) \mathbf{P}^*(x - \bar{\lambda} f \nu_U, y - \bar{\lambda} f \nu_V) \, dx \, dy}{\iint_{-\infty}^{\infty} |\mathbf{P}(x, y)|^2 \, dx \, dy}, \qquad (8.1\text{-}1)$$

where $\mathbf{P}$ is the complex pupil function and f is again the focal length of the lenses in Fig. 8-1.

If a screen with amplitude transmittance $\mathbf{t}_s(x, y)$ is placed in the pupil of this imaging system, the pupil function is modified, yielding a new pupil function $\mathbf{P}'(x, y)$ given by

$$\mathbf{P}'(x, y) = \mathbf{P}(x, y) \mathbf{t}_s(x, y). \qquad (8.1\text{-}2)$$

With a particular screen in place, therefore, the OTF becomes

$$\mathcal{H}(\nu_U, \nu_V) = \frac{\iint_{-\infty}^{\infty} \mathbf{P}(x, y) \mathbf{P}^*(x - \bar{\lambda} f \nu_U, y - \bar{\lambda} f \nu_V) \mathbf{t}_s(x, y) \mathbf{t}_s^*(x - \bar{\lambda} f \nu_U, y - \bar{\lambda} f \nu_V) \, dx \, dy}{\iint_{-\infty}^{\infty} |\mathbf{P}(x, y)|^2 |\mathbf{t}_s(x, y)|^2 \, dx \, dy}. \qquad (8.1\text{-}3)$$

At this point, further progress with a purely deterministic analysis is impossible, because of our lack of knowledge of the specific values of $\mathbf{t}_s(x, y)$ at each (x, y). The best we can hope to accomplish is to use knowledge of the statistics of $\mathbf{t}_s$ to calculate some measure of the *average* frequency response of the system, with the average being taken over an ensemble of screens. Of course, the *average* performance of the imaging system will in general not coincide with the actual performance with a particular screen in place. Nonetheless, lacking knowledge of the structure of the particular screen, we must resort to specifying average performance.

EFFECTS OF THIN RANDOM SCREENS ON IMAGE QUALITY   365

What might be a reasonable way to define an "average OTF" in a situation such as this? The most straightforward definition would be simply to take the expected value of $\mathcal{H}(\nu_U, \nu_V)$, using an appropriate statistical model for $\mathbf{t}_s$. Unfortunately, this straightforward definition often leads to complications. As examination of Eq. (8.1-3) shows, it requires that we take the expected value of the ratio of two correlated random variables, a task that is not easy to carry out.

Fortunately, an alternative definition exists that leads to more tractable results. We *define* the average OTF (which we represent by $\overline{\mathcal{H}}$) as

$$\overline{\mathcal{H}}(\nu_U, \nu_V) = \frac{E[\text{numerator of the OTF}]}{E[\text{denominator of the OTF}]}, \quad (8.1\text{-}4)$$

where $E[\cdot]$ is, as usual, an expectation operator.

It is possible to argue that the two definitions given above are nearly the same in most cases of interest. For example, in the most important case of a random phase screen (see Section 8.3), $|\mathbf{t}_s|^2 = 1$, and the two definitions are identical. In more general cases, it can be argued that, if the pupil function is much wider than the correlation width of the screen, the substantial amount of spatial averaging of the nonnegative integrand taking place in the denominator of $\overline{\mathcal{H}}$ yields a normalizing factor that is nearly constant and known, and as a consequence the two definitions are essentially the same.

In any case, it is quite arbitrary what we choose as a normalizing factor for the average optical transfer function, provided only that a value of unity results at the origin of the frequency plane. Again, both definitions satisfy this requirement. By adopting the second definition, we are choosing to specify system performance by describing the average weighting applied by the system to frequency component (ν_U, ν_V), normalized by the average weighting given to the zero-frequency component of intensity.

If the numerator and denominator of Eq. (8.1-3) are substituted in Eq. (8.1-4), interchanges of orders of integration and averaging yield

$$\overline{\mathcal{H}}(\nu_U, \nu_V) = \frac{\iint\limits_{-\infty}^{\infty} \mathbf{P}(x,y)\mathbf{P}^*(x - \bar{\lambda}f\nu_U, y - \bar{\lambda}f\nu_V) E[\mathbf{t}_s(x,y)\mathbf{t}_s^*(x - \bar{\lambda}f\nu_U, y - \bar{\lambda}f\nu_V)]\,dx\,dy}{\iint\limits_{-\infty}^{\infty} |\mathbf{P}(x,y)|^2 E[|\mathbf{t}_s(x,y)|^2]\,dx\,dy}.$$

$$(8.1\text{-}5)$$

If we assume that the spatial statistics of the screen are wide-sense stationary, the expected values are independent of x and y and can be factored outside the integrals. The result is an average optical transfer function given by

$$\overline{\mathcal{H}}(\nu_U, \nu_V) = \mathcal{H}_0(\nu_U, \nu_V)\overline{\mathcal{H}}_s(\nu_U, \nu_V) \tag{8.1-6}$$

where $\mathcal{H}_0$ is the OTF of the system in the absence of the screen [Eq. (8.1-1)], whereas $\overline{\mathcal{H}}_s(\nu_U, \nu_V)$ can be regarded as the average OTF of the screen and is given by

$$\overline{\mathcal{H}}_s(\nu_U, \nu_V) \triangleq \frac{\Gamma_t(\bar{\lambda}f\nu_U, \bar{\lambda}f\nu_V)}{\Gamma_t(0,0)}, \tag{8.1-7}$$

where Γ_t is the spatial autocorrelation function of the screen,

$$\Gamma_t(\Delta x, \Delta y) \triangleq E[\mathbf{t}_s(x,y)\mathbf{t}_s^*(x - \Delta x, y - \Delta y)]. \tag{8.1-8}$$

Equations (8.1-6) and (8.1-7) represent the important results of this section. We have proved that the average optical transfer function of an incoherent imaging system with a spatially stationary random screen in the pupil factors into the product of the OTF of the system without the screen, times an average OTF associated with the screen. The average OTF associated with the screen is simply the normalized spatial autocorrelation function of the amplitude transmittance of the screen.

8.1.3 The Average Point-Spread Function

Often it is convenient to refer to an "average point-spread function (PSF)" of a system under discussion. For simplicity we define this PSF as

$$\bar{s}(u,v) \triangleq \mathcal{F}^{-1}\{\overline{\mathcal{H}}(\nu_U, \nu_V)\}, \tag{8.1-9}$$

where $\mathcal{F}^{-1}\{\ \}$ signifies an inverse Fourier transform. The average PSF so defined is always nonnegative and real. Furthermore, since $\overline{\mathcal{H}}$ has been normalized to unity at the origin, $\bar{s}(u,v)$ defined in Eq. (8.1-9) will always have unit volume.

Since the average OTF is a product of the system OTF (without the screen) and an OTF associated with the screen, the average PSF must be expressible as a convolution of the PSF of the system with a PSF associated with the screen. Thus

$$\bar{s}(u,v) = s_0(u,v) * \bar{s}_s(u,v) \tag{8.1-10}$$

where

$$s_0(u,v) = \mathscr{F}^{-1}\{\mathscr{H}_0(\nu_U, \nu_V)\}$$

represents the spread function of the system without the screen, whereas

$$\bar{s}_s(u,v) = \mathscr{F}^{-1}\{\overline{\mathscr{H}}_s(\nu_U, \nu_V)\}$$

represents an average spread function associated with the screen.

8.2 RANDOM ABSORBING SCREENS

Suppose that the random screen placed in the system illustrated in Fig. 8-1 is a purely absorbing structure; that is, it introduces no appreciable phase shifts. The amplitude transmittance $\mathbf{t}_s(x, y)$ of such a screen is purely real and nonnegative and must have values lying between zero and unity. We may regard this screen as introducing a *random apodization* of the optical system, and we consider the effects of such an apodization on average image quality.

8.2.1 General Forms of the Average OTF and the Average PSF

The amplitude transmittance of a random absorbing screen may be written in the form

$$\mathbf{t}_s(x, y) = t_0 + r(x, y), \tag{8.2-1}$$

where t_0 is a real and nonnegative bias lying between zero and unity, whereas $r(x, y)$ is taken to be a spatially stationary, zero mean, real-valued random process, with values confined to the range

$$-t_0 \le r(x, y) \le 1 - t_0. \tag{8.2-2}$$

As stated previously in Eq. (8.1-6), the average OTF of the system with the screen in place is given by the product of the OTF without the screen and the normalized autocorrelation function of the screen. The autocorrelation function of the screen is easily seen to be

$$\Gamma_t(\Delta x, \Delta y) = E\{[t_0 + r(x, y)][t_0 + r(x - \Delta x, y - \Delta y)]\}$$
$$= t_0^2 + \Gamma_r(\Delta x, \Delta y), \tag{8.2-3}$$

where Γ_r is the autocorrelation function of $r(x, y)$. The normalizing constant needed is

$$\Gamma_t(0,0) = t_0^2 + \overline{r^2} = t_0^2 + \sigma_r^2. \tag{8.2-4}$$

With the normalized autocorrelation function of the random process $r(x, y)$ defined by

$$\gamma_r(\Delta x, \Delta y) = \frac{\Gamma_r(\Delta x, \Delta y)}{\sigma_r^2}, \tag{8.2-5}$$

the average OTF associated with the random screen takes the form

$$\overline{\mathcal{H}}_s(\nu_U, \nu_V) = \frac{t_0^2}{t_0^2 + \sigma_r^2} + \frac{\sigma_r^2}{t_0^2 + \sigma_r^2} \gamma_r(\bar{\lambda} f \nu_U, \bar{\lambda} f \nu_V). \tag{8.2-6}$$

The average OTF of the overall system is found by multiplying the above average OTF times the OTF of the system in the absence of the screen, yielding

$$\overline{\mathcal{H}}(\nu_U, \nu_V) = \frac{t_0^2}{t_0^2 + \sigma_r^2} \mathcal{H}_0(\nu_U, \nu_V)$$

$$+ \frac{\sigma_r^2}{t_0^2 + \sigma_r^2} \mathcal{H}_0(\nu_U, \nu_V) \gamma_r(\bar{\lambda} f \nu_U, \bar{\lambda} f \nu_V). \tag{8.2-7}$$

Several important properties of the average OTF of the screen $\overline{\mathcal{H}}_s(\nu_U, \nu_V)$ should be noted. First it is always nonnegative and real, because of the nonnegative and real character of the amplitude transmittance t_s. Second, for very high spatial frequencies (ν_U, ν_V), $\gamma_r(\bar{\lambda} f \nu_U, \bar{\lambda} f \nu_V) \to 0$, and the average OTF of the screen approaches the asymptote

$$\overline{\mathcal{H}}_s(\nu_U, \nu_V) \to \frac{t_0^2}{t_0^2 + \sigma_r^2}. \tag{8.2-8}$$

If $\sigma_r^2 \ll t_0^2$, this asymptotic value is close to unity, and the screen has little effect on image quality. If $\sigma_r^2 \gg t_0^2$, the asymptotic value is very small, and high spatial frequencies are strongly suppressed by the screen. For a screen with isotropic statistics (i.e., a circularly symmetrical autocorrelation function), the average OTF of the screen begins to approach its asymptotic value when

$$(\nu_U^2 + \nu_V^2)^{1/2} > \frac{\Delta l}{\bar{\lambda} f}, \tag{8.2-9}$$

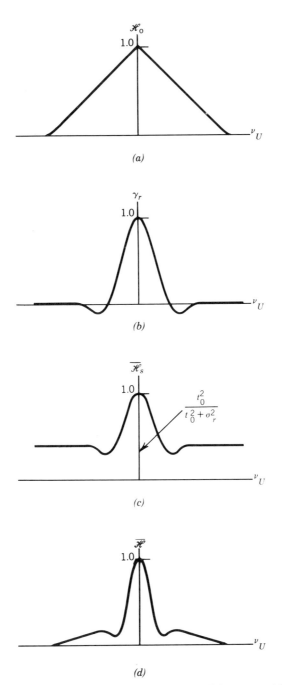

Figure 8-2. Effects of a random absorbing screen. (a) OTF of the system without screen; (b) normalized autocorrelation function of the screen; (c) average OTF of the screen; (d) average OTF of the system.

where Δl is the correlation length of the screen, defined for convenience by

$$\Delta l = \left[\frac{1}{\pi} \iint_{-\infty}^{\infty} |\gamma_r(\Delta x, \Delta y)|^2 \, d\Delta x \, d\Delta y \right]^{1/2}. \quad (8.2\text{-}10)$$

As an interesting exercise (see Problem 8-3), the reader may wish to prove that the minimum possible value of the asymptote $t_0^2/(t_0^2 + \sigma_r^2)$ for any random absorbing screen is identically equal to t_0, the average transmittance of the screen, a consequence of the fact that $\mathbf{t}_s$ is bounded between 0 and 1. Figure 8-2 illustrates the general character of the various transfer functions of concern for the case of a random absorbing screen.

The general character of the average PSF of the system can be understood by the following reasoning. To find the average PSF of the system with the screen present, we must convolve the PSF s_0 of the system without the screen and the average PSF $\bar{s}_s$ of the screen itself [cf. Eq. (8.1-10)]. To find the average PSF $\bar{s}_s$ of the screen, we must inverse Fourier transform the average OTF of the screen [Eq. (8.2-6)]. This latter operation yields

$$\bar{s}_s(u, v) = \frac{t_0^2}{t_0^2 + \sigma_r^2} \delta(u, v) + \frac{\sigma_r^2}{t_0^2 + \sigma_r^2} \bar{s}_r(u, v), \quad (8.2\text{-}11)$$

where

$$\bar{s}_r(u, v) = \mathscr{F}^{-1}\{\gamma_r(\bar{\lambda} f \nu_U, \bar{\lambda} f \nu_V)\}. \quad (8.2\text{-}12)$$

The average PSF of the system thus becomes

$$\bar{s}(u, v) = \frac{t_0^2}{t_0^2 + \sigma_r^2} s_0(u, v) + \frac{\sigma_r^2}{t_0^2 + \sigma_r^2} s_0(u, v) * \bar{s}_r(u, v). \quad (8.2\text{-}13)$$

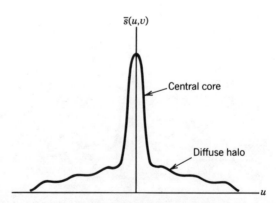

Figure 8-3. General character of the average PSF of a system with a random absorbing screen.

RANDOM ABSORBING SCREENS

The first term of this expression is often referred to as the "central core" of the average PSF. In most cases of interest, the second term is much wider than the first; this component is referred to as the "diffuse halo" in the average PSF. Figure 8-3 illustrates the general character of the average PSF.

8.2.2 A Specific Example

As a specific example of a random absorbing screen, we consider the "checkerboard" screen illustrated in Fig. 8-4. This structure consists of a multitude of contiguous $l \times l$ square cells, with randomly and independently chosen transmittance in each cell.

The screen itself may be regarded as being infinite in extent, although only a finite portion lies within the pupil of the optical system. To assure a model that is at least wide-sense stationary over space, the location of the screen with respect to the optical axis is taken to be random, with a uniform distribution of probability over an $l \times l$ square. This assumption simply implies a lack of knowledge of the exact location of the screen on the scale of a single cell. The transmittance t_s is assumed random and independent from cell to cell. For the moment we do not specify its exact probability distribution. The mean transmittance is represented by $\bar{t}_s$ or t_0 and the second moment by $\overline{t_s^2}$ or $t_0^2 + \sigma_r^2$.

We need to know the autocorrelation function of the screen

$$\Gamma_t(x_1, y_1; x_2, y_2) = \overline{t_s(x_1, y_1) t_s(x_2, y_2)} \qquad (8.2\text{-}14)$$

Figure 8-4. Random checkerboard absorbing screen.

in order to specify the average OTF of the system with the screen in place. Since different cells have statistically independent values of transmittance, the autocorrelation function can be written

$$\Gamma_t(x_1, y_1; x_2, y_2) = \overline{t_s^2} \cdot \text{Prob}\left\{\begin{array}{c}(x_1, y_1) \text{ and } (x_2, y_2) \text{ are}\\ \text{in the same cell}\end{array}\right\}$$

$$+ (\bar{t}_s)^2 \cdot \text{Prob}\left\{\begin{array}{c}(x_1, y_1) \text{ and } (x_2, y_2) \text{ are in}\\ \text{different cells}\end{array}\right\}.$$

(8.2-15)

Some reflection on the matter shows that, because of the uniform distribution of the absolute location of the screen,

$$\text{Prob}\left\{\begin{array}{c}(x_1, y_1) \text{ and } (x_2, y_2)\\ \text{are in the same cell}\end{array}\right\} = \Lambda\left(\frac{\Delta x}{l}\right)\Lambda\left(\frac{\Delta y}{l}\right), \quad (8.2\text{-}16)$$

where $\Delta x = x_1 - x_2$, $\Delta y = y_1 - y_2$, and $\Lambda(x) = 1 - |x|$ for $|x| \leq 1$, zero otherwise. It follows that

$$\Gamma_t(x_1, y_1; x_2, y_2) = \Gamma_t(\Delta x, \Delta y)$$

$$= \overline{t_s^2}\Lambda\left(\frac{\Delta x}{l}\right)\Lambda\left(\frac{\Delta y}{l}\right) + (\bar{t}_s)^2\left[1 - \Lambda\left(\frac{\Delta x}{l}\right)\Lambda\left(\frac{\Delta y}{l}\right)\right]$$

(8.2-17)

or

$$\Gamma_t(\Delta x, \Delta y) = \sigma_r^2 \Lambda\left(\frac{\Delta x}{l}\right)\Lambda\left(\frac{\Delta y}{l}\right) + t_0^2. \quad (8.2\text{-}18)$$

From this result we see that the normalized autocorrelation of the random portion of the screen amplitude transmittance is

$$\gamma_r(\Delta x, \Delta y) = \Lambda\left(\frac{\Delta x}{l}\right)\Lambda\left(\frac{\Delta y}{l}\right). \quad (8.2\text{-}19)$$

RANDOM ABSORBING SCREENS

The average OTF of an optical system with a screen such as this in place is now easily found by substituting (8.2-19) into (8.2-7), with the result

$$\overline{\mathcal{H}}(\nu_U, \nu_V) = \frac{t_0^2}{t_0^2 + \sigma_r^2} \mathcal{H}_0(\nu_U, \nu_V)$$

$$+ \frac{\sigma_r^2}{t_0^2 + \sigma_r^2} \mathcal{H}_0(\nu_U, \nu_V) \Lambda\left(\frac{\overline{\lambda} f \nu_U}{l}\right) \Lambda\left(\frac{\overline{\lambda} f \nu_V}{l}\right).$$

(8.2-20)

Noting that

$$\mathcal{F}^{-1}\left\{\Lambda\left(\frac{\overline{\lambda} f \nu_U}{l}\right) \Lambda\left(\frac{\overline{\lambda} f \nu_V}{l}\right)\right\} = \left(\frac{l}{\overline{\lambda} f}\right)^2 \text{sinc}^2\left(\frac{lu}{\overline{\lambda} f}\right) \text{sinc}^2\left(\frac{lv}{\overline{\lambda} f}\right),$$

(8.2-21)

we see that the average PSF of the system is

$$\bar{s}(u, v) = \frac{t_0^2}{t_0^2 + \sigma_r^2} s_0(u, v) + \frac{\sigma_r^2 (l/\overline{\lambda} f)^2}{t_0^2 + \sigma_r^2} s_0(u, v) * \text{sinc}^2\left(\frac{lu}{\overline{\lambda} f}\right) \text{sinc}^2\left(\frac{lv}{\overline{\lambda} f}\right).$$

(8.2-22)

The relative magnitudes of the terms in the average OTF and the average PSF can be assessed only if some specific assumptions are made regarding the statistics of the amplitude transmittance t_s. A case of some interest is that of a "black-and-white" checkerboard, for which $t_s = 1$ with probability p and $t_s = 0$ with probability $1 - p$. In this case

$$\bar{t}_s = t_0 = p$$

$$\overline{t_s^2} = t_0^2 + \sigma_r^2 = p \qquad (8.2\text{-}23)$$

$$\sigma_r^2 = p(1 - p).$$

The coefficients in Eq. (8.2-22) thus become

$$\frac{t_0^2}{t_0^2 + \sigma_r^2} = p, \qquad \frac{\sigma_r^2}{t_0^2 + \sigma_r^2} = 1 - p. \qquad (8.2\text{-}24)$$

Thus the average OTF is

$$\overline{\mathcal{H}}(\nu_U,\nu_V) = p\mathcal{H}_0(\nu_U,\nu_V) + (1-p)\mathcal{H}_0(\nu_U,\nu_V)\Lambda\left(\frac{\bar{\lambda}f\nu_U}{l}\right)\Lambda\left(\frac{\bar{\lambda}f\nu_V}{l}\right),$$
(8.2-25)

and the average PSF is given by

$$\bar{s}(u,v) = ps_0(u,v) + (1-p)\left(\frac{l}{\bar{\lambda}f}\right)^2 s_0(u,v) * \operatorname{sinc}^2\left(\frac{lu}{\bar{\lambda}f}\right)\operatorname{sinc}^2\left(\frac{lv}{\bar{\lambda}f}\right)$$

Clearly the more probable a clear cell (i.e., the larger p), the less effect the screen has on the performance of the system.

For consideration of a different distribution of the amplitude transmittance t_s, see Problem 8-4.

8.3 RANDOM-PHASE SCREENS

A second class of random screens, more important in practice than random absorbing screens, is the class of random-phase screens. A screen is called a *random-phase screen* if it changes the phase of the light transmitted in an unpredictable fashion but does not appreciably absorb light. The amplitude transmittance of such a screen takes the form

$$t_s(x,y) = \exp[j\phi(x,y)], \qquad (8.3\text{-}1)$$

where $\phi(x, y)$ is the (random) phase shift introduced at point (x, y).

As discussed in Section 7.1.1, the phase change ϕ can arise physically from changes of either the refractive index or the thickness of the screen (or both). Regardless of the physical origin of these phase changes, they are wavelength dependent, even in the absence of material dispersion, for they are proportional to the number of wavelengths of optical pathlength traveled by the wave as it passes through the screen. For a "thin" screen, therefore, the phase shift ϕ is taken to be

$$\phi(x,y) = \frac{2\pi}{\lambda}[L(x,y) - L_0], \qquad (8.3\text{-}2)$$

where $L(x, y)$ is the total optical path length (i.e., the product of refractive index and thickness) through the screen at (x, y), and L_0 is the mean pathlength associated with the screen.

In the analysis to follow, we replace the general wavelength λ by the mean wavelength $\bar{\lambda}$, thus neglecting the wavelength dependence of phase shift, in effect by assuming that the spectrum is sufficiently narrow to justify this approximation. As shown in Section 7.1.1 [in particular, in the development leading to Eq. (7.1-11)], this approximation is valid provided the pathlength differences through the screen do not exceed the coherence length of the illumination.

8.3.1 General Formulation

To understand the effects of a random phase screen on the performance of an incoherent imaging system, we must first find the spatial autocorrelation function of the amplitude transmittance t_s. Thus we must find the form of

$$\Gamma_t(x_1, y_1; x_2, y_2) = E[t_s(x_1, y_1) t_s^*(x_2, y_2)]. \quad (8.3\text{-}3)$$

Substituting (8.3-1) in (8.3-3), we see that

$$\Gamma_t(x_1, y_1; x_2, y_2) = E\{\exp[j\phi(x_1, y_1) - j\phi(x_2, y_2)]\}. \quad (8.3\text{-}4)$$

Two different interpretations of Eq. (8.3-4) are helpful in carrying the analysis further. First, the right-hand side of this equation can be recognized as being closely related to the second-order characteristic function of the joint random variables $\phi_1 = \phi(x_1, y_1)$ and $\phi_2 = \phi(x_2, y_2)$. With reference to Eq. (2.4-22), we see that

$$\Gamma_t(x_1, y_1; x_2, y_2) = \mathbf{M}_\phi(1, -1), \quad (8.3\text{-}5)$$

where

$$\mathbf{M}_\phi(\omega_1, \omega_2) = E[\exp(j\omega_1 \phi_1 + j\omega_2 \phi_2)] \quad (8.3\text{-}6)$$

is the characteristic function in question. Alternatively, (8.3-4) can be regarded as expressing Γ_t in terms of the first-order characteristic function of the phase difference $\Delta\phi = \phi_1 - \phi_2$,

$$\Gamma_t(x_1, y_1; x_2, y_2) = \mathbf{M}_{\Delta\phi}(1). \quad (8.3\text{-}7)$$

The two points of view are entirely equivalent, by virtue of the general fact that $\mathbf{M}_{\Delta\phi}(\omega) = \mathbf{M}_\phi(\omega, -\omega)$, proof of which is left as an exercise for the reader.

The general results (8.3-5) and (8.3-7) are as far as we can go without specific assumptions concerning the statistics of the phase $\phi(x, y)$. Accordingly, we turn to consideration of the most important special type of random-phase screen, namely, one with Gaussian statistics for the phase.

8.3.2 The Gaussian Random-Phase Screen

Let the random phase $\phi(x, y)$ be modeled as a zero mean Gaussian random process. Since both ϕ_1 and ϕ_2 are Gaussian, so is the phase difference $\Delta\phi$; further statistical properties include

$$\overline{\Delta\phi} = 0$$

$$\sigma_{\Delta\phi}^2 = \overline{(\phi_1 - \phi_2)^2} = D_\phi(x_1, y_1; x_2, y_2), \qquad (8.3\text{-}8)$$

where D_ϕ is the structure function of the random process $\phi(x, y)$ [cf. Eq. (3.4-17)]. Since the first-order characteristic function of the Gaussian random variable $\Delta\phi$ is

$$\mathbf{M}_{\Delta\phi}(\omega) = \exp\{-\tfrac{1}{2}\sigma_{\Delta\phi}^2\omega^2\}, \qquad (8.3\text{-}9)$$

appropriate substitutions yield

$$\Gamma_t(x_1, y_1; x_2, y_2) = \exp\{-\tfrac{1}{2}D_\phi(x_1, y_1; x_2, y_2)\} \qquad (8.3\text{-}10)$$

for the autocorrelation function of the amplitude transmittance of the screen.

If the random process $\phi(x, y)$ is *stationary in first increments*, the structure function of ϕ depends only on the coordinate differences $\Delta x = x_1 - x_2$ and $\Delta y = y_1 - y_2$. Thus

$$\Gamma_t(\Delta x, \Delta y) = \exp\{-\tfrac{1}{2}D_\phi(\Delta x, \Delta y)\}. \qquad (8.3\text{-}11)$$

The average OTF of the screen is thus given by

$$\mathcal{H}_s(\nu_U, \nu_V) = \exp\{-\tfrac{1}{2}D_\phi(\bar{\lambda}f\nu_U, \bar{\lambda}f\nu_V)\}. \qquad (8.3\text{-}12)$$

In the more restrictive case of a phase that is *wide-sense stationary*, the structure function can be expressed in terms of the normalized autocorrelation function $\gamma_\phi(\Delta x, \Delta y)$ of the phase [cf. Eq. (3.4-19)]

$$D_\phi(\Delta x, \Delta y) = 2\sigma_\phi^2[1 - \gamma_\phi(\Delta x, \Delta y)]. \qquad (8.3\text{-}13)$$

RANDOM-PHASE SCREENS

The average OTF of the screen then takes the form

$$\overline{\mathscr{H}}_s(\nu_U, \nu_V) = \exp\{-\sigma_\phi^2[1 - \gamma_\phi(\bar{\lambda}f\nu_U, \bar{\lambda}f\nu_V)]\}. \quad (8.3\text{-}14)$$

To understand the predicted behavior of the average OTF of an imaging system containing a random phase screen, it is first necessary to understand the behavior of the structure function D_ϕ. For a wide-sense stationary phase, we may consider the structure function of Eq. (8.3-13). Two important properties of this structure function are obvious:

(1) $\quad D_\phi(0,0) = 0$

(2) $\quad \left.\begin{array}{l} D_\phi(\infty, \Delta y) \\ D_\phi(\Delta x, \infty) \end{array}\right\} = 2\sigma_\phi^2$ $\quad (8.3\text{-}15)$

The latter property follows from the fact that the autocorrelation function γ_ϕ is expected to fall to zero as the separation between the two points becomes arbitrarily large. A typical behavior of the structure function is shown in Fig. 8-5 for three different variances σ_ϕ^2.

The typical behavior of the average OTF can now be sketched for the case under consideration. Figure 8-6 shows representative curves corresponding to $\mathscr{H}_0$, $\overline{\mathscr{H}}_s$, and $\overline{\mathscr{H}} = \mathscr{H}_0\overline{\mathscr{H}}_s$ for four values of phase variance. Note that for large ν_U or ν_V, the average OTF of the screen $\overline{\mathscr{H}}_s$ approaches

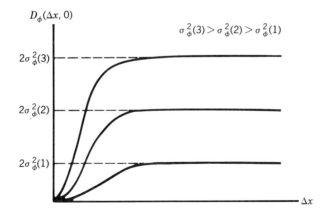

Figure 8-5. Typical behavior of the structure function of phase—wide-sense stationary case.

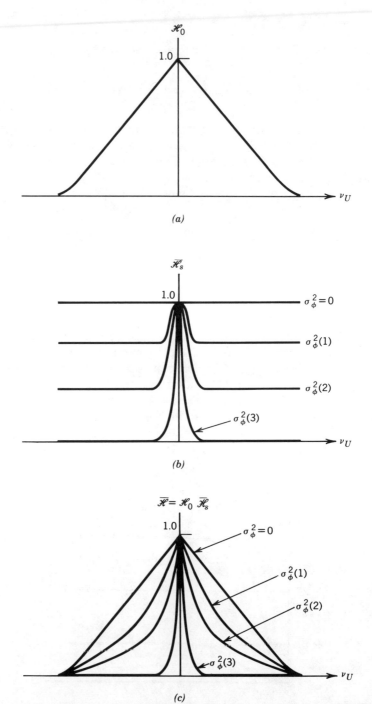

Figure 8-6. Typical OTF's for a system with a random phase screen: (*a*) diffraction-limited OTF; (*b*) average OTF of the screen; (*c*) average OTF of the system [$\sigma_\phi^2(1) < \sigma_\phi^2(2) < \sigma_\phi^2(3)$].

an asymptote

$$\left.\begin{array}{r}\overline{\mathcal{H}}_s(\nu_U,\infty)\\ \overline{\mathcal{H}}_s(\infty,\nu_V)\end{array}\right\} = \exp(-\sigma_\phi^2). \qquad (8.3\text{-}16)$$

For even modestly large σ_ϕ^2, this asymptote is extremely small.

The width of $\overline{\mathcal{H}}_s(\nu_U,\nu_V)$ will in general be considerably smaller than the width of the function $\gamma_\phi(\bar{\lambda}f\nu_U,\bar{\lambda}f\nu_V)$. This fact is perhaps most easily illustrated by considering a specific form for the autocorrelation function of the phase. For the purposes of illustration, therefore, consider a random-phase process with a circularly symmetric autocorrelation function

$$\gamma_\phi(r) = \exp\left\{-\left(\frac{r}{W}\right)^2\right\}, \quad r = \left[(\Delta x)^2 + (\Delta y)^2\right]^{1/2}. \qquad (8.3\text{-}17)$$

The average OTF of the screen thus becomes

$$\overline{\mathcal{H}}_s(\nu) = \exp\left\{-\sigma_\phi^2\left[1 - \exp\left(-\left(\frac{\bar{\lambda}f\nu}{W}\right)^2\right)\right]\right\},$$

$$\nu = \sqrt{\nu_U^2 + \nu_V^2}. \qquad (8.3\text{-}18)$$

For large phase variance, this OTF can be shown (see Problem 8-5) to fall to value $1/e$ when

$$\nu = \nu_{1/e} \cong \frac{W}{\bar{\lambda}f\sigma_\phi}. \qquad (8.3\text{-}19)$$

Thus the width of the average OTF of the screen varies directly with the width W of the phase autocorrelation function and inversely with the standard deviation σ_ϕ of the phase.

Returning to the case of a general phase autocorrelation function γ_ϕ, we now wish to find a useful approximate expression for the average OTF, which is valid when the OTF $\mathcal{H}_0$ of the original optical system is much wider than the average OTF of the screen. This approximation is found by rewriting the expression for $\overline{\mathcal{H}}_s$ as follows:

$$\overline{\mathcal{H}}_s = e^{-\sigma_\phi^2[1-\gamma_\phi]} = e^{-\sigma_\phi^2} + e^{-\sigma_\phi^2}\left[e^{\sigma_\phi^2\gamma_\phi} - 1\right]. \qquad (8.3\text{-}20)$$

The first term in this expression represents the asymptote to which the average OTF falls, whereas the second represents the rise above that asymptote. Now assuming that the OTF $\mathcal{H}_0$ of the original optical system is much wider than the second term in Eq. (8.3-20), we write

$$\overline{\mathcal{H}}(\nu_U, \nu_V) \cong \mathcal{H}_0(\nu_U, \nu_V) e^{-\sigma_\phi^2} + e^{-\sigma_\phi^2} \left[e^{\sigma_\phi^2 \gamma_\phi(\bar{\lambda} f \nu_U, \bar{\lambda} f \nu_V)} - 1 \right], \quad (8.3\text{-}21)$$

where we have replaced the factor $\mathcal{H}_0$ multiplying the second term by unity, its value at the origin.

Approximation (8.3-21) is particularly useful when we consider the average point-spread function of the overall system. With $s_0(u, v)$ again representing the PSF of the original system (without the screen), an inverse Fourier transformation of Eq. (8.3-21) yields an average PSF of the form

$$\bar{s}(u, v) \cong s_0(u, v) e^{-\sigma_\phi^2} + s_h(u, v), \quad (8.3\text{-}22)$$

where

$$s_h(u, v) = \mathcal{F}^{-1} \left\{ e^{-\sigma_\phi^2} \left[e^{\sigma_\phi^2 \gamma_\phi(\bar{\lambda} f \nu_U, \bar{\lambda} f \nu_V)} - 1 \right] \right\}. \quad (8.3\text{-}23)$$

We interpret the term $s_0(u, v) e^{-\sigma_\phi^2}$ to represent a diffraction-limited "core" of the PSF and the term $s_h(u, v)$ to represent a much broader "halo." Figure 8-7 illustrates this approximate form of the average PSF for three

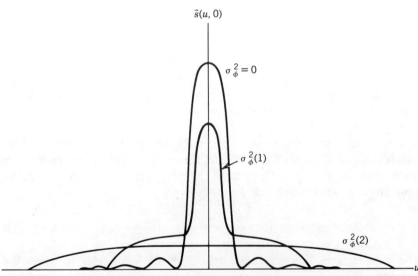

Figure 8-7. Average point-spread function for various phase variances $[\sigma_\phi^2(1) < \sigma_\phi^2(2)]$.

values of phase variance. It should be noted that the diffraction-limited "core" in the average PSF will very rapidly vanish with increasing phase variance and in many practical applications may be negligibly weak.

Finally, we note that, although the discussion of this section has been limited largely to the case of a wide-sense stationary screen, we can also draw some conclusions for the case of a screen that is stationary only in first increments. In this case, there may be no well-defined variance σ_ϕ^2 of the phase fluctuations, whereas the phase difference $\Delta\phi$ does have a well-defined variance for any separation. In general, the variance of $\Delta\phi$ will not approach an asymptote for large separations, but rather will increase indefinitely for increasing Δx or Δy. The structure function D_ϕ is again zero at the origin, but it need not approach an asymptotic value for large separations. Correspondingly modifications of Figs. 8-6b and 8-6c should thus be made. In particular, the average OTF of the screen need not contain a finite "plateau" at high frequencies, but rather will generally drop toward zero as the spatial frequencies increase. This, in turn, implies that the diffraction-limited "core" of the average PSF will be missing in this case.

8.3.3 Limiting Forms for Average OTF and Average PSF for Large Phase Variance

Certain limiting forms for the average OTF and the average PSF can be found when the phase fluctuations of a random-phase screen have a large variance. These limiting forms can provide useful approximations when the conditions for their validity are met.

Our attention is limited to phase screens that have the property that the first partial derivatives of the phase,

$$\phi_X(x, y) \triangleq \frac{\partial}{\partial x}\phi(x, y)$$
$$\phi_Y(x, y) \triangleq \frac{\partial}{\partial y}\phi(x, y),$$

(8.3-24)

are jointly stationary (in the strict sense) random processes. We begin the analysis with a slightly modified version of Eq. (8.3-4) for the autocorrelation function of the screen,

$$\Gamma_t(x_1, y_1; x_2, y_2) = E\{\exp[j(\phi(x_1, y_1) - \phi(x_1 - \Delta x, y_1 - \Delta y))]\}.$$

(8.3-25)

Now if the phase variance is large, we expect the value of the autocorrelation function Γ_t to drop toward zero in intervals Δx or Δy that are small compared with the correlation width of the phase $\phi(x, y)$. Accordingly, we can approximate the phase difference $\phi(x_1, y_1) - \phi(x_1 - \Delta x, y_1 - \Delta y)$ as having a linear dependence on Δx and Δy; thus

$$\phi(x_1, y_1) - \phi(x_1 - \Delta x, y_1 - \Delta y) \cong \Delta x \frac{\partial}{\partial x_1} \phi(x_1, y_1) + \Delta y \frac{\partial}{\partial y_1} \phi(x_1, y_1), \tag{8.3-26}$$

where we have dropped all higher-order terms in $(\Delta x, \Delta y)$.

With this approximation, the autocorrelation function of the screen becomes

$$\Gamma_t(x_1, y_1; x_1 - \Delta x, y_1 - \Delta y)$$

$$\cong E\left\{ \exp\left[j\Delta x \frac{\partial}{\partial x_1} \phi(x_1, y_1) + j\Delta y \frac{\partial}{\partial y_1} \phi(x_1, y_1) \right] \right\}$$

$$= \mathbf{M}_{\phi_X, \phi_Y}(\Delta x, \Delta y), \tag{8.3-27}$$

where $\mathbf{M}_{\phi_X, \phi_Y}$ represents the joint characteristic function of the two partial derivatives of the phase. The assumption that these partial derivatives are jointly stationary implies that the joint characteristic function is not a function of the coordinates (x_1, y_1).

The average OTF of the screen can now be expressed as a properly scaled version of the correlation function $\Gamma_t(\Delta x, \Delta y)$,

$$\overline{\mathcal{H}}_s(\nu_U, \nu_V) = \mathbf{M}_{\phi_X, \phi_Y}(\bar{\lambda} f \nu_U, \bar{\lambda} f \nu_V). \tag{8.3-28}$$

The average PSF associated with the screen is simply the inverse Fourier transform of this expression,

$$\bar{s}_s(u, v) = \frac{1}{(\bar{\lambda} f)^2} p_{\phi_X, \phi_Y}\left(\frac{u}{\bar{\lambda} f}, \frac{v}{\bar{\lambda} f} \right), \tag{8.3-29}$$

where $p_{\phi_X, \phi_Y}(\cdot, \cdot)$ signifies the joint probability density function of the partial derivatives ϕ_X and ϕ_Y.

Equation (8.3-29) has an interesting physical interpretation. In the limit of large phase variance, it is the slopes of the random-phase function that determine the distribution of energy in the average PSF. In effect, as the phase variance increases, the fluctuations of the slopes of the wavefront become so large that strictly geometric bending of the incident rays dominates over any diffraction effects that may be present.

In the special case of a zero-mean Gaussian random-phase screen, both partial derivatives are likewise Gaussian, and

$$\mathbf{M}_{\phi_X,\phi_Y}(\omega_X,\omega_Y) = \exp\left\{-\frac{1}{2}\left(\sigma_X^2\omega_X^2 + 2\sigma_X\sigma_Y\rho\omega_X\omega_Y + \sigma_Y^2\omega_Y^2\right)\right\}, \tag{8.3-30}$$

where σ_X^2 and σ_Y^2 represent the variances of ϕ_X and ϕ_Y, respectively, and ρ represents their correlation coefficient. The form of the average OTF of the screen thus becomes [with use of (8.3-28)]

$$\overline{\mathscr{H}}_s(\nu_U,\nu_V) = \exp\left\{-\frac{(\overline{\lambda}f)^2}{2}\left[\sigma_X^2\nu_U^2 + \sigma_Y^2\nu_V^2 + 2\sigma_X\sigma_Y\rho\nu_U\nu_V\right]\right\}. \tag{8.3-31}$$

Similarly, the average PSF of the screen takes a Gaussian form

$$\bar{s}_s(u,v) = \frac{1/(\overline{\lambda}f)^2}{2\pi\sigma_X^2\sigma_Y^2\sqrt{1-\rho^2}}\exp\left\{-\frac{\left[\left(\frac{u}{\sigma_X}\right)^2 + \left(\frac{v}{\sigma_Y}\right)^2 - 2\rho\left(\frac{u}{\sigma_X}\right)\left(\frac{v}{\sigma_Y}\right)\right]}{2(\overline{\lambda}f)^2(1-\rho^2)}\right\}. \tag{8.3-32}$$

With further specialization to the case of uncorrelated and identically distributed partial derivatives, the average PSF takes on the form of a circularly symmetrical Gaussian function,

$$\bar{s}_s(r) = \frac{1}{(\overline{\lambda}f)^2 \cdot 2\pi\sigma^4}\exp\left\{-\frac{r^2}{2(\overline{\lambda}f)^2\sigma^2}\right\} \tag{8.3-33}$$

with $r = [u^2 + v^2]^{1/2}$.

To summarize, we have found a limiting form for the average OTF of a phase screen when the phase variance is large. The result shows that the average OTF is given by a properly scaled version of the second-order characteristic function of the partial derivatives ϕ_X and ϕ_Y of the phase. When it is possible to hypothesize or derive a model for the joint statistics of ϕ_X and ϕ_Y, this limiting result can be useful. It also provides a demonstration that, for a *Gaussian* phase screen with large phase variance, the average OTF and average PSF are approximately Gaussian in shape.

8.4 EFFECTS OF AN EXTENDED RANDOMLY INHOMOGENEOUS MEDIUM ON WAVE PROPAGATION

In the previous sections of this chapter we considered the effects of thin random screens on the average performance of optical imaging systems. We now focus on the more important and more difficult case of an *extended* randomly inhomogeneous medium. As indicated in Fig. 8-8, the object of interest is an incoherent source. Between the imaging element and the object there exists an extended randomly inhomogeneous medium (e.g., the Earth's atmosphere). The image formed by such a system will be degraded by the presence of the inhomogeneous medium, and we seek a mathematical means for predicting these degradations.

As alluded to previously, the most important example of an extended randomly inhomogeneous medium is the Earth's atmosphere, which for centuries has limited the resolution of Man's images of the heavens above. Our analysis is directed toward this particular example from the start. As emphasized earlier in this chapter, our concern will be with the smooth and weak fluctuations of the refractive index of the clear air about us. We exclude from consideration the effects of particulate matter and aerosols on

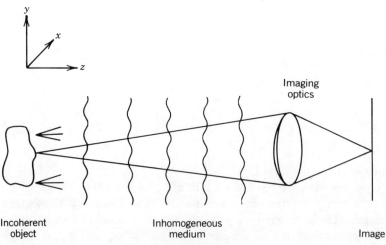

Figure 8-8. Imaging geometry.

optical propagation, these effects requiring the study of multiple scattering phenomena (see, e.g., Ref. 8-4, Vol. 2). We also limit our attention to optical propagation in a suitable spectral "window" of the atmosphere (such as the visible region of the spectrum), where atmospheric absorption is negligibly small. (For a detailed discussion of atmospheric absorption, see Ref. 8-5, Chapter 5.)

The literature on the subject of line-of-sight optical propagation through turbulence is vast. The reader may wish to consult any of a number of review articles and books on the subject (see, e.g., Refs. 8-4, 8-5 (Chapter 6), and 8-6 through 8-11). However, without doubt the most important and influential work on the subject is that of V. I. Tatarski, whose two books (Refs. 8-12 and 8-13) on the subject have been the foundation for most work that followed (see also Ref. 8-14).

Tatarski's work has had such an all-pervasive effect on this field that his notation is used in most work on this subject. To help the reader gain an ability to read the literature in this field, we abandon some of our earlier notation and adopt some of Tatarski's for the remainder of this chapter. To aid the reader in a transition to this new framework, we devote the subsection to follow to questions of definition and notation.

8.4.1 Notation and Definitions

The refractive index of the Earth's atmosphere varies over space, time, and wavelength. For our purposes, it is most convenient to express these dependences in the form

$$n(\vec{r}, t, \lambda) = n_0(\vec{r}, t, \lambda) + n_1(\vec{r}, t, \lambda), \tag{8.4-1}$$

where n_0 is the deterministic (nonrandom) portion of n, whereas n_1 represents random fluctuations of n about a mean value $\bar{n} = n_0 \cong 1$.

The deterministic changes in n are generally very slowly varying and macroscopic in spatial dimension. For example, n_0 contains the dependence of n on height above ground. Because of the comparatively long time scales associated with n_0, the time dependence of this term can be ignored in our discussions.

The random fluctuations n_1 arise in the presence of turbulence in the atmosphere. The turbulent eddies in the air have a range of scale sizes, varying from tens of meters or more to a few millimeters. The wavelength dependence of these random fluctuations can generally be ignored, with the result that we can rewrite Eq. (8.4-1) as

$$n(\vec{r}, t, \lambda) = n_0(\vec{r}, \lambda) + n_1(\vec{r}, t). \tag{8.4-2}$$

We further note that typical values of n_1 are several orders of magnitude smaller than unity (Ref. 8-15).

The time required for light to propagate through the atmosphere is only a small fraction of the "fluctuation time" of the random refractive index component n_1. For this reason, the time dependence of n_1 is often suppressed, with attention focused instead on the spatial properties. If temporal properties are of interest in a given problem, they can be introduced by invoking the "frozen turbulence" hypothesis (also known as Taylor's hypothesis), which assumes that a given realization of the random structure n_1 drifts across the measurement aperture with constant velocity (determined by the local wind conditions) but without any other change.

One of the most important statistical properties of the random process $n_1(\vec{r})$ is its spatial autocorrelation function

$$\Gamma_n(\vec{r}_1, \vec{r}_2) = E[n_1(\vec{r}_1)n_1(\vec{r}_2)]. \tag{8.4-3}$$

When n_1 is spatially stationary in three-dimensional space, we say that it is statistically *homogeneous*, and its autocorrelation function takes the simpler form

$$\Gamma_n(\vec{r}) = E[n_1(\vec{r}_1)n_1(\vec{r}_1 - \vec{r})], \tag{8.4-4}$$

where $\vec{r} = \vec{r}_1 - \vec{r}_2 = (\Delta x, \Delta y, \Delta z)$.

The power spectral density of n_1 is defined as the three-dimensional Fourier transform of $\Gamma_n(\vec{r})$ and in the notation to be used in the remainder of this chapter is written

$$\Phi_n(\vec{\kappa}) = \frac{1}{(2\pi)^3} \iiint_{-\infty}^{\infty} \Gamma_n(\vec{r}) e^{j\vec{\kappa}\cdot\vec{r}} d^3\vec{r}, \tag{8.4-5}$$

where $\vec{\kappa} = (\kappa_X, \kappa_Y, \kappa_Z)$ is called the wavenumber vector and may be regarded as a vector spatial frequency, with each component having units of radians per meter. Similarly, it is possible to express the autocorrelation function in terms of the power spectral density through

$$\Gamma_n(\vec{r}) = \iiint_{-\infty}^{\infty} \Phi_n(\vec{\kappa}) e^{-j\vec{\kappa}\cdot\vec{r}} d^3\vec{\kappa}. \tag{8.4-6}$$

If the refractive index fluctuations are further assumed to have an autocorrelation function with spherical symmetry, we say that n_1 is statistically *isotropic*, and the preceding three-dimensional Fourier transforms can be expressed in terms of single integrals through (see Ref. 8-16, pp. 251–253):

$$\Phi_n(\kappa) = \frac{1}{2\pi^2\kappa}\int_0^\infty \Gamma_n(r)\, r \sin(\kappa r)\, dr \qquad (8.4\text{-}7a)$$

$$\Gamma_n(r) = \frac{4\pi}{r}\int_0^\infty \Phi_n(\kappa)\,\kappa \sin(\kappa r)\, d\kappa, \qquad (8.4\text{-}7b)$$

where $\kappa = [\kappa_X^2 + \kappa_Y^2 + \kappa_Z^2]^{1/2}$ and $r = [(\Delta x)^2 + (\Delta y)^2 + (\Delta z)^2]^{1/2}$.

Occasionally it will be necessary to consider a two-dimensional autocorrelation function and a two-dimensional power spectral density of $n_1(\vec{r})$ across a plane with fixed axial z coordinate. This two-dimensional power spectral density is represented by $F_n(\kappa_X, \kappa_Y; z)$ and is related to the three-dimensional power spectral density through (see Problem 8-10)

$$F_n(\kappa_X, \kappa_Y; z) = \int_{-\infty}^\infty \Phi_n(\kappa_X, \kappa_Y, \kappa_Z)\, d\kappa_Z. \qquad (8.4\text{-}8)$$

The two-dimensional autocorrelation function $B_n(\vec{p}; z)$ is related to the two-dimensional power spectral density through

$$F_n(\vec{\kappa}; z) = \frac{1}{(2\pi)^2}\iint_{-\infty}^\infty B_n(\vec{p}; z)\, e^{j\vec{\kappa}\cdot\vec{p}}\, d^2\vec{p} \qquad (8.4\text{-}9a)$$

$$B_n(\vec{p}; z) = \iint_{-\infty}^\infty F_n(\vec{\kappa}; z)\, e^{-j\vec{\kappa}\cdot\vec{p}}\, d^2\vec{\kappa}, \qquad (8.4\text{-}9b)$$

where now $\vec{\kappa} = (\kappa_X, \kappa_Y)$ and $\vec{p} = (\Delta x, \Delta y)$. Note that by definition

$$B_n(\vec{p}; z) = E[n_1(\vec{p}_1; z) n_1(\vec{p}_1 - \vec{p}; z)] \qquad (8.4\text{-}10)$$

If the fluctuations of n_1 are statistically isotropic in the plane of constant z, then $B_n(\vec{p}; z)$ and $F_n(\vec{\kappa}; z)$ have circular symmetry, and Eqs. (8.4-9) can be reduced to

$$F_n(\kappa; z) = \frac{1}{2\pi}\int_0^\infty B_n(p; z) J_0(\kappa p)\, p\, dp \qquad (8.4\text{-}11a)$$

$$B_n(p; z) = 2\pi\int_0^\infty F_n(\kappa; z) J_0(\kappa p)\,\kappa\, d\kappa, \qquad (8.4\text{-}11b)$$

where $\kappa = [\kappa_X^2 + \kappa_Y^2]^{1/2}$ and $p = [(\Delta x)^2 + (\Delta y)]^{1/2}$.

With the establishment of the various notations above, we are now ready to consider the optical properties of the turbulent atmosphere in more detail.

8.4.2 Atmospheric Model

At optical frequencies the refractive index of air is given by (Ref. 8-15)

$$n = 1 + 77.6(1 + 7.52 \times 10^{-3}\lambda^{-2})\frac{P}{T} \times 10^{-6}, \qquad (8.4\text{-}12)$$

where λ is the wavelength of light in micrometers, P is the atmospheric pressure in millibars, and T is the temperature in Kelvins. The pressure variations of n are comparatively small and can be neglected, leaving temperature fluctuations as the dominant cause of fluctuations. For $\lambda = 0.5$ μm, the change dn of refractive index induced by an incremental change dT of temperature is

$$dn = \frac{79P}{T^2} \times 10^{-6} dT, \qquad (8.4\text{-}13)$$

again with P in millibars and T in Kelvins. For propagation near sea level, $|dn/dT|$ is on the order of 10^{-6}.

The random fluctuations n_1 of the refractive index are caused predominantly by random microstructure in the spatial distribution of temperature. The origin of this microstructure lies in extremely large scale temperature inhomogeneities caused by differential heating of different portions of the Earth's surface by the sun. These large-scale temperature inhomogeneities, in turn, cause large-scale refractive index inhomogeneities, which are eventually broken up by turbulent wind flow and convection, spreading the scale of the inhomogeneities to smaller and smaller sizes.

It is common to refer to the refractive index inhomogeneities as turbulent "eddies," which may be envisioned as packets of air, each with a characteristic refractive index. The power spectral density $\Phi_n(\vec{\kappa})$ of homogeneous turbulence may be regarded as a measure of the relative abundance of eddies with dimensions $L_X = 2\pi/\kappa_X$, $L_Y = 2\pi/\kappa_Y$, and $L_Z = 2\pi/\kappa_Z$. In the case of isotropic turbulence, $\Phi_n(\kappa)$ is a function of only one wavenumber κ, which may be considered as related to eddy size L through $L = 2\pi/\kappa$.

On the basis of classic work by Kolmogorov (Ref. 8-17) on the theory of turbulence, the power spectral density $\Phi_n(\kappa)$ is believed to contain three separate regions. For very small κ (very large scale sizes), we are concerned with the region in which most inhomogeneities originally arise. The mathematical form for Φ_n in this region is not predicted by the theory, for it depends on large-scale geographic and meteorological conditions. Furthermore, it is unlikely that the turbulence would be either isotropic or homogeneous at such scale sizes.

For κ larger than some critical wavenumber κ_0, the shape of $\Phi_n(\kappa)$ is determined by the physical laws that govern the breakup of large turbulent eddies into smaller ones. The scale size $L_0 = 2\pi/\kappa_0$ is called the *outer scale* of the turbulence. Near the ground, $L_0 \approx h/2$, where h is the height above ground. Typical numbers quoted for L_0 vary between 1 and 100 m, depending on atmospheric conditions and the geometry of the experiment in question.

For κ greater than κ_0, we enter the *inertial subrange* of the spectrum, where the form of Φ_n can be predicted from well-established physical laws governing turbulent flow. On the basis of Kolmogorov's work cited earlier, the form of Φ_n in the inertial subrange is given by

$$\Phi_n(\kappa) = 0.033 C_n^2 \kappa^{-11/3}, \qquad (8.4\text{-}14)$$

where C_n^2 is called the *structure constant* of the refractive index fluctuations and serves as a measure of the strength of the fluctuations.

When κ reaches another critical value κ_m, the form of Φ_n again changes. Turbulent eddies smaller than a certain scale size dissipate their energy as a result of viscous forces, resulting in a rapid drop in $\Phi_n(\kappa)$ for $\kappa > \kappa_m$. The scale size $l_0 \cong 2\pi/\kappa_m$ is referred to as the *inner scale* of the turbulence. A typical value for l_0 near the ground is a few millimeters. Tatarski includes the rapid decay of Φ_n for $\kappa > \kappa_m$ by use of the model

$$\Phi_n(\kappa) = 0.033 C_n^2 \kappa^{-11/3} \exp\left(\frac{-\kappa^2}{\kappa_m^2}\right). \qquad (8.4\text{-}15)$$

This equation is a reasonable approximation provided κ_m is chosen to equal $5.92/l_0$, and $\kappa > \kappa_0$.

Spectra (8.4-14) and (8.4-15) both have nonintegrable poles at the origin. In fact, since there is a finite amount of air associated with the Earth's atmosphere, the spectrum can not become arbitrarily large as $\kappa \to 0$. To overcome this defect in the model, a form known as the *von Kármán spectrum* is often adopted. The spectrum is then expressed approximately as

$$\Phi_n(\kappa) \cong \frac{0.033 C_n^2}{\left(\kappa^2 + \kappa_0^2\right)^{11/6}} \exp\left(\frac{-\kappa^2}{\kappa_m^2}\right). \qquad (8.4\text{-}16)$$

Note that for this spectrum,

$$\lim_{\kappa \to 0} \Phi_n(\kappa) = \frac{0.033 C_n^2}{\kappa_0^{11/3}}. \qquad (8.4\text{-}17)$$

It should be emphasized, however, that little is known about the actual shape of the spectrum in the very low wavenumber range, and Eq. (8.4-16) is only an artificial means for avoiding the pole at $\kappa = 0$. Furthermore, as we shall see, few optical experiments are significantly affected by eddies with scale sizes larger than the outer scale, so it is not really necessary to express a form for the spectrum in this range.

Figure 8-9 shows plots of $\Phi_n(\kappa)$ for the models (8.4-15) and (8.4-16), including indications of the locations of the wavenumbers κ_0 and κ_m between which the $\kappa^{-11/3}$ behavior holds.

In studying the effects of atmospheric turbulence on imaging systems, we shall find that it is the *structure function* of the refractive index fluctuations

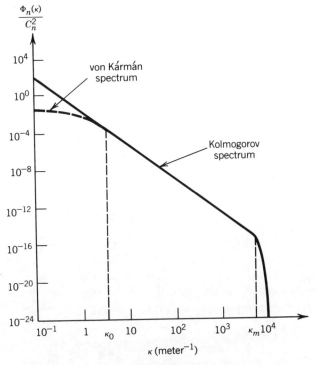

Figure 8-9. Power spectral densities of refractive index fluctuations.

EFFECTS OF AN INHOMOGENEOUS MEDIUM ON WAVE PROPAGATION 391

that influences the performance of such systems. By definition, this structure function is given by

$$D_n(\vec{r}_1, \vec{r}_2) = E\{[n_1(\vec{r}_1) - n_1(\vec{r}_2)]^2\}. \tag{8.4-18}$$

We wish to relate this structure function to the power spectral density Φ_n of the refractive index fluctuations.

If the turbulence is homogeneous and if $\Gamma_n(0)$ exists, it follows that [cf. Eq. (3.4-19)]

$$D_n(\vec{r}) = 2[\Gamma_n(0) - \Gamma_n(\vec{r})]. \tag{8.4-19}$$

Substituting (8.4-6) in (8.4-19), and taking account of the symmetry $\Phi_n(-\vec{\kappa}) = \Phi_n(\vec{\kappa})$ of any power spectral density, we find

$$D_n(\vec{r}) = 2 \iiint_{-\infty}^{\infty} [1 - \cos(\vec{\kappa} \cdot \vec{r})] \Phi_n(\vec{\kappa}) \, d^3\vec{\kappa}. \tag{8.4-20}$$

When the statistics of n_1 are isotropic, substitution of Eq. (8.4-7b) into (8.4-19) yields

$$D_n(r) = 8\pi \int_0^{\infty} \Phi_n(\kappa) \kappa^2 \left[1 - \frac{\sin \kappa r}{\kappa r}\right] d\kappa. \tag{8.4-21}$$

[To arrive at this expression, some care must be used in letting $r \to 0$ when expressing $\Gamma_n(0)$.]

A special advantage of using structure functions now becomes evident; namely, $D_n(r)$ is relatively insensitive to the behavior of $\Phi_n(\kappa)$ at very low wavenumbers. If we let κ become very small, the integrand of (8.4-21) is approximately $\kappa^4 \Phi_n(\kappa) r^2 / 3!$. Even though $\Gamma_n(0)$ may be infinite by virtue of behavior of $\Phi_n(\kappa)$ as κ^{-n} ($1 \leq n \leq 4$) as $\kappa \to 0$, nonetheless $D_n(r)$ will still be perfectly well defined. This indicates a certain insensitivity of $D_n(r)$ to the low-wavenumber portion of the spectrum. Furthermore, even if the very low wavenumber components of the refractive index power spectrum are nonhomogeneous and nonisotropic, n_1 may possess a homogeneous and isotropic structure function.

For the special case of interest here, the power spectral density behaves as $0.033 C_n^2 \kappa^{-11/3}$. Substitution of this form into Eq. (8.4-21) yields the integral

$$D_n(r) = 8\pi \times 0.033 C_n^2 \int_0^{\infty} \kappa^{-5/3} \left[1 - \frac{\sin \kappa r}{\kappa r}\right] d\kappa. \tag{8.4-22}$$

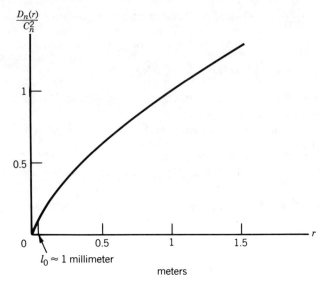

Figure 8-10. Refractive index structure function for Kolmogorov turbulence.

Now the integral identity (Ref. 8-18)

$$\int_0^\infty x^\nu \left(1 - \frac{\sin ax}{ax}\right) dx = \frac{-\Gamma(\nu)\sin\left(\frac{\pi\nu}{2}\right)}{a^{1+\nu}} \quad (-3 < \nu < -1)$$

(8.4-23)

can be used, where $\Gamma(\nu)$ represents the gamma function with argument ν. Noting that $\Gamma(-5/3) = 2.4110$, we obtain a structure function

$$D_n(r) = C_n^2 r^{2/3} \tag{8.4-24}$$

corresponding to the Kolmogorov spectrum of Eq. (8.4-14). The occurrence of a numerical coefficient of unity in this equation is no coincidence. C_n^2 is, in fact, defined in just such a way that this happens.

We should mention at this point that the value of the structure constant C_n^2 depends on both local atmospheric conditions and height above ground. Typical values near ground vary from 10^{-13} meter$^{-2/3}$ for strong turbulence to 10^{-17} meter$^{-2/3}$ for weak turbulence, with 10^{-15} meter$^{-2/3}$ often quoted as a typical "average" value.

EFFECTS OF AN INHOMOGENEOUS MEDIUM ON WAVE PROPAGATION

In closing this subsection, we note that the form of the structure function in Eq. (8.4-24) is valid only for $l_0 < r < L_0$, since the spectral density used in obtaining this result is valid only for $\kappa_0 < \kappa < \kappa_m$.

Figure 8-10 shows a plot of the structure function $D_n(r)$ expressed in Eq. (8.4-24).

8.4.3 Electromagnetic Wave Propagation Through the Inhomogeneous Atmosphere

Having characterized the statistical properties of the refractive index inhomogeneities in the atmosphere, we now consider the effects of these inhomogeneities on electromagnetic wave propagation. We consider a monochromatic electromagnetic wave, having time dependence $\exp(-j\omega t)$, propagating through the Earth's atmosphere. We have previously represented the refractive index of the atmosphere as

$$n(\vec{r}) = n_0(\vec{r}) + n_1(\vec{r}), \qquad (8.4\text{-}25)$$

where the time dependence has been suppressed. We suppose that the deterministic dependence $n_0(\vec{r})$ is essentially constant (independent of $\vec{r}$) over the region of our propagation experiment and thus represent the refractive index as

$$n(\vec{r}) = n_0 + n_1(\vec{r}). \qquad (8.4\text{-}26)$$

We assume that the atmosphere has constant magnetic permeability μ, but space-variant dielectric constant ε. Maxwell's equations then take the form

$$\nabla \cdot \vec{H} = 0$$

$$\nabla \times \vec{E} = j\omega\mu\vec{H}$$

$$\nabla \times \vec{H} = -j\omega\varepsilon\vec{E}$$

$$\nabla \cdot (\varepsilon\vec{E}) = 0, \qquad (8.4\text{-}27)$$

where $\vec{E}$ is the electric field, $\vec{H}$ is the magnetic field, and ∇ has vector components $(\partial/\partial x, \partial/\partial y, \partial/\partial z)$.

Applying a $\nabla \times$ operation to the second equation above, and substituting the third into the second, we obtain

$$\nabla \times (\nabla \times \vec{E}) = \omega^2 \mu \varepsilon \vec{E}. \qquad (8.4\text{-}28)$$

But

$$\nabla \times (\nabla \times \vec{E}) = -\nabla^2 \vec{E} + \nabla(\nabla \cdot \vec{E}), \qquad (8.4\text{-}29)$$

and from the last of Maxwell's equations,

$$\nabla \cdot (\varepsilon \vec{E}) = \varepsilon(\nabla \cdot \vec{E}) + \vec{E} \cdot \nabla \varepsilon = 0. \qquad (8.4\text{-}30)$$

Hence

$$\nabla \cdot \vec{E} = -\vec{E} \frac{\nabla \varepsilon}{\varepsilon} = -\vec{E} \cdot \nabla \ln \varepsilon, \qquad (8.4\text{-}31)$$

which, when substituted in Eqs. (8.4-29) and (8.4-28), yields

$$\nabla^2 \vec{E} + \omega^2 \mu \varepsilon \vec{E} + \nabla(\vec{E} \cdot \nabla \ln \varepsilon) = 0. \qquad (8.4\text{-}32)$$

Here, ln represents a natural (base e) logarithm.

The local velocity of propagation of the wave is $(\mu\varepsilon)^{-1/2}$, which is also equal to c/n, where c is the free-space velocity and n is the local refractive index. Hence

$$\mu\varepsilon = \frac{n^2}{c^2} \qquad (8.4\text{-}33)$$

and by the constancy of μ and c,

$$\nabla \ln \varepsilon = 2\nabla \ln n. \qquad (8.4\text{-}34)$$

Substituting these two equations into (8.4-32), we obtain

$$\nabla^2 \vec{E} + \frac{\omega^2 n^2}{c^2} \vec{E} + 2\nabla(\vec{E} \cdot \nabla \ln(n)) = 0, \qquad (8.4\text{-}35)$$

valid in any source-free region.

The last term in this equation introduces a coupling between the three components of $\vec{E}$ and thus corresponds to a depolarization term. It has been well established by past work that in the visible region of the spectrum, this term is completely negligible and can be replaced by zero (Ref. 8-19). Physically, depolarization effects are negligible because the inner scale of the turbulence l_0 is much larger than the wavelength λ. Thus the wave equation becomes

$$\nabla^2 \vec{E} + \frac{\omega^2 n^2}{c^2} \vec{E} = 0. \qquad (8.4\text{-}36)$$

EFFECTS OF AN INHOMOGENEOUS MEDIUM ON WAVE PROPAGATION 395

This equation differs from the conventional wave equation only through the fact that n^2 in the coefficient of the second term is a function of position $\vec{r}$.

Since all three components of the electric field obey the same wave equation, we can replace the vector equation by a scalar equation,

$$\nabla^2 U + \frac{\omega^2 n^2}{c^2} U = 0, \qquad (8.4\text{-}37)$$

where U can represent $\mathbf{E}_X$, $\mathbf{E}_Y$, or $\mathbf{E}_Z$.

Solution of this equation is accomplished by use of the method of small perturbations. Since $|n_1| \ll n_0$, it is reasonable to express the field U as the sum of a term $\mathbf{U}_0$ that would be obtained if the atmosphere had uniform refractive index n_0, plus a small correction term $\mathbf{U}_1$ that accounts for the effects of the index perturbations n_1. With this approximation, the wave equation becomes

$$\nabla^2(\mathbf{U}_0 + \mathbf{U}_1) + \frac{\omega^2}{c^2}(n_0 + n_1)^2(\mathbf{U}_0 + \mathbf{U}_1) = 0. \qquad (8.4\text{-}38)$$

Since $\mathbf{U}_0$ represents the unperturbed solution, it must satisfy

$$\nabla^2 \mathbf{U}_0 + k_0^2 \mathbf{U}_0 = 0, \qquad (8.4\text{-}39)$$

where $k_0^2 = \omega^2 n_0^2 / c^2$. Retention of only those terms that are first order in $\mathbf{U}_1$ and n_1 implies that $\mathbf{U}_1$ must satisfy

$$\nabla^2 \mathbf{U}_1 + k_0^2 \mathbf{U}_1 = \frac{-2k_0^2 n_1 \mathbf{U}_0}{n_0}. \qquad (8.4\text{-}40)$$

At this point and hereafter we assume that the mean refractive index n_0 is unity, an excellent approximation for the case of optical propagation.

Equation (8.4-40) is an inhomogeneous wave equation for $\mathbf{U}_1$, with a source term given by $-2k_0^2 n_1 \mathbf{U}_0$. Its solution can readily be expressed in terms of a convolution of the free-space Green's function (impulse response) $\exp(jk_0|\vec{r}|)/|\vec{r}|$ with the source term. The result is

$$\mathbf{U}_1(\vec{r}) = \frac{1}{4\pi} \iiint_V \frac{e^{jk|\vec{r}-\vec{r}'|}}{|\vec{r}-\vec{r}'|} \left[2k_0^2 n_1(\vec{r}') \mathbf{U}_0(\vec{r}') \right] d^3\vec{r}', \qquad (8.4\text{-}41)$$

where V is the scattering volume.

This expression for U_1 states that field perturbation U_1 can be found by summing a multitude of spherical waves generated at various points $\vec{r}'$ within the scattering volume V. The strength of the spherical wave generated at $\vec{r}'$ is proportional to the product of the incident unperturbed radiation and the refractive index perturbation at that point.

A further helpful approximation is obtained by use of the fact that the scattering angles involved in the propagation of visible light through the atmosphere are rather small. Since the smallest turbulent eddies are of the order of $l_0 \sim 2$ millimeters in size, whereas the wavelength is typically 0.5 micrometer, the scattering angles are no larger than $\lambda/l_0 \sim 2.5 \times 10^{-4}$ radians. Therefore, the maximum lateral displacement of a scatterer that contributes light to a given receiving point is much smaller than the axial distance of the scatterer from the receiver. As a result, the so-called *Fresnel approximation* (Ref. 8-20) can be applied to the integrand of Eq. (8.4-41), yielding

$$U_1(\vec{r}) = \frac{k_0^2}{2\pi} \iiint\limits_V \frac{\exp\left\{jk_0\left[(z-z') + \frac{|\vec{\rho}-\vec{\rho}'|^2}{2(z-z')}\right]\right\}}{z-z'} n_1(\vec{r}')U_0(\vec{r}')\, d^3\vec{r}', \qquad (8.4\text{-}42)$$

where $\vec{\rho}$ and $\vec{\rho}'$ represent the transverse displacement of $\vec{r}$ and $\vec{r}'$ from the z axis.

At this point we introduce a transformation used extensively by Tatarski (Refs. 8-12 and 8-13); namely, we define the complex quantity ψ as the natural logarithm of the field U:

$$\psi = \ln U. \qquad (8.4\text{-}43)$$

The reader may well wonder why such a transformation is useful; therefore we digress temporarily to address this question.

Our solution for the field was obtained by regarding the total field U as a sum of ever smaller contributions

$$U = U_0 + U_1 + U_2 + \cdots \qquad (8.4\text{-}44)$$

and by dropping those terms beyond the first perturbation U_1. This method of solution is referred to as the *Born approximation*, and it effectively neglects multiple scattering. There is, of course, a certain limited set of experimental conditions under which this solution can be expected to be accurate.

Tatarski introduced the transformation (8.4-43), called the *Rytov transformation*, at the very beginning of his analysis, rather than near the end, as we shall do. Such a substitution immediately transforms the wave equation (8.4-37) into the *Riccati equation*

$$\nabla^2 \psi(\vec{r}) + \nabla \psi(\vec{r}) \cdot \nabla \psi(\vec{r}) + \frac{\omega^2}{c^2} n^2(\vec{r}) = 0. \qquad (8.4\text{-}45)$$

Now the Riccati equation can be solved by assuming that

$$\psi = \psi_0 + \psi_1 + \psi_2 + \cdots \qquad (8.4\text{-}46)$$

and dropping all terms higher than ψ_1. This method of solution yields exactly the same result that we shall obtain for ψ_1, but it is thought to be valid under a wider range of conditions than apply for the Born approximation (see Ref. 8-4, Vol. 2, p. 349). The superiority of the Rytov method is borne out experimentally by the fact that, in the region of weak fluctuations, the statistics of the fluctuations of amplitude have been found to obey log-normal statistics. As we shall see, the solution for ψ implies that amplitude fluctuations obey log-normal statistics, whereas the solution for **U** does not imply statistics of this kind (Ref. 8-21) (see further discussion at the end of this subsection).

Returning to the transformation (8.4-43), we write

$$\mathbf{U} = \exp(\psi_0 + \psi_1)$$

$$\mathbf{U}_0 = \exp(\psi_0). \qquad (8.4\text{-}47)$$

In effect, we are representing **U** as a *multiplicatively* perturbed version of the free-space solution, rather than an *additively* perturbed version. [Note that $\mathbf{U}_1 \neq \exp(\psi_1)$.] We then have

$$\frac{\mathbf{U}}{\mathbf{U}_0} = 1 + \frac{\mathbf{U}_1}{\mathbf{U}_0} = e^{\psi_1}, \qquad (8.4\text{-}48)$$

and

$$\psi_1 = \ln\!\left(1 + \frac{\mathbf{U}_1}{\mathbf{U}_0}\right) \cong \frac{\mathbf{U}_1}{\mathbf{U}_0}. \qquad (8.4\text{-}49)$$

The last approximation is valid because $|\mathbf{U}_1| \ll |\mathbf{U}_0|$. Substituting Eq. (8.4-42) for U_1 in (8.4-49), we obtain

$$\psi_1(\vec{r}) = \frac{k_0^2}{2\pi U_0(\vec{r})} \iiint_V \frac{\exp\left\{jk_0\left[(z-z') + \frac{|\vec{\rho}-\vec{\rho}'|^2}{2(z-z')}\right]\right\}}{z-z'} n_1(\vec{r}') U_0(\vec{r}') \, d^3\vec{r}'.$$

(8.4-50)

This result is, in fact, identical with the result obtained by introducing the Rytov transformation at the very start.

With appropriate definitions, it is now possible to find expressions for the log-amplitude and phase of the wave perturbations. Let the amplitude and phase of the actual wave U be represented by A and S, while the amplitude and phase of the free-space solution are A_0 and S_0:

$$U = A\exp(jS)$$
$$U_0 = A_0\exp(jS_0). \quad (8.4\text{-}51)$$

Then

$$\psi_1 = \psi - \psi_0 = \ln\frac{A}{A_0} + j(S - S_0), \quad (8.4\text{-}52)$$

and defining

$$\chi \triangleq \ln\frac{A}{A_0} \quad \text{(the log-amplitude fluctuation)}$$

$$S_\delta \triangleq S - S_0 \quad \text{(the phase fluctuation)}, \quad (8.4\text{-}53)$$

we have

$$\psi_1 = \chi + jS_\delta. \quad (8.4\text{-}54)$$

Using (8.4-50), we conclude that

$$\begin{bmatrix} \chi \\ S_\delta \end{bmatrix} = \begin{bmatrix} \text{Re} \\ \text{Im} \end{bmatrix} \left\{ \frac{k_0^2}{2\pi U_0(\vec{r})} \iiint_V \frac{\exp\left\{jk_0\left[(z-z') + \frac{|\vec{\rho}-\vec{\rho}'|^2}{2(z-z')}\right]\right\}}{z-z'} \right.$$

$$\left. \times n_1(\vec{r}') U_0(\vec{r}') \, d^3\vec{r}' \right\}. \quad (8.4\text{-}55)$$

This equation represents the main result of the current subsection.

As mentioned earlier, the solutions obtained from the Born approximation and the Rytov approximation predict different probability density functions for amplitude A of the perturbed wave. In both cases the only random quantity present in the solution is the refractive index perturbation n_1. Equation (8.4-42) expresses the field perturbation $\mathbf{U}_1$ as a superposition of a vast number of independent contributions from different parts of the turbulent medium. According to the central limit theorem, we would expect the real and imaginary parts of $\mathbf{U}_1$ to obey Gaussian or normal statistics. The statistics predicted for the intensity of the total wave depend on the variances of the real and imaginary parts of $\mathbf{U}_1$ and on their correlation. When these variances are equal and the correlation is zero, the sum of $\mathbf{U}_0$ and $\mathbf{U}_1$ would be equivalent to the sum of a constant (nonrandom) phasor and a circular complex Gaussian phasor. The results in Section 2.9.4 imply that under these conditions $A = |\mathbf{U}|$ should obey Rician statistics. In general, however, the assumption of equal variances and zero correlation is not justified, and a more complicated solution for the intensity statistics would be expected.

On the other hand, Eq. (8.4-55) expresses the log-amplitude fluctuations χ as a superposition of a multitude of independent contributions. Again invoking the central limit theorem, the form of this solution leads us to predict Gaussian statistics for χ, which, in turn, implies that the amplitude A should be a log-normal variate.

The preponderance of experimental evidence favors the log-normal distribution under conditions of weak fluctuations, and it is generally agreed that under such conditions this statistical model is sufficiently accurate to justify its use in theoretical calculations.

8.4.4 The Log-Normal Distribution

Because the log-normal distribution plays an especially important role in the theory of propagation through turbulence, we now devote a short discussion to some of its properties.

The log amplitude χ is taken to be a Gaussian random variable with mean $\bar{\chi}$ and standard deviation σ_χ. Hence

$$p_\chi(\chi) = \frac{1}{\sqrt{2\pi}\,\sigma_\chi} \exp\left\{ -\frac{(\chi - \bar{\chi})^2}{2\sigma_\chi^2} \right\}. \tag{8.4-56}$$

To find the form of the probability density function for the amplitude

$$A = A_0 \exp(\chi), \tag{8.4-57}$$

we must introduce a probability transformation. Noting that

$$p_A(A) = p_\chi(\chi = \ln A)\left|\frac{d\chi}{dA}\right| \qquad (8.4\text{-}58)$$

and that $d\chi/dA = 1/A$, we find

$$p_A(A) = \frac{1}{\sqrt{2\pi}\,\sigma_\chi A}\exp\left\{-\frac{\left(\ln\frac{A}{A_0} - \bar{\chi}\right)^2}{2\sigma_\chi^2}\right\}, \qquad (A \geq 0). \quad (8.4\text{-}59)$$

In a similar fashion, the probability density function for the intensity $I = A^2$ is found to be

$$p_I(I) = \frac{1}{2\sqrt{2\pi}\,\sigma_\chi I}\exp\left\{-\frac{\left(\frac{1}{2}\ln\frac{I}{I_0} - \bar{\chi}\right)^2}{2\sigma_\chi^2}\right\}$$

$$= \frac{1}{2\sqrt{2\pi}\,\sigma_\chi I}\exp\left\{-\frac{\left(\ln\frac{I}{I_0} - 2\bar{\chi}\right)^2}{8\sigma_\chi^2}\right\}, \qquad (I \geq 0). \quad (8.4\text{-}60)$$

Equations (8.4-59) and (8.4-60) are both examples of the log-normal probability density function.

As they stand, these density functions have three independent parameters, $\bar{\chi}$, σ_χ, and I_0. However, if we fix the mean value of the log-normal variate, requiring, for example, that $\bar{I} = I_0$, we find that $\bar{\chi}$ and σ_χ can no longer be chosen independently.

To prove this assertion, let

$$\bar{I} = I_0 \overline{e^{2\chi}} = I_0. \qquad (8.4\text{-}61)$$

Now we use the following relationship, valid for any real-valued Gaussian random variable z and any complex constant $\mathbf{a}$:

$$E[e^{\mathbf{a}z}] = \exp\{\mathbf{a}\bar{z} + \tfrac{1}{2}\mathbf{a}^2\sigma_z^2\}. \qquad (8.4\text{-}62)$$

With $z = \chi$ and $a = 2$, we find

$$\bar{I} = I_0 e^{2\bar{\chi} + 2\sigma_\chi^2} = I_0, \qquad (8.4\text{-}63)$$

or equivalently,

$$\bar{\chi} = -\sigma_\chi^2. \qquad (8.4\text{-}64)$$

There are still two independent parameters here, but if we choose I_0 as one of them, $\bar{\chi}$ and σ_χ^2 cannot be chosen independently.

The preceding relationship is relevant to the propagation problem when we make the assumption that the waves propagate without any significant attenuation. Thus a unit intensity plane wave entering the atmosphere at $z = 0$ must, by energy conservation, still have unity *average* intensity when it reaches $z = L$.

The probability density function $p_I(I)$ is plotted in Fig. 8-11 for various values of the parameter σ_χ and subject to the energy conservation constraint. As can be seen, a wide variety of different shapes occur, even for a fixed mean I_0.

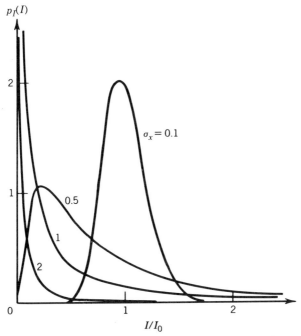

Figure 8-11. Log-normal probability density function for intensity when the mean intensity is unity.

The log-normal distribution has other properties that are rather unusual among distributions that describe real physical phenomena. First, it is known to violate the conditions necessary for a probability density function to be completely determined by its moments (Ref. 8-22). Second, it has been found to have a certain permanence in the sense that sums of independent log-normal variates tend toward Gaussian statistics very slowly (Ref. 8-23). However, these subjects take us a bit too far away from our prime goal, namely, determining the effects of turbulence on the performance of imaging systems.

8.5 THE LONG-EXPOSURE OTF

The inhomogeneities of the atmosphere are in constant turbulent motion, with the result that the instantaneous wavefront degradations fluctuate rapidly with time. To "freeze" the atmospheric degradations, thus eliminating any time averaging effects, it is necessary to use an exposure time from 0.01 to 0.001 second or shorter, depending on the effective wind velocity. Figure 8-12 shows examples of both long-exposure ($T \gg \frac{1}{100}$ second) and short-exposure ($T \ll \frac{1}{100}$ second) photographs of a star.

As might be expected from the appearance of the images in Fig. 8-12, there are distinct differences between the OTFs achieved with long and short exposures. In this section we consider only the long-exposure case, as appropriate, for example, in the imaging of faint astronomical objects, which may require seconds, minutes, or even hours of integration time. Underlying our analysis will be the assumption of temporal ergodicity, namely, that the long-time-average OTF—which is affected by many independent realizations of the atmospheric inhomogeneities—is identical with the ensemble-averaged OTF.

Recently there has arisen new interest in the short-exposure OTF, due partly to advances in adaptive optics (Ref. 8-24) and stellar speckle interferometry (Ref. 8-25). We discuss these subjects in later sections, but for the moment we limit our attention to the long-exposure case.

8.5.1 Long-Exposure OTF in Terms of the Wave Structure Function

With reference to Fig. 8-13, consider an extremely distant quasimonochromatic point source located on the optical axis of a simple imaging system. In the absence of atmospheric turbulence, this source would generate a plane wave normally incident on the lens. In the presence of the atmosphere, the plane wave incident on the inhomogeneous medium propagates into the

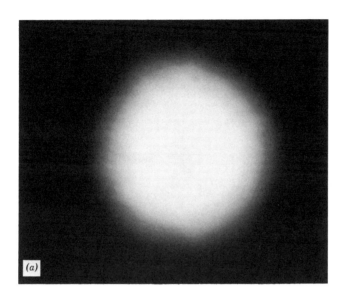

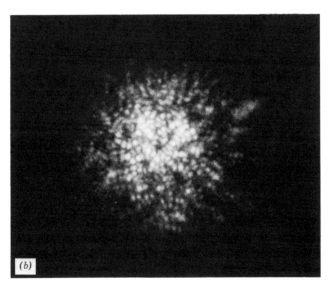

Figure 8-12. (*a*) Long- and (*b*) short-exposure photographs of the star *Lambda Cratis*. (Courtesy of Gerd Weigelt and Gerhard Baier, University of Erlangen.)

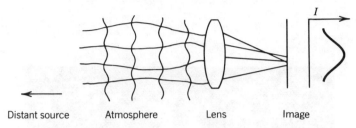

Figure 8-13. Imaging of a distant point source through the atmosphere.

medium, and ultimately a perturbed wave falls on the lens. The field distribution incident on the lens can be expressed as

$$U(x, y) = \sqrt{I_0} \exp\{\chi(x, y) + jS(x, y)\}, \qquad (8.5\text{-}1)$$

where I_0 is the intensity of the incident plane wave, and, as predicted by the Rytov solution, χ and S are Gaussian random variables.

Following the reasoning that led to Eq. (8.1-3), the instantaneous OTF of the system can be expressed as

$$\mathcal{H}(\nu_U, \nu_V) = \frac{\displaystyle\iint_{-\infty}^{\infty} \mathbf{P}(x, y)\mathbf{P}^*(x - \bar{\lambda}f\nu_U, y - \bar{\lambda}f\nu_V)\exp\{(\chi_1 + \chi_2) + j(S_1 - S_2)\}\,dx\,dy}{\displaystyle\iint_{-\infty}^{\infty} \mathbf{P}(x, y)\mathbf{P}^*(x, y)\exp\{2\chi\}\,dx\,dy},$$

$$(8.5\text{-}2)$$

where $\mathbf{P}(x, y)$ is the complex pupil function of the system in the absence of atmospheric turbulence, whereas

$$\chi_1 = \chi(x, y)$$
$$\chi_2 = \chi(x - \bar{\lambda}f\nu_U, y - \bar{\lambda}f\nu_V)$$
$$S_1 = S(x, y)$$
$$S_2 = S(x - \bar{\lambda}f\nu_U, y - \bar{\lambda}f\nu_V). \qquad (8.5\text{-}3)$$

Note that χ_1, χ_2, S_1, and S_2 are functions of time, but this time dependence has been suppressed in writing the instantaneous OTF.

THE LONG-EXPOSURE OTF

Under our assumption of ergodicity, the ensemble average OTF will be identical with the long-exposure OTF in the limit of infinitely long integration time. Thus we wish to take ensemble averages of the numerator and the denominator of Eq. (8.5-2). The result can be expressed as

$$\overline{\mathcal{H}}(\nu_U, \nu_V) = \mathcal{H}_0(\nu_U, \nu_V)\overline{\mathcal{H}}_L(\nu_U, \nu_V), \qquad (8.5\text{-}4)$$

where $\mathcal{H}_0$ is the OTF of the optical system in the absence of turbulence, whereas $\overline{\mathcal{H}}_L$ maybe regarded as the long-exposure OTF of the atmosphere, and is given by

$$\overline{\mathcal{H}}_L(\nu_U, \nu_V) = \frac{\overline{\Gamma}(\overline{\lambda}f\nu_U, \overline{\lambda}f\nu_V)}{\overline{\Gamma}(0,0)} \qquad (8.5\text{-}5)$$

with

$$\overline{\Gamma}(\Delta x, \Delta y) = E\left[\exp\{(\chi_1 + \chi_2) + j(S_1 - S_2)\}\right]. \qquad (8.5\text{-}6)$$

In writing $\overline{\Gamma}$ as a function of Δx and Δy, we have assumed that the wavefront perturbutions obey homogeneous statistics. Our ability to calculate the atmospheric OTF thus depends heavily on our knowledge of the statistical properties of χ and S. The fact that both χ and S obey Gaussian statistics is the key to success.

In general, we have no reason to suppose that χ and S are independent random processes, for their fluctuations both arise from the fluctuations of the refractive index. However, consider the following average:

$$\overline{(\chi_1 + \chi_2)(S_1 - S_2)} = \overline{\chi_1 S_1} - \overline{\chi_2 S_2} - \overline{\chi_1 S_2} + \overline{\chi_2 S_1}. \qquad (8.5\text{-}7)$$

If the refractive index fluctuations obey homogeneous statistics, χ and S must be jointly homogeneous, in which case

$$\overline{\chi_1 S_1} = \overline{\chi_2 S_2}. \qquad (8.5\text{-}8)$$

If in addition n obeys isotropic statistics, χ and S will be jointly isotropic, with the result

$$\overline{\chi_1 S_2} = \overline{\chi_2 S_1}. \qquad (8.5\text{-}9)$$

It follows that

$$\overline{(\chi_1 + \chi_2)(S_1 - S_2)} = 0, \qquad (8.5\text{-}10)$$

and we see that the random variables $(\chi_1 + \chi_2)$ and $(S_1 - S_2)$ are uncorrelated. Finally, as a result of the Gaussian statistics of χ and S, $(\chi_1 + \chi_2)$ and $(S_1 - S_2)$ are Gaussian, and their lack of correlation implies statistical independence. As a consequence, we see that

$$\overline{\Gamma}(\Delta x, \Delta y) = \overline{\exp(\chi_1 + \chi_2)} \,\overline{\exp[j(S_1 - S_2)]}. \quad (8.5\text{-}11)$$

From our earlier discussions of Gaussian phase screens, we know that

$$\overline{\exp[j(S_1 - S_2)]} = \exp[-\tfrac{1}{2} D_S(r)], \quad (8.5\text{-}12)$$

where $r = [(\Delta x)^2 + (\Delta y)^2]^{1/2}$ and $D_S(r)$ is the phase structure function,

$$D_S = \overline{(S_1 - S_2)^2}. \quad (8.5\text{-}13)$$

We must now calculate the average of $\exp(\chi_1 + \chi_2)$.

To aid in this calculation, we use the relation presented in Eq. (8.4-62), which is valid for any Gaussian random variable z:

$$E[e^{az}] = \exp\{a\bar{z} + \tfrac{1}{2} a^2 \sigma_z^2\} \quad (8.5\text{-}14)$$

Choosing $z = \chi_1 + \chi_2$ and $a = 1$, we obtain

$$E[e^{\chi_1 + \chi_2}] = e^{\tfrac{1}{2}\overline{(\chi_1 + \chi_2 - 2\bar{\chi})^2}} e^{2\bar{\chi}}. \quad (8.5\text{-}15)$$

Noting that

$$\tfrac{1}{2}\overline{(\chi_1 + \chi_2 - 2\bar{\chi})^2} = \tfrac{1}{2}\overline{[(\chi_1 - \bar{\chi}) + (\chi_2 - \bar{\chi})]^2}$$

$$= \tfrac{1}{2}\overline{(\chi_1 - \bar{\chi})^2} + \tfrac{1}{2}\overline{(\chi_2 - \bar{\chi})^2} + \overline{(\chi_1 - \bar{\chi})(\chi_2 - \bar{\chi})}$$

$$= C_\chi(0) + C_\chi(r), \quad (8.5\text{-}16)$$

where C_χ is the autocovariance of χ, we see that

$$E[e^{\chi_1 + \chi_2}] = \exp\{C_\chi(0) + C_\chi(r)\} e^{2\bar{\chi}}. \quad (8.5\text{-}17)$$

At this point we invoke conservation of energy to conclude that the mean intensity of an infinite plane wave propagating through a randomly inhomogeneous and lossless medium must remain constant. It follows [cf. Eq. (8.4-64)] that

$$\bar{\chi} = -\sigma_\chi^2 = -C_\chi(0). \quad (8.5\text{-}18)$$

Incorporating this fact in Eq. (8.5-17), we find

$$E[e^{\chi_1+\chi_2}] = \exp\{-C_\chi(0) + C_\chi(r)\}$$
$$= \exp\{-\tfrac{1}{2}D_\chi(r)\}, \quad (8.5\text{-}19)$$

where D_χ is the log-amplitude structure function,

$$D_\chi = \overline{(\chi_1 - \chi_2)^2}. \quad (8.5\text{-}20)$$

We conclude that

$$\bar{\Gamma}(r) = \exp\{-\tfrac{1}{2}D(r)\},$$
$$\bar{\Gamma}(0) = 1, \quad (8.5\text{-}21)$$

where $D = D_\chi + D_S$ is called the *wave structure function*. The total average OTF thus takes the form

$$\overline{\mathcal{H}}(\nu) = \mathcal{H}_0(\nu)\exp\{-\tfrac{1}{2}D(\bar{\lambda}f\nu)\}, \quad (8.5\text{-}22)$$

where $\nu = [\nu_U^2 + \nu_V^2]^{1/2}$ and a circularly symmetric OTF of the unperturbed optical system has been assumed for simplicity.

8.5.2 Near-Field Calculation of the Wave Structure Function

We now consider the problem of calculating a detailed expression for the wave structure function. Such a result will allow us to specify the form of the long-exposure atmospheric OTF in more detail.

In this first analysis of the problem, we adopt some rather severe simplifying assumptions that are only occasionally met in practice. Following this simplified analysis, however, we show how the validity of the results can be extended to a far wider range of conditions than might have been thought at the start (see Section 8.6).

The major assumptions adopted are listed below:

(1) The object of interest is at a very large distance from the lens, and its angular extent is so small that all parts of the object are affected identically by the atmosphere, at least over a long time average.
(2) The turbulence exists over a finite distance z in front of the imaging lens, and is homogeneous and isotropic in that region.

(3) The imaging system lies deep within the near field of the most important turbulent eddies, with the result that, to a good approximation, each ray incident on the inhomogeneous medium is simply delayed by that medium, with no significant bending of the rays. (This assumption is strictly valid only if $z \ll l_0^2/\bar{\lambda}$.)

Assumption 1 may be referred to as an "isoplanatic" assumption and is not a very restrictive one for long-time-average imaging. Assumption 2 is not valid for vertical viewing through the atmosphere, but it will be removed in Section 8.5.4. Assumption 3 is primarily a statement that the turbulence is so weak that no significant amplitude scintillation effects are present. Such a condition is generally not valid in practice; however, a possible exception is vertical viewing from a mountain-top observatory under conditions of good atmospheric "seeing." This assumption is removed in Section 8.6.1.

Figure 8-14 illustrates the geometry on which the calculation will be based. Assumption 1 allows us to consider only a single point source on the optical axis, producing a plane wave at the entrance to the region of turbulence. Assumption 3 is now used to express the phase delays S_1 and S_2, suffered by the two parallel rays shown, as

$$S_1 = \bar{k} \int_0^z [n_0 + n_1(\vec{r}_1)]\, dz'$$

$$S_2 = \bar{k} \int_0^z [n_0 + n_1(\vec{r}_2)]\, dz', \qquad (8.5\text{-}23)$$

where $\bar{k} = 2\pi/\bar{\lambda}$.

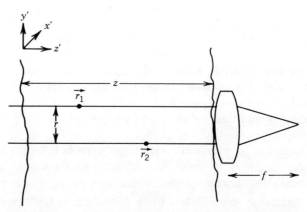

Figure 8-14. Geometry for phase structure function calculation.

THE LONG-EXPOSURE OTF

In addition, this same assumption implies that amplitude fluctuations are negligible, in which case $\chi_1 = \chi_2$ and $D_\chi(r) = 0$. Since the log-amplitude structure function is zero, the wave structure function equals the phase structure function, and the average atmospheric OTF is given by

$$\overline{\mathcal{H}}_L(\nu) = \exp\{-\tfrac{1}{2}D_S(\bar{\lambda}f\nu)\}. \tag{8.5-24}$$

In calculating the phase structure function, we choose the origin of our coordinate system to lie at the position where the lower ray in Fig. 8-14 enters the turbulent region. Then

$$n(\vec{r}_1) = n(z', r)$$
$$n(\vec{r}_2) = n(z', 0). \tag{8.5-25}$$

The phase structure function will be known if we can calculate the average

$$E\big[(S_1 - S_2)^2\big] = (\bar{k})^2 E\!\left[\left\{\int_0^z dz'\big[n_1(z', r) - n_1(z', 0)\big]\right\}^2\right]. \tag{8.5-26}$$

This quantity can be equivalently expressed as

$$\overline{(S_1 - S_2)^2} = (\bar{k})^2 E\!\left\{\int_0^z\!\int_0^z dz'\,dz''\big[n_1(z', r) - n_1(z', 0)\big]\right.$$

$$\left. \times \big[n_1(z'', r) - n_1(z'', 0)\big]\right\}$$

$$= (\bar{k})^2 \int_0^z dz' \int_0^z dz''\,\big[\overline{n_1(z', r)n_1(z'', r)} + \overline{n_1(z', 0)n_1(z'', 0)}$$

$$- \overline{n_1(z', r)n_1(z'', 0)} - \overline{n_1(z', 0)n_1(z'', r)}\big]. \tag{8.5-27}$$

The averages can be expressed in terms of covariance functions

$$D_S(r) = (\bar{k})^2 \int_0^z dz' \int_0^\infty dz''\Big[2C_n(z' - z'') - 2C_n\!\left(\sqrt{(z' - z'')^2 + r^2}\right)\Big]. \tag{8.5-28}$$

The difference of covariance functions can be expressed as a difference of structure functions,

$$2C_n(z' - z'') - 2C_n\left(\sqrt{(z' - z'')^2 + r^2}\right)$$

$$= \left[2C_n(0) - 2C_n\left(\sqrt{(z' - z'')^2 + r^2}\right)\right] - \left[2C_n(0) - 2C_n(z' - z'')\right]$$

$$= D_n\left(\sqrt{(z' - z'')^2 + r^2}\right) - D_n(z' - z''). \tag{8.5-29}$$

Combining (8.5-28) and (8.5-29), we find the phase structure function to be given by

$$D_S(r) = (\bar{k})^2 \int_0^z dz' \int_0^z dz'' \left[D_n\left(\sqrt{(z' - z'')^2 + r^2}\right) - D_n(z' - z'')\right]. \tag{8.5-30}$$

This expression can be simplified further by noting that the integrand is an even function of $z' - z''$; this fact allows us to reduce the double integral to a single integral, as we now demonstrate. Let $g(\cdot)$ be any even function of its argument. Letting $\Delta z = z' - z''$, we may write

$$\int_0^z dz' \int_0^z dz'' g(z' - z'') = \int_0^z dz'' \int_{-z''}^{z-z''} d\Delta z\, g(\Delta z), \tag{8.5-31}$$

where the area of integration is shown in Fig. 8-15. Since g depends only on

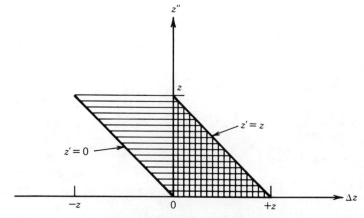

Figure 8-15. Region of integration.

THE LONG-EXPOSURE OTF

Δz and because it is an even function of Δz, the value of the integral can be obtained by doubling the integral over the doubly lined region, that is, the right-hand triangle. We conclude that

$$\int_0^z dz' \int_0^z dz'' g(z' - z'') = 2\int_0^z d(\Delta z) g(\Delta z) \int_0^{z-\Delta z} dz''$$

$$= 2\int_0^z (z - \Delta z) g(\Delta z) \, d(\Delta z). \quad (8.5\text{-}32)$$

Using this relationship, we obtain the following form for the phase structure function:

$$D_S(r) = 2(\bar{k})^2 \int_0^z (z - \Delta z) \left[D_n\left(\sqrt{(\Delta z)^2 + r^2}\right) - D_n(\Delta z) \right] d(\Delta z).$$

$$(8.5\text{-}33)$$

At this point we must adopt a specific form for the structure function of the refractive index fluctuations. According to the widely accepted Kolmogorov theory, this structure function is given by

$$D_n(r) = C_n^2 r^{2/3} \qquad l_0 < r < L_0. \quad (8.5\text{-}34)$$

Substitution of this form in the expression for the phase structure function yields

$$D_S(r) = 2(\bar{k})^2 C_n^2 \int_0^z (z - \Delta z) \left[(\Delta z^2 + r^2)^{1/3} - \Delta z^{2/3} \right] d\Delta z.$$

$$(8.5\text{-}35)$$

Note that, because of the restrictions on Eq. (8.5-34), this expression for $D_S(r)$ is strictly valid only for $\Delta z < L_0$, where L_0 is the outer scale of the turbulence. Thus it might be thought that the expression is valid only for pathlengths z shorter than the outer scale. Fortunately, when the path separation r is much less than L_0 (as it nearly always is, since the maximum separation of interest equals the diameter of the receiving optics), the integrand vanishes for large Δz, and the exact form of the structure function for $r > L_0$ is immaterial. In particular, for large Δz, the bracketed factor in the integrand of Eq. (8.5-35) behaves as

$$(\Delta z^2 + r^2)^{1/3} - \Delta z^{2/3} = \Delta z^{2/3}\left(1 + \frac{r^2}{\Delta z^2}\right)^{1/3} - \Delta z^{2/3}$$

$$\cong \Delta z^{2/3}\left[1 + \frac{1}{3}\frac{r^2}{\Delta z^2} - 1\right] = \frac{r^2}{3\Delta z^{4/3}} \to 0.$$

(8.5-36)

Hence for Δz larger than L_0 and r much smaller than L_0, the integrand is small enough to make a negligible contribution to the integral, and our expression for $D_S(r)$ can be used accurately for all pathlengths that do not violate our earlier, more basic assumptions.

The expression (8.5-35) may be regarded as involving a difference of two terms in the integrand, $z[(\Delta z^2 + r^2)^{1/3} - \Delta z^{2/3}]$ and $\Delta z[(\Delta z^2 + r^2)^{1/3} - \Delta z^{2/3}]$. Figure 8-16 shows both of these terms. The area under the first term (upper curve) is much greater than the area under the second term (lower curve). As a consequence, we shall entirely neglect the second term. Furthermore, the first term has dropped to essentially zero long before Δz reaches z, so we make little error in extending the upper limit of the integral to infinity. Thus

$$D_S(r) \cong 2(\bar{k})^2 C_n^2 z \int_0^\infty \left[(\Delta z^2 + r^2)^{1/3} - \Delta z^{2/3}\right] d\Delta z. \quad (8.5\text{-}37)$$

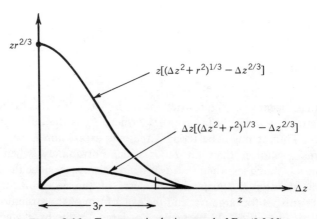

Figure 8-16. Two terms in the integrand of Eq. (8.5-35).

THE LONG-EXPOSURE OTF

With a change of variables $\Delta z = ru$, $d\Delta z = r\,du$, the expression becomes

$$D_S(r) = (\bar{k})^2 C_n^2 z r^{5/3} \int_{-\infty}^{\infty} \left[(u^2 + 1)^{1/3} - u^{2/3}\right] du. \quad (8.5\text{-}38)$$

The integral can be evaluated by numerical integration and is found to be equal to 2.91. Thus

$$D_S(r) = 2.91(\bar{k})^2 C_n^2 z r^{5/3}. \quad (8.5\text{-}39)$$

Having evaluated the phase structure function, we can now write expressions for the long-exposure atmospheric OTF [8-26]. Substituting (8.5-39) in (8.5-24), we obtain

$$\mathcal{H}_L(\nu) = \exp\left\{-\frac{1}{2} \times 2.91(\bar{k})^2 C_n^2 z (\bar{\lambda} f \nu)^{5/3}\right\}$$

$$= \exp\left\{-57.4 C_n^2 \frac{z f^{5/3}}{\bar{\lambda}^{1/3}} \nu^{5/3}\right\}. \quad (8.5\text{-}40)$$

A somewhat more convenient form that is independent of the optical system parameters is obtained if we express the OTF as a function of frequency Ω measured in cycles per radian of arc, rather than cycles per meter. The relationship between Ω and ν is $\Omega = f\nu$, with the result that

$$\mathcal{H}_L(\Omega) = \exp\left\{-\frac{1}{2} \times 2.91(\bar{k})^2 C_n^2 z (\bar{\lambda}\Omega)^{5/3}\right\}$$

$$= \exp\left\{-57.4 \frac{C_n^2 z}{\bar{\lambda}^{1/3}} \Omega^{5/3}\right\} \quad (8.5\text{-}41)$$

This equation represents the main result of our near-field analysis. Of course, to find the total OTF, the OTF given in (8.5-41) must be multiplied by the OTF of the optical system in the absence of atmospheric turbulence.

Special attention is called to the fact that the angular spatial frequency where $\mathcal{H}_L$ has dropped to value $1/e$ is given by

$$\Omega_{1/e} = \frac{\bar{\lambda}^{1/5}}{\left(57.4 C_n^2 z\right)^{3/5}}. \quad (8.5\text{-}42)$$

Thus the bandwidth of the OTF, defined in this particular way, depends on only the one-fifth power of wavelength, a very weak dependence.

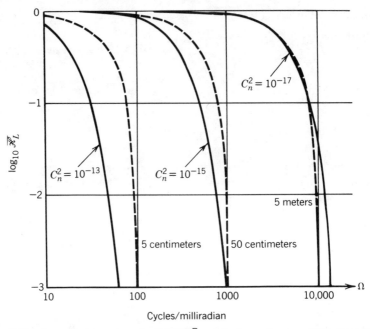

Figure 8-17. Long-exposure atmospheric OTF ($\bar{\lambda} = 0.5$ micrometers, $z = 100$ meters). Dotted curves represent the diffraction-limited OTFs of circular apertures.

Plots of the long-exposure atmospheric OTF are shown as solid lines in Fig. 8-17 for $\bar{\lambda} = 0.5$ micrometer, $z = 100$ meters, and various values of C_n^2. Shown dotted are the OTFs for diffraction-limited circular optics that have diameters 5 centimeters, 50 centimeters, and 5 meters. By comparing the solid and dotted curves, some feeling can be obtained for the effective reduction in aperture size that is comparable with the resolution-limiting effects of the turbulence.

8.6 GENERALIZATIONS OF THE THEORY

The mathematical expressions for the long-exposure OTF derived in Section 8.5 were obtained using a number of serious restrictions. In the sections to follow, we remove some of these restrictions.

First, we generalize our previous purely geometric derivation of the long-exposure OTF to include the effects of both bending of the rays and diffraction. Remarkably, the expression for the long-exposure OTF is found to be unchanged from that found with the more limited theory.

Second, we generalize our results to include the effects of slow changes of the structure constant C_n^2 along the propagation path. Such variations are particularly important for vertical imaging through the atmosphere (e.g., astronomical imaging), for the strength of the turbulence is a strong function of height above ground. Variations of C_n^2 along horizontal paths also occur frequently in practice.

Third, we introduce the concept of the atmospheric coherence diameter, represented by the symbol r_0, which is useful in understanding the limited resolution achievable in imagery gathered through the atmosphere, as well as in simplifying the mathematical form of the expressions for transfer functions.

Fourth and finally, we deal with the geometry of the imaging problem when the object lies at a finite distance from the imaging optics, rather than at infinite distance. Such considerations are important for imaging over horizontal paths, as well as for nonastronomical vertical viewing. In these cases, our attention must shift from plane wave propagation to spherical wave propagation.

8.6.1 Extension to Longer Propagation Paths—Amplitude and Phase Filter Functions

The calculation of the long-exposure atmospheric optical transfer function presented in Section 8.5 was based on the very restrictive assumption that, even for the smallest turbulent eddies, the effect of the refractive index perturbations is limited to a delay of the light rays passing through them. Thus, both geometric bending of rays and diffraction effects were ignored. The pathlengths for which this assumption is strictly valid are so short as to be of limited practical interest.

Here we generalize the analysis, taking more complete account of the effects of the inhomogeneous medium on waves propagating through it. The analysis is a modified version of that presented by Tatarski (Refs. 8-12 and 8-13). Surprisingly, we will find that the results of the more general theory are identical with those of the far simpler analysis done previously.

The geometry assumed for this calculation is illustrated in Fig. 8-18. The refractive index perturbations exist within a finite region lying between $z' = 0$ and $z' = z$ in the propagation path. Within this region, the fluctuations of $n_1(x', y', z')$ are assumed to be homogeneous. A plane wave enters the region of refractive index fluctuations at $z' = 0$, and the collecting aperture of the imaging system is assumed to lie in the plane $z' = z$.

The starting point for the analysis is Eq. (8.4-55), which relates the log-amplitude χ and the phase S_δ across the plane of the collecting aperture to the refractive index fluctuations n_1 of the inhomogeneous medium. For a

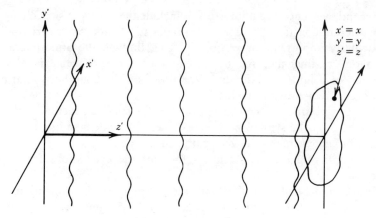

Figure 8-18. Propagation geometry.

unit intensity plane wave incident at $z' = 0$, the unperturbed solution within the medium takes the form

$$\mathbf{U}_0(\vec{r}') = \exp(j\bar{k}z'). \tag{8.6-1}$$

Substitution of this expression into Eq. (8.4-55) yields the following expressions for χ and S_δ:

$$\chi(x, y, z) = \frac{\bar{k}^2}{2\pi} \int_0^z dz' \int_{-\infty}^{\infty} dy' \int_{-\infty}^{\infty} dx' n_1(x', y', z')$$

$$\times \frac{\cos\left\{\dfrac{\bar{k}\left[(x - x')^2 + (y - y')^2\right]}{2(z - z')}\right\}}{z - z'}$$

$$S_\delta(x, y, z) = \frac{\bar{k}^2}{2\pi} \int_0^z dz' \int_{-\infty}^{\infty} dy' \int_{-\infty}^{\infty} dx' n_1(x', y', z')$$

$$\times \frac{\sin\left\{\dfrac{\bar{k}\left[(x - x')^2 + (y - y')^2\right]}{2(z - z')}\right\}}{z - z'}. \tag{8.6-2}$$

GENERALIZATIONS OF THE THEORY

For purposes of later analysis, it is convenient to rewrite (8.6-2) in the form

$$\chi(x, y, z) = \int_0^z q(x, y, z, z')\, dz'$$

$$S_\delta(x, y, z) = \int_0^z p(x, y, z, z')\, dz', \quad (8.6\text{-}3)$$

where

$$q(x, y, z, z') = \frac{\bar{k}^2}{2\pi} \iint_{-\infty}^{\infty} n_1(x', y', z') \frac{\cos\left\{\frac{\bar{k}\left[(x-x')^2 + (y-y')^2\right]}{2(z-z')}\right\}}{z-z'}\, dx'\, dy'$$

$$p(x, y, z, z') = \frac{\bar{k}^2}{2\pi} \iint_{-\infty}^{\infty} n_1(x', y', z') \frac{\sin\left\{\frac{\bar{k}\left[(x-x')^2 + (y-y')^2\right]}{2(z-z')}\right\}}{z-z'}\, dx'\, dy'.$$

(8.6-4)

Our goal in this analysis is to first calculate the two-dimensional power-spectral densities $F_\chi(\kappa_X, \kappa_Y; z)$ and $F_S(\kappa_X, \kappa_Y; z)$ of the log amplitude and phase in the plane $z' = z$. From these and related results we will be able to find the corresponding structure functions, and finally the long-exposure OTF will be calculated. The analysis is aided by recognizing that Eqs. (8.6-4) express q and p as two-dimensional convolution integrals in (x', y'). The integrations over z' in Eqs. (8.6-3) simply add the results of these convolutions for all distances z' along the propagation path. In all cases the pathlength z should be viewed as a fixed constant.

Since convolutions have been found to arise in the analysis, it is perhaps not unexpected that simplifications will occur if the analysis is performed in the frequency domain. The impulse responses in Eqs. (8.6-4) can be seen to be given by

$$h_\chi(x, y; z, z') = \frac{\bar{k}^2}{2\pi} \frac{\cos\left\{\frac{\bar{k}(x^2 + y^2)}{2(z-z')}\right\}}{z-z'}$$

$$h_S(x, y; z, z') = \frac{\bar{k}^2}{2\pi} \frac{\sin\left\{\frac{\bar{k}(x^2 + y^2)}{2(z-z')}\right\}}{z-z'}. \quad (8.6\text{-}5)$$

Defining corresponding two-dimensional transfer functions by

$$\mathbf{H}_{\underset{S}{\chi}}(\kappa_X, \kappa_Y; z, z') = \frac{1}{(2\pi)^2} \int\!\!\!\int_{-\infty}^{\infty} h_{\underset{S}{\chi}}(x, y; z, z') \exp[j(\kappa_X x + \kappa_Y y)]\, dx\, dy,$$

(8.6-6)

we use the Fourier transform relationship

$$\frac{1}{(2\pi)^2} \int\!\!\!\int_{-\infty}^{\infty} \exp\{j\alpha(x^2 + y^2)\}\exp\{j(\kappa_X x + \kappa_Y y)\}\, dx\, dy$$

$$= \frac{4\pi j}{\alpha} \exp\left\{-j\frac{\kappa_X^2 + \kappa_Y^2}{4\alpha}\right\},$$ (8.6-7)

together with the substitution $\alpha = \bar{k}/[2(z - z')]$, to find the following expressions for the transfer functions of interest:

$$\mathbf{H}_\chi(\kappa_X, \kappa_Y; z, z') = \bar{k}\sin\left[\frac{(\kappa_X^2 + \kappa_Y^2)}{2\bar{k}}(z - z')\right]$$

$$\mathbf{H}_S(\kappa_X, \kappa_Y; z, z') = \bar{k}\cos\left[\frac{(\kappa_X^2 + \kappa_Y^2)}{2\bar{k}}(z - z')\right].$$ (8.6-8)

We can now use the fundamental relationship of Eq. (3.3-12) governing the passage of random processes through linear, invariant systems to relate the two-dimensional power spectra of q and p to the two-dimensional power spectrum F_n of the refractive index fluctuations. The result is

$$F_q(\kappa_X, \kappa_Y; z, z') = |\mathbf{H}_\chi(\kappa_X, \kappa_Y; z, z')|^2 F_n(\kappa_X, \kappa_Y; z')$$

$$= \bar{k}^2 \sin^2\left[\frac{\kappa_t^2}{2\bar{k}}(z - z')\right] F_n(\kappa_X, \kappa_Y; z')$$ (8.6-9)

$$F_p(\kappa_X, \kappa_Y; z, z') = |\mathbf{H}_S(\kappa_X, \kappa_Y; z, z')|^2 F_n(\kappa_X, \kappa_Y; z')$$

$$= \bar{k}^2 \cos^2\left[\frac{\kappa_t^2}{2\bar{k}}(z - z')\right] F_n(\kappa_X, \kappa_Y; z'),$$

where $\kappa_t = (\kappa_X^2 + \kappa_Y^2)^{1/2}$ represents a radial wavenumber in the (κ_X, κ_Y) plane.

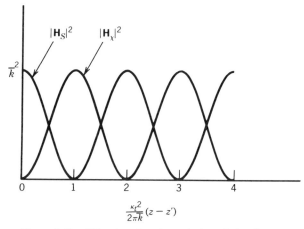

Figure 8-19. Filter functions for a single turbulent layer.

In Fig. 8-19 we show plots of $|H_\chi|^2$ and $|H_S|^2$ vs. normalized distance $(\kappa_t^2/2\pi\bar{k})(z - z')$ between the turbulent layer at z' and the collecting aperture at z. These curves may be interpreted physically as specifying how fluctuations of n_1 having wavenumber κ_t and lying in the plane z' are transferred into fluctuations of the log amplitude and phase of the wave incident on the collecting aperture of the imaging system. Note that fluctuations of n_1 with wavenumber κ_t make maximum contributions to fluctuations of the log amplitude χ when the distance $z - z'$ satisfies

$$z - z' = (2n + 1)\pi \frac{\bar{k}}{\kappa_t^2} \qquad n = 0, 1, \ldots, \qquad (8.6\text{-}10)$$

while maximum contributions to the fluctuations of the phase occur when

$$z - z' = 2n\pi \frac{\bar{k}}{\kappa_t^2} \qquad n = 0, 1 \ldots. \qquad (8.6\text{-}11)$$

There is a very close connection between results illustrated in Fig. 8-19 and the so-called Talbot effect, or the self-imaging properties of periodic gratings (Ref. 8-27).

Although we have found the influence of a particular turbulent layer located at z' on the log-amplitude and phase fluctuations of interest, we have not yet addressed the problem of adding the contributions from turbulent layers at all possible distances $z - z'$. Simple integration of F_q and F_p with respect to z' would yield correct results only if the correlation length of the turbulence in the z' direction were zero. Such is not the case, and a more careful analysis is necessary.

If we consider two random processes χ and q (or S_δ and p) related as in Eq. (8.6-3), it is a straightforward exercise to demonstrate that the autocorrelation functions of χ and S_δ are given by

$$\Gamma_\chi(\Delta x, \Delta y; z) = \iint_0^z \Gamma_q(\Delta x, \Delta y; z, z', z'')\, dz'\, dz''$$

$$\Gamma_S(\Delta x, \Delta y; z) = \iint_0^z \Gamma_p(\Delta x, \Delta y; z, z', z'')\, dz'\, dz'', \quad (8.6\text{-}12)$$

where Γ_q and Γ_p represent *cross*-correlation functions of $q(x, y, z, z')$ with $q(x, y, z, z'')$ in the first case and $p(x, y, z, z')$ with $p(x, y, z, z'')$ in the latter case. Fourier transforming both sides of Eq. (8.6-12) with respect to Δx and Δy, we obtain the relationships

$$F_\chi(\kappa_X, \kappa_Y; z) = \iint_0^z \tilde{F}_q(\kappa_X, \kappa_Y; z, z', z'')\, dz'\, dz''$$

$$F_S(\kappa_X, \kappa_Y; z) = \iint_0^z \tilde{F}_p(\kappa_X, \kappa_Y; z, z', z'')\, dz'\, dz'', \quad (8.6\text{-}13)$$

where $\tilde{F}_q$ and $\tilde{F}_p$ represent *cross*-spectral densities of $q(x, y, z, z')$ with $q(x, y, z, z'')$ on the one hand, and $p(x, y, z, z')$ with $p(x, y, z, z'')$ on the other.

At this point we use the fundamental relationship of Eq. (3.5-8) governing the passage of cross-spectral densities through linear invariant filters, coupled with the explicit expressions (8.6-8) for the transfer functions involved, to write

$$F_\chi(\kappa_t; z) = \iint_0^z \overline{k}^2 \sin\left[\frac{\kappa_t^2}{2k}(z - z')\right] \sin\left[\frac{\kappa_t^2}{2k}(z - z'')\right] \tilde{F}_n(\kappa_t; z' - z'')\, dz'\, dz''$$

$$F_S(\kappa_t; z) = \iint_0^z \overline{k}^2 \cos\left[\frac{\kappa_t^2}{2k}(z - z')\right] \cos\left[\frac{\kappa_t^2}{2k}(z - z'')\right] \tilde{F}_n(\kappa_t; z' - z'')\, dz'\, dz'',$$

$$(8.6\text{-}14)$$

where $\tilde{F}_n$ represents the cross-spectral density of the refractive index fluctuations in planes z' and z'' and isotropy in the transverse dimension has been assumed for simplicity.

GENERALIZATIONS OF THE THEORY

Further progress requires use of the trigonometric identities

$$\sin\left[\frac{\kappa_t^2}{2k}(z-z')\right]\sin\left[\frac{\kappa_t^2}{2k}(z-z'')\right]$$

$$=\frac{1}{2}\cos\left[\frac{\kappa_t^2}{2k}(z''-z')\right]-\frac{1}{2}\cos\left[\frac{\kappa_t^2}{2k}(2z-z''-z')\right]$$

(8.6-15)

and

$$\cos\left[\frac{\kappa_t^2}{2k}(z-z')\right]\cos\left[\frac{\kappa_t^2}{2k}(z-z'')\right]$$

$$=\frac{1}{2}\cos\left[\frac{\kappa_t^2}{2k}(z'-z'')\right]+\frac{1}{2}\cos\left[\frac{\kappa_t^2}{2k}(2z-z''-z')\right].$$

(8.6-16)

These relationships are now substituted into Eqs. (8.6-14), and a change of variables $\xi = z' - z''$ and $2\eta = z' + z''$ is made. The new region of integration in the (ξ, η) plane is illustrated in Fig. 8-20. Exploiting the symmetry of the integrand with respect to the variable ξ, our expressions for F_X and F_S

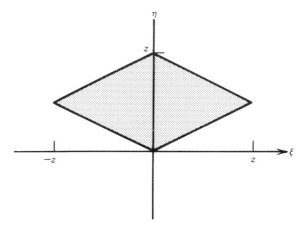

Figure 8-20. Region of integration.

become

$$\left.\begin{array}{l}F_X(\kappa_t; z)\\ F_S(\kappa_t; z)\end{array}\right\} = \bar{k}^2 \int_{-z}^{z} d\xi \tilde{F}_n(\kappa_t; \xi) \cos\left(\frac{\kappa_t^2}{2\bar{k}}\xi\right) \int_{+|\xi|/2}^{z-|\xi|/2} d\eta$$

$$\mp \bar{k}^2 \int_{-z}^{z} d\xi \tilde{F}_n(\kappa_t; \xi) \int_{+|\xi|/2}^{z-|\xi|/2} d\eta \cos\left[\frac{\kappa_t^2}{2\bar{k}}(2z - 2\eta)\right].$$

(8.6-17)

The integrals can be performed and yield

$$\left.\begin{array}{l}F_X(\kappa_t; z)\\ F_S(\kappa_t; z)\end{array}\right\} = \bar{k}^2 \int_{-z}^{z} \tilde{F}_n(\kappa_t; \xi)$$

$$\times \left[(z - |\xi|)\cos\frac{\kappa_t^2 \xi}{2\bar{k}} \pm \frac{\bar{k}}{\kappa_t^2}\sin\frac{\kappa_t^2|\xi|}{2\bar{k}} \mp \frac{\bar{k}}{\kappa_t^2}\sin\frac{\kappa_t^2}{2\bar{k}}(2z - |\xi|)\right] d\xi.$$

(8.6-18)

At this point it is necessary to make some approximations. The first uses the fact that the cross-spectral density $\tilde{F}_n(\kappa_t; \xi)$ falls rapidly toward zero for z separations ξ larger than $1/\kappa_t$. This property rests on the assumed statistical isotropy of the refractive index fluctuations, which suggests that there is little correlation between sinusoidal transverse components of identical wavenumbers in two planes separated axially by more than the reciprocal of the wavenumber in question. Since the important values of ξ are those no greater than $1/\kappa_t$, we see that over the significant range of integration we have

$$\frac{\kappa_t^2 \xi}{2\bar{k}} \le \frac{\kappa_t}{2\bar{k}} \ll 1, \qquad (8.6\text{-}19)$$

where the last inequality follows from the fact that the inner scale l_0 is much greater than the wavelength $\bar{\lambda}$. Furthermore, we are interested in the structure functions only for arguments that are small compared with the total pathlength L. Therefore, $1/\kappa_t \ll L$. In addition, from the previous discussion, the important range of ξ satisfies $\xi \le 1/\kappa_t$. It follows that $\xi \le 1/\kappa_t \ll 1/L$ and hence that $L - (\xi/2) \approx L$. In view of these facts, the following additional approximations can be made:

$$\cos\frac{\kappa_t^2 \xi}{2\bar{k}} \cong 1 \qquad \sin\frac{\kappa_t^2 \xi}{2\bar{k}} \cong \frac{\kappa_t^2 \xi}{2\bar{k}}$$

$$\sin\frac{\kappa_t^2(2z-\xi)}{2\bar{k}} \cong \sin\frac{\kappa_t^2 z}{\bar{k}}. \qquad (8.6\text{-}20)$$

We also note that, as a result of the rapid drop of $\tilde{F}_n(\kappa_t; \xi)$ with ξ, the limits of integration in Eq. (8.6-18) can be extended to infinity. With these simplifications we find

$$\left.\begin{array}{c}F_\chi(\kappa_t; z) \\ F_S(\kappa_t; z)\end{array}\right\} = \left[\bar{k}^2 z \mp \frac{\bar{k}^3}{\kappa_t^2}\sin\frac{\kappa_t^2 z}{\bar{k}}\right]\int_{-\infty}^{\infty}\tilde{F}_n(\kappa_t; \xi)\,d\xi. \qquad (8.6\text{-}21)$$

Finally, the relationship between the two-dimensional cross-spectral density $\tilde{F}_n(\kappa_t; \xi)$ and the three-dimensional spectrum $\Phi_n(\kappa_X, \kappa_Y, \kappa_Z)$ is introduced. From basic definitions we have (see Problem 8-10)

$$\Phi_n(\kappa_X, \kappa_Y, \kappa_Z) = \frac{1}{2\pi}\int_{-\infty}^{\infty}\tilde{F}_n(\kappa_t; \xi)\cos\kappa_Z\xi\,d\xi. \qquad (8.6\text{-}22)$$

It follows that

$$\int_0^\infty \tilde{F}_n(\kappa_t; \xi)\,d\xi = \pi\Phi_n(\kappa_X, \kappa_Y, 0). \qquad (8.6\text{-}23)$$

For isotropic turbulence,

$$\Phi_n(\kappa_X, \kappa_Y, \kappa_Z) = \Phi_n\left(\sqrt{\kappa_X^2 + \kappa_Y^2 + \kappa_Z^2}\right), \qquad (8.6\text{-}24)$$

and hence $\Phi_n(\kappa_X, \kappa_Y, 0) = \Phi_n(\kappa_t)$. Thus we arrive at the final expressions for the power spectra of the log amplitude and phase:

$$F_\chi(\kappa_t; z) = \pi\bar{k}^2 z\left(1 - \frac{\bar{k}}{\kappa_t^2 z}\sin\frac{\kappa_t^2 z}{\bar{k}}\right)\Phi_n(\kappa_t)$$

$$F_S(\kappa_t; z) = \pi\bar{k}^2 z\left(1 + \frac{\bar{k}}{\kappa_t^2 z}\sin\frac{\kappa_t^2 z}{\bar{k}}\right)\Phi_n(\kappa_t). \qquad (8.6\text{-}25)$$

Equations (8.6-25) represent a major result of our analysis. They provide us with knowledge of the relative magnitudes of the fluctuations of χ and S_δ as a function of pathlength z and wavenumber κ_t. The functions

$$|\mathcal{H}_\chi(\kappa_t; z)|^2 = \pi \bar{k}^2 z \left(1 - \frac{\bar{k}}{\kappa_t^2 z} \sin \frac{\kappa_t^2 z}{\bar{k}}\right)$$

$$|\mathcal{H}_S(\kappa_t; z)|^2 = \pi \bar{k}^2 z \left(1 + \frac{\bar{k}}{\kappa_t^2 z} \sin \frac{\kappa_t^2 z}{\bar{k}}\right) \quad (8.6\text{-}26)$$

are generally referred to as the "filter functions" for log amplitude and phase. They differ from the earlier filter functions of Eqs. (8.6-8) in that they apply for the entire integrated propagation path, whereas the earlier filter functions applied only for a single turbulent layer at distance $z - z'$ from the optical system.

The shapes of the log-amplitude and phase filter functions are illustrated in Fig. 8-21. Also shown in the same plot is the general form of the refractive index power spectrum Φ_n. In these plots z may be regarded as a fixed parameter and κ_t the variable of interest. Note that log-amplitude fluctuations are quite insensitive to the refractive index fluctuations at small wavenumbers (large scale sizes), whereas the phase fluctuations have their maximum sensitivity there.

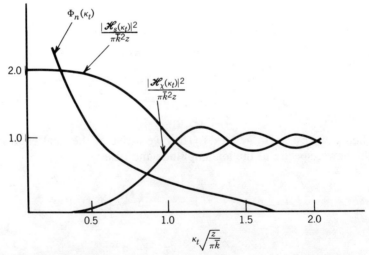

Figure 8-21. Log-amplitude and phase filter functions for an extended turbulent region—dependence on wavenumber.

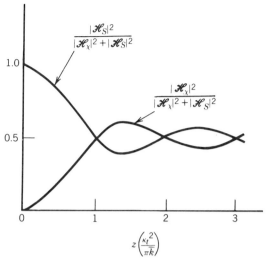

Figure 8-22. Log-amplitude and phase filter functions for an extended turbulent region—dependence on distance.

An alternative point of view is illustrated in Fig. 8-22. Here, κ_t should be regarded as a fixed parameter, whereas z is the variable of interest. The curves may be regarded as indicating the relative importance of log amplitude and phase fluctuations as a function of pathlength z. For very short paths ($z \ll \pi\bar{k}/\kappa_t^2$), the log-amplitude fluctuations are negligible, and essentially all fluctuations of the wave reside in the phase. For long paths ($z \gg \pi\bar{k}/\kappa_t^2$), the fluctuations residing in log amplitude and phase are essentially of equal importance. Note that for the particular distance $z = \pi\bar{k}/\kappa_t^2$, phase gratings at the far end of the path ($z' = 0$) are producing a pure amplitude effect in the observation plane, whereas phase gratings at the nearest end of the path ($z' = z$) are producing a pure phase effect in that plane. Hence an equal mixture of amplitude and phase effects is produced along the propagation path for this particular wavenumber.

Our ultimate interest in this treatment lies with the predictions of the general theory regarding the form of the long-time-average OTF. From Eq. (8.5-22), the form of this OTF depends entirely on the form of the wave structure function $D(r) = D_\chi(r) + D_S(r)$. The structure functions D_χ and D_S are related to the respective power spectra F_χ and F_S through

$$D_\chi(r) = 4\pi \int_0^\infty \left[1 - J_0(\kappa_t r)\right] F_\chi(\kappa_t; z) \kappa_t \, d\kappa_t$$

$$D_S(r) = 4\pi \int_0^\infty \left[1 - J_0(\kappa_t r)\right] F_S(\kappa_t; z) \kappa_t \, d\kappa_t. \qquad (8.6\text{-}27)$$

Thus the total wave structure function is

$$D(r) = 4\pi \int_0^\infty [1 - J_0(\kappa_t r)] [F_\chi(\kappa_t; z) + F_S(\kappa_t; z)] \kappa_t d\kappa_t$$

$$= 8\pi^2 \bar{k}^2 z \int_0^\infty [1 - J_0(\kappa_t r)] \Phi_n(\kappa_t) \kappa_t d\kappa_t. \quad (8.6\text{-}28)$$

With substitution of the Kolmogorov spectrum of Eq. (8.4-14), the wave structure function is seen to be given by

$$D(r) = 8\pi^2 \times (0.033) \bar{k}^2 z C_n^2 \int_0^\infty [1 - J_0(\kappa_t r)] \kappa_t^{-8/3} d\kappa_t. \quad (8.6\text{-}29)$$

Use of the integral identity† (Ref. 8-12, p. 269)

$$\int_0^\infty [1 - J_0(x)] x^{-p} dx = \pi \left\{ 2^p \Gamma^2\left(\frac{p+1}{2}\right) \sin\left[\frac{\pi(p-1)}{2}\right] \right\}^{-1}$$

$$1 < p < 3 \quad (8.6\text{-}30)$$

yields a final result

$$D(r) = 2.91 \bar{k}^2 z C_n^2 r^{5/3}, \quad (8.6\text{-}31)$$

which is exactly the same result obtained by the simpler analysis given in the previous subsection. Thus the forms of the OTF represented by (8.5-40) and (8.5-41) remain correct in this more general analysis.

The fundamental reason for the generality of the results of the simplified analysis can be deduced from the expressions for the amplitude and phase filter functions [Eqs. (8.6-26)]. For very short pathlengths, for which the simplified analysis is valid, we have

$$|\mathcal{H}_\chi|^2 \cong 0 \qquad |\mathcal{H}_S|^2 \cong \pi \bar{k}^2 z$$

It was in this regime that we neglected amplitude effects and retained only phase effects. From the more general results, however, we can see that at any pathlength (subject only to the restriction that it is not so long as to invalidate the perturbation analysis), the wave structure function depends on the sum of the two filter functions, and this sum is equal to $\pi \bar{k}^2 z$, exactly the same value ascribed to the phase filter function in the short-path regime. Hence the corrections to the amplitude and phase filter functions necessary in the long-path regime exactly cancel when the two transfer functions are added!

We conclude that the expressions for the atmospheric long-exposure OTF derived in Section 8.5.2 are valid under more general conditions than the initial analysis implied.

8.6.2 Effects of Smooth Variations of the Structure Constant C_n^2

Our goal in this section is to find an expression for the long-time-average OTF, analogous to Eq. (8.5-41), but valid when the strength of the turbulence varies slowly along the propagation path. Such variations are often encountered in atmospheric propagation and are particularly pronounced in vertical viewing through the atmosphere. The analysis presented here is an approximate one. Fortunately, it yields the same end result as a more careful examination of the problem. After the main result is derived, we point out the primary defect of the analysis and then discuss why this defect is unimportant in the calculation of the OTF of interest here. For a more complete study of the problem, the reader is referred to Ref. 8-12, Chapter 8.

The slow and smooth character of the variations of the strength of the turbulence suggests the use of a "quasihomogeneous" model for the structure function of the refractive index fluctuations,

$$D_n(\vec{r}_1, \vec{r}_2) = C_n^2\left(\frac{\vec{r}_1 + \vec{r}_2}{2}\right)|\vec{r}_1 - \vec{r}_2|^{2/3}, \tag{8.6-32}$$

when $l_0 < |r_1 - r_2| < L_0$. Implicit in this representation is the assumption that the significant changes of C_n^2 can occur only over distances comparable with L_0, or greater.

Our approximate model assumes that the atmosphere can be divided up into a series of layers or slabs of thickness Δz along the propagation path, and that Δz is chosen large enough so that, to a good approximation, the fluctuations of the log amplitude and phase introduced by different layers are uncorrelated. Such a model allows us to express the wave structure function after passage through N layers as the sum of the N wave structure functions associated with the individual layers,

$$D(r) = \sum_{i=1}^{N} D_i(r). \tag{8.6-33}$$

If z_i represents the z coordinate at the middle of the ith layer, we can then use Eq. (8.6-31) for the individual layers to write

$$D(r) = 2.91\bar{k}^2 \sum_{i=1}^{N} C_n^2(z_i) \Delta z \, r^{2/3}. \tag{8.6-34}$$

If we now further assume the variations of C_n^2 are slow by comparison with the length Δz, the finite sum can be replaced by an integral along the propagation path, yielding

$$D(r) = 2.91\bar{k}^2 r^{5/3} \int_0^z C_n^2(\xi) \, d\xi, \tag{8.6-35}$$

where z is the total pathlength. Finally, the form of the long-exposure OTF becomes [cf. Eq. (8.5-41)]

$$\mathscr{H}_L(\Omega) = \exp\left\{-57.4 \frac{\int_0^z C_n^2(\xi) \, d\xi}{\bar{\lambda}^{1/3}} \Omega^{5/3}\right\}. \tag{8.6-36}$$

The defect of the method of analysis used to obtain this result arises from our neglect of turbulent scale sizes larger than the outer scale L_0. The spectrum of the turbulence has maximum values at small wavenumbers (large scale sizes), yet in assuming that the refractive index fluctuations introduced by all slabs are uncorrelated, we have neglected the presence of these large-scale inhomogeneities. Nonetheless, the result we have derived agrees exactly with that obtained by the more thorough analysis referenced earlier. The reason for the success of the simplified and approximate analysis lies in the insensitivity of the particular quantity we calculated (i.e., the wave structure function) to large-scale turbulent structures. Such structures introduce neither significant amplitude variations nor significant phase-difference variations at the imaging aperture and hence have little influence on the wave structure function.

Having found the form of the wave structure function when the structure constant varies along the propagation path, we might naturally inquire as to what form these variations have. For horizontal imaging there is no analytical form that can be specified, since the variations are very much a function of the local terrain and wind conditions. For vertical viewing, the variations of C_n^2 are still subject to atmospheric conditions at the time of the experiment, but analytical approximations to C_n^2 have been proposed. One such analytical form is (Ref. 8-28)

$$C_n^2(\xi) = 4.2 \times 10^{-14} \xi^{-1/3} \exp\left(\frac{-\xi}{\xi_0}\right), \tag{8.6-37}$$

GENERALIZATIONS OF THE THEORY

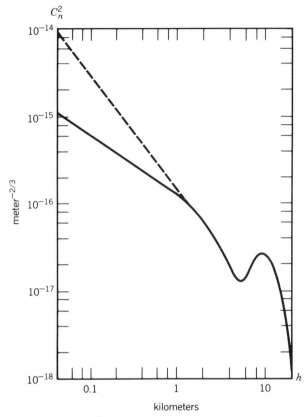

Figure 8-23. Structure constant C_n^2 versus height. Solid line, nighttime conditions. Dotted line, daytime conditions. (Courtesy of R. E. Hufnagel.)

with $\xi_0 = 3200$ meters. Figure 8-23 shows a typical dependence of C_n^2 on height above ground. The strength of the turbulence is seen to decrease with increasing height, a consequence of decreasing temperature fluctuations at higher altitudes. An increase of C_n^2 is observed in the region of the tropopause, shown in this figure in the vicinity of $h = 10$ kilometers.

8.6.3 The Atmospheric Coherence Diameter r_0.

A useful description of the image-degrading effects of the atmosphere is a parameter first introduced by Fried (Ref. 8-29) and represented by the symbol r_0. To introduce this parameter and to explain its significance, it is necessary to first consider a particular measure of resolution achieved by an imaging system.

Suppose that a transfer function $\mathcal{H}(\Omega)$ describes the performance of a particular imaging system. Further suppose that $\mathcal{H}(\Omega)$ is entirely real and circularly symmetric, as will be the case in all examples of interest to us here. Since $\mathcal{H}(0) = 1$, one possible measure of the resolution achieved by the system is the volume lying under the transfer function,

$$\mathcal{R} = 2\pi \int_0^\infty \Omega \mathcal{H}(\Omega) \, d\Omega. \tag{8.6-38}$$

We are interested here in the particular case of a long-exposure image through the atmosphere gathered by an otherwise perfect system that has a circular pupil with diameter D_0. The total average OTF in this case takes the form

$$\overline{\mathcal{H}}(\Omega) = \mathcal{H}_0(\Omega) \exp\left\{-57.4 \frac{\int_0^z C_n^2(\xi) \, d\xi}{\overline{\lambda}^{1/3}} \Omega^{5/3}\right\}, \tag{8.6-39}$$

where $\mathcal{H}_0(\Omega)$ is the OTF of the system in the absence of the atmosphere,

$$\mathcal{H}_0(\Omega) = \begin{cases} \dfrac{2}{\pi}\left[\cos^{-1}\left(\dfrac{\Omega}{\Omega_0}\right) - \dfrac{\Omega}{\Omega_0}\sqrt{1 - \left(\dfrac{\Omega}{\Omega_0}\right)^2}\right] & \Omega \leq \Omega_0 \\ 0 & \text{otherwise,} \end{cases} \tag{8.6-40}$$

and $\Omega_0 = D_0/\overline{\lambda}$ is the cutoff frequency (in cycles per radian of arc) of the optics. Our task now is to evaluate the expression (8.6-38) for the resolution $\mathcal{R}$ using the preceding transfer function.

With the change of variables $u = \Omega/\Omega_0 = \overline{\lambda}\Omega/D_0$, the required integral takes the form

$$\mathcal{R} = 4\left(\frac{D_0}{\overline{\lambda}}\right)^2 \int_0^1 u\left[\cos^{-1} u - u\sqrt{1 - u^2}\right]$$

$$\times \exp\left\{-57.4 \frac{\int_0^z C_n^2(\xi) \, d\xi}{\overline{\lambda}^{1/3}} \left(\frac{D_0}{\overline{\lambda}}\right)^{5/3} u^{5/3}\right\} du. \tag{8.6-41}$$

At this point we introduce the parameter r_0, which we shall call the *atmospheric coherence diameter*. For reasons that become apparent shortly,

GENERALIZATIONS OF THE THEORY

r_0 is defined as

$$r_0 \triangleq 0.185 \left[\frac{\bar{\lambda}^2}{\int_0^z C_n^2(\xi) \, d\xi} \right]^{3/5}. \quad (8.6\text{-}42)$$

When this expression is ultilized in Eq. (8.6-41), the resulting integral becomes

$$\mathcal{R} = 4\left(\frac{D_0}{\bar{\lambda}}\right)^2 \int_0^1 u\left[\cos^{-1}u - u\sqrt{1-u^2}\right] \exp\left\{-3.44\left(\frac{D_0}{r_0}\right)^{5/3} u^{5/3}\right\} du.$$

$$(8.6\text{-}43)$$

This integral has been evaluated numerically by Fried (Ref. 8-29) for various values of D_0/r_0, with the results shown in Fig. 8-24. Note that for $D_0/r_0 \ll 1$, the resolution $\mathcal{R}$ increases as the square of D_0/r_0, whereas for $D_0/r_0 \gg 1$, it approaches a constant asymptote, with value represented by $\mathcal{R}_{\max}$. The intersection of these two asymptotes occurs at $D_0 = r_0$, which is indeed the basis on which r_0 was defined.

The parameter r_0 is thus seen to provide a useful measure of the coherence diameter of the atmosphere. The resolution of a diffraction-limited system using a long exposure increases with aperture size until that size

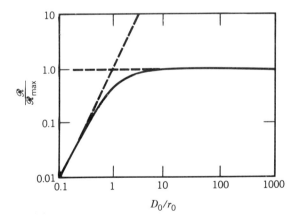

Figure 8-24. Dependence of normalized resolution $\mathcal{R}/\mathcal{R}_{\max}$ on the normalized diameter D_0/r_0 of the imaging optics for long-exposure imaging. (Courtesy of D. L. Fried and the Optical Society of America. From D. L. Fried, *J. Opt. Soc. Am.*, Vol. 56, p. 1378, 1966.)

reaches approximately r_0, beyond which the resolution stays roughly constant. The use of the parameter r_0 in expressions for atmospheric transfer functions makes them simpler in form and thereby aids in understanding their behavior.

Typical values of r_0 at a good mountain-top astronomical observatory might range from 5 centimeters under conditions of moderately poor seeing to 20 cm under conditions of exceptionally good seeing. An average value for good seeing might be 10 centimeters. Considerably smaller values can be expected over horizontal imaging paths.

8.6.4 Structure Function for a Spherical Wave

For astronomical objects viewed from the Earth, it is accurate to assume that any single object point generates a plane wave incident on the atmosphere. Hence the plane wave propagation results of earlier sections are directly applicable in such a situation. However, in most other applications such an assumption may be questionable. For systems gathering images of objects that lie within the Earth's atmosphere (e.g., for horizontal imaging or downward vertical imaging), the spherical nature of the ideal wavefronts generated by a single object point can not be neglected. Figure 8-25 illustrates the geometry of interest in such cases.

Expressions for the variance of the log amplitude and phase of a spherical wave propagating through a randomly inhomogeneous medium were derived by Tatarski (Ref. 8-12, Chapter 9). The wave structure function for this case was first derived by Fried (Ref. 8-28), based on results

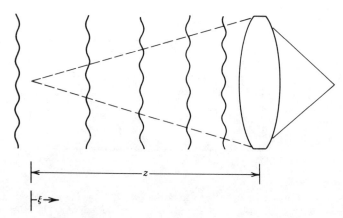

Figure 8-25. Geometry for spherical wave propagation.

of Schmeltzer (Ref. 8-30). We do not duplicate the analysis here, but the results are worth stating.

In the case of spherical wave propagation, the wave structure function is found to be given by

$$D(r) = 2.91(\bar{k})^2 r^{5/3} \int_0^z \left(\frac{\xi}{z}\right)^{5/3} C_n^2(\xi)\, d\xi. \qquad (8.6\text{-}44)$$

When the structure constant C_n^2 does not depend on distance along the path, this result becomes

$$D(r) = \tfrac{3}{8}\left[2.91\bar{k}^2 C_n^2 z r^{5/3}\right], \qquad (8.6\text{-}45)$$

which differs from the plane wave result only by a constant factor of $\tfrac{3}{8}$. As before, the long-exposure OTF of the atmosphere is related to the structure function through

$$\overline{\mathcal{H}}_L(\Omega) = \exp\{-\tfrac{1}{2}D(\bar{\lambda}\Omega)\}. \qquad (8.6\text{-}46)$$

8.7 THE SHORT-EXPOSURE OTF

Our analysis of the atmospheric OTF has concentrated exclusively on images collected with integration times much longer than the characteristic fluctuation time of the atmospherically induced wavefront deformations. The continuous evolution of independent realizations of atmospheric perturbations during the exposure time allowed the use of ensemble averaging to predict the time-averaged OTF. Our attention is now turned to the effects of atmospheric inhomogeneities on images obtained with integration times that are short compared with the characteristic fluctuation time of the atmosphere.

8.7.1 Long versus Short Exposures

The duration of the exposure time necessary to assure that an image has been gathered in the long-time-average regime is difficult to specify precisely. The first source of difficulty lies with the dependence of the required integration time on the particular atmospheric conditions present at the time the image is collected. A second source of difficulty lies in the dependence of the required time on the particular spatial frequencies of interest. If we adopt the "frozen turbulence" (i.e., Taylor's) hypothesis, we can assume that image degradations are caused by fixed patterns of refractive index perturbations drifting across the imaging path under the influence

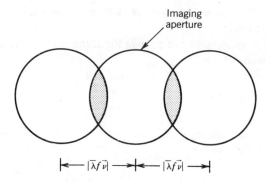

Figure 8-26. Areas on exit pupil that influence the spatial frequency $\vec{\nu} = \vec{s}/\lambda f$ of intensity.

of local wind conditions. If we consider a spatial frequency $\vec{\nu}$ corresponding to a fixed spacing $\vec{s} = \lambda f \vec{\nu}$ on the imaging aperture, we know that contributions to that particular spatial frequency come from only a limited area on the aperture, namely, the shaded area in Fig. 8-26. For high spatial frequencies, the regions on the aperture are small, and only a comparatively short time is required for a given set of wavefront deformations to drift out of these regions and to be replaced by new deformations. At lower spatial frequencies, the regions of importance on the aperture are larger, and hence longer times are required for replacement of the deformations.

To fully specify the time required to assure accuracy of the long-exposure model, it is necessary to specify the temporal power spectral densities associated with all frequency components of interest in an image. As a general rule of thumb, by no means universally applicable, it is often stated that exposure times substantially in excess of 0.01 second are required for accuracy of the long-exposure assumption.

In practice there arise many situations in which the long-exposure model is not accurate. For example, a motion picture camera mounted on an astronomical telescope can be run with frame exposure times less than 0.01 s, provided the brightness of the object under study is sufficiently great.

The PSFs and OTFs encountered for short-exposure images are markedly different from their long-exposure counterparts. As illustrated in Fig. 8-27 (cf. Fig. 8-12), the PSF of a long-exposure image is a smooth and broad function, and the corresponding OTF is narrow and smooth. On the other hand, the PSF for a short-exposure image is a jagged and narrower function, whereas the corresponding OTF has significant fluctuations of both magnitude and phase as a function of spatial frequency.

One of the most important facts about short-exposure images is that their quality is unaffected by the tilt component of the wavefront distortions. A

THE SHORT-EXPOSURE OTF

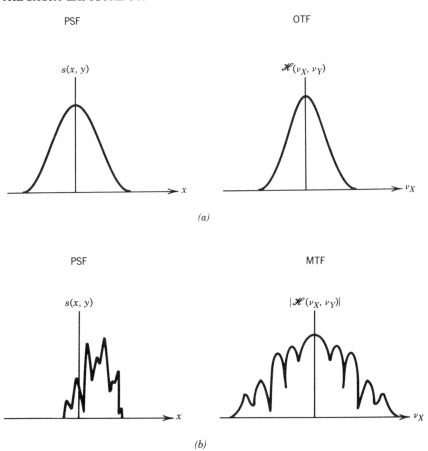

Figure 8-27. Typical point-spread functions and transfer functions for long- and short-exposure images: (*a*) long exposure; (*b*) short exposure.

tilt of the incident wavefront simply shifts the center of the image and affects the image in no other way. Provided the goal of the imaging experiment is to determine the structure of the object brightness distribution, but not its absolute position, tilt is of no consequence. On the other hand, for long-exposure images, changing tilt of the incident wavefront serves to broaden the PSF and narrow the OTF.

Since the structure of the OTF is statistical in nature for the short-exposure case, the best we can hope to do in mathematically describing it is to calculate some of its average properties. In Section 8.7.2 we find the expected or average OTF for short-exposure imagery, with the average over an ensemble of realizations of the atmospheric inhomogeneities.

8.7.2 Calculation of the Average Short-Exposure OTF

Our calculation closely follows that of Fried (Ref. 8-29). The optical transfer function of a single short-exposure image can be written as in Eq. (8.5-2), which we repeat here:

$$\mathcal{H}(\nu_u, \nu_v) = \frac{\displaystyle\iint_{-\infty}^{\infty} \mathbf{P}(x,y)\mathbf{P}^*(x - \bar{\lambda}f\nu_U, y - \bar{\lambda}f\nu_V)\exp\{(\chi_1 + \chi_2) + j(S_1 - S_2)\}\,dx\,dy}{\displaystyle\iint_{-\infty}^{\infty} |\mathbf{P}(x,y)|^2 e^{2\chi}\,dx\,dy}.$$

(8.7-1)

Averages of the numerator and denominator of this expression individually yielded our earlier expression for the long-exposure OTF. In this case we wish to take into account the fact that wavefront tilt has no effect on image quality in the short-exposure case. Thus our goal is to remove wavefront tilt from the phases S_1 and S_2 of Eq. (8.7-1) and then perform averaging.

The phase at point (x, y) within the collecting aperture of the imaging optics is represented by $S(x, y)$. Our goal is to find the least-squares fit of a planar wavefront to $S(x, y)$ and to subtract the phase associated with that planar wavefront, leaving a residual phase distribution that is free from tilt.

The linear component of $S(x, y)$ is taken to be of the form $a_X x + a_Y y$. We shall choose a_X and a_Y for any given $S(x, y)$ in such a way as to minimize the squared error,

$$\Delta = \iint_{-\infty}^{\infty} P(x,y)[S(x,y) - (a_X x - a_Y y)]^2\,dx\,dy, \qquad (8.7\text{-}2)$$

where the optical system has been assumed to be aberration-free and unapodized. Before undertaking the minimization, we simplify the expression for Δ somewhat. We have

$$\Delta = \iint_{-\infty}^{\infty} P(x,y)S^2(x,y)\,dx\,dy - 2\iint_{-\infty}^{\infty} P(x,y)(a_X x + a_Y y)S(x,y)\,dx\,dy$$

$$+ \iint_{-\infty}^{\infty}(a_X x - a_Y y)^2 P(x,y)\,dx\,dy. \qquad (8.7\text{-}3)$$

THE SHORT-EXPOSURE OTF   437

For a system with a circular, clear aperture of diameter D_0, we can easily show that the last term can be reduced to $\pi D_0^4(a_X^2 + a_Y^2)/64$. Now we find the partial derivatives $\partial \Delta/\partial a_X$ and $\partial \Delta/\partial a_Y$ and set them to zero. Interchanging orders of integration and differentiation, and solving for a_X and a_Y, we obtain

$$a_X = \frac{64}{\pi D_0^4} \int\!\!\!\int_{-\infty}^{\infty} xP(x,y)S(x,y)\,dx\,dy$$

$$a_Y = \frac{64}{\pi D_0^4} \int\!\!\!\int_{-\infty}^{\infty} yP(x,y)S(x,y)\,dx\,dy \qquad (8.7\text{-}4)$$

as the least-squares solution for a_X and a_Y. The fact that a_X and a_Y are linear functionals of $S(x,y)$ implies that, for Gaussian-distributed phase S, both tilt coefficients are also Gaussian random variables.

If we subtract the wavefront tilt from the phase distribution across the imaging aperture, the numerator of expression (8.7-1) for the OTF can be written

$$\text{Num} = \int\!\!\!\int_{-\infty}^{\infty} dx\,dy\, \mathbf{P}(x_1, y_1)\mathbf{P}^*(x_2, y_2)$$

$$\times \exp\{(\chi_1 + \chi_2) + j(S_1 - a_X x_1 - a_Y y_1) - j(S_2 - a_X x_2 - a_Y y_2)\},$$

$$(8.7\text{-}5)$$

where $(x_1, y_1) = (x, y)$, $(x_2, y_2) = (x - \bar{\lambda}f\nu_U, y - \bar{\lambda}f\nu_V)$. We must now average this expression over an ensemble of independent realizations of the atmospheric perturbations. In performing the required averages, it is helpful to note that, since S, a_X, and a_Y are all Gaussian random variables, so are $(S_1 - a_X x_1 - a_Y y_1)$ and $(S_2 - a_X x_2 - a_Y y_2)$.

The following assumptions are now adopted in order to simplify the evaluation of the averages involved:

(1) At any point (x, y), $S(x, y) - a_X x - a_Y y$ is assumed to be uncorrelated with (and, therefore, by virtue of the Gaussian statistics involved, independent of) a_X and a_Y. Equivalently, we are assuming that excursions of S about the tilt plane are not influenced by what tilt may be present. Detailed analysis shows (Ref. 8-31) that this is approximately, but not exactly, true.

(2) The difference of the residual phases after tilt removal $(S_1 - a_X x_1 - a_Y y_1) - (S_2 - a_X x_2 - a_Y y_2)$ is independent of the sum of the log

amplitudes $(\chi_1 + \chi_2)$. Since we previously established [see Eq. (8.5-10)] that $(\chi_1 + \chi_2)$ and $(S_1 - S_2)$ are independent, here we are effectively assuming that $E[(a_X\bar\lambda f\nu_U + a_Y\bar\lambda f\nu_V)(\chi_1 + \chi_2)] = 0$. Again, this must be regarded as an approximation.

If we now take the expected value of the numerator of the OTF, the orders of expectation and integration can be interchanged, leaving us with the evaluation of the average of the exponential in Eq. (8.7-5). Using assumption 2, above, we find with the help of Eqs. (8.5-19) and (8.5-14) that

$$E\{\exp[(\chi_1 + \chi_2) + j(S_1 - a_X x_1 - a_Y y_1) - j(S_2 - a_X x_2 - a_Y y_2)]\}$$

$$= \exp\Big\{-\tfrac{1}{2}D_\chi(\bar\lambda f\nu_U, \bar\lambda f\nu_V)$$

$$- \tfrac{1}{2}\overline{[(S_1 - a_X x_1 - a_Y y_1) - (S_2 - a_X x_2 - a_Y y_2)]^2}\Big\}. \quad (8.7\text{-}6)$$

Further simplification of this expression can be accomplished by use of the identity

$$[(S_1 - a_X x_1 - a_Y y_1) - (S_2 - a_X x_2 - a_Y y_2)]^2$$

$$= \{(S_1 - S_2)^2 + (a_X\bar\lambda f\nu_U + a_Y\bar\lambda f\nu_V)^2$$

$$+ 2[(S_1 - a_X x_1 - a_Y y_1) - (S_2 - a_X x_2 - a_Y y_2)]$$

$$\times [a_X\bar\lambda f\nu_U + a_Y\bar\lambda f\nu_V]\}. \quad (8.7\text{-}7)$$

Assumption 1 above, plus the reasonable assumption that excursions of phase about the tilt plane obey a symmetric probability density function, imply that the average of the last term is zero, leaving us with the result

$$E\{\exp[(\chi_1 + \chi_2) + j(S_1 - a_X x_1 - a_Y y_1) - j(S_2 - a_X x_2 - a_Y y_2)]\}$$

$$= \exp\Big\{-\tfrac{1}{2}D(\bar\lambda f\nu_U, \bar\lambda f\nu_V) + \tfrac{1}{2}(a_X\bar\lambda f\nu_U + a_Y\bar\lambda f\nu_V)^2\Big\}, \quad (8.7\text{-}8)$$

where $D = D_\chi + D_S$ is again the wave structure function.

At this point, to simply the results, we explicitly invoke an assumption of isotropic turbulence (such an assumption is already implicit in our previous assumption 2, above). In addition, we assume that the imaging optics have a

THE SHORT-EXPOSURE OTF

circular pupil, allowing us to express the results as a function of radial frequency $\nu = (\nu_U^2 + \nu_V^2)^{1/2}$. The expression for the expected short-exposure OTF of the atmosphere takes the form

$$\mathscr{H}_S(\nu) = \exp\left\{-\tfrac{1}{2}D(\bar\lambda f\nu) + \tfrac{1}{2}(\bar\lambda f\nu)^2\overline{\left(a_X^2 + a_Y^2\right)}\right\}. \qquad (8.7\text{-}9)$$

The average value of the quantity $a_X^2 + a_Y^2$ has been calculated by Fried (Ref. 8-32). The analysis is rather involved and is not repeated here. Fried finds that

$$(\bar\lambda f\nu)^2\overline{\left(a_X^2 + a_Y^2\right)} \cong 6.88\alpha\left(\frac{\bar\lambda f\nu}{r_0}\right)^{5/3}\left(\frac{\bar\lambda f\nu}{D_0}\right)^{1/3}, \qquad (8.7\text{-}10)$$

where α takes on the value unity for "near-field" propagation (valid when only phase effects are important), and value $\tfrac{1}{2}$ for "far field" propagation (valid when amplitude and phase effects are equally important). The symbol r_0 again represents the atmospheric coherence diameter, defined in Section 8.6.3, and D_0 is again the diameter of the entrance pupil of the imaging optics.

When the preceding expressions are incorporated in Eq. (8.7-9), together with the expression found in earlier sections for the wave structure function, the following form is obtained for the short-exposure OTF:

$$\mathscr{H}_S(\nu) = \exp\left\{-3.44\left(\frac{\bar\lambda f\nu}{r_0}\right)^{5/3}\left[1 - \alpha\left(\frac{\bar\lambda f\nu}{D_0}\right)^{1/3}\right]\right\}. \qquad (8.7\text{-}11)$$

Note that when the parameter α is set equal to zero, we obtain an expression equal to the long-exposure OTF,

$$\mathscr{H}_L(\nu) = \exp\left\{-3.44\left(\frac{\bar\lambda f\nu}{r_0}\right)^{5/3}\right\}. \qquad (8.7\text{-}12)$$

If the expression for $\mathscr{H}_S$ is rewritten in terms of frequency Ω measured in cycles per radian of arc, the corresponding result becomes

$$\mathscr{H}_S(\Omega) = \exp\left\{-3.44\left(\frac{\bar\lambda\Omega}{r_0}\right)^{5/3}\left[1 - \alpha\left(\frac{\Omega}{\Omega_0}\right)^{1/3}\right]\right\}, \qquad (8.7\text{-}13)$$

which is perhaps the most convenient form for expressing the results of our analysis.

Several comments about the preceding result are in order. First, we note that in the short exposure case the average OTF associated with the atmosphere depends on the diameter D_0 of the imaging optics, whereas in the long-exposure case the corresponding result is independent of the parameters of the imaging optics. The reason for the dependence on Ω_0 in the short-exposure case lies with the dependence of the mean-squared tilt on the reciprocal of $D_0^{1/3}$, as seen in Eq. (8.7-10). Thus the larger the aperture, the smaller the tilt component of wavefront distortion.

The difference between the long- and short-exposure results lies in the effect of the term $[1 - \alpha(\Omega/\Omega_0)^{1/3}]$. In the long-exposure case $\alpha = 0$, and this term reduces to unity. In the short-exposure case, a nonzero value for α results in a boost of the OTF, particularly as Ω approaches Ω_0. The different values for α in the near-field and far-field cases are simply a reflection of the fact that it is the tilt component of *phase* that has no effect on the OTF, and phase plays a less important role in the far-field case than in the near-field case. In the near-field case, all of the blur comes from phase effects, whereas in the far-field case, only half of the blur arises from phase perturbations, with the other half arising from amplitude effects.

Figure 8-28 shows plots of the combined system–atmosphere average OTFs for a telescope having a 1-m circular mirror and for an r_0 of 10

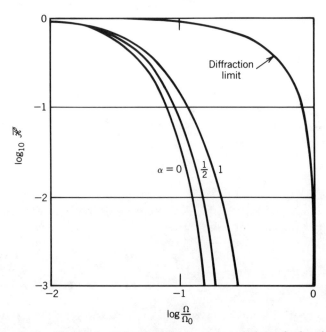

Figure 8-28. Combined system–atmosphere average optical transfer functions.

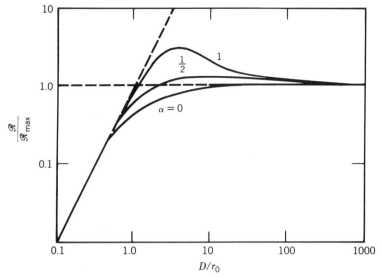

Figure 8-29. Normalized resolution $\mathscr{R}/\mathscr{R}_{max}$ versus normalized diameter of the imaging optics for both short- and long-exposure images. (Courtesy of D. L. Fried and the Optical Society of America. From D. L. Fried, *J. Opt. Soc. Am.*, Vol. 56, p. 1383, 1966.)

centimeters. The wavelength assumed is 0.5 μm. The curve labeled $\alpha = 0$ is the long-exposure OTF, whereas those labeled $\alpha = 1/2$ and 1 are for the short-exposure case. Also shown on the same graph is the diffraction-limited OTF of a system that has circular optics with a diameter of 1 meter.

Following the development of Section 8.6.3, we can again define the resolution of the system as the volume under the average OTF [Eq. (8.6-38)], this time using the short-exposure atmospheric OTF. Fried (Ref. 8-29) has numerically integrated the necessary equations, obtaining the results shown in Fig. 8-29. The limiting resolution for $D_0 \gg r_0$ is seen to be the same for all cases, because the tilt component of the wavefront distortions across the aperture diminishes as D_0 grows. Substantially higher resolution can be achieved in the short-exposure case when D_0 is nearly equal to r_0, particularly in the case of near-field propagation conditions.

8.8 STELLAR SPECKLE INTERFEROMETRY

In previous sections we studied the limitations to image quality caused by the presence of atmospheric inhomogeneities when long- and short-exposure images are gathered by an optical system. The effect of the atmosphere has been described by transfer functions that reduce the spatial frequency response at high frequencies, often limiting resolution to far smaller values than might be achieved by the same system operating in the absence of the atmosphere. Our attention is now turned to a novel and important data gathering and processing technique that allows information to be extracted from a series of short-exposure images at spatial frequencies far higher than would be passed by the average long- and short-exposure transfer functions studied earlier. The imaging technique of interest was invented by A. Labeyrie (Ref. 8-33) and was first demonstrated in astronomical observations by Gezari et al. (8-34).

In the section to follow, we discuss the basic principle of the method and the data processing operations involved. In Section 8.8.2 we present a heuristic analysis of the method. A more complete and rigorous analysis is outlined in Section 8.8.3, where the results of such an analysis are also presented. Finally, Section 8.8.4 discusses several extensions and alternate approaches.

8.8.1 Principle of the Method

We once more consider the different characters of the long- and short-exposure PSFs illustrated in Figs. 8-12 and 8-27. The short-exposure image of a point source is found to have a great deal of high-frequency structure, often referred to as "speckle," whereas the long-exposure image of a point is relatively smooth and regular. This fact suggests that the short-exposure OTF has greater high frequency response than the long-exposure OTF, as indeed is the case illustrated in that same figure.

An important distinction should be made between the OTF associated with a single short-exposure image and the expected or average OTF calculated in Section 8.7. The ensemble-averaging operation that leads to the latter OTF is itself an operation that suppresses high-frequency response, for at high frequencies the complex values of the OTF may vary wildly in both amplitude and phase from picture to picture. If we were to gather a large set of short-exposure photographs and center them all in such as way as to remove the effects of pure image shift from frame to frame, a sum of these aligned images would yield an image that closely agrees with the predictions of our average short-exposure OTF theory given in the previous section.

Given a set of short-exposure images, the procedure described earlier for combining them is not the only procedure that one can imagine. In fact, the method invented by Labeyrie rests on an alternate approach to extraction of information from such images. This alternate approach is motivated by the observation that, whereas the ensemble average of the short-exposure OTF falls off comparatively rapidly, the ensemble average of the squared modulus of the OTF has significant value out to much higher frequencies. The origin of this property is explained shortly, but first we describe Labeyrie's procedure in detail.

Assume that an astronomical telescope is used to collect a large number K of short-exposure photographs of an object of interest. A narrowband filter should be used in order to prevent blurring of the fine speckle-like image structure due to loss of temporal coherence of the light. This collection of images is now subjected to processing (by either digital or coherent optical means) of the following kind. Let the squared magnitude of the Fourier transform be calculated for each image. Thus if $I_i^{(k)}(u,v)$ represents the intensity associated with the kth image, we calculate its two-dimensional Fourier transform, given by

$$\mathcal{S}_i^{(k)}(\nu_U, \nu_V) = \iint_{-\infty}^{\infty} I_i^{(k)}(u,v) e^{j2\pi(u\nu_U + v\nu_V)} \, du \, dv. \qquad (8.8\text{-}1)$$

This image spectrum is, of course, related to the spectrum of the object (which does not change from frame to frame) and the OTF (which does change from frame to frame) through the usual product

$$\mathcal{S}_i^{(k)}(\nu_U, \nu_V) = \mathcal{H}^{(k)}(\nu_U, \nu_V) \mathcal{S}_0(\nu_U, \nu_V), \qquad (8.8\text{-}2)$$

where $\mathcal{S}_0$ is the Fourier transform of the object, whereas $\mathcal{H}^{(k)}$ is the OTF associated with the kth image. Now let the spectrum of each image be subjected to the squared modulus operation, generating a series of what we have called "energy spectra" in Chapter 3,

$$\mathcal{E}_i^{(k)}(\nu_U, \nu_V) = |\mathcal{S}_i^{(k)}(\nu_U, \nu_V)|^2. \qquad (8.8\text{-}3)$$

Finally we average these energy spectra by adding them together and dividing by the total number of images K. We assume that the number of images is sufficiently large that the finite average thus calculated is essentially the same as the ensemble average of the same quantity. Thus the procedure described has produced an estimate of the average energy spectrum of the image, which, in turn, depends on the average squared modulus of the short exposure OTF, or the average squared modulation transfer function (MTF),

$$\overline{|\mathcal{S}_i^{(k)}(\nu_U, \nu_V)|^2} = \overline{|\mathcal{H}^{(k)}(\nu_U, \nu_V)|^2} |\mathcal{S}_0(\nu_U, \nu_V)|^2. \qquad (8.8\text{-}4)$$

The procedure just described has been carried out by Labeyrie and others using a coherent optical system of the type illustrated in Fig. 8-30. The images reside on a film strip and are illuminated by coherent light from a laser. The film strip is drawn through the optical system, which produces at the output plane (the focal plane of the lens) a time-integrated intensity distribution that is the average of the energy spectra of the individual photographs. The fact that the *energy spectra* are averaged implies that the positions of the images withing their frames is unimportant, for shifts of the images result in linear phase shifts in the Fourier domain, and the processing system is insensitive to spectral phase distributions. Digital systems can also be used to perform the same operation, although the computational burden generally restricts their use to situations in which photon-counting detectors are used and the arrival rate of photons is rather small.

From Eq. (8.8-4) it is clear that if we were able to predict or measure the average squared MTF of the imaging system, and if that averaged quantity retained significant value out to frequencies higher than those present in the average short-exposure OTF, speckle interferometry would offer the possibility of extracting object information not retrievable from a single image or

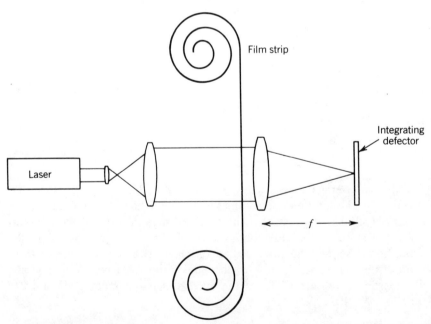

Figure 8-30. Coherent optical processing system used for averaging energy spectra of a set of short-exposure photographs.

STELLAR SPECKLE INTERFEROMETRY 445

from an image obtained by simply combining a number of centered short-exposure images.

However, it is also clear that the retrievable information about the object will in general not be complete, for it is the squared modulus of the object spectrum that can be obtained, not the complex spectrum itself. As in the case of the Michelson stellar interferometer working in the presence of atmospheric turbulence, as well as the intensity interferometer described in Section 6.3, no spectral phase information is available, and complete images of the object cannot be obtained in general.

Nonetheless, the less complete information can be exceedingly useful in many cases. The application to which Labeyrie first applied his technique was the measurement of the separation of binary stars. Of particular interest are those pairs of stars whose components are too close together to be resolved by a telescope in the presence of atmospheric degradations but are potentially resolvable by the diffraction-limited telescope aperture. If for convenience we consider the stars themselves to be ideal point sources, we can represent their brightness distribution by the intensity function

$$I_0(x, y) = I_1 \delta\left(x - \frac{\Delta x}{2}, y\right) + I_2 \delta\left(x + \frac{\Delta x}{2}, y\right), \qquad (8.8\text{-}5)$$

where x and y are to be interpreted as angular variables, Δx represents the angular separation of the components, and the possibility of different brightnesses for the two components has been retained. The squared modulus of the Fourier transform of this intensity distribution takes the form

$$|\mathscr{I}_0(\nu_X, \nu_Y)|^2 = \left(I_1^2 + I_2^2\right)\left[1 + \frac{2I_1 I_2}{I_1^2 + I_2^2}\cos(2\pi\Delta x \nu_X)\right]. \qquad (8.8\text{-}6)$$

Note in particular the sinusoidal "fringe" in this distribution, which will appear as a true optical fringe in the average intensity distribution at the output of the optical processing system shown in Fig. 8-30. The spatial frequency of this fringe identifies uniquely the separation Δx of the two components. To obtain an accurate measurement of the period of this fringe, we require that the average squared MTF of the imaging system have a component that has significant value out to frequencies much greater than a fringe period. Of course, our ability to recover this information depends entirely on the signal-to-noise ratio associated with the fringe, a subject we defer to Chapter 9. Figure 8-31 shows an experimentally recorded fringe pattern obtained in the manner described above.

We have asserted in several places that the average of the squared MTF will have significant value out to frequencies far higher than do the average

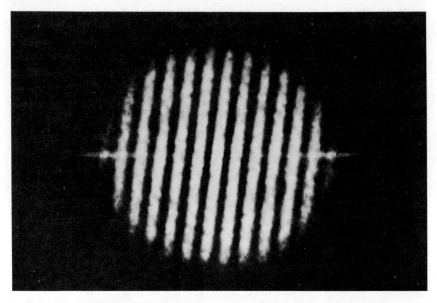

Figure 8-31. Average energy spectrum produced from 120 short-exposure images of the double star *9 Puppis*, after compensation for the speckle interferometry transfer function. Photograph supplied courtesy of Gerd Weigelt and Gerhard Baier, University of Erlangen.

long- or short-exposure OTFs. We turn now to a rather intuitive proof that this assertion is correct, following which we outline a more exact analysis.

8.8.2 Heuristic Analysis of the Method

The reason why the average of the squared short-exposure MTF of an optical system working within the Earth's atmosphere retains significant values out to relatively high frequencies can be understood with the help of intuitive reasoning and a minimum of mathematics. We wish to develop some understanding of the second-order statistics of the MTF. In accomplishing this goal, the interferometric view of the image-forming process is quite helpful.

Recall that, for a single short-exposure image, a particular spatial frequency component, having vector frequency $\vec{\nu}$, arises in the image plane by interference of light from points on the exit pupil separated by vector spacing $\vec{s} = \bar{\lambda} f \vec{\nu}$. As we slide such a vector spacing around the pupil, we gather a multitude of "elementary" contributions to this fringe, and the resulting contrast of this frequency component depends on the relative phases with which these contributions add, as well as on the amplitudes of the component fringes. See Section 7.4 for a more detailed discussion of this particular point of view.

The effects of the atmospheric distortions are to change the intensity and phase of the light incident on various parts of the pupil and thereby to change both the contrasts and the phases of the elementary fringes that compose any single frequency component of intensity. At low spatial frequencies, we are dealing with spacings that are very small; if the spacing of concern is smaller than the coherence diameter r_0 of the atmosphere, the log amplitudes and phase differences of the light incident at points a distance s apart are very small, with the result that such a frequency component is not affected by the presence of the atmospheric distortions. Such spatial frequencies lie within the low-frequency, high-response portion of the average short-exposure OTF.

If we now consider a spatial frequency sufficiently high that the corresponding spacing on the pupil is greater than r_0 but still considerably lower than the maximum spacing embraced by the pupil, the various elementary fringes will suffer both phase and contrast perturbations and will not perfectly reinforce one another when they are added together in the image plane. In fact, if we represent each sinusoidal fringe by a complex phasor, the addition of the various elementary fringes can be viewed as a form of random walk in the complex plane (see Fig. 2-10). The resultant phasor for any particular frequency has, after proper normalization, a complex value equal to the value of the OTF of the single short-exposure image under consideration at the particular frequency of interest.

Having built a random walk model for the "midrange" spatial frequency components, we can draw some conclusions about the statistical properties of the short-exposure OTF. As an approximation, let the exit pupil of the imaging system be imagined to consist of a multitude of independent correlation cells, each of diameter r_0. The number of such cells present in a pupil of diameter D_0 is simply

$$N_{\text{tot}} = \left(\frac{D_0}{r_0}\right)^2. \tag{8.8-7}$$

However, a particular spatial frequency $\vec{\nu}$ does not receive elementary fringe contributions from the entire exit pupil. Rather, contributions come only from the shaded regions of the exit pupil shown in Fig. 8-26. We represent the area of one of these shaded regions by the symbol $a(\vec{\nu})$. Now we recall that $a(\vec{\nu})$ is precisely equal to the numerator of the mathematical expression [Eq. (7.2-48)] for the diffraction-limited OTF at this particular spatial frequency. Hence the number of independent correlation cells contributing to this particular spatial frequency of the OTF is

$$N(\vec{\nu}) = \frac{a(\vec{\nu})}{\pi\left(\dfrac{r_0}{2}\right)^2} = \mathcal{H}_0(\nu)\left(\frac{D_0}{r_0}\right)^2, \tag{8.8-8}$$

where $\mathcal{H}_0$ is the diffraction-limited OTF of the optical system and the pupil has again been assumed circular with diameter D_0.

Knowing the number of independent phasors contributing to each spatial frequency component, we can now use our knowledge of the properties of random walks to draw some conclusions regarding the statistical properties of the OTF. First we note that, in the midfrequency range, where the number of contributing independent phasors is large, according to the arguments given in Section 2.9.2, the OTF must (to a good approximation) obey complex circular Gaussian statistics. As a corollary, the MTF must obey Rayleigh statistics and the square of the MTF must obey negative exponential statistics. These are powerful conclusions, but we emphasize that they are strictly true only in the midfrequency range, where the OTF has many independent randomly phased contributions.

Here we are particularly concerned about the mean-squared value of the MTF, for it is this quantity that plays a critical role in speckle interferometry [see Eq. (8.8-4)]. For the purpose of this simple argument, we suppose that the lengths of all phasors contributing to the random walk are identical and equal to β, which, in turn, we assume to be proportional to r_0^2,

$$\beta = \kappa r_0^2 \tag{8.8-9}$$

From Section 2.9, with the removal of the normalizing factor $\sqrt{N}$ in the definitions of the real and imaginary parts of the resultant, we see that the second moments of the real and imaginary parts of the numerator of the MTF are

$$\overline{r^2} = \overline{i^2} = N(\vec{\nu})\frac{\beta^2}{2}. \tag{8.8-10}$$

Using Eq. (8.8-9), we express the expected value of the square of the numerator of the MTF by

$$\overline{\mathrm{Num}^2} = \overline{r^2} + \overline{i^2} = N(\vec{\nu})\beta^2 = N(\vec{\nu})\kappa^2 r_0^4$$

$$= \mathcal{H}_0(\vec{\nu})\left(\frac{D_0}{r_0}\right)^2 \kappa^2 r_0^4. \tag{8.8-11}$$

In accord with the assumptions used in the earlier calculations of the long- and short-exposure cases, we treat the denominator of the OTF as approximately constant. Note that when the vector spacing on the aperture approaches zero, all phasors in the sum are fully correlated; that is, all have zero phase. As a consequence, the squared length of the resultant is just the square of the sum of the lengths of the individual phasors. Following the assumption of Eq. (8.8-9), we write

$$\text{Denom}^2 = \left(\kappa D_0^2\right)^2 \tag{8.8-12}$$

It follows that in the midfrequency region under consideration, the second moment of the MTF is equal to

$$\overline{|\mathcal{H}(\vec{\nu})|^2} = \frac{\overline{\text{Num}^2}}{\text{Denom}^2} = \left(\frac{r_0}{D_0}\right)^2 \mathcal{H}_0(\vec{\nu}). \tag{8.8-13}$$

Note in particular that in this midfrequency region, the second moment of interest is directly proportional to the MTF of the *diffraction-limited* optical system, and the proportionality constant is the ratio of the squares of r_0 and D_0. Since the diffraction-limited MTF does have significant value in this midfrequency region, so can the second moment of the MTF, provided the ratio of r_0 to D_0 is not too small.

At frequencies near the upper limit of the diffraction limited OTF, the overlap area on the pupil becomes comparatively small, implying that the number of independent phasors contributing to the OTF at such frequencies is small. Nonetheless, an examination of the arguments in Section 2.9.2 that led to the preceding expressions for the mean-square MTF shows that all the results we have used in arriving at Eq. (8.8-13) are valid for finite numbers of phasors. Although we cannot conclude that the squared MTF obeys negative-exponential statistics at such frequencies, we can nonetheless use the same expressions used previously for the second moment of the MTF.

Figure 8-32 illustrates the results of our approximate analysis. The second moment of the MTF behaves essentially the same as the average short-exposure OTF at low frequencies but falls to a value approximately equal to $(r_0/D_0)^2$ rather than to zero. At this point the behavior changes, and the function falls in proportion to the MTF of the diffraction-limited optical system, but with a scaling factor of $(r_0/D_0)^2$.

This concludes our heuristic discussion of stellar speckle interferometry. We turn next to the outline of a more complete and more accurate analysis.

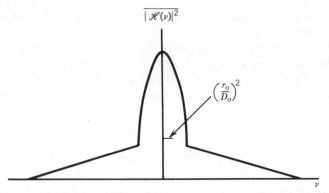

Figure 8-32. General behavior predicted for the mean-squared MTF of the combined optics and atmosphere.

8.8.3 A More Complete Analysis of Stellar Speckle Interferometry

A more complete analysis of the method described in Section 8.8.2 has been performed by Korff (Ref. 8-35). We outline his analysis here, omitting some details, which can be found in the reference cited.

As indicated in the previous section, the performance of the speckle interferometry method rests critically on the character of the second moment of the MTF of the optical system operating within the Earth's atmosphere. The starting point in this more complete analysis is thus an evaluation of the second moment $\overline{|\mathscr{H}(\vec{\nu})|^2}$. In evaluating this quantity, we wish to employ our detailed statistical knowledge concerning the amplitude and phase fluctuations introduced by the atmosphere.

Since the denominator of the MTF has been argued to be approximately constant, we can concentrate our attention on the properties of the numerator, which for the case at hand implies that we wish to find the second moment of

$$\text{Num} = \iint_{-\infty}^{\infty} \mathbf{P}(x, y)\mathbf{P}^*(x - \bar{\lambda}f\nu_U, y - \bar{\lambda}f\nu_V)$$

$$\times \mathbf{U}(x, y)\mathbf{U}^*(x - \bar{\lambda}f\nu_U, y - \bar{\lambda}f\nu_V)\, dx\, dy. \quad (8.8\text{-}14)$$

The moment in question can be written directly as

$$\overline{\text{Num}^2} = \iiiint_{-\infty}^{\infty} \mathbf{P}(\vec{r})\mathbf{P}^*(\vec{r} - \vec{s})\mathbf{P}^*(\vec{r}')\mathbf{P}(\vec{r}' - \vec{s})$$

$$\times \overline{\mathbf{U}(\vec{r})\mathbf{U}^*(\vec{r} - \vec{s})\mathbf{U}^*(\vec{r}')\mathbf{U}(\vec{r}' - \vec{s})}\, dx\, dy\, dx'\, dy', \quad (8.8\text{-}15)$$

where the vectors are defined by $\vec{r} \sim (x, y)$, $\vec{r}' \sim (x', y')$, $\vec{s} \sim (\bar{\lambda}f\nu_U, \bar{\lambda}f\nu_V)$ and the orders of averaging and integration have been interchanged. Expression of the field **U** in terms of a log amplitude χ and a phase S_δ [cf. Eq. (8.5-1)] yields an equivalent expression

$$\overline{\text{Num}^2} = I_0^2 \iiiint_{-\infty}^{\infty} \overline{\mathbf{P}(\vec{r})\mathbf{P}^*(\vec{r} - \vec{s})\mathbf{P}^*(\vec{r}')\mathbf{P}(\vec{r}' - \vec{s})}$$

$$\times \exp\{(\chi_1 + \chi_2 + \chi_3 + \chi_4) + j(S_1 - S_2 - S_3 + S_4)\}\, dx\, dy\, dx'\, dy',$$
(8.8-16)

where

$$\begin{aligned}
\chi_1 &= \chi(\vec{r}) & S_1 &= S_\delta(\vec{r}) \\
\chi_2 &= \chi(\vec{r} - \vec{s}) & S_2 &= S_\delta(\vec{r} - \vec{s}) \\
\chi_3 &= \chi(\vec{r}') & S_3 &= S_\delta(\vec{r}') \\
\chi_4 &= \chi(\vec{r}' - \vec{s}) & S_4 &= S_\delta(\vec{r}' - \vec{s}).
\end{aligned}$$
(8.8-17)

The reasoning underlying Eq. (8.5-10) can be extended to the case involving eight variables, rather than four, leading to the conclusion that the averages over the amplitude and phase terms can be computed individually (i.e., the sum of the log-amplitude terms is statistically independent of the sum of the phase terms). Recall that to arrive at this conclusion it was necessary to assume homogeneous and isotropic turbulence. Recognizing that both the sum of the log-amplitude terms and the sum of the phase terms obey Gaussian statistics, we can show [with the help of Eq. (8.5-14) and some algebra] that

$$\begin{aligned}
A_\chi &= \overline{\exp\{\chi_1 + \chi_2 + \chi_3 + \chi_4\}} \\
&= \exp[4C_\chi(0)] \exp\bigl[-D_\chi(|\vec{s}|) - D_\chi(|\vec{r} - \vec{r}'|) \\
&\quad - \tfrac{1}{2}D_\chi(|\vec{r} - \vec{r}' + \vec{s}|) - \tfrac{1}{2}D_\chi(|\vec{r} - \vec{r}' - \vec{s}|)\bigr],
\end{aligned}$$
(8.8-18)

$$\begin{aligned}
A_S &= \overline{\exp[j(S_1 - S_2 - S_3 + S_4)]} \\
&= \exp\bigl\{-D_S(|\vec{s}|) - D_S(|\vec{r} - \vec{r}'|) \\
&\quad + \tfrac{1}{2}D_S(|\vec{r} - \vec{r}' + \vec{s}|) \\
&\quad + \tfrac{1}{2}D_S(|\vec{r} - \vec{r}' - \vec{s}|)\bigr\}.
\end{aligned}$$
(8.8-19)

The product of A_χ and A_S can then be written in terms of the wave structure function as

$$A_\chi A_S = e^{4C_\chi(0)} Q(\vec{r},\vec{r}',\vec{s}),$$

where

$$Q(\vec{r},\vec{r}',\vec{s}) = \exp\{-D(|\vec{s}|) - D(|\vec{r}-\vec{r}'|)$$

$$- \tfrac{1}{2}D(|\vec{r}-\vec{r}'+\vec{s}|) - \tfrac{1}{2}D(|\vec{r}-\vec{r}'-\vec{s}|)\}$$

$$\times \exp\{D_S(|\vec{r}-\vec{r}'+\vec{s}|) + D_S(|\vec{r}-\vec{r}'-\vec{s}|)\}, \qquad (8.8\text{-}20)$$

yielding the following expression for the numerator of the mean-squared MTF:

$$\overline{\text{Num}^2} = \iiiint_{-\infty}^{\infty} \mathbf{P}(\vec{r})\mathbf{P}^*(\vec{r}-\vec{s})\mathbf{P}^*(\vec{r}')\mathbf{P}(\vec{r}'-\vec{s})$$

$$\times \exp[4C_\chi(0)] Q(\vec{r},\vec{r}',\vec{s})\, dx\, dy\, dx'\, dy'. \qquad (8.8\text{-}21)$$

The mean-squared denominator of the OTF is simply equal to the mean-squared numerator evaluated at zero spatial frequency. Thus we have

$$\overline{\text{Denom}^2} = \iiiint_{-\infty}^{\infty} |\mathbf{P}(\vec{r})|^2 |\mathbf{P}(\vec{r}')|^2 \exp[4C_\chi(0)] Q(\vec{r},\vec{r}',0)\, dx\, dy\, dx'\, dy'.$$

$$(8.8\text{-}22)$$

The mean-squared MTF can now be written as

$$\overline{|\mathcal{H}(\nu)|^2} = \frac{\displaystyle\iiiint_{-\infty}^{\infty} \mathbf{P}(\vec{r})\mathbf{P}^*(\vec{r}-\vec{s})\mathbf{P}^*(\vec{r}')\mathbf{P}(\vec{r}'-\vec{s})Q(\vec{r},\vec{r}',\vec{s})\, dx\, dy\, dx'\, dy'}{\displaystyle\iiiint_{-\infty}^{\infty} |\mathbf{P}(\vec{r})|^2 |\mathbf{P}(\vec{r}')|^2 Q(\vec{r},\vec{r}',0)\, dx\, dy\, dx'\, dy'}.$$

$$(8.8\text{-}23)$$

It remains to evaluate the preceding expression when the turbulence is assumed to obey Kolmogorov statistics. In this case the wave structure function takes the form

$$D(r) = 6.88\left(\frac{r}{r_0}\right)^{5/3}. \quad (8.8\text{-}24)$$

The phase structure function $D_S(r)$ depends on whether the propagation conditions are of the near-field or the far-field type,

$$D_S(r) = \begin{cases} D(r) & \text{(near field)} \\ \tfrac{1}{2}D(r) & \text{(far field)}. \end{cases} \quad (8.8\text{-}25)$$

Assuming near-field conditions, the expression for $Q(\vec{r}, \vec{r}', \vec{s})$ reduces to

$$Q(\vec{r}, \vec{r}', \vec{s}) = \exp\left\{-6.88\left(\frac{s}{r_0}\right)^{5/3}\right\}$$

$$\times \exp\left\{-6.88\left[\left(\frac{s}{r_0}\right)^{5/3} - \frac{1}{2}\left(\frac{|\vec{r} - \vec{r}' + \vec{s}|}{r_0}\right)^{5/3} - \frac{1}{2}\left(\frac{|\vec{r} - \vec{r}' - \vec{s}|}{r_0}\right)^{5/3}\right]\right\}.$$

$$(8.8\text{-}26)$$

Changing variables of integration to

$$\Delta \vec{r} = \vec{r} - \vec{r}' \sim (\Delta x, \Delta y)$$

$$\vec{\rho} = \vec{r} + \vec{r}' \sim (\rho_X, \rho_Y), \quad (8.8\text{-}27)$$

we obtain the following expression for the mean-squared MTF:

$$\overline{|\mathcal{H}(\nu)|^2} = \frac{\displaystyle\iint_{-\infty}^{\infty} Q(\Delta\vec{r}, \vec{s}) L(\Delta\vec{r}, \vec{s}) \, d\Delta x \, d\Delta y}{\left[\displaystyle\iint_{-\infty}^{\infty} |P(x, y)|^2 \, dx \, dy\right]^2}, \quad (8.8\text{-}28)$$

where $L(\Delta\vec{r}, \vec{s})$ represents the overlap integral

$$L(\Delta\vec{r}, \vec{s}) = \iint_{-\infty}^{\infty} P\left(\frac{\Delta\vec{r} + \vec{\rho} - 2\vec{s}}{2}\right) P^*\left(\frac{\Delta\vec{r} + \vec{\rho}}{2}\right)$$

$$\times P^*\left(\frac{\vec{\rho} - \Delta\vec{r} - 2\vec{s}}{2}\right) P\left(\frac{\vec{\rho} - \Delta\vec{r}}{2}\right) d\rho_X \, d\rho_Y. \quad (8.8\text{-}29)$$

Korff (Ref. 8-35) has evaluated the overlap integral $L(\Delta \vec{r}, \vec{s})$ for the case of a diffraction-limited system with a circular aperture and has evaluated the remaining integrals by numerical integration. His results are plotted in Fig. 8-33. A value of r_0 equal to 13 cm is assumed in this figure, and results are shown for telescope optics with diameters D_0 equal to 15 centimeters, 1.5 meters, and 5 meters, corresponding to ratios D_0/r_0 equal to 1.17, 11.7, and 38.4. Also shown on the same curve are the average long-exposure and average short-exposure OTFs for the same cases.

A comparison of Fig. 8-32 resulting from our approximate analysis and Fig. 8-33 shows that the conclusions of the earlier analysis are borne out by

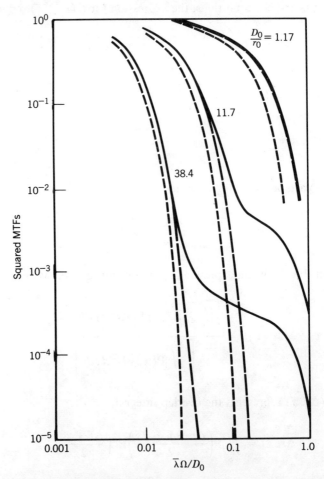

Figure 8-33. Theoretical prediction of the behavior of the mean-squared MTF for the combined optics and atmosphere. $\cdots = |\overline{\mathcal{H}}_L|^2$, $---=|\overline{\mathcal{H}}_S|^2$, $—=\overline{|\mathcal{H}|^2}$. (Courtesy of D. Korff and the Optical Society of America. From Ref. 8-35.)

the more exact calculation. At low frequencies the mean-squared MTF follows closely the shape of the average short-exposure OTF. There occurs a change of behavior when the mean-square MTF has fallen to a value of approximately $(D_0/r_0)^2$, at which time the shapes of the curves begin to follow an attenuated version of the diffraction-limited OTF of the optical system. It is precisely in this midfrequency region that the most useful information is derived from the speckle interferometry process.

8.8.4 Extensions

Our discussion of speckle interferometry would be incomplete if we did not refer the reader to some additional related developments. We broaden the discussion to include a number of related methods for extracting information from a series of short-exposure photographs taken in the presence of atmospheric distortions.

In 1971 McGlamery (Ref. 8-36) attempted to extract an image from a series of turbulence-degraded pictures using a certain averaging technique. His reasoning was that, if the pictures in the series of photographs were Fourier transformed (in this case by digital means) and the statistics of each frequency component were considered across the ensemble of pictures, the mean amplitudes and mean phases on a component-by-component basis ought to yield information about the amplitudes and phases of the undistorted object spectrum. Although the technique was sound in principle, its performance did not meet the expectations, as a result of the practical difficulties associated with the tracking of phases through multiples of 2π radians. That is, when the average phase of a given frequency component is calculated, it is necessary to perform the averaging on the "unwrapped" phase, with modulo 2π jumps removed. This proved extremely difficult, especially in the presence of typical amounts of noise.

There has been interest for some years in using the interferometric aspect of image formation to extract images from short-exposure recordings. Rhodes and Goodman (Ref. 8-37), following earlier leads by Jennison (Ref. 8-38) and Rogstad (Ref. 8-39), devised a scheme for breaking the telescope pupil up in such a way as to realize a series of triple interferometers simultaneously. The information extracted from the triple interferometers allows in principle the formation of images that are free from the effects of atmospherically induced phase distortions. This idea was tested in simulations by Brown (Ref. 8-40) for possible application to solar astronomy.

An important development occurred in 1974, when Knox and Thompson proposed a modification of Labeyrie's speckle interferometry technique that would allow full images (rather than simply autocorrelations of the object brightness distribution) to be recovered from a series of short-exposure images (Refs. 8-41 and 8-42). Their technique utilizes the fact that, over the image ensemble, correlations exist in the atmospherically induced phase perturbations of the Fourier components at closely spaced frequencies. Two Fourier components spaced by a distance smaller than $r_0/\bar{\lambda}f$ in the Fourier plane will have highly correlated atmospherically induced phase errors, whereas their phase components arising from the object may be quite different. If the phases of such components are subtracted, the atmospherically induced errors cancel, leaving a relatively error-free measurement of the phase difference between adjacent object frequency components. From a set of phase differences measured across the entire image spectrum, it is possible to determine a corresponding set of phases. With appropriate averaging of the modulus information and removal of its effects, that portion of the complex spectrum of the object lying within the OTF of the optics can be extracted, yielding a relatively distortion free image. This method has been applied by Stachnik et al. (Ref. 8-43) to obtain images of solar features.

We should also mention work of Worden et al. (Ref. 8-44), who made a somewhat different use of atmospherically induced speckle to extract images of astronomical objects. The speckle pattern observed in the presence of a single short-exposure image of a point source is equivalently the PSF of the imaging system at the particular instant when the picture was recorded. If this speckle pattern has one or a few widely separated peaks that are substantially higher than the surrounding intensity levels, convolution of that PSF with an intensity distribution associated with an object of small angular extent can yield a number of separated images of that object, one from each speckle peak, superimposed on a background. By shifting the image to line these subimages up, an image of the original object, blurred by an "average" speckle, is obtained. If the same procedure is performed on an image of a point source, a distribution of intensity associated with the average speckle is obtained. A deconvolution procedure is then used to remove the effect of the average speckle, leaving an improved image of the more extended object.

Finally it should be pointed out that in recent years an enormous amount of work has gone into the theory and practice of "adaptive optics" for obtaining diffraction-limited images in the presence of atmospheric turbulence. The block diagram of such a system is shown in Fig. 8-34. The wavefront sensor is an instrument that extracts from the arriving wave information regarding the atmospherically induced wavefront deformations suffered during propagation to the optical system. The wavefront computer combines these measurements in such a way as to yield an estimate of atmospherically induced wavefront errors across the pupil of the imaging optics. Finally, the deformable mirror is driven in real time to compensate

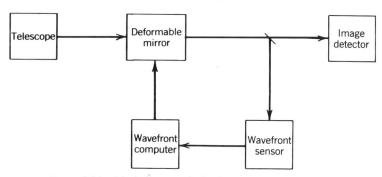

Figure 8-34. Block diagram of adaptive optic imaging system.

for the estimated distortions, yielding a diffraction-limited image at the image sensor. For an excellent review of this field, the reader can consult Ref. 8-24. In addition, a feature issue of the *Journal of the Optical Society of America* was devoted entirely to this subject area (Ref. 8-45).

8.9 GENERALITY OF THE THEORETICAL RESULTS

The vast majority of the theoretical results derived and presented here have been arrived at with the help of the Rytov approximation. It is natural to inquire as to their generality, since the Rytov solution is known to be limited to the case of weak fluctuations. It would be a mistake, however, to jump to the conclusion that results derived on the basis of a weak-fluctuation approximation must necessarily be invalid in the strong-fluctuation regime, as we discuss in what follows.

There have been numerous debates in the literature regarding the range of validity of the Rytov solution, based primarily on examinations of the magnitude of the terms that have been neglected (see, e.g., Refs. 8-46 and 8-47). Such a criterion leads one to the conclusion that the situations in which the Rytov solution is accurate are extremely limited. The assumptions of the weak-fluctuation theory are generally agreed to hold only in situations for which the log-amplitude variance is smaller than about 0.5. For a Kolmogorov spectrum of the turbulence, this criterion becomes (Ref. 8-4, Vol. 2, p. 445)

$$\sigma_{\ln I}^2 = 1.23 C_n^2 k^{7/6} z^{11/6} < 0.3 \tag{8.9-1}$$

When this condition is violated, the propagation is said to be occurring in the *strong* fluctuation regime. Nonetheless, there exist experimental results (e.g., Refs. 8-48 and 8-49) indicating that at least some of the predictions of the Rytov theory are correct in situations where it might have been argued that they should be substantially in error.

The understanding of this difficult problem was gradually enhanced after the discovery of a saturation phenomenon observed to occur during optical propagation over long paths (Ref. 8-50). Measurements of the variance of the intensity as a function of pathlength were observed to initially increase in accord with the weak-fluctuation theory but eventually to saturate at a value of unity for the ratio of the variance to the squared mean of the intensity.

A variety of theoretical methods have been developed to deal with propagation in the strong-fluctuation regime. These include the so-called "diagram method" (Ref. 8-51), the integral equation method (Ref. 8-52), the extended Huygens–Fresnel principle (Ref. 8-53), and the parabolic equation or moment equation method (Ref. 8-54). These various approaches are summarized in a review article by Strohbehn (Ref. 8-21).

The results of these mathematical analyses yield a remarkable and important result. The predictions of all methods for finding the mutual coherence function of a propagating wave yield identically the same result, namely, the result we obtained from the Rytov approximation. Thus our predictions for the optical transfer functions of imaging systems operating within the Earth's atmosphere can be used with confidence for both weak and strong fluctuations.

The predictions of the strong-fluctuation theories do differ markedly from those of the weak-fluctuation theory when higher-order coherence functions are calculated. Note in particular that the variance of the intensity involves fourth-order moments of the wave amplitude, and strong-fluctuation analysis is needed in order to explain the saturation phenomenon referred to above. In addition, it should be noted that the mean-squared MTF, which is of importance in stellar speckle interferometry (Section 8.8), involves fourth-order moments as well, and the range of validity of the solutions presented there is not well established. It seems likely, however, that the reasoning that led to the approximate solution in Section 8.8.2 will remain valid as long as the phase fluctuations are of the order of 2π radians or larger; therefore, the predictions of that theory in the midfrequency range should remain valid in the strong-fluctuation regime.

In conclusion, for the great majority of *imaging* problems, the predictions of weak-fluctuation and strong-fluctuation theories agree. Although the Rytov solution is strictly limited to small perturbations for the mathematics to be entirely correct, it is more physically oriented than the strong-fluctuation approaches and fortunately yields correct results in virtually all cases of interest to us here.

REFERENCES

8-1   H. G. Booker, J. A. Ratcliffe, and D. H. Shinn, *Phil. Trans. Roy. Soc. (Lond.), Ser. A,* **242**, 579 (1950).

8-2   J. A. Ratcliffe, *Rep. Progr. Phys.,* **19**, 188 (1956).

8-3   E. L. O'Neill, *Introduction to Statistical Optics,* Addison-Wesley, Reading, MA, pp. 99–101 (1963).

8-4   Akira Ishimaru, *Wave Propagation and Scattering in Random Media,* Vols. 1 and 2, Academic Press, New York (1978).

8-5   W. L. Wolfe and G. J. Zissis (editors), *The Infrared Handbook,* The Infrared Information and Analysis Center, Environmental Research Institute of Michigan, Ann Arbor, Michigan (1978). Note in particular Robert E. Hufnagel, Chapter 6, "Atmospheric Turbulence" and Benjamin Herman, Anthony J. LaRocca, and Robert E. Turner, Chapter 4, "Atmospheric Scattering."

8-6   J. W. Strohbehn, *Proc. IEEE,* **56**, 1301 (1968).

8-7   R. S. Lawrence and J. W. Strohbehn, *Proc. IEEE,* **58**, 1523 (1970).

8-8   R. W. Lee and J. C. Harp, *Proc. IEEE,* **57**, 375 (1969).

8-9   J. W. Strohbehn, "Optical Propagation Through the Turbulent Atmosphere," *Progress in Optics,* Vol. IX (E. Wolf, editor), North Holland Publishing Company, Amsterdam (1971).

8-10 *Radio Science,* Vol. 10, No. 1 (January 1975) (entire issue).

8-11 R. L. Fante, *Proc. IEEE,* **63**, 1669 (1975).

8-12 V. I. Tatarski, *Wave Propagation in a Turbulent Medium,* McGraw-Hill Book Company, New York (1961).

8-13 V. I. Tatarski, *The Effects of the Turbulent Atmosphere on Wave Propagation,* National Science Foundation Report TT-68-50464 (1968).

8-14 L A. Chernoff, *Wave Propagation in a Random Medium,* McGraw-Hill Book Company, New York (1960).

8-15 S. F. Clifford, "The Classical Theory of Wave Propagation in a Turbulent Medium," in *Laser Beam Propagation in the Atmosphere* (J. W. Strohbehn, editor), Springer-Verlag, Heidelberg (1978).

8-16 R. N. Bracewell, *The Fourier Transform and its Applications,* McGraw-Hill Book Company, New York (1965).

8-17 A. Kolmogorov, in *Turbulence, Classic Papers on Statistical Theory* (S. K. Friedlander and L. Topper, editors), Wiley-Interscience, New York (1961).

8-18 This identity can be obtained by analytic continuation of formula 6.5(1), p. 317, of A. Erdelyi, W. Magnus, F. Oberhettinger, and F. G. Tricomi, *Table of Integral Transforms,* Vol. 1, McGraw Hill Book Company, New York (1954).

8-19 S. F. Clifford and J. W Strohbehn, *IEEE Transact. Antennas Propagation,* **AP-15**, 416 (1967).

8-20 J. W. Goodman, *Introduction to Fourier Optics,* McGraw-Hill Book Company, New York, Chapter 4 (1968).

8-21 J. W. Strohbehn, "Modern Theories in the Propagation of Optical Waves," in *Laser Beam Propagation in the Atmosphere* (J. W. Strohbehn, editor) Springer-Verlag, Heidelberg (1978).
8-22 R. Barakat, *J. Opt. Soc. Am.*, **66**, 211 (1966).
8-23 R. L. Mitchell, *J. Opt. Soc. Am.*, **58**, 1267 (1968).
8-24 John W. Hardy, *Proc. IEEE*, **66**, 651 (1978).
8-25 J. C. Dainty, "Stellar Speckle Interferometry," in *Laser Speckle and Related Phenomena* (J. C. Dainty, editor), Springer-Verlag, Heidelberg (1975).
8-26 R. E. Hufnagel and N. R. Stanley, *J. Opt. Soc. Am.*, **54**, 52 (1964).
8-27 F. Talbot, *Phil. Mag.*, **9**, 401 (1836).
8-28 D. L. Fried, *J. Opt. Soc. Am.*, **56**, 1380 (1966).
8-29 D. L. Fried, *J. Opt. Soc. Am.*, **56**, 1372 (1966).
8-30 R. A. Schmeltzer, *Quart. Appl. Math.*, **24**, 339 (1967).
8-31 G. R. Heibreider, *IEEE Transact. Antennas Propagation*, **AP-15**, 90 (1967).
8-32 D. L. Fried, *J. Opt. Soc. Am.*, **55**, 1427 (1965).
8-33 A. Labeyrie, *Astron. Astrophys.*, **6**, 85 (1970).
8-34 D. Y. Gezari, A. Labeyrie, and R. V. Stachnik, *Astrophys. J.*, **173**, L1 (1972).
8-35 D. Korff, *J. Opt. Soc. Am.*, **63**, 971 (1973).
8-36 B. L. McGlamery, *Astronomical Use of Television Image Sensors*, NASA Technical Report SP-256, (1971).
8-37 W. T. Rhodes and J. W. Goodman, *J. Opt. Soc. Am.*, **63**, 647 (1973).
8-38 R. C. Jennison, *Monthly Notices Roy. Astron. Soc.*, **118**, 276 (1958).
8-39 D. H. Rogstad, *Appl. Opt.*, **7**, 585 (1968).
8-40 T. M. Brown, *J. Opt. Soc. Am.*, **68**, 883 (1978).
8-41 K. T. Knox and B. Thompson, *Astrophys. J.*, **193**, L45 (1974).
8-42 K. T. Knox, *J. Opt. Soc. Am.*, **66**, 1236 (1976).
8-43 R. V. Stachnik, P. Nisenson, D. C. Ehn, R. H. Hudgin, and V. E. Scherf, *Nature*, **266**, 149 (1977).
8-44 S. P. Worden, C. R. Lynds, and J. W. Harvey, *J. Opt. Soc. Am.*, **66**, 1243 (1976).
8-45 Topical Issue on Adaptive Optics, *J. Opt. Soc. Am.*, **67** (Mach 1977) (entire issue).
8-46 D. A. de Wolf, *J. Opt. Soc. Am.*, **55**, 812 (1965).
8-47 W. P. Brown, Jr., *J. Opt. Soc. Am.*, **56**, 1045 (1966).
8-48 M. E. Gracheva, A. S. Gurvich, S. S. Kashkarov, and V. V. Pokasov, "Similarity Relations and Their Experimental Verification for Strong Intensity Fluctuations of Laser Radiation," in *Laser Beam Propagation in the Atmosphere* (J. W. Strohbehn, editor), Springer-Verlag, Heidelberg (1978).
8-49 A. S. Gurvich, M. A. Kallistratova, and N. S. Time, *Radiophys. Quant. Electron.*, **11**, 771 (1968).
8-50 M. E. Gracheva and A. S. Gurvich, *Izv. VUZ, Radiofiz*, **8**, 717 (1965), [English translation: *Radiophys. Quant. Electron.*, **8**, 511 (1965)].
8-51 R. C. Bourret, *Can. J. Phys.*, **40**, 782 (1962); *Nuovo Cimento*, **26**, 1 (1962).
8-52 W. P. Brown, Jr., *J. Opt. Soc. Am.*, **61**, 1051 (1971); **62**, 45 (1972).
8-53 R. Lutomirski and H. Yura, *J. Opt. Soc. Am.*, **61**, 482 (1971).
8-54 T. L. Ho and M. Beran, *J. Opt. Soc. Am.*, **58**, 1335 (1968).

ADDITIONAL READING

F. Roddier, "The Effects of Atmospheric Turbulence in Optical Astronomy," *Progress in Optics*, Vol. XIX (E. Wolf, editor), North-Holland Publishing Company, Amsterdam (1981).

PROBLEMS

8-1 Consider a random screen placed in the geometry of Fig. 8-1, but with an amplitude transmittance $t_s(x, y; \xi, \eta)$ that depends on which object point (ξ, η) is considered. Assuming that the statistical autocorrelation function $E[t_s(x, y; \xi, \eta) t_s^*(x - x_0, y - y_0; \xi, \eta)]$ is independent of (ξ, η), find the average OTF of the system.

8-2 A random screen with amplitude transmittance $t_s(x, y)$ is illuminated in the geometry shown in Fig. 8-2p. The amplitude transmittance of the screen is $t_s(x, y)$, and its spatial autocorrelation function (assumed spatially stationary) is $\Gamma_t(\Delta x, \Delta y)$. Assuming unit amplitude quasimonochromatic illumination of the screen and neglecting the finite extent of the screen, find the autocorrelation function of the fields in a plane at distance z_2 from the screen.

Hint: The spatial transfer function for fields propagating through free space is (Ref. 5-24, p. 54)

$$\mathcal{H}(\nu_X, \nu_Y) = \begin{cases} \exp\left[j2\pi \frac{z}{\lambda}\sqrt{1 - (\lambda \nu_X)^2 - (\lambda \nu_Y)^2}\right] & \text{for } \sqrt{\nu_X^2 + \nu_Y^2} < \frac{1}{\lambda} \\ 0 & \text{otherwise.} \end{cases}$$

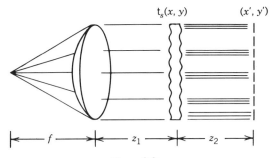

Figure 8-2p.

8-3 Prove that for a random absorbing screen, and for frequencies corresponding to spacings sufficiently large that $\gamma_r \approx 0$, the value of the average OTF of the screen can never be less than $\bar{t}_0$, the average value of the amplitude transmittance of the screen.

Hint: Remember that $0 \le t_s \le 1$.

8-4 Find the average OTF and the average PSF of a random checkerboard absorbing screen when the amplitude transmittance t_s is uniformly distributed between zero and unity.

8-5 Show that, for a stationary Gaussian phase screen, with the phase having a circularly symmetric, Gaussian-shaped autocorrelation function [Eq. (8.3-17)], in the limit of large phase variance the $1/e$ frequency of the average OTF is given by Eq. (8.3-19); thus,

$$\nu_{1/e} = \frac{W}{\bar{\lambda} f \sigma_\phi}.$$

8-6 For Problem 8-5, show that the $1/e$ frequency of the average OTF is in fact independent of the average wavelength $\bar{\lambda}$.

8-7 Consider a random-phase screen in which the phase $\phi(x, y)$ is given by

$$\phi(x, y) = \phi_0 \cos\left(2\pi \frac{x}{l} + \theta\right),$$

where θ is a random variable, uniformly distributed on $(-\pi, \pi)$. Find the average of OTF associated with such a screen.

8-8 The intensity $I = I_0 \exp(2\chi)$ of a plane wave after propagation through the turbulent atmosphere has mean $\bar{I}$ and standard deviation σ_I. Assuming that there is no loss of energy during propagation, find an expression for $\sigma_I/\bar{I}$ in terms of σ_χ^2. What is the possible range of $\sigma_I/\bar{I}$ as σ_χ^2 ranges from 0 to ∞?

8-9 A set of experimental measurements performed with the use of an astronomical telescope shows that the long-exposure OTF drops to value $1/e$ at an angular frequency of 80 cycles per milliradian under typical conditions. The measurements were made with light of mean wavelength 0.550 μm.

(a) Find the value of $\int_0^\infty C_n^2(\xi) d\xi$ looking up through the atmosphere.

(b) Imagine the atmosphere to be a uniformly turbulent medium with $C_n^2 = 10^{-15}$ m$^{-2/3}$ independent of height. What is the effective thickness of the atmosphere?

8-10 Consider a refractive index fluctuation $n_1(r)$ that may be regarded as statistically homogeneous.

(a) Show that the two-dimensional power spectral density F_n of n_1 in a plane of fixed z is related to the three-dimensional spectral density Φ_n through

$$F_n(\kappa_X, \kappa_Y; z) = \int_{-\infty}^{\infty} \Phi_n(\kappa_X, \kappa_Y, \kappa_Z) \, d\kappa_Z.$$

(b) Show that the two-dimensional cross-spectral density $\tilde{F}_n(\kappa_X, \kappa_Y; \Delta z)$ of the fluctuations of n_1 in two transverse planes separated by Δz is related to the three-dimensional power spectral density by

$$\tilde{F}_n(\kappa_X, \kappa_Y; \Delta z) = \int_{-\infty}^{\infty} \Phi_n(\kappa_X, \kappa_Y, \kappa_Z) \cos(\kappa_Z \Delta z) \, d\kappa_Z.$$

8-11 Two monochromatic, equal-intensity point sources are separated by distance s in a plane P_1. At distance z_1 from this plane there exists a thin, statistically homogeneous, Gaussian phase screen having phase variance σ_ϕ^2 and normalized autocorrelation function γ_ϕ. The structure of the screen is changing with time, but its statistics are ergodic. At a distance z_2 behind the screen we place an observing screen on which fringes can be seen. Find the long-time-average fringe visibility on the observing screen using the following assumptions:

(a) Spherical waves can be represented by their paraxial approximations.
(b) The rays are delayed but not appreciably bent by passage through the screen, and diffraction effects can be ignored.
(c) The finite size of the screen can be ignored.
(d) Temporal coherence effects can*not* be neglected.
Express the observed fringe visibility as a function of all relevant parameters.

Comment: You are examining the possibility of recording holograms through a time-varying distorting medium.

8-12 Find the filter function that relates the two-dimensional cross-spectral density of S and χ to the three-dimensional spectral density Φ_n of the refractive index fluctuations.

8-13 The purpose of this problem is to gain some insight into the relationship between the Born approximation and the Rytov approximation.

(a) Show that for the case of plane wave propagation, the Born approximation yields exactly the same expression for the additive field perturbation U_1 as the Rytov approximation yields for the exponent ψ_1 of the multiplicative field perturbation, aside from a constant multiplier.

(b) On the basis of the above observation, find filter functions that relate the two-dimensional power spectral densities of components of U_1 that are in phase and in quadrature with U_0 to the three-dimensional power spectral density Φ_n of the refractive index fluctuations.

(c) Assess the validity of the assertion often made in the literature that the Born approximation yields a prediction of Rician statistics for the amplitude $|U_0 + U_1|$ of the total field.

9

Fundamental Limits in Photoelectric Detection of Light

Light interacts with matter in a fundamentally random or stochastic way. As a consequence, any measurement of light will be accompanied by certain unavoidable fluctuations. We attribute these fluctuations to quantum effects; that is, light can be absorbed only in small discrete energy "packets," or quanta (Ref. 9-1). The goal of this chapter is to develop a statistical model for such fluctuations and to explore the associated limitations on the extraction of information from optical waves.

The most fundamental approach to understanding such phenomena would be through the theory of quantum electrodynamics (QED). Thus the electromagnetic fields would be quantized, and the implications of the basic postulates of quantum mechanics would be explored in the context of the detection problem. Such an approach is most fundamental but is also comparatively difficult, for it requires a thorough knowledge of the mathematics of quantum mechanics and rests very little on physical intuition.

As a consequence of the obstacles associated with the rigorous approach, an alternative formalism has been chosen for this chapter. We shall deal with the so-called semiclassical theory of photodetection. Such an approach has the benefit of being comparatively simple in terms of the mathematical background required, as well as allowing a greater use of physical intuition. Fortunately it has been shown (Refs. 9-2, 9-3) that the predictions of the semiclassical theory are in complete agreement with the predictions of the more rigorous quantum mechanical approach for all detection problems involving the *photoelectric effect*. Since the vast majority of optical detection problems do indeed rest on the photoelectric effect, there is relatively little loss of generality by making this assumption at the start.

There are a number of excellent general references to the semiclassical theory of detection that the reader may wish to consult, either for alternative explanations or in some cases for more detailed discussions. We mention in particular Refs. 9-4 through 9-6. For a discussion of the relative merits of the semiclassical and QED treatments, see Ref. 9-7. Those particularly interested in the rigorous QED approach should consult the work of R. Glauber (Ref. 9-8).

9.1 THE SEMICLASSICAL MODEL FOR PHOTOELECTRIC DETECTION

The semiclassical approach provides a highly physical means for describing the interaction of light and matter. The distinguishing characteristic of this formalism is the fact that electromagnetic fields are treated in a completely classical manner until they interact with the atoms of the photosensitive material on which they are incident. Thus there is no necessity to deal with quantization of the electromagnetic field; only the interaction of the classical field and matter is quantized.

When electromagnetic fields are incident on a photosurface, a complex set of events can transpire. The major steps in this process can be identified as (1) absorption of a quantum of light energy (i.e., a photon) and the transfer of that energy to an excited electron, (2) transport of the excited electron to the surface, and finally (3) release of the electron from the surface. We shall refer to the release of such an electron from the photosurface as a *photoevent*. The number K such events occurring in a given time interval is referred to as the number of *photocounts*.

The semiclassical theory is based on the following three fundamental assumptions concerning the statistical properties of photoevents. First, it is assumed that the probability of occurrence of a single photoevent from an area of the photosurface that is small compared with the coherence area of the incident light, in a time period shorter than the coherence time of the light (but much longer than the period of the optical vibration), is proportional to the *intensity* of the incident wave, the length of the time interval itself, and the size of the area in question on the photosurface. Stated mathematically, the probability of observing one photoevent in time Δt from area ΔA is taken to be

$$P(1; \Delta t, \Delta A) = \alpha \Delta t \Delta A \, I(x, y; t), \qquad (9.1\text{-}1)$$

where α is a proportionality constant and $I(x, y; t)$ represents the intensity of the wave at time t and at coordinates (x, y) on the photosurface. Second, it is assumed that the probability of more than one photoevent occurring in such a short time interval and in such a small area is vanishingly small compared with the probability of one or zero photoevents. (Thus no multiple events are allowed.) Finally, it is assumed that the numbers of photoevents occurring in any two nonoverlapping time intervals are statistically independent. (The photoemission process has no memory.)

MODEL FOR PHOTOELECTRIC DETECTION

The alert reader may recall that these three assumptions are identical to those used in Section 3.7.2 in the discussion of Poisson impulse processes and were seen there to lead directly to Poisson statistics for the number of pulses occurring in a fixed time interval. If each photoevent is represented by a unit area Dirac δ function in time and space, the resulting random process is a space–time Poisson impulse process, with rate equal to the intensity of the light times the proportionality constant α. In accord with Eq. (3.7-8), therefore, the probability of observing K photoevents in the time interval $(t, t + \tau)$ can be expressed as

$$P(K) \triangleq P(K; t, t + \tau) = \frac{(\overline{K})^K}{K!} e^{-\overline{K}}, \tag{9.1-2}$$

where, if $\mathscr{A}$ represents the illuminated area of the photosurface, the mean number of photoevents $\overline{K}$ is given by

$$\overline{K} = \alpha \iint_{\mathscr{A}} \int_{t}^{t+\tau} I(x, y; \xi) \, d\xi \, dx \, dy. \tag{9.1-3}$$

It is generally convenient to express this result in terms of a quantity we refer to as *integrated intensity* W having the dimensions of energy and defined by

$$W = \iint_{\mathscr{A}} \int_{t}^{t+\tau} I(x, y; \xi) \, d\xi \, dx \, dy. \tag{9.1-4}$$

Note that simpler forms for the integrated intensity are possible when the intensity incident on the photosurface is constant in time and/or over space. Thus when the intensity has constant value I_0 (independent of both time and space), the expression for W simplifies to

$$W = I_0 A \tau, \tag{9.1-5}$$

whereas when only time variations are allowed, the expression becomes

$$W = A \int_{t}^{t+\tau} I(\xi) \, d\xi. \tag{9.1-6}$$

When expressed in terms of integrated intensity, the probability of observing K photoevents is given as

$$P(K) = \frac{(\alpha W)^K}{K!} e^{-\alpha W}. \tag{9.1-7}$$

It is possible, and indeed useful, to express the constant α in terms of other, better known, physical constants. Since the integrated intensity W is in fact the energy incident on the photosurface during the measurement time of interest, and since each photon of light carries energy $h\bar{\nu}$, the mean number of photoevents in time τ can be expressed as

$$\overline{K} = \alpha W = \frac{\eta W}{h\bar{\nu}}, \qquad (9.1\text{-}8)$$

where h is Planck's constant (6.626196×10^{-34} joule-s); $\bar{\nu}$ is the mean optical frequency of the radiation; and η, called the *quantum efficiency*, represents the average number of photoevents produced by each incident photon ($\eta \leq 1$). We conclude that the proportionality constant α is equivalently given by

$$\alpha = \frac{\eta}{h\bar{\nu}}. \qquad (9.1\text{-}9)$$

At this point the initial introduction to the semiclassical theory is complete. However, the discussion has implicitly used an important assumption, namely, that the space-time variations of intensity are entirely deterministic, or equivalently known a priori. Attention is now turned to more realistic cases that involve random fluctuations of the classical intensity.

9.2 EFFECTS OF STOCHASTIC FLUCTUATIONS OF THE CLASSICAL INTENSITY

When light having a deterministic variation of intensity over space and time is incident on a photodetector, we have seen that the fluctuations of the photocounts obey Poisson statistics. In most problems of real interest, however, the light wave incident on the photosurface has stochastic attributes; that is, it is not possible to predict exactly what the fluctuations of the light wave will be. As we shall see, any stochastic fluctuations of the classical intensity can influence the statistical properties of the photoevents that are observed. For this reason, it is necessary to regard the Poisson distribution of Eq. (9.1-7) as a *conditional* probability distribution; the conditioning is based on knowledge of the exact value of the integrated intensity W.

In practice, it is the *unconditional* probability distribution of the photoevents that is of interest. To obtain this distribution, it is necessary to average the conditional statistics of Eq. (9.1-7) over the statistics of the integrated intensity. It is helpful to explicitly represent the fact that the Poisson distribution (9.1-7) is a conditional distribution, by expressing it in the form $P(K|W)$. Here, as usual, the vertical bar indicates that the distribution is conditioned on knowledge of the quantity that follows it. The unconditional probability of observing K photoevents can now be expressed as

$$P(K) = \int_0^\infty P(K|W) p_W(W) \, dW \qquad (9.2\text{-}1)$$

$$= \int_0^\infty \frac{(\alpha W)^K}{K!} e^{-\alpha W} p_W(W) \, dW,$$

where $p_W(W)$ is the probability density function of the integrated intensity. This equation will serve as the basis for all future calculations of photoevent statistics. It is called *Mandel's formula* after the individual who first derived it (Ref. 9-10). It is also common to refer to a function $P(K)$ defined by this formula as the *Poisson transform* of the probability density $p_W(W)$.

It should be evident from Eq. (9.2-1) that, in spite of the underlying conditional Poisson nature of the photoevents, the statistics are in general *not* Poisson when the classical intensity has random fluctuations of its own. In effect, we see photocount fluctuations that are the compounded consequence of both the fundamental uncertainties associated with the interaction of light and matter and the classical fluctuations of the light incident on the photosurface. Thus the photoevents form a doubly stochastic Poisson process (see Section 3.7.5).

Before turning to calculations of photoevent statistics in some specific cases, it is worthwhile stating some general relationships that follow directly from Mandel's formula. In particular, we wish to calculate the nth factorial moment of the distribution $P(K)$,

$$E[K(K-1)\cdots(K-n+1)] = \sum_{K=0}^\infty K(K-1)\cdots(K-n+1) P(K)$$

$$= \sum_{K=0}^\infty K(K-1)\cdots(K-n+1)$$

$$\times \int_0^\infty \frac{(\alpha W)^K}{K!} e^{-\alpha W} p_W(W) \, dW. \qquad (9.2\text{-}2)$$

Interchanging the orders of summation and integration, we can now recognize the inner sum as the nth factorial moment of a Poisson random variate with mean αW. According to Eq. (3.7-3), the value of the sum is simply $(\alpha W)^n$. Thus evaluation of the unconditioned moment in question is accomplished as

$$E[K(K-1) \cdots (K-n+1)] = \int_0^\infty (\alpha W)^n p_W(W) \, dW = \alpha^n \overline{W^n}.$$

(9.2-3)

From this result, the mean and variance of K are easily found to be

$$\overline{K} = \alpha \overline{W}, \qquad \sigma_K^2 = \alpha \overline{W} + \alpha^2 \sigma_W^2.$$ (9.2-4)

Note in particular that the variance σ_K^2 of K consists of two distinct terms, each of which has a physical interpretation. The first, which is proportional to the total energy incident during the measurement, can be interpreted as representing the effects of pure Poisson noise introduced by the random interaction of light and matter. The second, because it is proportional to the variance of the fluctuations of the incident intensity, is the classically expected result in the absence of any noise associated with the interaction of light and matter.

The reader is encouraged to consider the relationship between the characteristic functions of the photocounts and the integrated intensity in Problem 9-1.

9.2.1 Photocount Statistics for Well-Stabilized, Single-Mode Laser Radiation

Consider a single-mode laser operating well above threshold. The light from this source falls on a photosurface, and we wish to determine the statistical distribution of the number of photoevents observed in any τ-second interval. Assume that to an excellent approximation the intensity of the incident light may be considered constant over both space and time and represent that intensity by the symbol I_0. The integrated intensity can then be seen to be given in this simple case by

$$W = I_0 A \tau,$$ (9.2-5)

and as a consequence, the probability density function of the integrated intensity takes the form

$$p_W(W) = \delta(W - I_0 A \tau).$$ (9.2-6)

With substitution of this equation into Mandel's formula [Eq. (9.2-1)], the integration is trivially performed, yielding the following result:

$$P(K) = \int_0^\infty \frac{(\alpha W)^K}{K!} e^{-\alpha W} \delta(W - I_0 A \tau) \, dW$$

$$= \frac{(\alpha I_0 A \tau)^K}{K!} e^{-\alpha I_0 A \tau}. \tag{9.2-7}$$

Finally, noting that the mean number of photoevents is $\overline{K} = \alpha I_0 A \tau$, we can equivalently write the probability of obtaining K photoevents,

$$P(K) = \frac{(\overline{K})^K}{K!} e^{-\overline{K}}. \tag{9.2-8}$$

Following warnings that the statistics of the photocounts will in general not be Poisson, we have discovered a case in which they are exactly so. This should not be surprising, however, for the particular case examined here has been one in which there are absolutely no classical fluctuations of the intensity. Thus there are no "excess fluctuations" of the photocounts above and beyond the basic Poisson statistics associated with the interaction of light and matter.

As a reminder, recall that the factorial moments of the Poisson distribution are given by [cf. Eq. (3.7-3)]

$$E[K(K-1)\cdots(K-n+1)] = \alpha^n (\overline{W})^n. \tag{9.2-9}$$

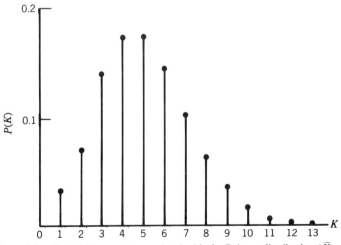

Figure 9-1. Probability masses associated with the Poisson distribution ($\overline{K} = 5$).

Similarly, the variance can be expressed in terms of the mean as

$$\sigma_K^2 = \overline{K}. \qquad (9.2\text{-}10)$$

It is worth noting that the signal-to-noise ratio associated with this distribution, as defined by the ratio of mean to standard deviation, is given by

$$\frac{S}{N} = \frac{\overline{K}}{\sigma_K} = \sqrt{\overline{K}} \qquad (9.2\text{-}11)$$

and is seen to increase in proportion to the square root of the mean number of photocounts.

The model used for light from a single-mode, amplitude-stabilized laser is clearly an idealization. However, with sufficient care this idealization can be closely approached in practice, and it is important to understand the character of the count fluctuations in this limiting case. Figure 9-1 illustrates the probability masses associated with various values of K when the count statistics are Poisson.

9.2.2 Photocount Statistics for Polarized Thermal Radiation with a Counting Time Much Shorter Than the Coherence Time

We now consider the case of thermal radiation and the associated photocount distribution. For the present, consideration is restricted to the simplest case from the analytical point of view, namely, completely polarized radiation and a counting interval that is short compared with the coherence time of the light. In practice, a counting time this short would be extremely difficult to achieve with true thermal light, for a 1 nm bandwidth at a wavelength of 500 nm would require a counting time much smaller than about one picosecond (10^{-12} second)! However, with pseudothermal light the condition can be met easily.

For such a short counting time, the value of the incident intensity $I(t)$ is approximately constant over the entire counting interval. As a consequence, the integrated intensity is simply equal to the product of the intensity, the counting time, and the detector area,

$$W = I(t)A\tau. \qquad (9.2\text{-}12)$$

However, the value of the intensity within that interval is random and obeys negative exponential statistics; from (9.2-12) it follows that the same is true of the integrated intensity,

$$p_W(W) = \frac{1}{\overline{W}}\exp\left\{\frac{-W}{\overline{W}}\right\}, \qquad W \geq 0. \qquad (9.2\text{-}13)$$

The photocount statistics can now be found by substituting Eq. (9.2-13) in Mandel's formula and performing the required integration. The steps are outlined in the following equation, and the result is shown:

$$P(K) = \int_0^\infty \frac{(\alpha W)^K}{K!} e^{-\alpha W} \cdot \frac{1}{\overline{W}} e^{-W/\overline{W}} dW$$

$$= \frac{\alpha^K}{K!\overline{W}} \int_0^\infty W^K \exp\left[-W\left(\alpha + \frac{1}{\overline{W}}\right)\right] dW = \frac{1}{1 + \alpha\overline{W}} \left(\frac{\alpha\overline{W}}{1 + \alpha\overline{W}}\right)^K. \tag{9.2-14}$$

With the substitution $\overline{K} = \alpha\overline{W}$, the equivalent expression

$$P(K) = \frac{1}{1 + \overline{K}} \left(\frac{\overline{K}}{1 + \overline{K}}\right)^K \tag{9.2-15}$$

is obtained. This probability distribution is called the *Bose–Einstein* distribution (or, in statistics, the *geometric* distribution), and it plays an extremely important role in the statistical physics of indistinguishable particles (bosons). For our purposes, it suffices to note that the factorial moments of this distribution are given by

$$E[K(K-1)\cdots(K-n+1)] = n!(\overline{K})^n. \tag{9.2-16}$$

It follows that the variance can be expressed in terms of the mean as

$$\sigma_K^2 = \overline{K} + (\overline{K})^2. \tag{9.2-17}$$

Note that of the two terms composing the variance, the first again represents the Poisson noise associated with the basic interaction of light and matter, whereas the second represents the classical fluctuations of the integrated intensity, which will be very significant in this case if $\overline{K} \gg 1$. The signal-to-noise ratio associated with Bose–Einstein counts is readily seen to be

$$\frac{S}{N} = \frac{\overline{K}}{\sigma_K} = \sqrt{\frac{\overline{K}}{1 + \overline{K}}}. \tag{9.2-18}$$

This expression asymptotically approaches *unity* as the mean number of counts increases, indicating that the count fluctuations are always very substantial indeed.

When the mean number of counts $\overline{K}$ becomes much smaller than unity, it can readily be shown that the differences between the Poisson distribution and the Bose–Einstein distribution become small. For such small means, only one event and zero events have significant probability. The following equations indicate how these respective probabilities are asymptotically the same for the two distributions:

$$\text{Poisson} \quad \begin{cases} P(0) = e^{-\overline{K}} \cong 1 - \overline{K} \\ P(1) = \overline{K} e^{-\overline{K}} \cong \overline{K} \end{cases} \quad (9.2\text{-}19)$$

$$\text{Bose–Einstein} \quad \begin{cases} P(0) = \dfrac{1}{1 + \overline{K}} \cong 1 - \overline{K} \\ P(1) = \dfrac{\overline{K}}{(1 + \overline{K})^2} \cong \overline{K}. \end{cases} \quad (9.2\text{-}20)$$

Finally, in Fig. 9-2 the probability masses associated with the Bose–Einstein distribution are shown for the same value of mean used in Fig. 9-1. Comparison of these two figures shows that when the mean number of counts is greater than unity, the spread of the Bose–Einstein distribution is far greater than that of the Poisson distribution, and as a consequence far greater fluctuations of the photocounts are expected for the distribution of Fig. 9-2 than for that of Fig. 9-1.

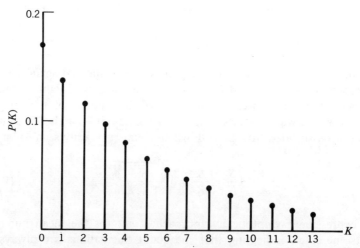

Figure 9-2. Probability masses associated with the Bose–Einstein distribution ($\overline{K} = 5$).

EFFECTS OF STOCHASTIC FLUCTUATIONS OF INTENSITY 475

The discussion of the count statistics for polarized thermal light with a short counting interval is now complete. The next section deals with the more general case of an unrestricted counting interval.

9.2.3 Photocount Statistics for Polarized Thermal Light and an Arbitrary Counting Interval

As pointed out earlier, it is extremely difficult to realize an experiment with true thermal light in which the counting interval is much shorter than the coherence time of the incident light. For this reason it is important to investigate the count statistics for counting times that are comparable with or longer than that coherence time. The assumption that the incident light is fully polarized is retained for the time being. The procedure for finding the photocount statistics is similar to that used before. First we must find the probability density function $p_W(W)$ for the integrated intensity, then substitute that density function in Mandel's formula, and finally perform the required integration.

Determination of the statistics of the integrated intensity is a nontrivial problem. However, we have met this problem before, and solutions have already been found. The reader is referred to Section 6.1 (in Chapter 6), where the statistics of the time-integrated intensity were considered. There an approximate solution for $p_W(W)$ was obtained (Section 6.1.2) and an exact solution was also outlined (Section 6.1.3). Here we concentrate on the use of the approximate expression for $p_W(W)$ and the resulting predictions for photocount statistics. For a discussion of the exact solution, the reader may consult Ref. 9-11.

Assume for the time being that the wave incident on the photosurface has a coherence area that far exceeds the area of the detector. With this assumption, attention can be concentrated completely on temporal coherence effects. It is then possible to directly utilize the approximate solution for $p_W(W)$ presented in Eq. (6.1-31) (i.e., the gamma probability density function) and repeated below:

$$p_W(W) = \begin{cases} \left(\dfrac{\mathcal{M}}{\overline{W}}\right)^{\mathcal{M}} \dfrac{W^{\mathcal{M}-1}\exp\left(-\mathcal{M}\dfrac{W}{\overline{W}}\right)}{\Gamma(\mathcal{M})}, & W \geq 0 \\ 0 & \text{otherwise.} \end{cases} \quad (9.2\text{-}21)$$

Here the parameter $\mathcal{M}$ represents the number of "degrees of freedom" of the intensity included within the measurement interval and when purely temporal degrees of freedom are involved, is given by [cf. Eq. (6.1-30)]

$$\mathcal{M} = \left[\frac{1}{T}\int_{-\infty}^{\infty} \Lambda\left(\frac{\tau}{T}\right)|\gamma(\tau)|^2 d\tau\right]^{-1}, \qquad (9.2\text{-}22)$$

where $\gamma(\tau)$ is the complex degree of coherence of the incident wave.

Two limiting cases are worth attention, namely, the cases of an integration time very long and very short compared with the coherence time of the light. These cases have been dealt with in Chapter 6, specifically in Eqs. (6.1-18) and (6.1-19), to which the reader is referred. Note that, no matter how short the time, the number of degrees of freedom can never be less than 1, and in this limiting case the gamma density of Eq. (9.2-21) reduces to the negative exponential distribution. When the integration time is much longer than the coherence time, the number of degrees of freedom is equal to the number of correlation intervals embraced by the measurement interval. In addition, it is not difficult to show that as the number of degrees of freedom grows large, the gamma density function asymptotically approaches a Gaussian density function (see Problem 9-2).

When the approximate form of the probability density function for integrated intensity is known, the problem remains to calculate the probability distribution function of the number of photocounts occurring in a time interval of arbitrary length. This calculation is carried out with the help of Mandel's formula, which in this case is given by

$$P(K) = \int_0^\infty \left[\frac{(\alpha W)^K}{K!} e^{-\alpha W}\right] \frac{\left(\frac{\mathcal{M}}{\overline{W}}\right)^{\mathcal{M}} W^{\mathcal{M}-1} \exp\left(-\mathcal{M}\frac{W}{\overline{W}}\right)}{\Gamma(\mathcal{M})} dW.$$

$$(9.2\text{-}23)$$

The integration can be performed without great difficulty and yields the following probability distribution for the number of counts K observed in a τ-second interval:

$$P(K) = \frac{\Gamma(K+\mathcal{M})}{\Gamma(K+1)\Gamma(\mathcal{M})}\left[1+\frac{\mathcal{M}}{\overline{K}}\right]^{-K}\left[1+\frac{\overline{K}}{\mathcal{M}}\right]^{-\mathcal{M}}, \qquad (9.2\text{-}24)$$

where $\overline{K} = \alpha\overline{W}$. This distribution is known as the *negative binomial* distribution and has been found to be a rather good approximation to the photocount distribution of interest. Note that the number of degrees of freedom $\mathcal{M}$ is a parameter of the distribution, as indeed we might have

EFFECTS OF STOCHASTIC FLUCTUATIONS OF INTENSITY 477

anticipated. When the integration time τ is very small, the number of degrees of freedom is essentially unity, and the negative binomial distribution can readily be shown to reduce to the Bose–Einstein distribution (see Problem 9-3).

Having found an approximate expression for the photocount distribution for polarized thermal light with an arbitrary counting time and with perfect spatial coherence, we now very briefly discuss the modifications of the results that are needed when the wave is not perfectly polarized.

9.2.4 Polarization Effects

The preceding discussion assumed that the light incident on the photosurface is perfectly polarized. The case of thermal light with an arbitrary degree of polarization is also of interest. To derive the probability distribution of the photocounts in this general case, we first note that, when the light is partially polarized, the total integrated intensity may be regarded as the sum of two statistically independent integrated intensity components, one for each of the polarization components of the wave after passage through a polarization instrument that diagonalizes the coherency matrix [cf. Eq. (4.3-38)]. Thus

$$W = W_1 + W_2, \quad (9.2\text{-}25)$$

where

$$\overline{W} = \overline{W}_1 + \overline{W}_2$$

$$\overline{W}_1 = \frac{\overline{W}}{2}(1 + \mathscr{P}) \quad (9.2\text{-}26)$$

$$\overline{W}_2 = \frac{\overline{W}}{2}(1 - \mathscr{P}).$$

Since W_1 and W_2 are statistically independent for thermal light, it follows that

$$p_W(W) = p_1(W_1) * p_2(W_2), \quad (9.2\text{-}27)$$

where p_1 and p_2 are the probability density functions of W_1 and W_2, respectively.

To make further progress we must call on a result that holds for all photoelectric counting distributions. In words, this result states that, when the probability density function of the integrated intensity can be expressed as the (continuous) convolution of two probability density functions, the corresponding probability distribution of the counts can be expressed as a (discrete) convolution of two count probability distributions, one arising from each of the individual continuous density functions. Thus if $p_1(W)$ and $p_2(W)$ are density functions as in Eq. (9.2-27), and if $P_1(n)$ and $P_2(n)$ are the associated discrete counting probability distribution functions (derivable from Mandel's formula applied to each of the continuous densities), then

$$P(K) = \sum_{k=0}^{K} P_1(k) P_2(K-k). \tag{9.2-28}$$

The reader is asked to prove this general result in Problem 9-5.

Applying the preceding result in the present case and taking account of the negative binomial distributions associated with the counts generated by each of the independent polarization components, we obtain an expression for the probability distribution of the counts for partially polarized thermal light,

$$P(K) = \left[1 + \frac{2\mathcal{M}}{\overline{K}(1+\mathcal{P})}\right]^{-K} \left[1 + \frac{\overline{K}}{\mathcal{M}} + \left(\frac{\overline{K}}{2\mathcal{M}}\right)^2 (1-\mathcal{P})\right]^{-\mathcal{M}}$$

$$\times \sum_{k=0}^{K} \frac{\Gamma(K-k+\mathcal{M})}{\Gamma(K-k+1)\Gamma(\mathcal{M})}$$

$$\times \frac{\Gamma(k+\mathcal{M})}{\Gamma(k+1)\Gamma(\mathcal{M})} \left[\frac{\overline{K}(1-\mathcal{P}^2) + 2\mathcal{M}(1-\mathcal{P})}{\overline{K}(1-\mathcal{P}^2) + 2\mathcal{M}(1+\mathcal{P})}\right]^k.$$

(9.2-29)

This discrete convolution can be carried out numerically to determine the count probability distribution for any given degree of polarization. When one of the polarization components has zero intensity, the discrete convolution reduces to the negative binomial distribution associated with the counts generated by the single remaining component. As should be expected, when the light is completely unpolarized, the convolution reduces to a single negative binomial distribution with $2\mathcal{M}$ degrees of freedom.

In closing this section, it should be reemphasized that the solutions presented here for the cases involving thermal light have been approximate ones, with the approximations arising from the approximate forms used for $p_W(W)$. More exact treatments are possible, based on the exact formulation

EFFECTS OF STOCHASTIC FLUCTUATIONS OF INTENSITY

for the statistics for W presented in Section 6.1.3. For discussions of such an approach and comparisons with the approximate results derived here, the reader may consult, for example, Refs. 9-6 and 9-11.

9.2.5 Effects of Incomplete Spatial Coherence

Our previous discussion assumed that the thermal light striking the photosurface was completely coherent in a spatial sense. In such a case the number of degrees of freedom $\mathcal{M}$ is determined strictly by temporal effects. When the wave is not spatially coherent, its spatial structure can affect the number of degrees of freedom; at any given time, different parts of the photosensitive surface may experience different levels of incident intensity. In such cases a modification of the concept of degrees of freedom is required so as to include the possibility of both temporal and spatial degrees of freedom, as we outline in the following paragraphs.

To make the analysis tractable, we make a number of assumptions about the nature of the light falling on the photosensitive surface. We assume the light to be of thermal origin and to be completely polarized. In addition, the light is assumed to be cross-spectrally pure, in which case the complex degree of coherence can be factored into temporal and spatial components. Finally, both the temporal and spatial fluctuations of the intensity are assumed to be at least wide-sense stationary,[†] in which case

$$\gamma_{12}(\Delta x, \Delta y; \tau) = \mu_{12}(\Delta x, \Delta y)\gamma(\tau). \qquad (9.2\text{-}30)$$

With reference to Fig. 6-2, just as we divided time up into approximately independent temporal correlation cells, so we must divide the detector *area* up into approximately independent *spatial* correlation cells. The total integrated intensity is then regarded as the sum of many independent exponentially distributed random variables, one from each of the time–space correlation cells.

If the mean and variance of the gamma distribution approximating the statistics of the temporally and spatially integrated intensity are to match the true mean and variance of W, the parameter $\mathcal{M}$ of the gamma distribution must be properly chosen. In view of the factorization evident in Eq. (9.2-30), the number of degrees of freedom required to match the true mean and variance will be expressible as a product of a number of temporal

---

[†]Strictly speaking, we require only that the modulus of the complex degree of coherence depend only on differences of spatial and temporal coordinates. This requirement is weaker than wide-sense stationarity and is satisfied, for example, by spatial coherence effects predicted by the Van Cittert–Zernike theorem.

degrees of freedom $\mathcal{M}_t$ times a number of spatial degrees of freedom $\mathcal{M}_s$,

$$\mathcal{M} = \mathcal{M}_t \mathcal{M}_s. \tag{9.2-31}$$

As demonstrated in Section 6.1-1, when only time variations are important, the number of temporal degrees of freedom can be expressed (after suitable manipulation) in terms of a single integral,

$$\mathcal{M}_t = \left[\frac{1}{T} \int_{-\infty}^{\infty} \Lambda\left(\frac{\tau}{T}\right) |\gamma(\tau)|^2 d\tau \right]^{-1}. \tag{9.2-32}$$

It is also possible to reduce the expression for the number of spatial degrees of freedom to a double integral,

$$\mathcal{M}_s = \left[\frac{1}{A} \iint_{-\infty}^{\infty} \mathcal{R}(\Delta x, \Delta y) |\mu_{12}(\Delta x, \Delta y)|^2 d\Delta x \, d\Delta y \right]^{-1}, \tag{9.2-33}$$

where the function $\mathcal{R}$ represents the normalized autocorrelation function of an effective "pupil function" $P(x, y)$ associated with the photosensitive area and A is the area of the photosensitive surface; that is, if

$$P(x, y) = \begin{cases} 1 & (x, y) \text{ within photosensitive area} \\ 0 & \text{otherwise,} \end{cases} \tag{9.2-34}$$

then

$$\mathcal{R}(\Delta x, \Delta y) = \frac{\iint_{-\infty}^{\infty} P\left(x + \frac{\Delta x}{2}, y + \frac{\Delta y}{2}\right) P\left(x - \frac{\Delta x}{2}, y - \frac{\Delta y}{2}\right) dx \, dy}{\iint_{-\infty}^{\infty} P^2(x, y) \, dx \, dy}. \tag{9.2-35}$$

When the photosensitive area A is much smaller than the coherence area A_c of the incident radiation, it is easy to show that the number of spatial degrees of freedom $\mathcal{M}_s$ reduces to unity. When the photosensitive area is much larger than the coherence area, it can be shown (see Problem 9-4) that the number of spatial degrees of freedom reduces to the ratio of the detector area to the coherence area (or equivalently to the number of spatial coherence areas of the light embraced by the photosensitive surface),

$$\mathcal{M}_s \cong \frac{A}{A_c}. \tag{9.2-36}$$

This expression can be further modified when the incident wave arises from a spatially incoherent source. In this case the Van Cittert–Zernike theorem and Eq. (5.6-12) lead us to the following equivalent expression for the number of spatial degrees of freedom, valid when $A \gg A_c$:

$$\mathcal{M}_s \cong \frac{A\Omega_S}{(\bar{\lambda})^2}. \tag{9.2-37}$$

Here, as in Eq. (5.6-12), Ω_S represents the angular size of the source, when viewed from the detector.

The discussion of the spatial aspects of the problem is now complete, and attention will be turned to an introduction to the concept of the degeneracy parameter of the light and the role it plays in determining the photocount statistics for thermal light.

9.3 THE DEGENERACY PARAMETER

At this point the reader should be convinced that there is a fundamental difference between the statistical properties of photocounts generated by highly stable, single-mode laser light and the more chaotic light associated with thermal sources. Indeed, this difference shows itself particularly clearly when one examines in more detail the fluctuations of the photocounts generated by both types of radiation, as we do in the section to follow. However, the situation is more complicated than might be imagined at first glance. The differences between the statistics of the photocounts for the two types of light are not always great. In fact, in the visible region of the electromagnetic spectrum, under most circumstances it is very difficult to distinguish which kind of radiation is present, based on measurements of the photocount statistics. The key parameter that determines the distinguishability of these two types of radiation will be shown to be the *degeneracy parameter*, which we soon define.

In the section to follow we consider the fluctuations of photocounts when different kinds of light are incident on a photosurface. Such considerations lead us to the definition of the degeneracy parameter. In Section 9.3.2 this parameter is considered for the particular case of blackbody radiation. The importance of the degeneracy parameter is further emphasized when we consider applications in the final sections of this chapter.

9.3.1 Fluctuations of Photocounts

Our purpose in this section is to examine the variance of the photocounts produced by thermal light and to consider when this variance is significantly different from that produced by stabilized single-mode laser light. We first re-emphasize the direct link between the variance of the counts and the variance of the classical fluctuations of intensity incident on the photosurface.

To calculate the variance of the count fluctuations, we must first find the second moment of the counts $\overline{K^2}$. A simple approach to this problem is to first note that, when conditioned by knowledge of the integrated intensity W, the number of counts K is a Poisson variate with mean αW [see Eq. (9.1-7)]. Therefore, the conditional second moment of K is given by

$$E_{K|W}[K^2] = (\alpha W)^2 + \alpha W. \qquad (9.3\text{-}1)$$

To determine the unconditioned second moment of K, we must simply average Eq. (9.3-1) with respect to W. The result is

$$\overline{K^2} = \alpha^2 \overline{W^2} + \alpha \overline{W}$$

$$= \alpha^2 \sigma_W^2 + (\alpha \overline{W})^2 + \alpha \overline{W}. \qquad (9.3\text{-}2)$$

To find the variance of K, we need only subtract the square of the mean of K, or equivalently subtract $(\alpha \overline{W})^2$. The result is the following expression for the variance of the counts [cf. Eq. (9.2-4)]:

$$\sigma_K^2 = \overline{K} + \alpha^2 \sigma_W^2. \qquad (9.3\text{-}3)$$

Note that it has not been necessary to make *any* assumptions about the statistics of the classical fluctuations of integrated intensity in arriving at Eq. (9.3-3). The result is completely general; that is, it applies for any type of radiation falling on the surface of the photodetector. Furthermore, each term of this equation has a physical interpretation. The first term, $\overline{K}$, is simply the variance of the counts that would be observed if the classical intensity were constant and the photocounts were purely Poisson. We refer to this contribution to the count fluctuations as "shot noise," by analogy with the Poisson-distributed shot noise observed, for example, in an electronic vacuum diode (Ref. 9-12). The second term, $\alpha^2 \sigma_W^2$, is clearly zero if there are no fluctuations of the classical intensity. Therefore, it is the component of count variance caused by fluctuations of the classical intensity. In the case of stabilized single-mode laser light, this component would be identically zero, and the count variance would be simply that arising from Poisson-distributed counts. When thermal light is incident on the photosurface, the classical fluctuations are nonzero, and the variance of the photocounts is larger than that expected for a Poisson distribution by an amount that is proportional to the variance of the integrated intensity. This extra component of variance of the counts is often referred to as "excess noise," meaning that it is above and beyond that expected for pure Poisson fluctuations.

At this point we introduce the assumption that the light incident on the photosurface is polarized thermal light. For such light, a combination of Eqs. (6.1-10) and (6.1-17) shows that

$$\sigma_W^2 = \frac{(\overline{W})^2}{\tau} \int_{-\infty}^{\infty} \Lambda\left(\frac{\xi}{\tau}\right) |\gamma(\xi)|^2 d\xi = \frac{(\overline{W})^2}{\mathcal{M}}. \quad (9.3\text{-}4)$$

Hence

$$\sigma_K^2 = \overline{K} + \alpha^2 \frac{\overline{W}^2}{\mathcal{M}} = \overline{K} + \frac{(\overline{K})^2}{\mathcal{M}} \quad (9.3\text{-}5)$$

or

$$\sigma_K^2 = \overline{K}\left(1 + \frac{\overline{K}}{\mathcal{M}}\right). \quad (9.3\text{-}6)$$

Note in particular that the ratio of classically induced fluctuations to shot noise fluctuations is simply $\overline{K}/\mathcal{M}$. The important role of this parameter is emphasized by giving it a name of its own. Therefore, we define the *count degeneracy parameter* as follows:

$$\delta_c = \frac{\overline{K}}{\mathcal{M}} \quad (9.3\text{-}7)$$

Physically speaking, the count degeneracy parameter can be interpreted as the average number of counts that occur in a single coherence interval of the incident radiation. It can also be described as the average number of counts per "degree of freedom" or per "mode" of the incident wave. When $\delta_c \ll 1$, it is highly probable that there will be no more than one count per coherence interval of the wave, with the result that shot noise predominates over classically induced noise. On the other hand, when $\delta_c \gg 1$, there are many photoevents present in each coherence interval of the wave. The result is a "bunching" of the photoevents by the classical intensity fluctuations, and an increase of the variance of the counts to the point where the classically induced fluctuations are far stronger than the shot noise variations.

Because the count degeneracy parameter is proportional to $\overline{K}$, it is also proportional to the quantum efficiency of the photosurface. Sometimes it is useful to remove this dependence on the characteristics of the particular detector that may be present, and to deal with a degeneracy parameter that is a property of the incident wave itself. We thus define the *wave degeneracy parameter* as

$$\delta_w = \frac{\delta_c}{\eta}. \tag{9.3-8}$$

This new degeneracy parameter may be considered to be the count degeneracy parameter that would be obtained with an ideal detector having a quantum efficiency of unity.

The distribution of photocounts obtained with polarized thermal light is determined by the combination of the parameters $\overline{K}$ and δ_c, as can be seen by rewriting the negative binomial distribution of Eq. (9.2-24) in the form

$$P(K) = \frac{\Gamma\left(K + \frac{\overline{K}}{\delta_c}\right)}{K!\,\Gamma\left(\frac{\overline{K}}{\delta_c}\right)} \left[(1 + \delta_c)^{\overline{K}/\delta_c}\left(1 + \frac{1}{\delta_c}\right)^K\right]^{-1}. \tag{9.3-9}$$

We now prove a very important fact—when the count degeneracy parameter approaches zero, the photocount distribution $P(K)$, which is given by a negative binomial distribution, becomes indistinguishable from a Poisson distribution. For proof of this assertion, several approximations are necessary. First, when the degeneracy parameter is much less than 1, the gamma functions in Eq. (9.3-9) can be replaced using Stirling's approximation (Ref. 9-13):

$$\Gamma\left(\frac{\overline{K}}{\delta_c}\right) \cong \sqrt{2\pi}\left(\frac{\overline{K}}{\delta_c}\right)^{\overline{K}/\delta_c - 1/2} e^{-\overline{K}/\delta_c}$$

$$\Gamma\left(K + \frac{\overline{K}}{\delta_c}\right) \cong \sqrt{2\pi}\left(K + \frac{\overline{K}}{\delta_c}\right)^{K + \overline{K}/\delta_c - 1/2} e^{-K - \overline{K}/\delta_c}. \tag{9.3-10}$$

Second, for δ_c much less than 1, the following approximations are valid:

$$\left(1 + \frac{1}{\delta_c}\right)^K \cong \left(\frac{1}{\delta_c}\right)^K$$

$$(1 + \delta_c)^{\overline{K}/\delta_c} \cong e^{\overline{K}}. \tag{9.3-11}$$

Combining the above approximations, we find that the probability of observing K counts in time τ is given approximately by

$$P(K) = \frac{(\overline{K})^K}{K!} e^{-\overline{K}} \left[\left(1 + \frac{K}{\mathcal{M}}\right)^{K+\mathcal{M}-1/2} e^{-K} \right] \qquad (9.3\text{-}12)$$

where we have noted explicitly that $\overline{K}/\delta_c$ is identically $\mathcal{M}$, the number of degrees of freedom. For a final approximation we note that if the average number of photoevents per coherence interval, δ_c, is small, there is a high degree of probability that the actual number of photoevents occurring in one coherence interval, $K/\mathcal{M}$, is likewise small. It follows that

$$\left[1 + \frac{K}{\mathcal{M}}\right]^{\mathcal{M}} \cong e^K, \qquad (9.3\text{-}13)$$

and that the quantity in brackets in Eq. (9.3-12) has a value very close to unity. Thus the probability of obtaining K counts is, to a good approximation, given by the Poisson distribution. Note that the approximations used in arriving at this final result are more and more accurate as the degeneracy parameter becomes smaller and smaller. Therefore, we might more properly state this result in the form of a limit,

$$\lim_{\delta_c \to 0} P(K) = \frac{(\overline{K})^K}{K!} e^{-\overline{K}}. \qquad (9.3\text{-}14)$$

To emphasize the important result we have just obtained, we restate it once more in words:

For polarized thermal radiation, when the count degeneracy parameter approaches zero, the probability distribution of the photocounts approaches a Poisson distribution.

A physical understanding of this result can be gained from the following considerations. If the count degeneracy parameter is much less than 1, it is highly probable that there will be either zero or one counts in each separate coherence interval of the incident classical wave. In such a case the classical intensity fluctuations have a negligible "bunching" effect on the photoevents, for (with high probability) the light is simply too weak to generate multiple events in a single coherence cell. If negligible bunching of the events takes place, the count statistics will be indistinguishable from those produced by stabilized single-mode laser radiation, for which no bunching occurs.

Note that whereas we have specifically assumed that the incident light is polarized, a similar result holds for partially polarized thermal light, provided the independent intensities of the polarization components, obtainable by passage of the light through a polarization instrument that diagonalizes the coherency matrix, each have small degeneracy parameters.

The significance of the result we have obtained cannot be fully appreciated until we determine typical values of the degeneracy parameter that can be expected in practice. Such values are dramatically different in the microwave and visible regions of the electromagnetic spectrum, as discussed in the section that follows.

9.3.2 The Degeneracy Parameter for Blackbody Radiation

In statistical physics and thermodynamics it is common to introduce the idealized concept of a *blackbody*. A blackbody is defined as an object that perfectly absorbs all the radiant energy incident on it. If that body is in thermal equilibrium with its surroundings, then in addition to being a perfect absorber, it must also be a perfect radiator, in the sense that it must reradiate as much energy as it absorbs; otherwise, it could not remain in thermal equilibrium. The ideal nature of a blackbody facilitates the calculation of a (temperature-dependent) spectral distribution that is characteristic of such radiation. Many radiation emitters encountered in practice can be regarded as blackbodies or approximations to blackbodies. For example, the gross features of the spectrum of the sun are approximately those of a blackbody at a temperature of 6000 K.

Calculation of the spectral distribution of energy radiated by a blackbody is a problem that occupied the attention of many physicists in the nineteenth century. Among the most famous of these studies were those by Lord Rayleigh and Sir James Jean, who derived an expression for the spectral distribution of blackbody radiation with the help of the classical law of equipartition of energy. They discovered that the predictions thus derived agreed with experiment only in the long wavelength limit, and that in the short-wavelength limit, the results lead to the famous "ultraviolet catastrophe," a spectral distribution that increased without bound as wavelength approached zero.

The dilemmas associated with blackbody radiation were resolved only by the introduction of a hypothesis that departed markedly from the precepts of classical physics. In 1900 Max Planck published a new derivation of the blackbody radiation law that incorporated the radical assumption that energy could be radiated and absorbed only in discrete packets or quanta. The law predicted by this theory was in agreement with all the experimental results known at that time. With this work, the quantum theory of radiation was born.

Planck's theory of blackbody radiation is relevant to us in that it provides a very specific prediction as to the magnitudes of the degeneracy parameter expected for thermal light in various parts of the electromagnetic spectrum. To apply Planck's results, we must regard each degree of freedom of the incident radiation as being analogous to a harmonic oscillator. Such a picture can be arrived at explicitly by applying the frequency-domain sampling theorem to the finite-time waveform that is incident on the photodetector in the problem of interest to us. The number of degrees of freedom of the waveform is the same regardless of whether time or frequency samples are considered. Indeed, the energy falling on the photosurface can be regarded as the sum of the energies carried by either the temporal samples or the frequency samples; both sums yield the same result.

The energies associated with each such harmonic oscillator are assumed to be quantized, with allowable discrete energy states given by

$$E_n = nh\nu, \qquad (9.3\text{-}15)$$

where n is an integer, the constant h is now known as Planck's constant ($h = 6.626196 \times 10^{-34}$ joule-second), and ν is the frequency of the oscillator. In an experiment involving a large number of such oscillators, the numbers of oscillators in each of the possible energy states is assumed to follow a Maxwell–Boltzmann distribution. That is, the number N_n of oscillators with energy E_n is assumed to be given by

$$N_n = N_0 \exp\left\{-\frac{E_n}{kT}\right\} = N_0 \exp\left\{-\frac{nh\nu}{kT}\right\}, \qquad (9.3\text{-}16)$$

where N_0 is a constant, E_n is the nth allowable energy level, k is Boltzmann's constant ($k = 1.38 \times 10^{-23}$ joule/Kelvin), and T is the temperature in Kelvin. To obtain the probability that a given oscillator is in the nth energy state (or equivalently has "occupation number" n), we must normalize Eq. (9.3-16) to yield a set of numbers that sum over n to yield unity. Such a normalization can be performed (see Problem 9-7). The result is

$$P(n) = (1 - e^{-h\nu/kT})(e^{-h\nu/kT})^n. \qquad (9.3\text{-}17)$$

Comparison of (9.3-17) with (9.2-15) shows that energy states obey a Bose–Einstein distribution, with mean occupation number per mode

$$\bar{n} = \frac{1}{e^{h\nu/kT} - 1} \qquad (9.3\text{-}18)$$

or equivalently, mean energy per mode

$$\overline{E} = \frac{h\nu}{e^{h\nu/kT} - 1}. \tag{9.3-19}$$

The wave degeneracy parameter is simply the average number of photons per mode. This is precisely the quantity represented by Eq. (9.3-18). If the radiation of concern is narrowband, the frequency ν in that equation can be replaced by the frequency $\bar{\nu}$ at the center of the spectrum. Hence the degeneracy parameter for blackbody radiation from a narrowband source in thermal equilibrium is given by

$$\delta_W = \frac{1}{e^{h\bar{\nu}/kT} - 1}. \tag{9.3-20}$$

This equation forms the basis for our understanding of the differences between radiation in the optical region of the spectrum and radiation at lower frequencies.

Consider first the case of radiation with a sufficiently low frequency that $h\bar{\nu} \ll kT$. Under such a condition, the degeneracy parameter of Eq. (9.3-20) is well approximated by

$$\delta_W \cong \frac{kT}{h\bar{\nu}}. \tag{9.3-21}$$

In this regime, the degeneracy parameter is inversely proportional to frequency and is a very large number. At the opposite extreme $h\bar{\nu} \gg kT$, however, the degeneracy parameter decreases exponentially with increasing frequency and is a very small number. Figure 9.3 shows contours of constant wave degeneracy parameter in a plane with mean wavelength as one coordinate and source temperature as the other coordinate. From Fig. 9-3 it is quite clear that in the microwave region of the spectrum ($\bar{\lambda} \cong 10^{-1}$ meters) any source temperature greater than a fraction of a degree produces a wave degeneracy parameter that is much greater than unity. Hence in this region of the spectrum we expect classically induced fluctuations of photocounts to have a far stronger effect than pure shot noise fluctuations. On the other hand, in the visible region of the spectrum ($\bar{\lambda} \cong 5 \times 10^{-7}$ meters), source temperatures in excess of 20,000 K are required to produce a wave degeneracy parameter greater than unity. Since the sun has an effective blackbody temperature of only 6000 K, we can conclude that in the visible region of the spectrum, the vast majority of sources encountered produce radiation with a wave degeneracy parameter that is small, and hence noise

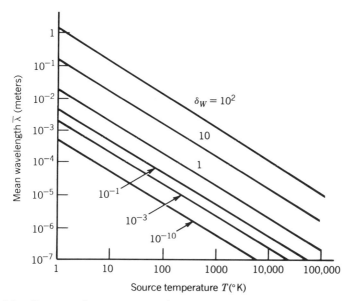

Figure 9-3. Contours of constant wave degeneracy parameter as a function of source temperature and wavelength.

produced by the quantized nature of the radiation is far greater than the noise produced by classical fluctuations of the intensity.

A few additional comments are needed in closing this section. First, we have concentrated our attention on the wave degeneracy parameter, which is a property of the radiation falling on the photodetector. That detector invariably has a quantum efficiency that is less than unity. Hence the count degeneracy parameter will be even smaller than the wave degeneracy parameter, and in the visible region of the spectrum the likelihood of encountering true thermal radiation for which classical fluctuations of the intensity will dominate the count statistics is made even more remote. (However, it should be noted that *pseudothermal* sources can produce radiation with very high degeneracy parameter, and in such cases the classical fluctuations of the intensity can be the dominant source of fluctuations of the photocounts.) In addition, it is possible that the photodetector or collecting optics may intercept only a fraction of one spatial mode from the source. (In practice, a great many temporal modes will be captured during a measurement interval of any reasonable duration.) In such a case the count degeneracy parameter may again be smaller than the wave degeneracy parameter, as a result of the incomplete capture of a spatial mode. Although the minimum value of $\mathcal{M}$ is unity, the reduction of the energy striking the photosurface must be taken into account. In such a case, the count degeneracy parameter must be reduced from its normal value by a factor of the ratio of the effective measurement area to the coherence area of the incident light. For the case of an extended incoherent source, this possibility is covered by expressing the count degeneracy parameter in the form that follows:

$$\delta_c = \begin{cases} \dfrac{\eta}{e^{h\bar{\nu}/kT} - 1} & A > A_c \\ \dfrac{A}{A_c} \dfrac{\eta}{e^{h\bar{\nu}/kT} - 1} = \dfrac{A\Omega}{(\bar{\lambda})^2} \dfrac{\eta}{e^{h\bar{\nu}/kT} - 1} & A < A_c, \end{cases} \quad (9.3\text{-}22)$$

where we have used the fact that the coherence area of the light incident from an incoherent source is given by the ratio of the square of the wavelength to the solid angle subtended by the source at the detector (cf. Problem 5-15).

Finally, for the reader interested in studying the problem of blackbody radiation further, we mention specifically Refs. 9-14 through 9-16.

9.4 NOISE LIMITATIONS OF THE AMPLITUDE INTERFEROMETER AT LOW LIGHT LEVELS

Here and in the following sections we discuss applications of the photoelectron counting theory developed in previous sections of this chapter. There are, in fact, many applications that could be discussed, for virtually any optical experiment is limited in its accuracy most fundamentally by the finite amount of light that is utilized in the measurements involved. We have chosen for emphasis here experiments aimed at measuring the parameters of simple fringe patterns. The reasons for this emphasis are several. First, fringe parameter measurement provides a relatively well defined and tractable example of application of the theory. The desired parameters are easily defined, and methods for their measurement are readily devised based on common sense. Second, we have seen throughout this book that fringe parameter measurement is central to all problems involving coherence. The fundamental descriptors of light waves utilized in coherence theory are in fact measurable parameters of fringes. By examining the limitations to fringe parameter measurement, we are actually examining the limitations to the measurability of coherence itself.

We shall deal with two different approaches to the measurement of fringe parameters. In this section we consider what can reasonably be called an

LIMITATIONS OF THE AMPLITUDE INTERFEROMETER

"amplitude interferometry" or a "predetection correlation" method. Such methods are used, for example, in the Michelson stellar interferometer. More generally, we noted in Chapter 7 that any system that forms images by directly focusing light onto a detector can be viewed as an interferometric imaging system; each Fourier component of the object can be regarded as the superposition of a multitude of fringes of the same spatial frequency arising from a fixed spacing embraced in many ways by the exit pupil of the system. In Section 9.5 we consider a different approach to fringe parameter measurement, namely, the use of "intensity interferometry" or equivalently a "postdetection correlation" method. This type of measurement was discussed from a purely classical point of view in Chapter 6. Finally, in Section 9.6, we discuss the noise limitations of the stellar speckle interferometer introduced earlier in Section 8.8. Our goal in all cases is to discover the limiting sensitivity of the measurement techniques considered. More specifically, how does the accuracy of the measurement depend on the number of photoevents participating in that measurement?

9.4.1 The Measurement System and the Quantities to Be Measured

The amplitude interferometry method will be assumed to utilize a measurement system of the kind illustrated in Fig. 9-4. A sinusoidal distribution of intensity, representing an ideal fringe, falls on a detector consisting of N discrete elements arranged in a closely packed linear array. This fringe may have originated, for example, from a Michelson stellar interferometer, which is attempting to determine the diameter of a distant stellar source. Each element of the detector is assumed to be followed by a separate counter.[†] The counter associated with the nth detector produces at the end of a τ-second counting period a number $K(n)$ representing the number of photoevents produced by that detector element during the measurement interval. All counters in the array are gated simultaneously, so that at the end of the common counting period there is produced a "count vector" $\vec{K}$ of length N, where each component of that vector is the number of counts generated by a different element of the array.

Several assumptions are made about the character of the fringe pattern. First, the spatial frequency of the fringe pattern is assumed to be known a priori. In practice, this is a good assumption. For example, if the fringe is generated by a Michelson stellar interferometer, the fringe period is de-

---

[†]We are rather casually avoiding some complicated electronics issues here. To count the discrete photoevents produced by the array, substantial amplification of the detected signals will be required, and a suitable electronic threshold mechanism must be introduced. In practice, not every count will be detected, and some false counts will register. We are not considering such details because our interest is in the fundamental aspects of the problem.

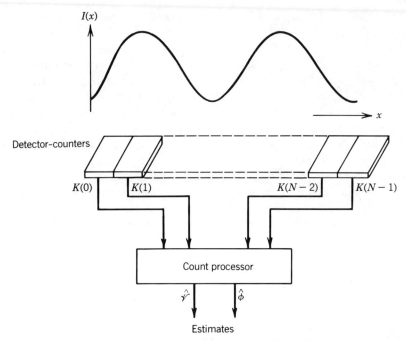

Figure 9-4. Detection and estimation system assumed for the amplitude interferometer.

termined by the subaperture spacing, the wavelength, and the focal length, all of which are presumed to be known. Second, the amplitude of the fringe is assumed to be constant across the detector array. In effect, we are assuming that the light involved is quasimonochromatic and that the time-averaged intensities of the two beams are constant across the array. Third, the spatial period of the fringe is assumed to be large compared to the size of a single detector in the array. This assumption allows us to approximate the intensity across any one detector element as constant. Finally, we make the somewhat artificial assumption that an integer number of periods of the fringe exist across the array. This latter assumption allows us to simplify the problem (in a way that will be made clearer later), yet still allows us to find the fundamental limits to the accuracy of the measurement of interest.

The intensity distribution of the fringe incident on the detector array is represented mathematically as

$$I(x, y) = (I_1 + I_2)\left[1 + \mathscr{V}\cos\left(\frac{2\pi x}{L} + \phi\right)\right], \qquad (9.4\text{-}1)$$

LIMITATIONS OF THE AMPLITUDE INTERFEROMETER   493

where I_1 and I_2 are the (constant) time-average intensities of the two interfering beams at the detector array, $\mathcal{V}$ is the visibility of the fringe, whereas L and ϕ are the spatial period and spatial phase of the fringe pattern, respectively.

The purpose of the measurement is presumably to determine $\mathcal{V}$ and ϕ. In some experiments, such as one in which a Michelson stellar interferometer is being used to gather information in the presence of time-varying atmospheric inhomogeneities, the phase of the fringe may be fluctuating rapidly as a function of time. We assume that the counting interval used in the measurement is sufficiently short to assure that the fringe is "frozen" in time on the detector array, in which case no reduction of fringe visibility will be expected as a result of fringe motion. Our goal is to find how accurately $\mathcal{V}$ and ϕ can be determined as a function of the number of photoevents detected by the array.

9.4.2 Statistical Properties of the Count Vector

It will be helpful in our analysis to have at our disposal certain information regarding the statistical properties of the counts $K(n)$. The statistics of concern are a function of the particular kind of light that is taking part in the interference experiments. For example, if the light is single-mode amplitude-stabilized laser light, each component of the count vector will be a Poisson variate. On the other hand, if the two light beams are polarized and thermal in origin, the counts obey negative-binomial statistics. We shall assume that thermal light is involved, since this is the case in virtually all experiments aimed at forming images from interferometric data. We further assume that the light is polarized. The first statistical quantity of interest is the mean count vector. Of course, the expected value of the count from the nth detector is simply proportional to the intensity of the portion of the fringe incident on that detector. Thus

$$\overline{K}(n) = \alpha A\tau(I_1 + I_2)\left[1 + \mathcal{V}\cos\left(\frac{2\pi n p_0}{N} + \phi\right)\right], \quad (9.4\text{-}2)$$

where α is given by Eq. (9.1-9), p_0 is the number of periods of the fringe embraced by the detector array, τ is the integration time, and A is the area of a single detector element. Also of interest is the second moment of the nth count, $\overline{K^2(n)}$. From Eqs. (9.3-2) and (9.3-5) we can easily show that

$$\overline{K^2}(n) = \overline{K}(n) + [\overline{K}(n)]^2\left(1 + \frac{1}{\mathcal{M}}\right), \quad (9.4\text{-}3)$$

where $\mathscr{M}$ is the number of temporal degrees of freedom in the measurement interval. In a similar fashion (see Problem 9-8) it is possible to show that the correlation between the counts registered by the mth and nth detectors ($m \neq n$) is given by

$$\overline{K(m)K(n)} = \overline{K(m)}\,\overline{K(n)}\left(1 + \frac{1}{\mathscr{M}}\right) \qquad (9.4\text{-}4)$$

With these results at our disposal we can now evaluate a specific method for measuring the fringe parameters of interest.

9.4.3 The Discrete Fourier Transform as an Estimation Tool

For estimation of the parameters of the fringe, some specific estimation procedure must be adopted. We choose here the discrete Fourier transform (DFT) (Ref. 9-17, Chapter 6) of the count vector as the primary tool for this task. By the DFT of the count vector, we mean explicitly the complex sequence $\mathscr{X}(p)$ given by

$$\mathscr{X}(p) = \frac{1}{N}\sum_{n=0}^{N-1} K(n) e^{j(2\pi np/N)}. \qquad (9.4\text{-}5)$$

If we evaluate the component of the DFT with index p_0 (where p_0 is again the number of periods of the fringe across the array), the amplitude and the phase of that component will be seen to yield information regarding the amplitude and phase of the fringe pattern of interest. This method of estimating fringe parameters under photon-limited conditions has been investigated in detail in Ref. 9-18, where it is shown to be an *optimum* procedure, in the maximum likelihood sense, when the fringe visibility is small. When the fringe visibility is large, the method is not strictly optimum, but it is extremely practical and yields respectable performance.

Before proceeding further, a short comment on our assumption that an integer number of fringe periods exist across the array is in order. The statement that the DFT coefficient with index p_0 has an amplitude and phase that provide estimates of the amplitude and phase of the incident fringe is true only if the above condition is satisfied. When a noninteger number of periods are Fourier transformed, the phenomenon known as "leakage" (Ref. 9-17, Section 9-5) leads to a spreading of the fringe-parameter information over several DFT coefficients and a change of the value of the p_0 coefficient. Since the fringe period is precisely known in advance, however, there is no reason why the system cannot be designed to capture an integer number of fringe periods.

LIMITATIONS OF THE AMPLITUDE INTERFEROMETER

Exactly how will we estimate the parameters $\mathscr{V}$ and ϕ from the p_0th component of the DFT? To answer this question we must first consider the character of the mean of this DFT component. To do so, let Eq. (9.4-5) be written in terms of two equations, one each for the real and imaginary parts of $\mathscr{X}(p_0)$. Since the count vector $\vec{K}$ is entirely real, the resulting equations are

$$\mathscr{X}_R \triangleq \text{Re}\{\mathscr{X}(p_0)\} = \frac{1}{N} \sum_{n=0}^{N-1} K(n) \cos \frac{2\pi n p_0}{N} \quad (9.4\text{-}6)$$

$$\mathscr{X}_I \triangleq \text{Im}\{\mathscr{X}(p_0)\} = \frac{1}{N} \sum_{n=0}^{N-1} K(n) \sin \frac{2\pi n p_0}{N}, \quad (9.4\text{-}7)$$

where $\mathscr{X}_R$ and $\mathscr{X}_I$ are the real and imaginary parts of $\mathscr{X}(p_0)$, respectively.

The means values of $\mathscr{X}_R$ and $\mathscr{X}_I$ can easily be evaluated by using the expression (9.4-2) for $\bar{K}(n)$. Straightforward manipulations yield the following expressions for these means:

$$\bar{\mathscr{X}}_R = \frac{\alpha A \tau (I_1 + I_2)}{2} \mathscr{V} \cos \phi$$

$$\bar{\mathscr{X}}_I = \frac{\alpha A \tau (I_1 + I_2)}{2} \mathscr{V} \sin \phi. \quad (9.4\text{-}8)$$

It is now possible to describe our strategy for estimating the parameters of interest. If there were no noise associated with the photodetection process, the actual values of $\mathscr{X}_R$ and $\mathscr{X}_I$ would be the mean values given in Eq. (9.4-8). In such a case the detected fringe amplitude, which we represent by C, could be obtained by simply taking the square root of the sum of the squares of the two expressions in that equation. Similarly, the phase of the fringe could be obtained by taking the arc tangent of the ratio of $\mathscr{X}_I$ to $\mathscr{X}_R$. In the absence of noise, such a strategy would yield error-free estimates of the detected fringe amplitude and phase. In the presence of photocount fluctuations, the strategy is not perfect, in the sense that there will always be some difference between our estimates of the parameters and the true parameters. Nonetheless, the performance of such an estimator has been found to be excellent (Ref. 9-17), and we adopt it here. Thus our estimates $\hat{C}$ and $\hat{\phi}$ of the detected fringe amplitude and phase are given as follows:

$$\hat{C} = \left(\mathscr{X}_R^2 + \mathscr{X}_I^2\right)^{1/2}$$

$$\hat{\phi} = \tan^{-1}(\mathscr{X}_I / \mathscr{X}_R). \quad (9.4\text{-}9)$$

To estimate the visibility of the fringe, again in the absence of noise, one further step is required. Since the noise-free fringe amplitude is

$$\left(\bar{\mathcal{X}}_R^2 + \bar{\mathcal{X}}_I^2\right)^{1/2} = \frac{\alpha A \tau (I_1 + I_2)}{2} \mathcal{V}, \qquad (9.4\text{-}10)$$

It is necessary to divide by $\alpha A \tau (I_1 + I_2)/2$ to obtain $\mathcal{V}$. Note that the parameters α, τ, and A are all known in advance. However, in general the sum of the two incident intensities, $I_1 + I_2$ is not known. Lack of knowledge of the total intensity incident on the array does not impede our ability to estimate the phase of the fringe pattern, but it does impede our ability to estimate its visibility. The only solution is to obtain a separate estimate of the value of this sum. Such an estimate could be obtained from the DC component of the DFT, $\mathcal{X}(0)$, which has a mean value

$$\overline{\mathcal{X}}(0) = \frac{1}{N} \sum_{n=0}^{N-1} \overline{K}(n) = \alpha A \tau (I_1 + I_2). \qquad (9.4\text{-}11)$$

Often in practice a sequence of measurements is to be performed for fringes of different spatial frequencies (different subaperture spacings in the case of the Michelson stellar interferometer). During the course of this sequence it is possible to obtain many independent estimates of the total incident intensity, one for each different fringe measured. Presumably this total intensity does not change with time or with the spatial frequency of the fringe of interest, and hence from the sequence of measurements an estimate of the total intensity can be obtained that is much more accurate than any one estimate of fringe amplitude. For this reason we assume that the sum of intensities is, for all practical purposes, a known quantity. Thus the visibility of the fringes can be estimated simply as

$$\hat{\mathcal{V}} = \frac{2\left(\mathcal{X}_R^2 + \mathcal{X}_I^2\right)^{1/2}}{\alpha A \tau (I_1 + I_2)}. \qquad (9.4\text{-}12)$$

9.4.4 Accuracy of the Visibility and Phase Estimates

The rules for estimating the fringe visibility and phase from the DFT were derived by considering the case in which the light level is so high that photocount fluctuations can be neglected. Attention is now turned to the crucial question of how accurately these parameters can be measured by this technique when the fluctuations of the photocounts can*not* be neglected. As

LIMITATIONS OF THE AMPLITUDE INTERFEROMETER

an aid in answering this question, we consider the variances and covariance of the real and imaginary parts of the DFT coefficient $\mathcal{X}(p_0)$. To illustrate, consider

$$\sigma_R^2 = \overline{\mathcal{X}_R^2} - (\overline{\mathcal{X}_R})^2. \tag{9.4-13}$$

Now with the help of Eqs. (9.4-3) and (9.4-4),

$$\overline{\mathcal{X}_R^2} = \frac{1}{N^2} \sum_{m=0}^{N-1} \sum_{n=0}^{N-1} \overline{K(m)K(n)} \cos\frac{2\pi m p_0}{N} \cos\frac{2\pi n p_0}{N}$$

$$= \frac{1}{N^2} \sum_{n=0}^{N-1} \overline{K^2(n)} \cos^2\frac{2\pi n p_0}{N} \quad \bigg\} \begin{array}{l} m = n \\ \text{terms} \end{array}$$

$$+ \left(\frac{1}{N} \sum_{n=0}^{N-1} \overline{K(n)} \cos\frac{2\pi n p_0}{N} \right)^2 \left(1 + \frac{1}{\mathcal{M}}\right) \quad \bigg\} \begin{array}{l} m \neq n \\ \text{terms.} \end{array} \tag{9.4-14}$$

$$- \left(\frac{1}{N} \sum_{n=0}^{N-1} [\overline{K(n)}]^2 \cos\frac{2\pi n p_0}{N} \right)\left(1 + \frac{1}{\mathcal{M}}\right)$$

Using Eq. (9.4-3) for the second moment of $K(n)$, we obtain

$$\sigma_R^2 = \frac{1}{N^2} \sum_{n=0}^{N-1} \overline{K(n)} \cos^2\frac{2\pi n p_0}{N} + \frac{1}{\mathcal{M}} \left(\frac{1}{N} \sum_{n=0}^{N-1} \overline{K(n)} \cos\frac{2\pi n p_0}{N} \right)^2. \tag{9.4-15}$$

Finally, with the application of Eq. (9.4-2) for $\overline{K}(n)$, and without loss of generality, with selection of the phase reference such that ϕ is zero (we are always able to choose any convenient phase reference we wish), the summations can be performed and yield

$$\sigma_R^2 = \frac{\alpha A \tau (I_1 + I_2)}{2N} \left[1 + \frac{\alpha A \tau (I_1 + I_2)}{2} \frac{N \mathcal{V}^2}{\mathcal{M}} \right]. \tag{9.4-16}$$

Defining $\overline{K}_1$ and $\overline{K}_2$ to be the average numbers of photoevents generated by the two interfering beams across the *entire array*, that is,

$$\overline{K}_1 = \alpha A \tau N I_1, \qquad \overline{K}_2 = \alpha A \tau N I_2, \tag{9.4-17}$$

we can express the variance of $\mathscr{X}_R$ as

$$\sigma_R^2 = \frac{\overline{K}_1 + \overline{K}_2}{2N^2}\left(1 + \frac{\overline{K}_1 + \overline{K}_2}{2\mathscr{M}}\mathscr{V}^2\right). \tag{9.4-18}$$

Application of an identical analysis to find the variance of $\mathscr{X}_I$ yields

$$\sigma_I^2 = \frac{\alpha A\tau(I_1 + I_2)}{2N} = \frac{\overline{K}_1 + \overline{K}_2}{2N^2}. \tag{9.4-19}$$

Finally, the covariance $E[(\mathscr{X}_R - \overline{\mathscr{X}}_R)(\mathscr{X}_I - \overline{\mathscr{X}}_I)]$ can be evaluated and is found to be identically zero.

The preceding analysis has shown that the real and imaginary parts of $\mathscr{X}(p_0)$ are uncorrelated and in general have different means and variances. Figure 9-5 illustrates the various quantities of interest. The true value of the fringe phasor is represented by **C** and is seen to be directed along the positive real axis, in accord with our assumption that the true phase of the fringe is zero. About the point defined by the tip of **C** there exists a "noise cloud," the contours of which can be regarded as contours of constant probability density. These contours are broader in the direction of the real axis than they are in the direction of the imaginary axis. More

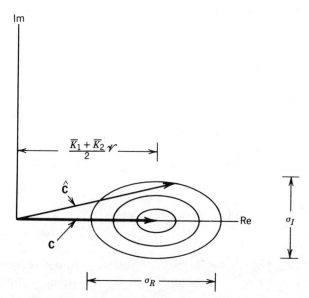

Figure 9-5. Phasor diagram for noisy fringe estimation.

generally, when the true phase of the fringe is not zero, the contours are elongated in the direction of **C**.

Referring back to Eqs. (9.4-18) and (9.4-19), it can be seen that the two variances differ only though the second term in parentheses in the former equation. This term can be recognized as being the product of the squared visibility with the arithmetic average of the degeneracy parameters of the two incident beams. As noted in Section 9.3.2, these degeneracy parameters are almost always much less than unity for visible thermal radiation. In addition, the fringe visibility can never exceed unity. Hence this second term, which causes the asymmetry of the noise cloud in Fig. 9-5, is generally negligible, provided we are dealing with thermal light in the visible region of the spectrum. Accordingly, hereafter we shall assume that the two variances are identical.

With the above analysis as background, we can now arrive at expressions indicating the accuracies that can be achieved for estimates of fringe visibility and phase. Two different approaches to achieving useful results can be taken. One is to assume that the total number of photocounts across the array is sufficiently great to apply the central limit theorem to the real and imaginary parts of $\mathcal{X}(p_0)$. In such a case, the problem of determining fringe amplitude and phase is identical to that of determining the amplitude and phase of a constant phasor in circular complex Gaussian noise. Such an approach was taken in Ref. 9-18. Here we choose to take a somewhat simpler approach, which requires a different assumption. Rather than assuming that the central limit theorem can be invoked, instead we assume that the width of the noise cloud in Fig. 9-5 is much smaller than the length of the true phasor along the real axis. (See Sections 2.9.5 and 6.2.3 for similar analyses carried out with this same high signal-to-noise ratio assumption.) Referring to Fig. 9-5, the mathematical statement of this assumption is

$$\frac{\bar{K}_1 + \bar{K}_2}{2N} \mathcal{V} \gg \sqrt{\frac{\bar{K}_1 + \bar{K}_2}{2N^2}}, \quad \text{or} \quad \mathcal{V} \gg \sqrt{\frac{2}{\bar{K}_1 + \bar{K}_2}}. \quad (9.4\text{-}20)$$

Thus the fringe visibility must be greater than a certain limit, where the limit decreases as the number of photoevents detected by the array increases. Under such a condition, the errors in estimation of the fringe amplitude are caused almost entirely by the noise component that is in phase with the true phasor (variance σ_R^2 in this case), whereas errors in the estimated phase are caused almost entirely by the noise component in quadrature with the true phasor (variance σ_I^2 in this case). The signal-to-noise ratio associated with the estimate of fringe amplitude (and also fringe visibility, since I_1 and I_2 are assumed accurately known) then takes the from

$$\left(\frac{S}{N}\right)_{rms} \cong \frac{\bar{\mathcal{X}}_R}{\sigma_R} = \sqrt{\frac{\bar{K}_1 + \bar{K}_2}{2}} \cdot \mathcal{V}, \qquad (9.4\text{-}21)$$

whereas the rms error associated with measurement of fringe phase is given by

$$\sigma_\phi \cong \frac{\sigma_I}{\bar{\mathcal{X}}_R} = \sqrt{\frac{2}{\bar{K}_1 + \bar{K}_2}} \cdot \frac{1}{\mathcal{V}}. \qquad (9.4\text{-}22)$$

Equations (9.4-21) and (9.4-22) are the primary results of this section. Some discussion of their implications is in order. Considering the problem of measuring fringe visibility first, the key facts to note in Eq. (9.4-22) are that the signal-to-noise ratio depends (1) on the square root of the total number of photoevents detected by the entire array (as might be expected for a Poisson-type phenomenon) and (2) linearly on the visibility of the fringe. One important implication of these results concerns the integration time required to achieve a predetermined signal-to-noise ratio. Since $\bar{K}_1$ and $\bar{K}_2$ are linearly proportional to the integration time τ, we can state a third important conclusion, namely, to hold the signal-to-noise ratio constant while decreasing the visibility, $\bar{K}_1 + \bar{K}_2$ must be increased in proportion to $1/\mathcal{V}^2$.

For some purposes it is convenient to express the result (9.4-21) in a different form. Assuming that the average intensities of the two interfering beams are equal ($\bar{K}_1 = \bar{K}_2$), and noting that $\delta_c = \bar{K}/\mathcal{M}$ and $\mathcal{M} \cong \tau/\tau_c$, where τ_c is the coherence time of the light, we can write

$$\left(\frac{S}{N}\right)_{rms} = \sqrt{\frac{\tau}{\tau_c}} \sqrt{\delta_c} \, \mathcal{V}, \qquad (9.4\text{-}23)$$

thereby explicitly noting the role of the count degeneracy parameter.

Turning to the problem of measuring fringe phase, the primary conclusions to be noted are that (1) the rms error in measurement of phase is inversely proportional to the square root of total number of photoevents produced by the array and (2) the rms phase error is inversely proportional to the visibility of the fringe.

With the preceding results in hand, it should now be possible to estimate, for example, the observation time required to determine the visibility of a fringe formed in a Michelson stellar interferometer. To achieve a given

LIMITATIONS OF THE INTENSITY INTERFEROMETER   501

signal-to-noise ratio Eq. (9.4-23) implies that the ratio of the observation time to the coherence time of the light must satisfy

$$\frac{\tau}{\tau_c} = \left[\left(\frac{S}{N}\right)_{\text{rms}}\right]^2 \cdot \frac{1}{\delta_c \mathcal{V}^2}, \qquad (9.4\text{-}24)$$

where we have assumed that the two interfering beams have the same count degeneracy parameter δ_c. Suppose that we wish to achieve a signal-to-noise ratio of 10 in the visibility measurement using light with a degeneracy parameter of 10^{-3}. The averaging time required to achieve this accuracy will depend on the magnitude of the visibility we are measuring. If the visibility of interest is close to unity, appropriate substitutions in (9.4-24) show that the observation time required is about 10^5 coherence times. On the other hand, if the visibility of the fringe of interest is only 0.1, the observation time required is about 10^7 coherence times. The coherence time itself depends on the spectral width of the light that is incident on the photodetector. If the spectral width is, for example, 0.001 micrometers (10 Å) and the mean wavelength is 0.5 micrometers, the coherence time is approximately 10^{-12} seconds. Hence a measurement time of 10^7 coherence times is still extremely short. This example is somewhat artificial in that in real astronomical problems the effective degeneracy parameter is likely to be much smaller than that assumed here. Nonetheless, the numbers will prove useful for comparison purposes when we consider the sensitivity of the intensity interferometer in the section that follows.

9.5 NOISE LIMITATIONS OF THE INTENSITY INTERFEROMETER AT LOW LIGHT LEVELS

In the preceding section we considered one method for measuring the visibility of an incident fringe pattern, or equivalently, the complex coherence factor μ_{12} of the light incident on two spatially separated apertures. (Since the incident intensities were assumed perfectly known, the complex coherence factor can be determined from the visibility and will, in fact, equal the visibility when the average intensities of the two interfering beams are the same.) That method superimposed the two beams before detection.

In this section we reconsider the intensity interferometer discussed in Section 6.3. In this case the light incident on two spatially separated apertures is detected directly, without bringing together the two optical beams. The detected photocurrents are then correlated, and the fringe visibility is determined from that correlation. The reader may wish to reread

Section 6.3 at this time. In that section, the limitations posed by purely classical noise arising from the fluctuations of the intensity of the thermal radiation striking the photodetectors were investigated. Here we concentrate on the noise limitations of the intensity interferometer caused by the discrete nature of the photoevents generated at each of the two detectors. In general, both types of noise will be present. As we shall see, however, in the visible region of the spectrum, photocount fluctuations are the primary limitation to the sensitivity and accuracy of the intensity interferometer.

9.5.1 The Counting Version of the Intensity Interferometer

The version of the the intensity interferometer depicted in Fig. 6-15 is appropriate for a purely classical situation in which the continuous currents generated by the two photodetectors are operated on and combined using analog filters and devices. In the case of interest here, the analysis will be simplified if we assume a slightly different form of the interferometer, as depicted in Fig. 9-6. The light beams collected by two large mirrors are focused onto two separate photodetectors. Each such photodetector is assumed to be followed by a counter, which counts the number of photoevents observed in a time interval of duration τ_0 seconds. From the numbers of counts K_1 and K_2 produced in the two arms of the interferometer, the mean numbers of counts $\overline{K}_1$ and $\overline{K}_2$ expected in the two arms are subtracted, respectively. The resulting "count fluctuations" ΔK_1 and ΔK_2 are then multiplied together and passed to an averaging accumulator, where

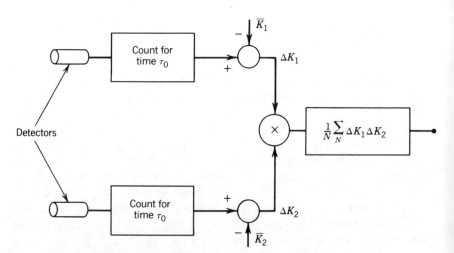

Figure 9-6. Counting version of the intensity interferometer.

LIMITATIONS OF THE INTENSITY INTERFEROMETER

this count product is added to the count products produced in earlier τ_0-second periods, and the total sum is divided by the number of accumulated count products. The output of interest is the average count product, which will be seen to yield information about the visibility of the fringe that would have been formed if the two optical beams had been directly interfered. Our goal is to find the relationship between the expected value of the averaged count product and the fringe visibility or complex coherence factor. In addition, we wish to find the variance of this quantity, so that the rms signal-to-noise ratio associated with the measurement can be determined and compared with the similar quantities found in the previous section.

9.5.2 The Expected Value of the Count-Fluctuation Product and Its Relationship to Fringe Visibility

By the "count fluctuations" we mean explicitly the differences between the actual numbers of counts obtained in a τ_0-second interval at detectors 1 and 2 and the expected values of these two numbers of counts. Thus

$$\Delta K_1 = K_1 - \overline{K}_1, \qquad \Delta K_2 = K_2 - \overline{K}_2. \qquad (9.5\text{-}1)$$

The averaging accumulator at the output of the system depicted in Fig. 9-6 in effect produces an estimate of the expected value of the product of the two count fluctuations. Thus we are interested in the statistical properties of the quantity $\overline{\Delta K_1 \Delta K_2}$, and in particular its mean and variance. As an aid to this analysis, we first calculate the expected value of the product of K_1 and K_2,

$$\overline{K_1 K_2} = \sum_{K_1=0}^{\infty} \sum_{K_2=0}^{\infty} K_1 K_2 P(K_1, K_2), \qquad (9.5\text{-}2)$$

where $P(K_1, K_2)$ represents the joint probability distribution of K_1 and K_2. First note that, from the basic properties of conditional probabilities,

$$P(K_1, K_2) = \iint_0^{\infty} P(K_1, K_2 | W_1, W_2) p_W(W_1, W_2) \, dW_1 \, dW_2. \qquad (9.5\text{-}3)$$

In addition, since K_1 and K_2 are independent when conditioned by the integrated intensities W_1 and W_2, respectively, we can write

$$P(K_1, K_2 | W_1, W_2) = P(K_1 | W_1) P(K_2 | W_2)$$

$$= \frac{(\alpha W_1)^{K_1}}{K_1!} e^{-\alpha W_1} \frac{(\alpha W_2)^{K_2}}{K_2!} e^{-\alpha W_2}, \qquad (9.5\text{-}4)$$

where we have utilized the fact that both K_1 and K_2 are conditionally Poisson. Using these facts, the following expression for the average of the count product can now be written

$$\overline{K_1 K_2} = \sum_{K_1=0}^{\infty} \sum_{K_2=0}^{\infty} K_1 K_2 \iint_0^{\infty} \frac{(\alpha W_1)^{K_1}}{K_1!} e^{-\alpha W_1} \frac{(\alpha W_2)^{K_2}}{K_2!} e^{-\alpha W_2}$$

$$\times p_W(W_1, W_2)\, dW_1\, dW_2. \qquad (9.5\text{-}5)$$

At this point we interchange the orders of summation and integration in (9.5-5). Using the relationship

$$\sum_{K_1=0}^{\infty} K_1 \frac{(\alpha W_1)^{K_1}}{K_1!} e^{-\alpha W_1} = \alpha W_1, \qquad \sum_{K_2=0}^{\infty} K_2 \frac{(\alpha W_2)^{K_2}}{K_2!} e^{-\alpha W_2} = \alpha W_2,$$

$$(9.5\text{-}6)$$

we can express the average of the count product in terms of the average of the classical integrated intensities at the two detectors,

$$\overline{K_1 K_2} = \alpha^2 \overline{W_1 W_2}. \qquad (9.5\text{-}7)$$

Some investigation of the average of the products of the integrated intensities is needed. Substitution of the definitions of the two integrated intensities into the average and interchange of orders of integration and averaging yields

$$\overline{W_1 W_2} = E\left[A^2 \int_t^{t+\tau_0} I_1(\xi_1)\, d\xi_1 \int_t^{t+\tau_0} I_2(\xi_2)\, d\xi_2 \right]$$

$$= \iint_t^{t+\tau_0} \Gamma_I(P_1, P_2; \xi_1 - \xi_2)\, d\xi_1\, d\xi_2, \qquad (9.5\text{-}8)$$

where P_1 and P_2 represent the centers of the two collecting apertures of the interferometer, whereas Γ_I is the cross-correlation of the intensities incident at those two points.

To make further progress, it is necessary to incorporate some specific assumptions about the nature of the light involved in the measurement. We assume that the light is (1) polarized and *thermal* in origin and (2) *cross-spectrally pure*, allowing us to separate the temporal and spatial

LIMITATIONS OF THE INTENSITY INTERFEROMETER

aspects of coherence. With these assumptions, the cross-correlation function of the two intensities can be reduced to the form

$$\Gamma_I(P_1, P_2; \xi_1 - \xi_2) = \bar{I}_1 \bar{I}_2 \left[1 + |\mu_{12}|^2 |\gamma(\xi_1 - \xi_2)|^2 \right]. \quad (9.5\text{-}9)$$

It is worth further emphasis that this expression is valid only for thermal or pseudothermal light. It specifically is not valid for single-mode, amplitude-stabilized laser light. The symmetry properties of $\gamma(\xi_1 - \xi_2)$ allow further simplification of the integrals of concern. By use of a reduction similar to one used on several occasions before [see, e.g., Eq. (6.2-18)], the double integral is reduced to a single integral

$$\overline{W_1 W_2} = 2\tau_0 A^2 \int_0^{\tau_0} \left(1 - \frac{\eta}{\tau_0}\right) \Gamma_I(P_1, P_2; \eta) \, d\eta$$

$$= 2 A^2 \tau_0 \bar{I}_1 \bar{I}_2 \int_0^{\tau_0} \left(1 - \frac{\eta}{\tau_0}\right) d\eta$$

$$+ 2 A^2 \tau_0 \bar{I}_1 \bar{I}_2 |\mu_{12}|^2 \int_0^{\tau_0} \left(1 - \frac{\eta}{\tau_0}\right) |\gamma(\eta)|^2 \, d\eta. \quad (9.5\text{-}10)$$

Substituting this result back into Eq. (9.5-7), and taking note of the definition (9.2-22) for the number of degrees of freedom, we find

$$\overline{\Delta K_1 \Delta K_2} = \frac{\bar{K}_1 \bar{K}_2}{\mathcal{M}} \mu_{12}^2, \quad (9.5\text{-}11)$$

where $\mu_{12} = |\mu_{12}|$. Thus, knowing K_1, K_2, and $\mathcal{M}$, we can determine μ_{12} from $\overline{\Delta K_1 \Delta K_2}$. Finally, this result can equally well be expressed in terms of fringe visibility $\mathcal{V}$. Since

$$\mathcal{V} = \frac{2\sqrt{\bar{I}_1 \bar{I}_2}}{\bar{I}_1 + \bar{I}_2} \mu_{12} = \frac{2\sqrt{\bar{K}_1 \bar{K}_2}}{\bar{K}_1 + \bar{K}_2} \mu_{12}, \quad (9.5\text{-}12)$$

the average of the count-fluctuation product is given by

$$\overline{\Delta K_1 \Delta K_2} = \left(\frac{\bar{K}_1 + \bar{K}_2}{2}\right)^2 \frac{\mathcal{V}^2}{\mathcal{M}}. \quad (9.5\text{-}13)$$

Equation (9.5-13), relating the average count-fluctuation product to the fringe visibility that would be observed if the two beams of light were to interfere, is an important milestone in our analysis. It shows that, if a sufficiently large number of count fluctuation products are entered into the averaging accumulator, so that the estimate of the true statistical average is reasonably accurate, the visibility $\mathscr{V}$ can be estimated from the information present at the output of the interferometer. Note that, as in the case of the classical intensity interferometer, no information about the phase of the fringe is available. What we do not know at this point is how many count-fluctuation products must be averaged for the estimate of visibility to be an accurate one. This brings us to a consideration of the fluctuations of the estimate of $\mathscr{V}$, for which we must turn our attention to the noise present at the output of the interferometer.

9.5.3 The Signal-to-Noise Ratio Associated with the Visibility Estimate

A fully general study of the noise fluctuations associated with the output of the counting interferometer illustrated in Fig. 9-6 would be nontrivial. The difficulty arises in simultaneously including the effects on that noise of both the classically induced fluctuations and the shot-noise fluctuations of the counts. Whereas the shot-noise fluctuations of the counts at the outputs of the two detectors are statistically independent, the classically induced fluctuations are not. In fact, it is precisely the statistical dependence of those counts that allows us to extract information about the fringe visibility. Not only does the "signal" portion of the interferometer output depend on this statistical relationship between the counts, but also the noise at the output is influenced by it. The full analysis of the interferometer, including both of these effects, is a very difficult analytical problem.

Fortunately, in the particular case of most interest, namely, light originating from a true thermal source in the visible region of the spectrum, a much simplified analysis will suffice. We know that for such sources, because of the very small degeneracy parameter of the light they emit, the fluctuations of the photocounts are strongly dominated by pure shot noise. We cannot neglect the classically induced fluctuations of the counts when we calculate the signal component of the output, but we can neglect them when we calculate the noise, simply because their contribution to the noise is so small.

In the analysis that follows, we consider first the signal-to-noise ratio associated with the measurement of a single count product. It is defined as follows:

$$\left(\frac{S}{N}\right)_1 = \frac{\overline{\Delta K_1 \Delta K_2}}{\left[\overline{(\Delta K_1 \Delta K_2)^2} - \overline{(\Delta K_1 \Delta K_2)}^2\right]^{1/2}}. \qquad (9.5\text{-}14)$$

After calculating this quantity, we then specify the signal-to-noise ratio at the output of the averaging accumulator simply by multiplying the signal-to-noise ratio for one count product by the square root of the number of independent measurements averaged in the accumulator. The only requirement for accuracy of this procedure is that the count fluctuations be uncorrelated from counting interval to counting interval, a property that does hold for the Poisson shot noise we have assumed to be the main limit to accuracy.

The quantity inside the brackets of the denominator of Eq. (9.5-14) can be evaluated by noting that, for the purpose of noise calculations, the count fluctuations as the two detectors are those of statistically independent Poisson variates. It follows that

$$\overline{(\Delta K_1 \Delta K_2)^2} - \overline{(\Delta K_1 \Delta K_2)}^2 = \overline{\Delta K_1^2} \, \overline{\Delta K_2^2} - \overline{(\Delta K_1 \Delta K_2)}^2 = \overline{\Delta K_1^2} \, \overline{\Delta K_2^2} = \overline{K_1} \overline{K_2},$$
(9.5-15)

where the properties $\overline{\Delta K} = 0$ (by definition) and $\overline{\Delta K^2} = \overline{K}$ (by the Poisson assumption) have been used. Substitution of Eqs. (9.5-13) and (9.5-15) in (9.5-14) shows that the signal-to-noise ratio associated with a single count product is given by

$$\left(\frac{S}{N}\right)_1 = \frac{\left(\frac{\overline{K_1} + \overline{K_2}}{2}\right)^2 \mathscr{V}^2 \mathscr{M}}{\sqrt{\overline{K_1}\overline{K_2}}}.$$
(9.5-16)

When the average intensities incident on the two detectors are equal, the expression for the signal-to-noise ratio reduces to the useful form

$$\left(\frac{S}{N}\right)_1 = \delta_c \mathscr{V}^2,$$
(9.5-17)

where δ_c is the count degeneracy parameter of the light incident on either detector.

It should be emphasized once more that Eq. (9.5-17) represents the signal-to-noise ratio only for a single count-fluctuation product, based on the counts in a single counting interval of length τ_0 seconds. Even a casual inspection of this result indicates that a problem exists. Since the degeneracy parameter has been assumed to be much less than unity, and since the visibility of the fringe can never exceed unity, we see that the signal-to-noise ratio of Eq. (9.5-17) is always much less than unity! Note that this expression is independent of the length τ_0 of the single counting period used. Therefore, the signal-to-noise ratio is not improved by counting longer with the counters that follow the photodetectors. We conclude that it is impossible to extract information about the fringe visibility from a single count-product measurement, for the noise dominates the signal by a large margin.

To obtain a more accurate estimate of the visibility, we must resort to averaging the count products obtained in many independent counting intervals. This is precisely the function of the averaging accumulator shown at the output of the interferometer in Fig. 9-6. Assuming that the count-product fluctuations are independent from counting interval to counting interval, we see that the rms signal-to-noise ratio associated with averaged results from N counting intervals is given by

$$\left(\frac{S}{N}\right)_N = \sqrt{N}\,\delta_c \mathscr{V}^2. \tag{9.5-18}$$

If τ_0 is the basic counting interval, and if the counters can be reset instantaneously, the total measurement time is $\tau = N\tau_0$. Hence the expression (9.5-18) can be restated with total measurement time as a parameter,

$$\left(\frac{S}{N}\right)_N = \sqrt{\frac{\tau}{\tau_0}}\,\delta_c \mathscr{V}^2. \tag{9.5-19}$$

Note that it is, in fact, advantageous to have the basic counting interval τ_0 as *small* as possible, for then the number of independent count products averaged in a fixed total measurement time is maximized.

To obtain a more concrete feeling for the implications of this analysis, we examine exactly the same example that was treated in Section 9.4 for the amplitude interferometer. The ratio of the total measurement time to the basic counting time required in order to achieve a predetermined signal-to-noise ratio can be written, from (9.5-19), as

$$\frac{\tau}{\tau_0} = \frac{\left[\left(\frac{S}{N}\right)_N\right]^2}{\delta_c^2 \mathscr{V}^4}. \tag{9.5-20}$$

Let the degeneracy parameter of the light be 10^{-3}, the basic counting interval τ_0 be 10^{-7} seconds, and the required signal-to-noise ratio be 10 (as in the previous example). The required measurement time now depends on the *fourth* power of visibility. If the visibility of the fringe is unity, the required measurement time in this case is at least 10 seconds (compared with a small fraction of a second for the amplitude interferometer). If the visibility of the fringe drops to 0.1, the measurement time increases to 10^8 seconds, or about 28 hours (again to be compared with a small fraction of a second)!

If the sensitivity of the intensity interferometer is really as poor as indicated above, why has this instrument been of practical interest? The answer lies partly in the fact that the collecting apertures of the intensity interferometer can be far larger than the collecting apertures of an amplitude interferometer, and thus a larger fraction of a single coherence cell can be covered by the collecting aperture in the present case. Our assumption that the degeneracy parameter of the counts is the same for the two interferometers when utilizing light from the same source is not really true. When the collecting aperture in an arm of an interferometer is smaller than the size of a single coherence cell, then the count degeneracy parameter at the photodetector for that arm is proportional to the area of that aperture [cf. Eq. (9.3-22)]. The diameter of the largest allowable collector in a Michelson interferometer operating within the Earth's atmosphere is of the order of 10 centimeters (or perhaps somewhat smaller); larger apertures result in a loss of fringe visibility due to the presence of more than one atmospheric coherence cell in the measurement. On the other hand, the intensity interferometer, which is insensitive to atmospheric distortions of the phase of the light impinging on the detectors, can use collecting apertures that are far larger than those mentioned previously. For example, the intensity interferometer at Narrabri, Australia, has collectors that are about 7 meters in diameter. Thus the effective count degeneracy parameter of the detected light is about $(70)^2$ times larger for this intensity interferometer than for a comparable amplitude interferometer.

There are a variety of other reasons for interest in the intensity interferometer, in spite of its comparatively low sensitivity. First, the pathlengths in the two arms of such an interferometer need only be equalized and maintained to a fraction of c/B, where c is the velocity of light and B is the *electrical* bandwidth of the postdetection electronics. For an amplitude interferometer, the corresponding requirement is equalization to a fraction of $c/\Delta\nu$, where $\Delta\nu$ is the *optical* bandwidth of the interferometer. It would not be uncommon to have several orders of magnitude difference between the electrical and optical bandwidths. Hence the alignment tolerances are greatly relaxed with the intensity interferometer.

A second advantage of the intensity interferometer is that relatively imperfect collectors can be used, whereas an amplitude interferometer requires high-precision optical components.

A third advantage is that inhomogeneities of the atmosphere have comparatively little effect on the performance of an intensity interferometer but a profound effect on an amplitude interferometer. The detectors in the intensity interferometer are completely insensitive to any phase errors of the optical waves falling on them. The only significant effect comes from scintillation induced by the atmosphere, and these effects are often comparatively minor. An amplitude interferometer, by comparison, is exceedingly sensitive to atmospherically induced phase perturbations, even when the aperture sizes are small. In such a case the fringe that is to be detected is actually running back and forth on the detector array, as a result of the constantly changing relative phase shifts introduced in the two atmospheric paths intercepted by the two arms of the interferometer. In practice, such wandering of the fringe negates the possibility of extracting phase information about the complex coherence factor (the intensity interferometer also cannot determine this phase). If also makes the job of extracting visibility information more difficult than it would be if the fringe were completely stationary.

In summary, the lower inherent sensitivity of the intensity interferometer is at least partially compensated for by the increased areas possible for the collecting apertures, the lower precision required of those collecting elements, the relative insensitivity to atmospheric effects, and the much relaxed tolerances on alignment of the system. However, the majority of current research in interferometric imaging is directed toward the amplitude interferometer, precisely because of its superior noise performance. The reader interested in pursuing the subject of intensity interferometry further is referred to Ref. 6-24 for a detailed discussion of the history and performance capabilities of the intensity interferometer.

9.6 NOISE LIMITATIONS IN SPECKLE INTERFEROMETRY

Our final analysis considers the noise limitations encountered in stellar speckle interferometry, particularly the fundamental limitations that arise from the finite number of photoevents participating in any measurement. The reader may wish to review Section 8.8, which introduced the basic ideas behind stellar speckle interferometry, before proceeding further. Here it suffices to remind the reader that, by averaging the squared moduli of the Fourier spectra of an ensemble of detected short-exposure images, an estimate of the squared modulus of the object spectrum can be obtained, free from degrading effects of the atmosphere. The ability to extract such Fourier information about the object is limited, however, particularly for the weak objects of greatest interest in astronomy, by the noise that is inherent in the photodetection process.

Our attention is first focused on the analytical model that will be used to study the sensitivity of this imaging method. A calculation of the spectral density of the detected imagery is presented next, followed by a calculation of the fluctuations that are encountered in one possible estimation procedure used to determine the squared modulus of the object spectrum. Finally, the signal-to-noise ratio (S/N) achieved by the process will be calculated. Some general comments will then complete the final section. Alternative discussions of this problem can be found in Refs. 9-19 through 9-21.

9.6.1 A Continuous Model for the Detection Process

The analyses of amplitude and intensity interferometry presented in the previous sections used discrete models for the detection process. By this we mean that the analysis of amplitude interferometry assumed a discrete set of small photodetectors, each of which produced an element of a count vector, whereas the detectors used by the intensity interferometer were assumed to be gated in discrete time intervals, each producing a discrete sequence of counts for further processing. The reader may appreciate being introduced to another method of analysis here, namely, one that uses a spatially *continuous* model of the photodetection process.

In this case we suppose that the detector is continuous in space and capable of registering not only the occurrence of a photoevent anywhere on its sensitive surface, but also the location of that photoevent. The detected signal is then represented in the form

$$d(x, y) = \sum_{n=1}^{K} \delta(x - x_n, y - y_n), \quad (9.6\text{-}1)$$

where $d(x, y)$ represents the detected signal as a function of the two spatial coordinates and $\delta(x - x_n, y - y_n)$ represents a particular photoevent occurring at spatial coordinates (x_n, y_n) in terms of a two-dimensional δ function centered at those coordinates. There is a total of K such photoevents, at different locations on the photosurface, during the time that this single image is detected. In this representation, K, x_n, and y_n are all to be regarded as random variables, with statistical properties to be described in the following paragraphs.

The model described in the preceding paragraph is that of a compound or inhomogeneous Poisson impulse process, of the kind discussed in Section 3.7. In accord with the semiclassical theory of photodetection, the probability that K photoevents occur in an area A on the photodetector is taken to be Poisson, under the assumption that the incident light is thermal in origin and has very small degeneracy parameter. Thus the probability of detecting

K photoevents in area A is written

$$P(K) = \frac{\left[\iint\limits_{-\infty}^{\infty} \lambda(x, y)\, dx\, dy\right]^K}{K!} \exp\left[-\iint\limits_{-\infty}^{\infty} \lambda(x, y)\, dx\, dy\right], \quad (9.6\text{-}2)$$

where the "rate" $\lambda(x, y)$ of the Poisson impulse process is related to the classical intensity $I(x, y)$ falling on the photosurface through

$$\lambda(x, y) = \alpha I(x, y)\tau, \quad (9.6\text{-}3)$$

and the finite extent of the detector has been incorporated in $I(x, y)$. Here α is given by Eq. (9.1-9), and τ represents the detector integration time for this particular image. Since the distribution $I(x, y)$ of classical intensity is unknown a priori, we first treat $\lambda(x, y)$ as a given known function and then later average over the statistics of λ. This procedure is entirely consistent with rules of conditional statistics. We further note, for future use, that, when conditioned by knowledge of the number K of events, the event locations (x_n, y_n) are independent random variables, with common probability density function [cf. Eq. (3.7-14)]

$$p(x_n, y_n) = \frac{\lambda(x_n, y_n)}{\iint\limits_{-\infty}^{\infty} \lambda(x, y)\, dx\, dy}, \quad (9.6\text{-}4)$$

where the rate function $\lambda(x, y)$, which is proportional to the classical image intensity, is a nonnegative function.

Figure 9-7 shows a typical classical intensity distribution and the corresponding typical detected image; the illustration is one-dimensional for simplicity.

9.6.2 The Spectral Density of the Detected Imagery

The speckle interferometry method rests on the generation of an accurate estimate of the spectral density of the detected imagery. It is thus important to consider the statistical properties of such spectral estimates. In this section we focus on the mean or expectation of the spectral density estimate. In Section 9.6.3 we consider the fluctuations of the estimate about the mean.

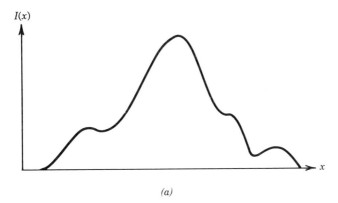

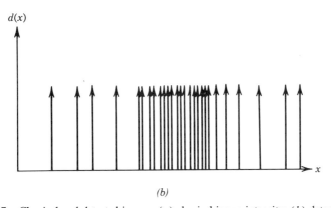

Figure 9-7. Classical and detected images: (*a*) classical image intensity; (*b*) detected image.

The classical intensity $I(x, y)$ falling on the photosurface is not representable as a sample function of a stationary random process. The finite area of the photosensitive surface in effect supplies a "window" through which the incident images must be measured, and independent of whether the images incident on this window are stationary, they certainly are non-stationary after windowing. In fact, the integral (over the detector area) of each sample function of the windowed intensity process is finite, a consequence of the fact that each image contains finite optical power. Hence the Fourier transform of each sample function exists, and it is appropriate to deal with the *energy* spectral density of the images, rather than the power spectral density. This conclusion is also true for the detected image $d(x, y)$. Our interest lies in determining the expected value of the squared modulus for the Fourier transform $\mathbf{D}(\nu_X, \nu_Y)$ of the detected image data, where

$$\mathbf{D}(\nu_X, \nu_Y) = \iint_{-\infty}^{\infty} d(x, y) \exp[j2\pi(\nu_X x + \nu_Y y)] \, dx \, dy. \quad (9.6\text{-}5)$$

As a first step, expression (9.6-1) for $d(x, y)$ is substituted into the Fourier transform of (9.6-5), yielding

$$\mathbf{D}(\nu_X, \nu_Y) = \sum_{n=1}^{K} \exp[j2\pi(\nu_X x_n + \nu_Y y_n)]. \quad (9.6\text{-}6)$$

The squared modulus of this quantity is given by

$$|\mathbf{D}(\nu_X, \nu_Y)|^2 = \sum_{n=1}^{K} \sum_{m=1}^{K} e^{j2\pi[\nu_X(x_n - x_m) + \nu_Y(y_n - y_m)]}. \quad (9.6\text{-}7)$$

It remains to find the expected value of $|\mathbf{D}|^2$ over the statistics of K, (x_n, y_n), and λ. It is convenient to first regard K and $\lambda(x, y)$ as known quantities, average over the conditional statistics of (x_n, y_n) and (x_m, y_m), and then average over K and λ. Thus our first goal is to compute

$$E_{nm}[|\mathbf{D}(\nu_X, \nu_Y)|^2] = \sum_{n=1}^{K} \sum_{m=1}^{K} E_{nm}[e^{j2\pi[\nu_X(x_n - x_m) + \nu_Y(y_n - y_m)]}], \quad (9.6\text{-}8)$$

where E_{nm} signifies an average over (x_n, y_n) and (x_m, y_m).

Two classes of terms can be identified: (1) K terms for which $n = m$, each of which yields unity and (2) $K^2 - K$ terms for which $n \ne m$. For the latter terms we know that (x_n, y_n) and (x_m, y_m) are independent random variables and, therefore, that

$$p(x_n, y_n; x_m, y_m) = \frac{\lambda(x_n, y_n)}{\iint_{-\infty}^{\infty} \lambda(x, y) \, dx \, dy} \frac{\lambda(x_m, y_m)}{\iint_{-\infty}^{\infty} \lambda(x, y) \, dx \, dy}. \quad (9.6\text{-}9)$$

For these $K^2 - K$ terms, the result of the averaging process is

$$E_{nm}[e^{j2\pi[\nu_X(x_n - y_m) + \nu_Y(y_n - y_m)]}] = \frac{\left|\iint_{-\infty}^{\infty} \lambda(x, y) e^{j2\pi(\nu_X x + \nu_Y y)} \, dx \, dy\right|^2}{\iint_{-\infty}^{\infty} \lambda(x, y) \, dx \, dy}.$$

$$(9.6\text{-}10)$$

Thus the result of averaging $|\mathbf{D}|^2$ over the statistics of (x_n, y_n) and (x_m, y_m) becomes

$$E_{nm}\left[|\mathbf{D}(\nu_X, \nu_Y)|^2\right] = K + (K^2 - K)\left|\frac{\Lambda(\nu_X, \nu_Y)}{\Lambda(0, 0)}\right|^2, \quad (9.6\text{-}11)$$

where $\Lambda(\nu_X, \nu_Y)$ is the Fourier transform of the rate function $\lambda(x, y)$.

At this point a few comments are in order regarding the number of counts K in a single frame. This number varies from frame to frame. In some applications, particularly those where accurate photon-counting equipment is used, it is possible to measure K for each detected frame. In such a case it would not be appropriate to treat K as a random variable, since it is completely known for each measurement. In other cases it is not possible to measure K, such as when the detector is photographic film. Under such circumstances K must be treated as a random variable. Here we assume the latter case, although later comments will address the changes necessary when K can, in fact, be measured for each frame.

Continuing our averaging process, we next find the expected value of Eq. (9.6-11) over the random variable K, assuming $\lambda(x, y)$ to be known. Representing the conditional mean of K (given λ) by $\overline{K}_{(\lambda)}$, and noting that, for Poisson statistics,

$$E[K^2 - K] = \left[\overline{K}_{(\lambda)}\right]^2, \quad (9.6\text{-}12)$$

we find that

$$E_{n,m,K}\left[|\mathbf{D}(\nu_X, \nu_Y)|^2\right] = \overline{K}_{(\lambda)} + |\Lambda(\nu_X, \nu_Y)|^2. \quad (9.6\text{-}13)$$

Finally, averaging over the statistics of $\lambda(x, y)$, we obtain

$$\mathscr{E}_d(\nu_X, \nu_Y) = E\left[|\mathbf{D}(\nu_X, \nu_Y)|^2\right] = \overline{K} + \mathscr{E}_\lambda(\nu_X, \nu_Y), \quad (9.6\text{-}14)$$

where $\overline{K}$ is the unconditional mean of K and $\mathscr{E}_\lambda(\nu_X, \nu_Y) = E[|\Lambda(\nu_X, \nu_Y)|^2]$.

Thus the spectral density of the detected image is the sum of a constant spectral level $\overline{K}$, plus the spectral density of the rate function. This result is in agreement with Eq. (3.7-32), which was obtained by a related argument. Alternate forms of this result are also useful. First, if we define a normalized energy spectral density

$$\hat{\mathscr{E}}_\lambda(\nu_X, \nu_Y) = \frac{\mathscr{E}_\lambda(\nu_X, \nu_Y)}{\mathscr{E}_\lambda(0, 0)}, \quad (9.6\text{-}15)$$

we have

$$\mathcal{E}_d(\nu_X, \nu_Y) = \overline{K} + (\overline{K})^2 \hat{\mathcal{E}}_\lambda(\nu_X, \nu_Y). \tag{9.6-16}$$

Furthermore, since $\lambda(x, y)$ is proportional to the classical intensity $I(x, y)$, we must have

$$\hat{\mathcal{E}}_\lambda(\nu_X, \nu_Y) = \hat{\mathcal{E}}_i(\nu_X, \nu_Y), \tag{9.6-17}$$

where $\hat{\mathcal{E}}_i(\nu_X, \nu_Y)$ is the normalized energy spectral density of the classical image intensity incident on the detector. The result is illustrated in Fig. 9-8.

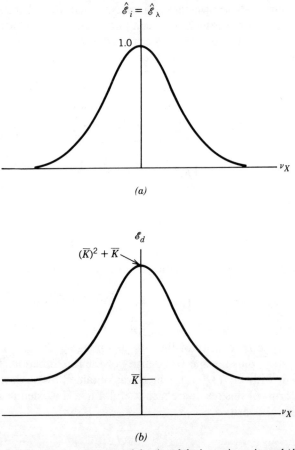

Figure 9-8. (a) Normalized energy spectral density of the image intensity and (b) corresponding energy spectral density of the detected image.

NOISE LIMITATIONS IN SPECKLE INTERFEROMETRY

Having calculated the mean or expected distribution of the squared modulus of the Fourier spectrum of the image, we next consider the more difficult problem of calculating the fluctuations associated with the estimate.

9.6.3 Fluctuations of the Estimate of Image Spectral Density

In the imaging problem of interest here, the initial desired result is an accurate estimate of the normalized energy spectral density $\hat{\mathscr{E}}_i$ of the classical image intensity falling on the detector. Because of the simple relationship that exists between $\mathscr{E}_d$ and the energy spectral density $\hat{\mathscr{E}}_i$ of the detected image [Eq. (9.6-17)], a reasonable approach is to first estimate $\mathscr{E}_d$ and the express $\hat{\mathscr{E}}_i$ as

$$\hat{\mathscr{E}}_i = \frac{\mathscr{E}_d - \overline{K}}{(\overline{K})^2}. \tag{9.6-18}$$

The quantity $\overline{K}$ is simply a measure of the total image brightness, which we assume is either known a priori or can be determined accurately by a suitable photometric measurement. (An alternate estimation procedure, described in Ref. 9-21, in which $\overline{K}$ is replaced by the actual number K of photoevents detected in the picture, is discussed briefly later.) The fluctuations in our estimate of $\hat{\mathscr{E}}_i$ are determined by the fluctuations in our measurement of $\mathscr{E}_d$. It is these fluctuations that we wish to find here.

An estimate of $\mathscr{E}_d$ can be made by measuring $|\mathbf{D}|^2$ for a single image. The expected value of this estimate is, of course, $\mathscr{E}_d(\nu_X, \nu_Y)$. But how far from this expected value is a single measurement likely to be? To answer this question, it is necessary to find the second moment of $|\mathbf{D}|^2$; that is, we must calculate

$$E\left[|\mathbf{D}|^4\right] = \sum_{n=1}^{K} \sum_{m=1}^{K} \sum_{p=1}^{K} \sum_{q=1}^{K} E\left[\exp\left\{j2\pi\left[\nu_X(x_n - x_m + x_p - x_q)\right.\right.\right.$$
$$\left.\left.\left. + \nu_Y(y_n - y_m + y_p - y_q)\right]\right\}\right]. \tag{9.6-19}$$

This calculation is a lengthy one and is presented in Appendix C. The result is found to be

$$E\left[|\mathbf{D}|^4\right] = \overline{K} + 2\overline{K}^2 + 4(1 + \overline{K})\mathscr{E}_\lambda(\nu_X, \nu_Y)$$
$$+ \mathscr{E}_\lambda(2\nu_X, 2\nu_Y) + 2\mathscr{E}_\lambda^2(\nu_X, \nu_Y). \tag{9.6-20}$$

If we subtract the square of the mean of $|\mathbf{D}|^2$, that is, the square of Eq.

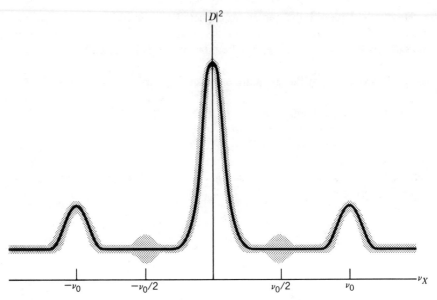

Figure 9-9. Spectral density estimate for a sinusoidal image. The solid line represents the mean, and the shaded area represents the standard deviation of the estimate at each frequency.

(9.6-14), we obtain the variance of $|\mathbf{D}|^2$,

$$\sigma^2_{|\mathbf{D}|^2} = \overline{K} + (\overline{K})^2 + 2(2 + \overline{K})\mathscr{E}_\lambda(\nu_X, \nu_Y) + \mathscr{E}_\lambda(2\nu_X, 2\nu_Y) + \mathscr{E}^2_\lambda(\nu_X, \nu_Y). \tag{9.6-21}$$

Equivalently, using the proportionality between λ and I, we have

$$\sigma^2_{|\mathbf{D}|^2} = \overline{K} + (\overline{K})^2 + 2(2 + \overline{K})(\overline{K})^2 \hat{\mathscr{E}}_i(\nu_X, \nu_Y)$$
$$+ (\overline{K})^2 \hat{\mathscr{E}}_i(2\nu_X, 2\nu_Y) + (\overline{K})^4 \hat{\mathscr{E}}^2_i(\nu_X, \nu_Y). \tag{9.6-22}$$

This equation represents the main result of this section. It is sufficiently interesting to deserve some comment. Note in particular that the fluctuations of the spectral density of the detected image at frequency (ν_X, ν_Y) depend not only on the spectral density of the classical intensity at the same frequency, but *also on the spectral density at frequency* $(2\nu_X, 2\nu_Y)$! Stated in other words, a frequency component of the classical intensity at $(2\nu_X, 2\nu_Y)$ induces fluctuations of the spectral estimate at frequency (ν_X, ν_Y). This "half-frequency" phenomenon is a fundamental property of photon-limited images. It has previously been noted in another context by Walkup (Ref. 9-22). The character of the spectral density estimate of a photon-limited image consisting of a single spatial sinusoid (of finite extent) is illustrated in Fig. 9-9. Note the fluctuations of the spectral estimate at one-half the frequency of the sinusoid actually present.

Having found both the mean and the variance of the spectral estimate, we can now consider the signal-to-noise ratio associated with the measurement.

9.6.4 Signal-to-Noise Ratio for Stellar Speckle Interferometry

On the basis of calculations performed in the preceding two sections, it is now possible to express the rms signal-to-noise ratio associated with a single-frame estimate of the normalized energy spectral density of the image at frequency (ν_X, ν_Y). Following subtraction of the bias $\overline{K}$ associated with the mean of $\mathscr{E}_d$ [cf. Eq. (9.6-16)], the signal-to-noise ratio takes the form

$$\left(\frac{S}{N}\right)_1 = \frac{(\overline{K})^2 \hat{\mathscr{E}}_i(\nu_X, \nu_Y)}{\left\{ \overline{K}^4 \hat{\mathscr{E}}_i^2(\nu_X, \nu_Y) + 2\overline{K}^3 \hat{\mathscr{E}}_i(\nu_X, \nu_Y) + \overline{K}^2 [1 + 4\hat{\mathscr{E}}_i(\nu_X, \nu_Y) + \hat{\mathscr{E}}_i(2\nu_X, 2\nu_Y)] + \overline{K} \right\}^{1/2}}$$

(9.6-23)

A more useful form of the result is obtained if we remember that it is really the spectral density of the *object* that we ultimately seek, not that of the image. It is now necessary to incorporate the relationship between these two spectral densities, taking into account the effects of atmospheric turbulence. Such calculations have been carried out in Section 8.8. Best suited for our use is the result of the heuristic analysis in Section 8.8.2. Noting that the normalized spectral densities of the image and object are related by [cf. Eq. (8.8-4)]

$$\hat{\mathscr{E}}_i(\nu_X, \nu_Y) = \overline{|\mathscr{H}(\nu_X, \nu_Y)|^2} \hat{\mathscr{E}}_o(\nu_X, \nu_Y), \tag{9.6-24}$$

we can use the result (8.8-13) to express the average squared short-exposure OTF,

$$\overline{|\mathscr{H}(\nu_X, \nu_Y)|^2} = \left(\frac{r_0}{D_0}\right)^2 \mathscr{H}_0(\nu_X, \nu_Y). \tag{9.6-25}$$

Here, as before, r_0 is the atmospheric coherence diameter, D_0 is the diameter of the telescope collecting aperture, and $\mathcal{H}_0$ is the diffraction-limited OTF of the telescope in the absence of the atmosphere. With the substitution

$$\hat{\mathscr{E}}_i(\nu_X, \nu_Y) = \left(\frac{r_0}{D_0}\right)^2 \mathcal{H}_0(\nu_X, \nu_Y) \hat{\mathscr{E}}_o(\nu_X, \nu_Y) \qquad (9.6\text{-}26)$$

and the definition

$$\bar{k} = \left(\frac{r_0}{D_0}\right)^2 \bar{K} \qquad (9.6\text{-}27)$$

representing the average number of photoevents *per speckle*, the signal-to-noise ratio for a single frame takes the form

$$\left(\frac{S}{N}\right)_1 =$$

$$\frac{\bar{k}\mathcal{H}_0(\nu_X,\nu_Y)\hat{\mathscr{E}}_o(\nu_X,\nu_Y)}{\left\{[1+\bar{k}\mathcal{H}_0(\nu_X,\nu_Y)\hat{\mathscr{E}}_o(\nu_X,\nu_Y)]^2 + \frac{1}{\bar{K}}[1+4\bar{k}\mathcal{H}_0(\nu_X,\nu_Y)\hat{\mathscr{E}}_o(\nu_X,\nu_Y)+\bar{k}\mathcal{H}_0(2\nu_X,2\nu_Y)\hat{\mathscr{E}}_o(2\nu_X,2\nu_Y)]\right\}^{1/2}}.$$

$$(9.6\text{-}28)$$

In practice, the measured data are obtained not from one frame of imagery, but from a large number of frames taken in time sequence. With the assumption that the realizations of the state of the atmosphere are independent from frame to frame, the signal-to-noise ratio associated with an average over N frames of imagery is

$$\left(\frac{S}{N}\right)_N = \sqrt{N}\left(\frac{S}{N}\right)_1. \qquad (9.6\text{-}29)$$

We now have in hand expressions for the signal-to-noise ratio that should suffice in any assessment of the limitations of stellar speckle interferometry. These results require some further discussion, however, which we take up in Section 9.6.5.

9.6.5 Discussion of the Results

The expression for the single-frame signal-to-noise ratio [Eq. (9.6-28)] reveals some interesting and important properties of the stellar speckle interferometry technique. Most important is the fact that, as the number of photoevents per speckle $\bar{k}$ grows large without bound, the signal-to-noise ratio approaches unity. Thus it is not possible to achieve a signal-to-noise ratio greater than unity by use of a single frame for the estimate of the image spectral density. This property is shared by all spectral estimates that rely on a Fourier transform of a single sample function of a random process (see, e.g., the discussion of "periodograms" in Ref. 9-12, Section 6-6). The only way to increase the signal-to-noise ratio is to average the single-frame estimates over many frames, yielding the behavior described in Eq. (9.6-29).

There is a striking resemblance between the intensity interferometer and the stellar speckle interferometer in this regard. The signal-to-noise ratio associated with any single count product in the intensity interferometer was found to be bounded from above by unity. Only by averaging over many independent count products could the performance be improved. The analogy does not stop here. In the case of the intensity interferometer, the critical parameter determining performance is the count degeneracy parameter, or the average number of photoevents produced in a single coherence interval of the incident light. In the case of the stellar speckle interferometer, a similar role is played by the parameter $\bar{k}$, the average number of photoevents occurring in a single spatial coherence cell of the atmosphere.

The expression (9.6-28) for the single-frame signal-to-noise ratio is complicated by the fact it depends on the spectral content of the object at both (ν_X, ν_Y) and $(2\nu_X, 2\nu_Y)$. This complication can be removed if we restrict attention to frequencies that lie between one-half the diffraction-limited cutoff frequency and the cutoff frequency of the telescope, for then we know that the double frequency term can make no contribution to the noise at the frequency of interest. In this case the single-frame signal-to-noise can be written

$$\left(\frac{S}{N}\right)_1 = \frac{\bar{k}\mathscr{H}_0\hat{\mathscr{E}}_o}{\left\{(1+\bar{k}\mathscr{H}_0\hat{\mathscr{E}}_o)^2 + \frac{1}{\bar{K}}(1+4\bar{k}\mathscr{H}_0\hat{\mathscr{E}}_o)\right\}^{1/2}}. \quad (9.6\text{-}30)$$

There are three limiting regions of interest, each with a different dependence on $\bar{k}$:

(1) For $\bar{k}\mathscr{H}_0\hat{\mathscr{E}}_o \gg 1$, $(S/N)_1 \approx 1$ (independent of $\bar{k}$),
(2) For $\bar{k}\mathscr{H}_0\hat{\mathscr{E}}_o \ll 1$ but $\bar{K} \gg 1$, $(S/N)_1 \approx \bar{k}\mathscr{H}_0\hat{\mathscr{E}}_o$,
(3) For $\bar{k}\mathscr{H}_0\hat{\mathscr{E}}_o \ll 1$ but $\bar{K} \ll 1$, $(S/N)_1 \approx \bar{k}^{3/2}(D_0/r_0)\mathscr{H}_0\hat{\mathscr{E}}_o$.

Figure 9-10 shows a typical dependence of the single-frame signal-to-noise ratio on $\bar{k}$. The three different regions of performance are indicated. Note the asymptotic approach to a signal-to-noise ratio of unity at large $\bar{k}$ (on the right). In this region $(S/N)_1$ is relatively independent of the brightness of the object. In the middle range, $(S/N)_1$ increases in proportion to the average number of photoevents per coherence area. Only in the third range, when the total number of photoevents *per picture* is much less than unity, does increase of the telescope aperture improve the signal-to-noise ratio. This latter range is generally not important in practice, since the resulting signal-to-noise ratio is so small.

Dainty and Greenaway (Ref. 9-21) have shown that if the actual number of photoevents in a given frame is known, then in (9.6-18) that number should be subtracted, rather than $\bar{K}$, the expected number. Then the performance described in the middle region extends through the third region on the left. Thus the signal-to-noise ratio remains proportional to $\bar{k}$

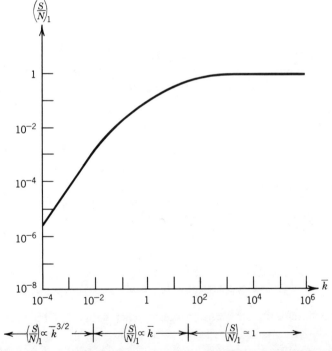

Figure 9-10. Single-image rms signal-to-noise ratio for speckle interferometry, as a function of $\bar{k}$, the mean number of photoevents per speckle. Assumptions: spatial frequency 0.8 of the cutoff, telescope mirror diameter 1.5 meters, atmospheric coherence diameter 10 centimeters normalized object energy spectral density 1.0 (appropriate for a point-source object).

in all regions for which $\bar{k} \ll 1$. They have also examined so-called q-clipped estimates, in which all frames having fewer than q photoevents are discarded and do not take part in the averaging process. The results in this case are more complicated, but it is shown that by discarding frames containing 0 or 1 photoevents, best performance is achieved, and that performance again yields a single-frame signal-to-noise ratio that increases in proportion to $\bar{k}$ when $\bar{k} \ll 1$, as in the middle region of Fig. 9-10.

REFERENCES

9-1   A. Einstein, *Ann. Physik*, **17**, 132 (1905).
9-2   L. Mandel, E. C. G. Sudarshan, and E. Wolf, *Proc. Phys. Soc.* (London), **84**, 435 (1964).
9-3   W. E. Lamb, Jr. and M. O. Scully, in *Polarization: Matiere et Rayonnement*, Press Universitaires de France, Paris, p. 363 (1969).
9-4   L. Mandel, "Fluctuations of Light Beams," in *Progress in Optics*, Vol. II (E. Wolf, editor), North Holland Publishing Company, Amsterdam, p. 181 (1963).
9-5   J. R. Klauder and E. C. G. Sudarshan, *Fundamentals of Quantum Optics*, W. A. Benjamin, New York (1968).
9-6   B. Saleh, *Photoelectron Statistics*, Springer-Verlag, Berlin (1978).
9-7   L. Mandel, "The Case for and against Semiclassical Radiation Theory," in *Progress in Optics*, Vol. XIII (E. Wolf, editor), North Holland Publishing Company, Amsterdam, pp. 27–68 (1976).
9-8   R. J. Glauber, *Phys. Rev.*, **130**, 2529; **131**, 2766 (1963).
9-9   P. J. Vernier, "Photoemission," in *Progress in Optics*, Vol. XIV (E. Wolf, editor), North Holland Publishing Company, Amsterdam, pp. 245–325 (1976).
9-10 L. Mandel, *Proc. Phys. Soc.* (*London*), **74**, 233 (1959).
9-11 C. L. Mehta, "Theory of Photoelectric Counting," in *Progress in Optics*, Vol. VIII (E. Wolf, editor), North Holland Publishing Company, Amsterdam, pp. 373–440 (1970).
9-12 W. B. Davenport, Jr. and W. L. Root, *Random Signals and Noise*, McGraw-Hill Book Company, New York, (1958).
9-13 M. Abramowitz and I. A. Stegan, *Handbook of Mathematical Functions*, Dover Publications, New York, p. 257 (1972).
9-14 Edwin Goldin, *Waves and Photons. An Introduction to Quantum Optics*, John Wiley & Sons, New York, Section 5.3 (1982).
9-15 R. W. Boyd, *Radiometry and the Detection of Optical Radiation*, John Wiley & Sons, New York, Chapter 3 (1983).
9-16 A. E. Siegman, *An Introduction to Lasers and Masers*, McGraw-Hill Book Company, New York, Chapter 11 (1971).
9-17 E. O. Brigham, *The Fast Fourier Transform*, Prentice-Hall, Englewood Cliffs, NJ (1974).
9-18 J. F. Walkup and J. W. Goodman, *J. Opt. Soc. Am.* **63**, 399–407 (1973).
9-19 F. Roddier, "Signal-to-Noise Ratio in Speckle Interferometry," in *Imaging in Astronomy*, AAS/SAO/OSA/SPIE Topical Meeting, Paper ThC6, Boston (1975).
9-20 J. W. Goodman and J. F. Belsher, "Photon Limited Images and Their Restoration," Technical Report RADC-TR-76-50 (March 1976); "Precompensation and Postcom-

pensation of Photon Limited Degraded Images," Technical Report RADC-TR-76-382 (December 1976); "Photon Limitations in Imaging and Image Restoration, Technical Report RADC-TR-77-175 (May 1977) (all available from Rome Air Development Center, Griffiss AFB, NY 13441).

9-21   J. C. Dainty and A. H. Greenaway, *J. Opt. Soc. Am.*, **69**, 786–790 (1979).

9-22   John F. Walkup, *Limitations in Interferometric Measurements and Image Restoration at Low Light Levels*, Ph.D. Dissertation, Department of Electrical Engineering, Stanford University (July 1971).

ADDITIONAL READING

Donald L. Snyder, *Random Point Processes*, John Wiley & Sons, New York (1975).

B. E. A. Saleh and M. C. Teich, "Multiplied Poisson Noise in Pulse, Particle, and Photon Detection," *Proc. IEEE*, **69**, 229–245 (1981).

R. M. Gagliardi and S. Karp, *Optical Communications*, John Wiley & Sons, New York (1976).

PROBLEMS

9-1   Consider the characteristic function of the number of photoevents occurring in a τ-second interval when light falls on a photosurface. Express this characteristic function in terms of the characteristic function of the integrated intensity of the incident light. Note that to do so requires extension of the characteristic function of W to a function of a complex variable.

9-2   Using characteristic functions, show that the gamma density in Eq. (9.2-21) asymptotically approaches a Gaussian density as the parameter $\mathcal{M}$ grows arbitrarily large.

9-3   Show that the negative binomial distribution in Eq. (9.2-24) reduces to the Bose–Einstein distribution when the number of degrees of freedom is unity.

9-4   Demonstrate that, when the photosensitive area of a detector is much larger than the coherence area of an incident cross-spectrally pure wave, the number of spatial degrees of freedom reduces to the ratio of the detector area to the coherence area of the incident wave.

9-5   A partially polarized thermal light wave is incident on a photosurface. The total incident integrated intensity of the wave can be regarded to consist of two statistically independent components, W_1 (mean $\overline{W}_1$) and W_2 (mean $\overline{W}_2$). Thus the probability density of W

PROBLEMS 525

can be expressed as the convolution of the probability densities of W_1 and W_2. Under such circumstances show that the probability distribution $P(K)$ of the total number of photoevents observed can be expressed as a discrete convolution of the distribution functions $P_1(K)$ and $P_2(K)$ for the numbers of photoevents that would be observed if W_1 and W_2 were incident individually.

9-6 Equation (9.2-29) presents a general expression for the probability distribution of the photocounts when the light is thermal in origin and partially polarized. Show that when the degree of polarization is 0, this expression reduces to a negative binomial distribution with $2\mathcal{M}$ degress of freedom. Repeat for the case of a perfectly polarized wave, showing that the distribution becomes a negative binomial distribution with $\mathcal{M}$ degrees of freedom.

9-7 Given the assumption that the energy levels of an harmonic oscillator can take on only the values $nh\bar{\nu}$, and given the Maxwell–Boltzmann distribution of occupation numbers (Eq. 9.3-16), show that the probability distribution associated with occupation number is Bose–Einstein, and derive the mean occupation number.

9-8 Show that for the detector array depicted in Fig. 9-4, and for the assumptions adopted in connection with the problem treated there (including the assumption of thermal light), the correlation between the counts registered on the jth and kth detector elements is given by

$$E[K(j)K(k)] = \begin{cases} \overline{K}(j)\overline{K}(k)\left(1 + \dfrac{1}{\mathcal{M}}\right) & k \neq j \\ \overline{K}(j) + [\overline{K}(j)]^2\left(1 + \dfrac{1}{\mathcal{M}}\right) & k = j. \end{cases}$$

9-9 Consider the noise performance of the system of Fig. 9-4 when the light impinging on the detector array is pseudothermal and has a *large* degeneracy parameter (i.e., $\delta_c \gg 1$). Make the following assumptions about the relative values of the various parameters:

$$\overline{K}_1 = \overline{K}_2, \qquad \delta_c \mathcal{V}^2 \gg 1.$$

Find an expression for the rms signal-to-noise ratio associated with the measurement of fringe visibility, and compare the result with that obtained in Eq. (9.4-23).

9-10 A single-mode, amplitude-stabilized laser is intensity modulated to produce an intensity as a function of time given by

$$I(t) = \frac{I_0}{2}[1 + \cos(2\pi\nu_m t + \theta)],$$

Where I_0 and ν_m are known constants, but θ is a random variable uniformly distributed on 0 to 2π.

(a) Find the mean number $\overline{K}$ of photoevents detected by a photodetector in a time interval of length τ when such light is incident.

(b) Find the variance σ_K^2 of the number of photocounts observed in τ seconds.

9-11 The field emitted by a laser oscillating in N equal strength, independent modes falls on a photodetector. The photoevents are counted over an interval that is sufficiently short that the incident light stays constant (but random) over this interval.

(a) Find the mean $\overline{K}$ and the variance σ_K^2 of the number of photocounts observed, expressing the latter result as a function of N and $\overline{K}$.

(b) Express the ratio of the classically induced component of count variance to the shot-noise-induced component of count variance, stating the answer in terms of N and $\overline{K}$.

9-12 A certain fluorescence process produces extremely short pulses of light, each carrying a known classical energy W_0. The number of pulses incident on a photoelectric detector per second is Poisson distributed, with the mean λ pulses per second known. We count the number K of photoelectrons released from the photosurface in a measurement time τ.

(a) Express the variance σ_K^2 of the number of photoevents K in terms of the mean number of counts $\overline{K}$ in time τ and the mean number of counts per pulse $\overline{N}$.

(b) On the basis of the result of part (a), under what condition do you expect classically induced fluctuations to exceed the shot-noise fluctuations due to the photoemission process itself?

9-13 A certain photodetector generates a finite pulse of constant known area for every photoevent that is detected. To a first approximation,

PROBLEMS

the pulses generated may be assumed to be rectangular in form, to have a duration τ, and to have peak voltage V_0.

(a) Suppose that single-mode, amplitude-stabilized laser light (intensity I_0) is incident on the photosensitive surface (area = A) of the detector. Find the probability distribution of the voltage V observed at the detector output at any arbitrarily selected time t_0.

(b) Repeat part (a) for the case of polarized thermal light, with no restrictions on the coherence time of the light.

Appendix A

The Fourier Transform

The Fourier transform is perhaps the most important analytical tool needed for work in statistical optics, or for that matter in the field of modern optics in general. For this reason we present here a brief summary of the most important Fourier transform theorems and Fourier transform pairs needed in practice. No attempt at derivation of these properties or relationships is made here. Rather, for such information the reader is encouraged to consult any of a number of excellent books on the subject (see, e.g., Refs. A-1 through A-4).

A.1 FOURIER TRANSFORM DEFINITIONS

In this book we have chosen to use definitions of the forward Fourier transform that have a *positive* exponential kernel. Thus our definitions of the one-dimensional and two-dimensional Fourier transforms of (generally complex valued) functions $\mathbf{f}(x)$ and $\mathbf{f}(x, y)$ are:

$$\mathbf{F}(\nu) = \int_{-\infty}^{\infty} \mathbf{f}(x) e^{j2\pi\nu x} \, dx \tag{A-1}$$

and

$$\mathbf{F}(\nu_X, \nu_Y) = \iint_{-\infty}^{\infty} \mathbf{f}(x, y) e^{j2\pi(\nu_X x + \nu_Y y)} \, dx \, dy. \tag{A-2}$$

Along with these definitions are corresponding definitions of the one-dimensional and two-dimensional *inverse* Fourier transforms,

$$\mathbf{f}(x) = \int_{-\infty}^{\infty} \mathbf{F}(\nu) e^{-j2\pi\nu x} \, d\nu \tag{A-3}$$

and

$$\mathbf{f}(x, y) = \iint_{-\infty}^{\infty} \mathbf{F}(\nu_X \nu_Y) e^{-j2\pi(\nu_X x + \nu_Y y)} \, d\nu_X \, d\nu_Y. \tag{A-4}$$

BASIC PROPERTIES OF THE FOURIER TRANSFORM 529

The reader may be more accustomed to using one of several other possible definitions of these transforms. For example, often a negative sign is used in the exponential kernel of the forward transform (and a positive sign in the kernel of the inverse). The question then naturally arises as to how a Fourier transform defined with the positive kernel [call it $F^+(\nu)$] can be related to the transform defined with the negative kernel [$F^-(\nu)$]. Just a single line of algebra shows that the desired relationship is

$$F^+(\nu) = F^-(-\nu) \tag{A-5}$$

in the one-dimensional case and

$$\mathbf{F}^+(\nu_X, \nu_Y) = \mathbf{F}^-(-\nu_X, -\nu_Y) \tag{A-6}$$

in the two-dimensional case. Thus tables of Fourier transforms valid for one definition can easily be converted into corresponding tables for the other.

A.2 BASIC PROPERTIES OF THE FOURIER TRANSFORM

A number of relationships that are useful in manipulating Fourier transforms are presented now without proof, for both the one-dimensional and two-dimensional cases. Throughout this appendix, **g** and **h** represent functions (generally complex valued) of one or two variables, and **G** and **H** represent their Fourier transforms, defined in accord with Eqs. (A-1) or (A-2). In all cases, a script $\mathscr{F}\{\ \}$ represents a Fourier transform operator in either one or two dimensions. The dimensionality should be clear from the context. When only one form of the relationship is shown, it holds for both the one-dimensional and the two-dimensional cases.

Linearity. If **a** and **b** represent arbitrary complex constants, then, for both the one- and two-dimensional cases,

$$\mathscr{F}\{\mathbf{ag} + \mathbf{bh}\} = \mathbf{aG} + \mathbf{bH} \tag{A-7}$$

Similarity. If a and b are real-valued constants, then

$$\mathscr{F}\{\mathbf{g}(ax)\} = \frac{1}{|a|}\mathbf{G}\left(\frac{\nu}{a}\right) \tag{A-8}$$

and

$$\mathscr{F}\{g(ax,by)\} = \frac{1}{|ab|}G\left(\frac{\nu_X}{a},\frac{\nu_Y}{b}\right). \tag{A-9}$$

Shift. If a and b are real-valued constants, then

$$\mathscr{F}\{g(x-a)\} = e^{j2\pi\nu a}G(\nu) \tag{A-10}$$

and

$$\mathscr{F}\{g(x-a, y-b\} = e^{j2\pi(\nu_X a + \nu_Y b)}G(\nu_X,\nu_Y). \tag{A-11}$$

Parseval's Theorem. In one dimension

$$\int_{-\infty}^{\infty} |g(x)|^2 dx = \int_{-\infty}^{\infty} |G(\nu)|^2 d\nu, \tag{A-12}$$

and in two dimensions

$$\iint_{-\infty}^{\infty} |g(x,y)|^2 dx\, dy = \iint_{-\infty}^{\infty} |G(\nu_X,\nu_Y)|^2 d\nu_X d\nu_Y. \tag{A-13}$$

Convolution Theorem. In one dimension

$$\mathscr{F}\left\{\int_{-\infty}^{\infty} g(\xi)h(x-\xi)\,d\xi\right\} = G(\nu)H(\nu), \tag{A-14}$$

and in two dimensions

$$\mathscr{F}\left\{\iint_{-\infty}^{\infty} g(\xi,\eta)h(x-\xi, y-\eta)\,d\xi\,d\eta\right\} = G(\nu_X,\nu_Y)H(\nu_X,\nu_Y). \tag{A-15}$$

Autocorrelation Theorem. In one dimension

$$\mathscr{F}\left\{\int_{-\infty}^{\infty} h(\xi)h^*(\xi-x)\,d\xi\right\} = |H(\nu)|^2, \tag{A-16}$$

and in two dimensions

$$\mathscr{F}\left\{\iint_{-\infty}^{\infty} h(\xi,\eta)h^*(\xi-x,\eta-y)\,d\xi\,d\eta\right\} = |H(\nu_X,\nu_Y)|^2. \tag{A-17}$$

ONE-DIMENSIONAL FOURIER TRANSFORMS

Fourier Integral Theorem. At each point of continuity of g, a Fourier transform followed by an inverse Fourier transform yields the value of the original function g again. At a point of discontinuity of g, the succession of a transform and an inverse transform yields: (1) in one dimension, the arithmetic mean of the values of the function on both sides of the discontinuity, and (2) in two dimensions, the angular average of the function about the point of discontinuity.

A.3 TABLE OF ONE-DIMENSIONAL FOURIER TRANSFORMS

Table A-1 presents a collection of one-dimensional Fourier transform pairs that is fully adequate as an aid for the material presented in this book

Table A-1 One-Dimensional Fourier Transform pairs

Function	Transform
$e^{-\pi x^2}$	$e^{-\pi \nu^2}$
1	$\delta(\nu)$
$\delta(x)$	1
$\cos \pi x$	$\frac{1}{2}\delta(\nu - \frac{1}{2}) + \frac{1}{2}\delta(\nu + \frac{1}{2})$
$\sin \pi x$	$\frac{j}{2}\delta(\nu - \frac{1}{2}) - \frac{j}{2}\delta(\nu + \frac{1}{2})$
rect x	sinc ν
$\Lambda(x)$	$\text{sinc}^2 \nu$
$e^{-|x|}$	$\dfrac{2}{1+(2\pi\nu)^2}$
$\dfrac{1}{1+(2\pi x)^2}$	$\frac{1}{2}e^{-|\nu|}$
$J_0(2\pi x)$	$\dfrac{\text{rect}(\nu/2)}{\pi(1-\nu^2)^{1/2}}$
$\dfrac{J_1(2\pi x)}{2x}$	$(1-\nu^2)^{1/2}\text{rect}(\nu/2)$
$-\dfrac{j}{\pi x}$	sgn ν
$e^{j\pi x^2}$	$e^{j(\pi/4)}e^{-j\pi\nu^2}$
$\dfrac{m^m x^{m-1}\exp(-mx)}{\Gamma(m)}$	$\left(1 - j\dfrac{2\pi\nu}{m}\right)^{-m}$

Table A-2 Two-Dimensional Fourier Transform Pairs

Function	Transform
rect x rect y	$\text{sinc}\, \nu_X \,\text{sinc}\, \nu_Y$
$\delta(x, y)$	1
$e^{-\pi(x^2+y^2)} = e^{-\pi r^2}$	$e^{-\pi(\nu_X^2+\nu_Y^2)} = e^{-\pi \rho^2}$
circ r	$\dfrac{J_1(2\pi\rho)}{\rho}$
$\delta(r-a)$	$2\pi a J_0(2\pi a \rho)$
$\dfrac{1}{r}$	$\dfrac{1}{\rho}$
$\Lambda(x)\Lambda(y)$	$\text{sinc}^2 \nu_X \,\text{sinc}^2 \nu_Y$
$\left[\dfrac{J_1(2\pi r)}{r}\right]^2$	$2\left[\cos^{-1}\dfrac{\rho}{2} - \dfrac{\rho}{2}\sqrt{1-\dfrac{\rho^2}{4}}\right]\text{rect}\dfrac{\rho-1}{2}$
$e^{j\pi(x^2+y^2)} = e^{j\pi r^2}$	$je^{-j\pi(\nu_X^2+\nu_Y^2)} = je^{-j\pi\rho^2}$

A.4 TABLE OF TWO-DIMENSIONAL FOURIER TRANSFORM PAIRS

Table A-2 presents a collection of two-dimensional Fourier transform pairs that may prove useful in the study of this book. The symbol r represents radius in the (x, y) plane, whereas the symbol ρ represents radius in the two-dimensional spatial frequency plane (ν_X, ν_Y).

REFERENCES

A-1 R. N. Bracewell, *The Fourier Transform and its Applications*, 2nd ed., McGraw-Hill Book Company, New York (1978).

A-2 A. Papoulis, *The Fourier Integral and its Applications*, McGraw-Hill Book Company, New York (1962).

A-3 A. Papoulis, *Systems and Transforms with Applications to Optics*, McGraw-Hill Book Company, New York (1968).

A-4 J. D. Gaskill, *Linear Systems, Fourier Transforms, and Optics*, John Wiley and Sons, New York (1978).

Appendix B

Random Phasor Sums

Because of the great importance of the random walk problem in statistical optics, we present in this appendix a generalization of the theory discussed in Chapter 2, Section 2.9. There the assumption was made that the phases of the individual phasors contributing to the sum are independent and uniformly distributed on the interval $(-\pi, \pi)$. Here we derive results valid when the phases take on an arbitrary probability density function $p_\phi(\phi)$ while remaining identically distributed and independent. The characteristic function corresponding to the probability density function of the phase is represented by $\mathbf{M}_\phi(\omega)$.

As in Section 2.9, we consider the sum

$$\mathbf{a} = ae^{j\theta} = \frac{1}{\sqrt{N}} \sum_{k=1}^{N} \alpha_k e^{j\phi_k}, \tag{B-1}$$

where N represents the number of independent phasors contributing to the random walk and α_k represents the length of the kth phasor. The α_k are assumed to be independent of each other and of the phases and identically distributed. The real and imaginary parts of this sum are

$$r = \operatorname{Re}\{ae^{j\theta}\} = \frac{1}{\sqrt{N}} \sum_{k=1}^{N} \alpha_k \cos \phi_k$$

$$i = \operatorname{Im}\{ae^{j\theta}\} = \frac{1}{\sqrt{N}} \sum_{k=1}^{N} \alpha_k \sin \phi_k. \tag{B-2}$$

Clearly, the means of the real and imaginary parts of the resultant can be expressed as

$$\bar{r} = \frac{1}{\sqrt{N}} \sum_{k=1}^{N} \overline{\alpha_k \cos \phi_k}$$

$$\bar{i} = \frac{1}{\sqrt{N}} \sum_{k=1}^{N} \overline{\alpha_k \sin \phi_k}. \tag{B-3}$$

Further simplification can be made by noting that the means of the cosine and sine functions can be related to the characteristic function of the random variables ϕ_k. Expanding the cosine and sine functions using Euler's formulas, we can express the means as

$$\bar{r} = \frac{\sqrt{N}\,\bar{\alpha}}{2}\left[\mathbf{M}_\phi(1) + \mathbf{M}_\phi(-1)\right]$$

$$\bar{i} = \frac{\sqrt{N}\,\bar{\alpha}}{2j}\left[\mathbf{M}_\phi(1) - \mathbf{M}_\phi(-1)\right]. \tag{B-4}$$

These results are the most general ones consistent with our previous assumptions but can be simplified if further constraints are applied. For example, if the probability density function of the random phases is assumed to be *symmetrical* about zero, the characteristic function is entirely real and even in the variable ω (Ref. B-1, p. 14). It follows that the means of the real and imaginary parts of the random walk become

$$\bar{r} = \sqrt{N}\,\bar{\alpha}_k \mathbf{M}_\phi(1)$$

$$\bar{i} = 0. \tag{B-5}$$

Having found the first-order moments of the real and imaginary parts, we turn to second-order moments, with the goal of specifying the variances of the real and imaginary parts and their covariance. General expressions for the second-order moments of r and i are

$$\overline{r^2} = \frac{1}{N}\sum_{k=1}^{N}\sum_{n=1}^{N}\overline{\alpha_k\alpha_n\cos\phi_k\cos\phi_n}$$

$$\overline{i^2} = \frac{1}{N}\sum_{k=1}^{N}\sum_{n=1}^{N}\overline{\alpha_k\alpha_n\sin\phi_k\sin\phi_n}$$

$$\overline{ri} = \frac{1}{N}\sum_{k=1}^{N}\sum_{n=1}^{N}\overline{\alpha_k\alpha_n\cos\phi_k\sin\phi_n}. \tag{B-6}$$

Again using Euler's formulas, we obtain the following general expressions for the moments of the trigonometric functions of concern:

$$\overline{\cos\phi_k \cos\phi_n} = \begin{cases} \frac{1}{4}[2\mathbf{M}_\phi(1)\mathbf{M}_\phi(-1) + \mathbf{M}_\phi^2(1) + \mathbf{M}_\phi^2(-1)] & k \neq n \\ \frac{1}{4}[2 + \mathbf{M}_\phi(2) + \mathbf{M}_\phi(-2)] & k = n \end{cases}$$

$$\overline{\sin\phi_k \sin\phi_n} = \begin{cases} \frac{1}{4}[2\mathbf{M}_\phi(1)\mathbf{M}_\phi(-1) - \mathbf{M}_\phi^2(1) - \mathbf{M}_\phi^2(-1)] & k \neq n \\ \frac{1}{4}[2 - \mathbf{M}_\phi(2) - \mathbf{M}_\phi(-2)] & k = n \end{cases}$$

(B-7)

$$\overline{\cos\phi_k \sin\phi_n} = \begin{cases} \frac{1}{4j}[\mathbf{M}_\phi^2(1) - \mathbf{M}_\phi^2(-1)] & k \neq n \\ \frac{1}{4j}[\mathbf{M}_\phi(2) - \mathbf{M}_\phi(-2)] & k = n. \end{cases}$$

Combining these expressions, we find the second-order moments of interest to be given by

$$\overline{r^2} = \frac{\overline{\alpha^2}}{4}[2 + \mathbf{M}_\phi(2) + \mathbf{M}_\phi(-2)]$$

$$+ \frac{(N-1)(\overline{\alpha})^2}{4}[2\mathbf{M}_\phi(1)\mathbf{M}_\phi(-1) + \mathbf{M}_\phi^2(1) + \mathbf{M}_\phi^2(-1)]$$

$$\overline{i^2} = \frac{\overline{\alpha^2}}{4}[2 - \mathbf{M}_\phi(2) - \mathbf{M}_\phi(-2)]$$

$$+ \frac{(N-1)(\overline{\alpha})^2}{4}[2\mathbf{M}_\phi(1)\mathbf{M}_\phi(-1) - \mathbf{M}_\phi^2(1) - \mathbf{M}_\phi^2(-1)]$$

$$\overline{ri} = \frac{\overline{\alpha^2}}{4j}[\mathbf{M}_\phi(2) - \mathbf{M}_\phi(-2)] + \frac{(N-1)(\overline{\alpha})^2}{4j}[\mathbf{M}_\phi^2(1) - \mathbf{M}_\phi^2(-1)].$$

(B-8)

Subtracting the squares of the means from $\overline{r^2}$ and $\overline{i^2}$ and the product of the means from $\overline{ri}$, we obtain the variances and the covariance,

$$\sigma_r^2 = \frac{\overline{\alpha^2}}{4}\left[2 + \mathbf{M}_\phi(2) + \mathbf{M}_\phi(-2)\right]$$

$$- \frac{(\bar{\alpha})^2}{4}\left[2\mathbf{M}_\phi(1)\mathbf{M}_\phi(-1) + \mathbf{M}_\phi^2(1) + \mathbf{M}_\phi^2(-1)\right]$$

$$\sigma_i^2 = \frac{\overline{\alpha^2}}{4}\left[2 - \mathbf{M}_\phi(2) - \mathbf{M}_\phi(-2)\right]$$

$$- \frac{(\bar{\alpha})^2}{4}\left[2\mathbf{M}_\phi(1)\mathbf{M}_\phi(-1) - \mathbf{M}_\phi^2(1) - \mathbf{M}_\phi^2(-1)\right]$$

$$\mathrm{cov}(r,i) = \frac{\overline{\alpha^2}}{4j}\left[\mathbf{M}_\phi(2) - \mathbf{M}_\phi(-2)\right] - \frac{(\bar{\alpha})^2}{4j}\left[\mathbf{M}_\phi^2(1) - \mathbf{M}_\phi^2(-1)\right].$$

(B-9)

For the special case of a probability density function for the phases that is even about the origin, we obtain the simpler expressions

$$\sigma_r^2 = \frac{\overline{\alpha^2}}{2}\left[1 + \mathbf{M}_\phi(2)\right] - (\bar{\alpha})^2 \mathbf{M}_\phi^2(1)$$

$$\sigma_i^2 = \frac{\overline{\alpha^2}}{2}\left[1 - \mathbf{M}_\phi(2)\right]$$

$$\mathrm{cov}(r,i) = 0.$$

(B-10)

Finally, when the phases are uniformly distributed, we have

$$\mathbf{M}_\phi(1) = \mathbf{M}_\phi(2) = 0,$$

(B-11)

and the variances and covariances reduce to

$$\sigma_r^2 = \sigma_i^2 = \frac{\overline{\alpha^2}}{2}$$

$$\mathrm{cov}(r,i) = 0,$$

(B-12)

which is identical with the results obtained in Section 2.9.

RANDOM PHASOR SUMS

There is one further subtlety regarding the random walk problem that we wish to clarify here. It was argued in Section 2.9 that, when the number of terms in the sums of Eq. (B-2) grows large, the central limit theorem implies that the statistics of the real and imaginary parts of the sum tend asymptotically toward Gaussian. Such an argument is valid regardless of whether the phases associated with the individual contributions have uniform statistics. However, the assumption has been made that the real and imaginary parts are asymptotically *jointly* Gaussian random variables; that is, they satisfy together a second-order Gaussian probability density function [cf. Eq. (2.9-5)]. Whereas the Gaussian character of their *marginal* densities is implied by the central limit theorem, their joint Gaussian character is less obvious.

To prove joint Gaussianity, we make the simplifying assumption that the phases ϕ_k are uniformly distributed, independent random variables. We retain the assumption that the amplitudes α_k are independent of the phases and of each other. The joint characteristic function of the real and imaginary parts r and i is given by

$$\mathbf{M}_{ri}(\omega_1, \omega_2) = E[e^{j(\omega_1 r + \omega_2 i)}]. \quad (B\text{-}13)$$

We define polar coordinate variables in the (ω_1, ω_2) plane, through

$$\omega_1 = \Omega \cos \chi$$

$$\omega_2 = \Omega \sin \chi. \quad (B\text{-}14)$$

With substitution of Eqs. (B-2) and (B-14) in (B-13) and use of the trigonometric identity $\cos A \cos B + \sin A \sin B = \cos(A - B)$, the characteristic function becomes

$$\mathbf{M}_{ri}(\omega_1, \omega_2) = E\left[\exp\left\{\frac{1}{\sqrt{N}} \sum_{k=1}^{N} \alpha_k \Omega \cos(\chi - \phi_k)\right\}\right]. \quad (B\text{-}15)$$

For the moment we refrain from averaging over the α_k and take a conditional average over the ϕ_k, with the result

$$\mathbf{M}_{ri}(\omega_1, \omega_2) = \prod_{k=1}^{N} E_\alpha\left[J_0\left(\frac{\alpha_k \Omega}{\sqrt{N}}\right)\right], \quad (B\text{-}16)$$

where J_0 is a Bessel function of the first kind, order zero. As N grows large, the argument of the Bessel function grows small, allowing that function to be approximated by the first two terms of its power series expansion about the origin,

$$\mathbf{M}_{ri}(\omega_1, \omega_2) = \prod_{k=1}^{N} E_\alpha \left[1 - \left(\frac{\alpha_k \Omega}{\sqrt{N}} \right)^2 \right]. \tag{B-17}$$

At this point we perform the averaging operation over the amplitudes α_k, yielding

$$\mathbf{M}_{ri}(\omega_1, \omega_2) = \left[1 - \frac{\overline{\alpha^2} \Omega^2}{N} \right]^N. \tag{B-18}$$

If the number of terms N in the random walk is allowed to increase without bound, the joint characteristic function of the real and imaginary parts asymptotically approaches a circularly symmetrical Gaussian function,

$$\lim_{N \to \infty} \mathbf{M}_{ri}(\omega_1, \omega_2) = e^{-\overline{\alpha^2} \Omega^2}. \tag{B-19}$$

Finally, an inverse Fourier transform of this characteristic function yields the two-dimensional Gaussian joint probability density function,

$$p_{ri}(r, i) = \frac{1}{4\pi \overline{\alpha^2}} \exp\left\{ -\frac{r^2 + i^2}{4\overline{\alpha^2}} \right\}. \tag{B-20}$$

Hence we have proved that the real and imaginary parts of the random walk are *joint* Gaussian random variables.

Whereas the preceding argument assumed that the phases of the individual components of the random walk are uniformly distributed, a more complex argument allows one to demonstrate that joint Gaussianity holds asymptotically even when the phases are nonuniformly distributed.

REFERENCE

B-1   R. N. Bracewell, *The Fourier Transform and its Applications*, 2nd ed., McGraw-Hill Book Company, New York (1978).

Appendix C

Fourth-Order Moment of the Spectrum of a Detected Speckle Image

In this appendix the fourth-order moment of the Fourier transform of an image used in speckle interferometry is calculated. The goal is to derive the result presented in Eq. (9.6-20).

Our starting point is Eq. (9.6-19), which we repeat here for convenience:

$$E[|\mathbf{D}|^4] = \sum_{n=1}^{K} \sum_{m=1}^{K} \sum_{p=1}^{K} \sum_{q=1}^{K} E\left[\exp\{j2\pi[\nu_X(x_n - x_m + x_p - x_q) + \nu_Y(y_n - y_m + y_p - y_q)]\}\right]. \tag{C-1}$$

The K^4 terms in this summation can be placed in 15 different classes as follows:

(1)	$n = m = p = q$	K terms
(2)	$n = m, p = q, n \neq p$	$K(K-1)$ terms
(3)	$n = m, p \neq q \neq n$	$K(K-1)(K-2)$ terms
(4)	$n = p, m = q, n \neq m$	$K(K-1)$ terms
(5)	$n = p, m \neq q \neq n$	$K(K-1)(K-2)$ terms
(6)	$n = q, m = p, n \neq m$	$K(K-1)$ terms
(7)	$n = q, m \neq p \neq n$	$K(K-1)(K-2)$ terms
(8)	$n = m = p, n \neq q$	$K(K-1)$ terms
(9)	$n = p = q, n \neq p$	$K(K-1)$ terms
(10)	$n = p = q, n \neq m$	$K(K-1)$ terms
(11)	$p = q = m, n \neq m$	$K(K-1)$ terms
(12)	$n \neq m \neq p \neq q$	$K(K-1)(K-2)(K-3)$ terms
(13)	$p = q, n \neq m \neq p$	$K(K-1)(K-2)$ terms
(14)	$m = q, n \neq m \neq p$	$K(K-1)(K-2)$ terms
(15)	$m = p, n \neq m \neq q$	$K(K-1)(K-2)$ terms

For the moment, condition the statistics by a known rate function $\lambda(x, y)$; we shall later average over the statistics of λ. Thus we first average over the $2K + 1$ random variables $(x_1, y_1), (x_2, y_2), \ldots, (x_K, y_K)$, K. Noting that for a Poisson random variable K, the following expectation over K, conditioned on a known λ, is given by

$$E[K(K-1)\cdots(K-k+1)] = [\bar{K}_{(\lambda)}]^k, \quad \text{(C-2)}$$

where $\bar{K}_{(\lambda)}$ represents the conditional mean of K. The contributions of the 15 sets of terms identified above can now be written as follows:

(1) $\bar{K}_{(\lambda)}$
(2) $[\bar{K}_{(\lambda)}]^2$
(3) $[\bar{K}_{(\lambda)}]^3 |\hat{\Lambda}(\nu_X, \nu_Y)|^2$
(4) $[\bar{K}_{(\lambda)}]^2 |\hat{\Lambda}(2\nu_x, 2\nu_y)|^2$
(5) $[\bar{K}_{(\lambda)}]^3 \hat{\Lambda}(2\nu_X, 2\nu_Y)[\hat{\Lambda}^*(\nu_X, \nu_Y)]^2$
(6) $[\bar{K}_{(\lambda)}]^2$
(7) $[\bar{K}_{(\lambda)}]^3 |\hat{\Lambda}(\nu_X, \nu_Y)|^2$
(8) $[\bar{K}_{(\lambda)}]^2 |\hat{\Lambda}(\nu_X, \nu_Y)|^2$
(9) $[\bar{K}_{(\lambda)}]^2 |\hat{\Lambda}(\nu_X, \nu_Y)|^2$
(10) $[\bar{K}_{(\lambda)}]^2 |\hat{\Lambda}(\nu_X, \nu_Y)|^2$
(11) $[\bar{K}_{(\lambda)}]^2 |\hat{\Lambda}(\nu_X, \nu_Y)|^2$
(12) $[\bar{K}_{(\lambda)}]^4 |\hat{\Lambda}(\nu_X, \nu_Y)|^4$
(13) $[\bar{K}_{(\lambda)}]^3 |\hat{\Lambda}(\nu_X, \nu_Y)|^2$
(14) $[\bar{K}_{(\lambda)}]^3 \hat{\Lambda}^*(2\nu_X, 2\nu_Y)[\hat{\Lambda}(\nu_X, \nu_Y)]^2$
(15) $[\bar{K}_{(\lambda)}]^3 |\hat{\Lambda}(\nu_X, \nu_Y)|^2$

Here, the definition

$$\hat{\Lambda}(\nu_X, \nu_Y) = \frac{\iint\limits_{-\infty}^{\infty} \lambda(x, y) e^{j2\pi(\nu_X x + \nu_Y y)}\, dx\, dy}{\iint\limits_{-\infty}^{\infty} \lambda(x, y)\, dx\, dy} \quad \text{(C-3)}$$

has been used. Noting further that

$$\Lambda(\nu_X, \nu_Y) = \bar{K}_{(\lambda)} \hat{\Lambda}(\nu_X, \nu_Y) \quad \text{(C-4)}$$

and combining these results, we obtain

$$E[|\mathbf{D}|^4] = \overline{K}_{(\lambda)} + 2[\overline{K}_{(\lambda)}]^2 + 4[1 + \overline{K}_{(\lambda)}]|\Lambda(\nu_X,\nu_Y)|^2$$

$$+ \Lambda(2\nu_X, 2\nu_Y)[\Lambda^*(\nu_X,\nu_Y)]^2 + \Lambda^*(2\nu_X, 2\nu_Y)[\Lambda(\nu_X,\nu_Y)]^2$$

$$+ |\Lambda(2\nu_X, 2\nu_Y)|^2 + |\Lambda(\nu_X,\nu_Y)|^4. \quad \text{(C-5)}$$

To simplify the result further, express $\Lambda(\nu_X, \nu_Y)$ in terms of its modulus and phase,

$$\Lambda(\nu_X, \nu_Y) = |\Lambda(\nu_X, \nu_Y)| e^{j\theta(\nu_X, \nu_Y)} \quad \text{(C-6)}$$

and write the fourth moment of D as

$$E[|\mathbf{D}|^4] = \overline{K}_{(\lambda)} + 2[\overline{K}_{(\lambda)}]^2 + 4[1 + \overline{K}_{(\lambda)}]|\Lambda(\nu_X,\nu_Y)|^2$$

$$+ 2|\Lambda(2\nu_X, 2\nu_Y)||\Lambda(\nu_X,\nu_Y)|^2 \cos[\theta(2\nu_X, 2\nu_Y) - 2\theta(\nu_X,\nu_Y)]$$

$$+ |\Lambda(2\nu_X, 2\nu_Y)|^2 + |\Lambda(\nu_X,\nu_Y)|^4. \quad \text{(C-7)}$$

The task now remains to average over the statistics of $\lambda(x, y)$. If the image intensity distribution extends over a finite region of size $L \times L$, then under rather general conditions, for $\nu_X \gg 1/L$ and $\nu_Y \gg 1/L$, $\Lambda(\nu_X, \nu_Y)$ is approximately a circular complex Gaussian random process, with correlation extending over a region of dimensions approximately $2/L \times 2/L$ in the frequency domain. It follows that the phase θ is uniformly distributed on $(-\pi, \pi)$ and that $|\Lambda|^2$ obeys negative exponential statistics. Furthermore, for such frequencies $\theta(2\nu_X, 2\nu_Y)$, $\theta(\nu_X, \nu_Y)$, $|\Lambda(2\nu_X, 2\nu_Y)|$, and $|\Lambda(\nu_X, \nu_Y)|$ are all approximately independent. Using these facts and averaging over λ, we find

$$E[\overline{K}_{(\lambda)}] = \overline{K}$$

$$E[2|\Lambda(2\nu_X, 2\nu_Y)||\Lambda(\nu_X,\nu_Y)|^2 \cos[\theta(2\nu_X, 2\nu_Y) - 2\theta(\nu_X,\nu_Y)]] = 0$$

$$E[|\Lambda(\nu_X, \nu_Y)|^2] = \mathscr{E}_\lambda(\nu_X, \nu_Y)$$

$$E[|\Lambda(\nu_X, \nu_Y)|^4] = 2\mathscr{E}_\lambda(\nu_X, \nu_Y). \quad \text{(C-8)}$$

Substituting these relations in Eq. (C-7), we obtain the final result,

$$E[|\mathbf{D}|^4] = \bar{K} + 2\bar{K}^2 + 4(1 + \bar{K})\mathscr{E}_\lambda(\nu_X, \nu_Y)$$

$$+ \mathscr{E}_\lambda(2\nu_X, 2\nu_Y) + 2\mathscr{E}_\lambda^2(\nu_X, \nu_Y), \qquad (\text{C-9})$$

which agrees with Eq. (9.6-20). Thus the proof is complete.

Index

Aberrations, 336
Abramowitz, M., 523
Adaptive optics, 402, 456, 457
Airy pattern, 328
Alford, W. P., 191
Allen, L., 152
Amplitude grating, 328
Amplitude interferometry, 272, 490-501
Amplitude splitting, 157
Amplitude transfer function, 322
Amplitude transmittance, 222, 289
Analytic signal, 101, 104-108, 344
Angular spectrum, 324
Aperture synthesis, 336
Apparent transfer function, 329
Armstrong, J. A., 152
Astronomical telescope, 443
Asymmetry factor, 264
Atmosphere:
  coherence diameter of, 429-432
  effects on fringe measurement, 340
  inhomogeneities of, 361, 362
  model for, 388-393
  propagation through, 4
Autocorrelation function, 73, 321
Autocorrelation theorem, 530
Autocovariance function, 78
Average:
  statistical, 15
  time, 66

Baier, G., 403, 446
Barakat, R., 460
Bayes' rule, 14
Becherer, R. J., 330, 356

Beckmann, P., 56, 357
Bell, R. J., 229
Belsher, J. F., 523
Beran, M. J., 228, 356, 460
Berek, M., 157, 228
Bias, 265, 266
Binary stars, 445
Binomial distribution, 90
Blackbody radiation, 486
Blanc-Lapierre, A., 157, 228
Boileau, E, 281
Booker, H. G., 459
Born, M., 1, 152, 228, 356
Born approximation, 396, 397, 399, 464
Bose-Einstein distribution, 473, 474
Bourret, R. C., 460
Boyd, R. W., 523
Boxcar approximation, 244, 245
Bracewell, R. N., 56, 229, 356, 459, 532, 538
Brames, B. J., 358
Brigham, E. O., 523
Broadband light, 231
Brown, R. Hanbury, 271, 282, 357
Brown, T. M., 455, 460
Brown, W. P., Jr., 460
Bruck, Yu M., 357
Bunching, 485
Burckhardt, C. B., 356
Bures, J., 281

Carter, W. H., 229
Central limit theorem, 31-33
Characteristic function:
  definition, 19
  for Gaussian random variable, 20

Characteristic function (*Continued*)
  joint, 20
  for Poisson random variable, 20
Checkerboard random phase screen, 371-374
Chernoff, L. A., 459
Circular source, 211-218
Circular statistics, 42-44, 108, 109, 122
Clifford, S. F., 459
Coherence:
  area, 210, 221, 235, 308, 323
  complex degree of, 162, 164, 175
  high order, 237, 238
  length, 296, 299
  mutual, 174
  self, 161
  spatial, 170-187
  temporal, 158-170
  time, 158, 167
Coherence relationships:
  exit pupil to image plane, 300-303
  focal plane to focal plane, 292-296
  object to image, 296-300
Coherency matrix, 130-134
Coherent field, 202-205
Coherent imaging, 321, 322
Coherent limit, 321
Coherent transfer function, 322
Collier, R. J., 356
Complex coherence factor, 181, 183
Complex degree of coherence, 162, 164, 175, 183
Complex degree of spectral coherence, 202
Complex envelope, 103, 104
Complex Gaussian moment theorem, 260, 274, 354
Complex Gaussian random process, 108, 109, 350
Complex valued random variables, 40-44
Condie, M. A., 281
Convolution, 312, 530
Cooley, J. W., 229
Correlation, 17
Count degeneracy parameter, 483
Count fluctuation product, 503
Count vector, 491
Covariance, 17
Crimmins, T. R., 358
Critical illumination, 306, 307

Cross-correlation function, 79
Cross-spectral density:
  definition of, 80
  propagation of, 201
Cross-spectral purity, 187-195, 233, 260, 275
Currie, D. G., 357

Daino, B., 152
Dainty, J. C., 348, 357, 358, 460, 522, 524
Davenport, W. B., Jr., 56, 523
Degeneracy parameter, 481-490
  for blackbody radiation, 486-490
Degree of polarization, 134-136
Degrees of freedom, 475, 476
Delisle, C., 281
Density function, 11, 12
Depolarization, 394
Detector, 4
de Wolf, D. A., 460
Dialetis, D., 357
Diameter, stellar, 340
Diffraction, 222-228
Diffuser, 151, 233, 349
  moving, 193-195
Diffusion, 142
Dillon, H. S., 229
Discrete Fourier transform, 494
Distribution function:
  definition, 9, 10
  joint, 13
Double star, 446
Duffieux, P. M., 321, 356
Dugundji, J., 111
Dumontet, P., 157, 228
Dutta, K., 356

Ehn, D. C., 460
Eigenvalues, 135
Einstein, A., 523
Electromagnetic propagation, 393-399
Energy spectral density, 68
Ensemble average coherence, 351-356
Entrance pupil, 300, 301, 323
Erdelyi, A., 459
Ergodicity, 64-68
Evanescent wave, 206
Excess fluctuations, 471, 483
Exit pupil, 300, 301

Factorial moment, 469
Fante, R. L., 459
Far field, 225
Fast Fourier transform, 170
Feller, W., 56
Fellgett, P., 170, 229
Fiddy, M. A., 358
Fienup, J. R., 346, 347, 357, 358
Filter function:
  in atmospheric propagation, 415-427
  for log amplitude and phase, 424, 425
Fizeau, H., 357
Fizeau stellar interferometer, 336-338
Flügge, S., 356
Focal length, 291
Fourier spectroscopy, 169, 170, 232, 344
Fourier transform:
  definitions, 528, 529
  four-dimensional, 295
  one-dimensional pairs, 531
  properties of, 529-531
  two-dimensional pairs, 532
Fractional bias, 265, 266, 283
Francon, M., 229
Fraunhofer diffraction, 210, 226
Fresnel diffraction, 396
Fried, D. L., 429, 431, 432, 436, 441, 460
Frieden, B. R., 111, 282
Friedlander, S. K., 459
Fringe measurement, 500

Gabor, D., 111
Gagliardi, R. M., 524
Gamma probability density, 246, 475
Gas discharge:
  high pressure, 166
  low pressure, 164, 165
Gaskill, J. D., 532
Gaussian:
  circular statistics, 42-44
  complex random variable, 41-44
  density function, 12
  line shape, 165, 166
  moment theorem, 39, 44, 84, 85, 278
  random phase screen, 376-384
  random process, 82-85, 108, 109
  random variable, 33-40

reference sphere, 310
spectrum, 242, 243, 283
Generalized Van Cittert-Zernike theorem, 218-222
Geometric distribution, 473
Gezari, D. Y., 442, 460
Glauber, R. J., 152, 466, 523
Gold, A., 191
Goldin, E., 523
Goodman, J. W., 152, 229, 281, 282, 356, 357, 455, 459, 460, 523
Gordon, E. I., 357
Gracheva, M. E., 460
Grating spectroscopy, 170
Greenaway, A. H., 522, 524
Green's function, 395
Grimes, D., 326, 356
Gurvich, A. S., 460

Half-frequency phenomenon, 518
Hardy, J. W., 460
Harmonic oscillator, 487
Harp, J. C., 459
Harvey, J. W., 460
Heibreider, G. R., 460
Helmholtz equation, 200, 201, 234
Herman, B., 459
Hilbert transform, 105, 199, 344, 345
Ho, T. L., 460
Hodara, H., 147, 152
Hologram, 463
Holsztnski, W., 358
Homogeneous turbulence, 386
Hopkins, H. H., 157, 228, 287, 356
Hudgin, R. H., 460
Hufnagel, R. E., 429, 459, 460
Huiser, A. M. J., 357
Huygens-Fresnel principle, 117, 196

Image formation:
  analyzed by four-dimensional linear systems approach, 312-320
  analyzed with incident mutual intensity function, 307-312
  analyzed by integration over source, 303-307
  as interferometric process, 331-335
Imaging with partially coherent light, 286-331
Incoherence, 205-207, 322-324

Incoherent imaging, 320, 321
Incoherent object, 320
Independence, 15, 18
Inertial subrange, 389
Inhomogeneities of atmosphere, 384-393
Inner scale, 389
Instantaneous frequency, 141
Integrated intensity, 238-256, 467
  approximate probability density for, 244-250
  exact probability density for, 250-256
  measured with finite time, 256-271
Integration time:
  in intensity interferometer, 508
  in Michelson stellar interferometer, 501
Intensity:
  definition, 123
  instantaneous, 123
Intensity interferometer:
  advantages of, 273, 509, 510
  classical analysis of, 271-281
  counting version, 502
  ideal output of, 277
  noise limitations of, 277, 501-510
  output signal-to-noise ratio, 281, 519-523
Interferogram, 159, 160, 168
Interferometer:
  Fizeau stellar, 336-338
  Michelson, 158
  Michelson stellar, 338-340
  stellar speckle, 441-455, 510-523
Ishimaru, A., 459
Isoplanatic assumption, 408
Isoplanatic spread function, 313
Isotropic turbulence, 386, 387

Jean, Sir James, 486
Jennison, R. C., 455, 460
Joint random variables, 12-15
Jones, D. G. C., 152
Jones, R. C., 127, 152
Joyeux, Y., 281

Kallistratova, M. A., 460
Karhunen-Loève expansion, 109-111, 250-252
Karp, S., 524

Kashkarov, S. S., 460
Kingslake, R., 356
Kinzley, R. E., 356
Klauder, J. R., 523
Klein, M. V., 152
Knapp, S. L., 357
Knox, K. T., 455, 460
Kohler, D., 357
Köhler illumination, 224, 306, 307
Kolmogorov, A., 388, 389, 459
Kolmogorov spectrum, 389, 392, 426, 457
Korff, D., 450, 454, 460
Kulagin, E. S., 357

Labeyrie, A., 357, 442, 443, 444, 445, 455, 460
Lamb, W. E., Jr., 523
Lambda Cratis, 403
Laplace transform, 344
La Rocca, A. J., 459
Laser:
  gas, 230
  modes, 229
  multimode, 145-151
  single mode, 282
Laser light, 4, 121, 138-151, 193
Lawrence, R. S., 459
Lee, R. W., 459
Lens law, 298
Liewer, K. M., 357
Lin, L. H., 356
Linear filtering, 71-73
Linear system, four dimensional, 312
Lloyd's mirror, 230
Loève, M., 111
Log amplitude fluctuation, 398
Log normal:
  distribution, 399-402
  statistics, 397
Long-exposure image, 433-435
Long-exposure OTF, 402-414
Lord Rayleigh, see Strutt, J. W. (Lord Rayleigh)
Lorentzian spectrum, 165, 166, 242, 243, 252, 284
Lowenthal, S., 281
Luneburg, R. K., 356
Lutomirski, R., 460
Lynds, C. R., 460

McGlamery, B. L., 455, 460
Magnus, W., 459
Mandel, L., 1, 147, 152, 157, 203, 228, 229, 244, 281, 357, 523
Mandel's formula, 469
Marathay, A. S., 229, 357
Martienssen, W., 152
Maxwell-Boltzmann distribution, 487
Maxwell's equations, 393
Mehta, C. L., 281, 523
Mertz, L., 229
Michelson, A. A., 229, 338, 357
Michelson interferometer, 158, 232
Michelson stellar interferometer, 338-340, 359, 360, 445, 491
Microdensitometer, 324
Middleton, D., 56
Miller, R. H., 357
Mitchell, A. C. G., 229
Mitchell, R. L., 460
Moments:
    definition, 16
    joint, 17
    obtained from characteristic function, 19
Moment theorem:
    complex Gaussian, 237, 260, 274
    for laser light, 283
    for Poisson variates, 57
    real Gaussian, 39, 44, 278
Monochromatic light, propagation of, 117
Monotonic transformations of probability, 23-25
Multivariate transformations of probability, 23-25, 27-29
Murdoch, D. C., 152
Mutual coherence:
    definition, 174
    function, 183
    propagation of, 195-202
Mutual intensity:
    definition, 180, 181
    statistical properties of, 256-271

Narrowband light, propagation of, 120, 196
Negative binomial distribution, 476
Negative exponential statistics, 123, 124, 350, 448
Nisenson, P., 460

Nonmonotonic transformations of probability, 25-27
Nussenzvieg, H. M., 357

Oberhettinger, F., 459
Occupation number, 487
Oliver, B. M., 357
O'Neill, E. L., 152, 357, 459
Optical axis, 293
Optical transfer function, 321, 335
    average, 364, 368, 369, 373, 374, 376-380, 382, 383
Outer scale, 389

Paley-Wiener condition, 345
Papoulis, A., 56, 532
Paraxial approximation, 178, 209, 219, 220, 224, 291, 294, 302
Parrent, G. B., 152, 228, 330, 356
Parseval's theorem, 114, 530
Partial polarization, 127-138, 247-250
Parzen, E., 56
Pearson, K., 357
Pease, F. G., 338, 357
Peřina, J., 228, 356
Periodogram, 521
Phase information, importance of, 340-343
Phase retrieval, 343-347
Phase screen, Gaussian, 462
Phase unwrapping, 455
Phasor, 44-56
Photocount:
    definition, 466
    fluctuations, 481-486
Photocount statistics:
    for laser light, 470-472
    polarization effects, 477-479
    for polarized thermal light, 472-477
    spatial coherence effects, 479-481
Photoelectric effect, 465
Photoemission, 466
Photoevent, 466
Photosurface, 466
Piazzolla, S., 152
Picinbono, B., 281
Pinhole size effects, 183-187
Planck, M., 486
Planck's constant, 468, 487

Point-spread function:
  amplitude, 298
  atmospheric, 403, 435
  average, 366, 370, 371, 374, 380, 383
  intensity, 320
Poisson:
  density function, 12
  impulse process, 85-98, 467
  statistics, 88-91
  transform, 469
Pokasov, V. V., 460
Polarization:
  analyzer, 129
  circular, 131
  degree of, 134-136
  matrix, 128
  partial, 240, 241
Pollak, H. O., 282
Power spectral density, 68, 164-166
Probability:
  axioms obeyed by, 8
  conditional, 14
  definition of, 7, 8
  marginal, 14
Prolate spheroidal functions, 254
Propagation:
  of mutual coherence, 195-202
  of mutual intensity, 198, 200
  through inhomogeneous atmosphere, 393-399
Pseudothermal light, 150-152, 161, 240, 243, 283
Pupil function, 224, 225, 315, 325, 333, 364

Quantum efficiency, 468
Quantum electrodynamics, 465
Quantum mechanics, 465
Quasi-homogeneous source, 219
Quasimonochromatic conditions, 180

Radio astronomy, 331
Random:
  absorbing screen, 367-374, 461
  checkerboard, 462
  experiment, 7
  phase screen, 374-384, 406
  phasor sum, 44-56, 264, 447, 448, 533-536
Random process:
  complex, 99-109
  definition, 60
  Gaussian, 82-85
  Poisson impulse, 85-99
Random variable:
  complex, 40-44
  continuous, 9-12
  definition, 8
  discrete, 9-12
  Gaussian, 33-40
  independent, 31
  mixed, 9-12
  sums of, 29-33, 37-39
  transformations of, 21-29
Random walk, 44-56, 447
Ratcliffe, J. A., 459
Rayleigh:
  limit, 326
  Lord, *see* Strutt, J. W. (Lord Rayleigh)
  probability density function, 50
  separation, 359
  statistics, 123, 448
Rectangular spectrum, 243
Reducibility, 187, 189, 192
Redundancy of vector spacings, 336
Refractive index of atmosphere, 385
Relative frequency, 7
Rhodes, W. T., 455, 460
Riccati equation, 397
Rice, S. O., 244, 281
Rician statistics, 52, 399, 464
Rigden, J. D., 357
Risken, H., 143, 152
Roberts, J. A., 358
Robinson, S. R., 357
Roddier, F., 461, 523
Rogers, G. L., 331, 356
Rogstad, D. H., 455, 460
Roman, P., 152, 357
Root, W. L., 56, 523
Rough surface, 349
Russell, F. D., 357
Rytov:
  approximation, 399, 457, 464
  transformation, 397

Sakai, H., 229
Saleh, B. E. A., 358, 523, 524
Sample function, 60, 62

INDEX 549

Scalar theory, 116
Schell, A. C., 229
Schell's theorem, 225, 226, 228
Scherf, V. E., 460
Schmeltzer, R. A., 433, 460
Schwartz's inequality, 17, 74, 204
Screen:
 random, 362-384
 thin, 362
Scully, M. D., 523
Self coherence, 161, 183
Semiclassical theory of photodetection, 465-468
Shinn, D. H., 459
Shore, R. A., 229
Short-exposure:
 image, 433-435
 OTF, 402, 433-441
Shumaker, J. B., 229
Siegman, A. E., 523
Silver, S., 152
Singh, K., 229
Slepian, D., 282
Small perturbations, 395
Smith, A. W., 152
Snyder, D. L., 524
Sodin, L. G., 357
Solar astronomy, 455, 456
Spano, P., 152
Sparrow condition, 328
Speckle:
 atmospheric, 442
 in coherent imaging, 347-356
 contrast, 350
 first-order statistics, 348-350
 power spectral density of, 353-356
Spectral coherence, 202
Spectral density estimate, 517-519
Spiller, E., 152
Spizzichino, A., 56, 282, 357
Spontaneous emission, 139, 142, 146
Spread function, *see* Point-spread function
Stachnik, R. V., 456, 460
Standard deviation, 16
Stanley, N. R., 460
Stationarity:
 in increments, 63, 141
 strict sense, 63, 84
 wide sense, 63, 84

Steel, W. H., 229
Stegan, I. A., 523
Stellar speckle interferometry, 402, 441-455
 basic principle of, 442-446
 complete analysis of, 450-455
 heuristic analysis of, 446-450
 noise limitations of, 510-523
 signal-to-noise ratio in, 519-523
Stimulated emission, 139, 142
Strohbehn, J. W., 458, 459, 460
Strong fluctuation regime, 458
Structure constant, 389
 dependence on height, 428, 429
 effects of smooth variations of, 427-429
 typical values of, 392
Structure function:
 definition, 79
 of refractive index, 390-393
Strutt, J. W. (Lord Rayleigh), 1, 348, 357, 486
Sudarshan, E. C. G., 523
Sums of random variables, 29-33, 37-39
Sun, 235
Swing, R. E., 356

Tamburrini, M., 152
Tango, W. J., 358
Tatarski, V. I., 385, 396, 397, 432, 459
Talbot effect, 419
Talbot, F., 460
Taylor's hypothesis, 386, 433
Teich, M. C., 524
Telescope, 336, 338, 339
Thermal light, 4, 121, 240, 250
 partially polarized, 127-138
 polarized, 121-124
 unpolarized, 124-127
Thermal source, 4
Thin lens, 290-292
Thin object, 287
Thomas, J. B., 52, 53, 56
Thompson, B. J., 229, 286, 326, 356, 455, 460
Tichenor, D. A., 356
Time, N. S., 460
Topper, L., 459
Transfer function, 314, 315

Transformations of random variables, 21-29
Transmission cross-coefficient, 319
Tricomi, F. G., 459
Triple interferometer, 455
Tukey, J. W., 229
Turbulence, atmospheric, 388-393
Turbulent eddies, 388
Turner, R. E., 459
Twiss, R. Q., 271, 282, 358

Ultraviolet catastrophe, 486
Unitary matrix, 129, 133, 135

Vanasse, G. A., 229
Van Cittert, P. H., 157, 207
Van Cittert-Zernike theorem, 207-222, 235, 236, 308-310, 334, 336, 352, 353, 481
  generalized, 311, 358
Van der Pol oscillator, 142
Van Schooneveld, C., 358
Van Toorn, P., 357
Variance, 16
Vector spacing, 333, 446
Verdet, E., 157, 228, 348, 357
Vernier, P. J., 523
Visibility, 163, 182
Von Kármán spectrum, 389, 390
Von Laue, M., 157, 228, 348, 357

Walkup, J. F., 523, 524
Walther, A., 357
Wave degeneracy parameter, 484
Wave equation, 199
Wave front splitting, 157
Wavefront tilt, 436
Wave structure function, 407
  near field calculation of, 407-413
  for spherical wave propagation, 432, 433
Weigelt, G., 403, 446
Whitney, R. E., 229
Wiener, N., 152
Wiener-Khinchin theorem, 74
Wolf, E., 1, 152, 157, 199, 203, 228, 229, 281, 282, 343, 356, 357, 358, 459, 461, 523
Wolfe, W. L., 459
Worden, S. P., 456, 460

Young, T., 170, 229
Young's experiment, 170-177, 196, 231, 234, 272, 331, 351
Young's fringes, 336
Yura, H., 460

Zardecki, A., 281
Zemansky, M. W., 229
Zernike, F., 157, 207, 228, 311, 356
Zissis, G. J., 459